Communications
in Computer and Information Science   2150

Rationale

The CCIS series is devoted to the publication of proceedings of computer science conferences. Its aim is to efficiently disseminate original research results in informatics in printed and electronic form. While the focus is on publication of peer-reviewed full papers presenting mature work, inclusion of reviewed short papers reporting on work in progress is welcome, too. Besides globally relevant meetings with internationally representative program committees guaranteeing a strict peer-reviewing and paper selection process, conferences run by societies or of high regional or national relevance are also considered for publication.

Topics

The topical scope of CCIS spans the entire spectrum of informatics ranging from foundational topics in the theory of computing to information and communications science and technology and a broad variety of interdisciplinary application fields.

Information for Volume Editors and Authors

Publication in CCIS is free of charge. No royalties are paid, however, we offer registered conference participants temporary free access to the online version of the conference proceedings on SpringerLink (http://link.springer.com) by means of an http referrer from the conference website and/or a number of complimentary printed copies, as specified in the official acceptance email of the event.

CCIS proceedings can be published in time for distribution at conferences or as post-proceedings, and delivered in the form of printed books and/or electronically as USBs and/or e-content licenses for accessing proceedings at SpringerLink. Furthermore, CCIS proceedings are included in the CCIS electronic book series hosted in the SpringerLink digital library at http://link.springer.com/bookseries/7899. Conferences publishing in CCIS are allowed to use Online Conference Service (OCS) for managing the whole proceedings lifecycle (from submission and reviewing to preparing for publication) free of charge.

Publication process

The language of publication is exclusively English. Authors publishing in CCIS have to sign the Springer CCIS copyright transfer form, however, they are free to use their material published in CCIS for substantially changed, more elaborate subsequent publications elsewhere. For the preparation of the camera-ready papers/files, authors have to strictly adhere to the Springer CCIS Authors' Instructions and are strongly encouraged to use the CCIS LaTeX style files or templates.

Abstracting/Indexing

CCIS is abstracted/indexed in DBLP, Google Scholar, EI-Compendex, Mathematical Reviews, SCImago, Scopus. CCIS volumes are also submitted for the inclusion in ISI Proceedings.

How to start

To start the evaluation of your proposal for inclusion in the CCIS series, please send an e-mail to ccis@springer.com.

Andrew M. Olney · Irene-Angelica Chounta ·
Zitao Liu · Olga C. Santos · Ig Ibert Bittencourt
Editors

Artificial Intelligence in Education

Posters and Late Breaking Results, Workshops and Tutorials, Industry and Innovation Tracks, Practitioners, Doctoral Consortium and Blue Sky

25th International Conference, AIED 2024
Recife, Brazil, July 8–12, 2024
Proceedings, Part I

Springer

Editors
Andrew M. Olney 🆔
University of Memphis
Memphis, TN, USA

Irene-Angelica Chounta 🆔
University of Duisburg-Essen
Duisburg, Germany

Zitao Liu 🆔
Jinan University
Guangzhou, China

Olga C. Santos 🆔
UNED
Madrid, Spain

Ig Ibert Bittencourt 🆔
Universidade Federal de Alagoas
Maceio, Brazil

ISSN 1865-0929   ISSN 1865-0937 (electronic)
Communications in Computer and Information Science
ISBN 978-3-031-64314-9   ISBN 978-3-031-64315-6 (eBook)
https://doi.org/10.1007/978-3-031-64315-6

This Springer imprint is published by the registered company Springer Nature Switzerland AG
The registered company address is: Gewerbestrasse 11, 6330 Cham, Switzerland

If disposing of this product, please recycle the paper.

Preface

The 25th International Conference on Artificial Intelligence in Education (AIED 2024) was hosted by Centro de Estudos e Sistemas Avançados do Recife (CESAR), Brazil from July 8 to July 12, 2024. It was set up in a face-to-face format but included an option for an online audience. AIED 2024 was the next in a longstanding series of annual international conferences for the presentation of high-quality research on intelligent systems and the cognitive sciences for the improvement and advancement of education. Note that AIED is ranked A in CORE (top 16% of all 783 ranked venues), the well-known ranking of computer science conferences. The AIED conferences are organized by the prestigious International Artificial Intelligence in Education Society, a global association of researchers and academics, which has already celebrated its 30th anniversary, and aims to advance the science and engineering of intelligent human-technology ecosystems that support learning by promoting rigorous research and development of interactive and adaptive learning environments for learners of all ages across all domains.

The theme for the AIED 2024 conference was "AIED for a World in Transition". The conference aimed to explore how AI can be used to enhance the learning experiences of students and teachers alike when disruptive technologies are turning education upside down. Rapid advances in Artificial Intelligence (AI) have created opportunities not only for personalized and immersive experiences but also for ad hoc learning by engaging with cutting-edge technology continually, extending classroom borders, from engaging in real-time conversations with large language models (LLMs) to creating expressive artifacts such as digital images with generative AI or physically interacting with the environment for a more embodied learning. As a result, we now need new approaches and measurements to harness this potential and ensure that we can safely and responsibly cope with a world in transition. The conference seeks to stimulate discussion of how AI can shape education for all sectors, how to advance the science and engineering of AI-assisted learning systems, and how to promote broad adoption.

AIED 2024 attracted broad participation. We received 334 submissions for the main program, of which 280 were submitted as full papers, and 54 were submitted as short papers. Of the full paper submissions, 49 were accepted as full papers, and another 27 were accepted as short papers. The acceptance rate for full papers and short papers together was 23%. These accepted contributions are published in the Springer proceedings volumes LNAI 14829 and 14830.

The submissions underwent a rigorous double-masked peer-review process aimed to reduce evaluation bias as much as possible. The first step of the review process was done by the program chairs, who verified that all papers were appropriate for the conference and properly anonymized. Program committee members were asked to declare conflicts of interest. After the initial revision, the program committee members were invited to bid on the anonymized papers that were not in conflict according to their declared conflicts of interest. With this information, the program chairs made the review assignment, which consisted of three regular members to review each paper plus a senior member to provide a meta-review. The management of the review process (i.e., bidding, assignment, discussion, and meta-review) was done with the EasyChair platform, which was configured so that reviewers of the same paper were anonymous to each other. A subset of the program committee members were not included in the initial assignment but were asked to be ready to do reviews that were not submitted on time (i.e., the emergency review period). To avoid a situation where program committee members would be involved in too many submissions, we balanced review assignments and then rebalanced them during the emergency review period.

As a result, each submission was reviewed anonymously by at least three Program Committee (PC) members and then a discussion was led by a Senior Program Committee (SPC) member. PC and SPC members were selected based on their authorship in previous AIED conferences, their experience as reviewers in previous AIED editions, their h-index as calculated by Google Scholar, and their previous positions in organizing and reviewing related conferences. Therefore all members were active researchers in the field, and SPC members were particularly accomplished on these metrics. SPC members served as meta-reviewers whose role was to seek consensus to reach the final decision about acceptance and to provide the corresponding meta-review. They were also asked to check and highlight any possible biases or inappropriate reviews. Decisions to accept/reject were taken by the program chairs. For borderline cases, the contents of the paper were read in detail before reaching the final decision. In summary, we are confident that the review process assured a fair and equal evaluation of the submissions received without any bias, as far as we are aware.

Beyond paper presentations, the conference included a Doctoral Consortium Track, Late-Breaking Results, a Workshops and Tutorials Track, and an Industry, Innovation and Practitioner Track. There was a WideAIED track, which was established in 2023, where opportunities and challenges of AI in education were discussed with a global perspective and with contributions coming also from areas of the world that are currently under-represented in AIED. Additionally, a BlueSky special track was included with contributions that reflect upon the progress of AIED so far and envision what is to come in the future. The submissions for all these tracks underwent a rigorous peer review process. Each submission was reviewed by at least three members of the AIED community, assigned by the corresponding track organizers who then took the final decision about acceptance. The participants of the conference had the opportunity to attend three keynote talks: "Navigating Strategic Challenges in Education in the Post-Pandemic AI Era" by Blaženka Divjak, "Navigating the Evolution: The Rising Tide of Large Language Models for AI and Education" by Peter Clark, and "Artificial Intelligence in Education and Public Policy: A Case from Brazil" by Seiji Isotani. These contributions are published in the Springer proceedings volumes CCIS 2150 and 2151.

The conference also included a Panel with experts in the field and the opportunity for the participants to present a demonstration of their AIED system in a specific session of Interactive Events. A selection of the systems presented is included as showcases on the web page of the IAIED Society[1]. Finally, there was a session with presentations of papers published in the International Journal of Artificial Intelligence in Education[2],

---

[1] https://iaied.org/showcase.

[2] https://link.springer.com/journal/40593.

the journal of the IAIED Society indexed in the main databases, and a session with the best papers from conferences of the International Alliance to Advance Learning in the Digital Era (IAALDE)[3], an alliance of research societies that focus on advances in computer-supported learning, to which the IAIED Society belongs.

For making AIED 2024 possible, we thank the AIED 2024 Organizing Committee, the hundreds of Program Committee members, the Senior Program Committee members, and the AIED proceedings chairs Paraskevi Topali and Rafael D. Araújo. In addition, we would like to thank the Executive Committee of the IAIED Society for their advice during the conference preparation, and specifically two of the working groups, the Conference Steering Committee, and the Diversity and Inclusion working group. They all gave their time and expertise generously and helped with shaping a stimulating AIED 2024 conference. We are extremely grateful to everyone!

July 2024

Andrew M. Olney
Irene-Angelica Chounta
Zitao Liu
Olga C. Santos
Ig Ibert Bittencourt

---

[3] https://alliancelss.com/.

Organization

Conference General Co-chairs

Olga C. Santos   UNED, Spain
Ig Ibert Bittencourt   Universidade Federal de Alagoas, Brazil

Program Co-chairs

Andrew M. Olney   University of Memphis, USA
Irene-Angelica Chounta   University of Duisburg-Essen, Germany
Zitao Liu   Jinan University, China

Doctoral Consortium Co-chairs

Yu Lu   Beijing Normal University, China
Elaine Harada T. Oliveira   Universidade Federal do Amazonas, Brazil
Vanda Luengo   Sorbonne Université, France

Workshop and Tutorials Co-chairs

Cristian Cechinel   Federal University of Santa Catarina, Brazil
Carrie Demmans Epp   University of Alberta, Canada

Interactive Events Co-chairs

Leonardo B. Marques   Federal University of Alagoas, Brazil
Ben Nye   University of Southern California, USA
Rwitajit Majumdar   Kumamoto University, Japan

Industry, Innovation and Practitioner Co-chairs

Diego Dermeval   Federal University of Alagoas, Brazil
Richard Tong   IEEE Artificial Intelligence Standards
                               Committee, USA
Sreecharan Sankaranarayanan Amazon, USA
Insa Reichow   German Research Center for Artificial
                               Intelligence, Germany

Posters and Late Breaking Results Co-chairs

Marie-Luce Bourguet   Queen Mary University of London, UK
Qianru Liang   Jinan University, China
Jingyun Wang   Durham University, UK

Panel Co-chairs

Julita Vassileva   University of Saskatchewan, Canada
Alexandra Cristea   Durham University, UK

Blue Sky Co-chairs

Ryan S. Baker   University of Pennsylvania, USA
Benedict du Boulay   University of Sussex, UK
Mirko Marras   University of Cagliari, Italy

WideAIED Co-chairs

Isabel Hilliger   Pontificia Universidad Católica, Chile
Marco Temperini   Sapienza University of Rome, Italy
Ifeoma Adaji   University of British Columbia, Canada
Maomi Ueno   University of Electro-Communications, Japan

Local Organising Co-chairs

Rafael Ferreira Mello	Universidade Federal Rural de Pernambuco, Brazil
Taciana Pontual	Universidade Federal Rural de Pernambuco, Brazil

AIED Mentoring Fellowship Co-chairs

Amruth N. Kumar	Ramapo College of New Jersey, USA
Vania Dimitrova	University of Leeds, UK

Diversity and Inclusion Co-chairs

Rod Roscoe	Arizona State University, USA
Kaska Porayska-Pomsta	University College London, UK

Virtual Experiences Co-chairs

Guanliang Chen	Monash University, Australia
Teng Guo	Jinan University, China
Eduardo A. Oliveira	University of Melbourne, Australia

Publicity Co-chairs

Son T. H. Pham	Nha Viet Institute, USA
Pham Duc Tho	Hung Vuong University, Vietnam
Miguel Portaz	UNED, Spain

Volunteer Co-chairs

Isabela Gasparini	Santa Catarina State University, Brazil
Lele Sha	Monash University, Australia

Proceedings Co-chairs

Paraskevi Topali	National Education Lab AI, Radboud University, Netherlands
Rafael D. Araújo	Federal University of Uberlândia, Brazil

Awards Co-chairs

Ning Wang	University of Southern California, USA
Beverly Woolf	University of Massachusetts, USA

Sponsorship Chair

Tanci Simões Gomes	CESAR, Brazil

Scholarship Chair

Patrícia Tedesco	Universidade Federal de Pernambuco, Brazil

Steering Committee

Noboru Matsuda	North Carolina State University, USA
Eva Millan	Universidad de Málaga, Spain
Sergey Sosnovsky	Utrecht University, Netherlands
Ido Roll	Israel Institute of Technology, Israel
Maria Mercedes T. Rodrigo	Ateneo de Manila University, Philippines

Blue Sky Track Program Committee Members

Giora Alexandron	Weizmann Institute of Science, Israel
Gautam Biswas	Vanderbilt University, USA
Mingyu Feng	WestEd, USA
Neil Heffernan	Worcester Polytechnic Institute, USA
Sharon Hsiao	Santa Clara University, USA
Kenneth Koedinger	Carnegie Mellon University, USA
Roberto Martinez-Maldonado	Monash University, Australia

Gordon McCalla   University of Saskatchewan, Canada
Agathe Merceron   BHT Berlin, Germany
Tanja Mitrovic   University of Canterbury, New Zealand
Luc Paquette   University of Illinois at Urbana-Champaign, USA
Anna Rafferty   Carleton College, USA
Ana Serrano Mamolar   University of Burgos, Spain
Beverly Woolf   University of Massachusetts, USA

WideAIED Track Program Committee Members

Ifeoma Adaji   University of British Columbia, Canada
Gabriel Astudillo   Pontificia Universidad Católica de Chile, Chile
Paolo Fantozzi   LUMSA University, Italy
Kazuma Fuchimoto   University of Electro-Communications, Japan
Shili Ge   Guangdong University of Foreign Studies, China
Isabel Hilliger   Pontificia Universidad Católica de Chile, Chile
Zuzana Kubincová   Comenius University, Slovakia
Carla Limongelli   Roma Tre University, Italy
Ivana Marenzi   L3S Research Center, Germany
Ronald Pérez-Álvarez   Universidad de Costa Rica, Costa Rica
Mar Perez-Sanagustin   Université Paul Sabatier Toulouse III, France
Valerio Rughetti   Università Telematica Internazionale Uninettuno,
                               Italy
Daniele Schicchi   ITD, National Council of Research, Italy
Filippo Sciarrone   Universitas Mercatorum, Italy
Andrea Sterbini   Sapienza University of Rome, Italy
Davide Taibi   ITD, National Council of Research, Italy
Juan Andrés Talamás   Tecnológico de Monterrey, Mexico
Marco Temperini   Sapienza University of Rome, Italy
Maomi Ueno   University of Electro-Communications, Japan
Giacomo Valente   University of L'Aquila, Italy
Ignacio Villagran   Pontificia Universidad Católica de Chile, Chile
Esteban Villalobos   Université Paul Sabatier, France

Late-Breaking Results Track Program Committee Members

Ifeoma Adaji   University of British Columbia, Canada
Mohammad Alshehri   University of Jeddah, Saudi Arabia
Pablo Arnau González   Universidad de Valencia, Spain
Yuya Asano   University of Pittsburgh, USA

Gabriel Astudillo	Pontificia Universidad Católica de Chile, Chile
Francisco Leonardo	.
Aureliano Ferreira	UFERSA, Brazil
Shiva Baghel	Extramarks Education India Pvt. Ltd., India
Rabin Banjade	University of Memphis, USA
Sami Baral	Worcester Polytechnic Institute, USA
Hemilis Joyse Barbosa Rocha	UFPE, Brazil
Luca Benedetto	University of Cambridge, UK
Raul Benites Paradeda	State University of Rio Grande do Norte, Brazil
Jiang Bo	East China Normal University, China
Diana Borza	Babeş-Bolyai University, Romania
Matthieu Branthome	Université de Bretagne Occidentale CREAD, France
Minghao Cai	University of Alberta, Canada
May Kristine Jonson Carlon	RIKEN Center for Brain Science, Japan
Liangyu Chen	East China Normal University, China
Jade Cock	EPFL, Switzerland
Clayton Cohn	Vanderbilt University, USA
Syaamantak Das	Indian Institute of Technology Bombay, India
Deep Dwivedi	Indraprastha Institute of Information Technology, India
Yo Ehara	Tokyo Gakugei University, Japan
Abul Ehtesham	The Davey Tree Expert Company, USA
Lara Ceren Ergenç	King's College London, UK
Luyang Fang	University of Georgia, USA
Zechu Feng	University of Hong Kong, China
Rafael Ferreira Mello	CESAR School, Brazil
Ritik Garg	Extramarks Education Pvt. Ltd., India
Sahin Gokcearslan	Gazi University, Turkey
Sara Guerreiro-Santalla	Universidade da Coruña, Spain
Shivang Gupta	Carnegie Mellon University, USA
Songhee Han	University of Texas at Austin, USA
John Hollander	University of Memphis, USA
Xinying Hou	University of Michigan, USA
Mingliang Hou	Dalian University of Technology, China
Rositsa V. Ivanova	University of St. Gallen, Switzerland
Yuan-Hao Jiang	East China Normal University, China
Stamos Katsigiannis	Durham University, UK
Ekaterina Kochmar	MBZUAI, UAE
Elizabeth Koh	Nanyang Technological University, Singapore
Tatsuhiro Konishi	Shizuoka University, Japan

Marco Kragten	Amsterdam University of Applied Science, Netherlands
Ehsan Latif	University of Georgia, USA
Morgan Lee	Worcester Polytechnic Institute, USA
Junyoung Lee	Nanyang Technological University, Singapore
Liang-Yi Li	National Taiwan Normal University, Taiwan
Mengting Li	University of Hong Kong, China
Jianwei Li	Beijing University of Posts and Telecommunications, China
Jeongki Lim	Parsons School of Design, USA
Yu Lu	Beijing Normal University, China
Pascal Muam Mah	AGH University of Krakow, Poland
Leonardo Brandão Marques	University of São Paulo, Brazil
Paola Mejia Domenzain	EPFL, Switzerland
Fatma Miladi	MIRACL, Tunisia
Yoshimitsu Miyazawa	National Center for University Entrance Examinations, Japan
Yasuhiro Noguchi	Shizuoka University, Japan
Ange Adrienne Nyamen Tato	École de Technologie Supérieure, France
Chaohua Ou	Georgia Institute of Technology, USA
Qianqian Pan	Nanyang Technological University, Singapore
Andrew Petersen	University of Toronto, Canada
Vijay Prakash	Indian Institute of Technology Bombay, India
Ethan Prihar	École polytechnique fédérale de Lausanne, Switzerland
Kun Qian	Columbia University, USA
Emanuel Queiroga	Federal University of Pelotas, Brazil
Krishna Chaitanya Rao Kathala	University of Massachusetts Amherst, USA
Vasile Rus	University of Memphis, USA
Martin Ruskov	University of Milan, Italy
Sabrina Schork	UAS Aschaffenburg, Germany
Sabine Seufert	University of St. Gallen, Switzerland
Mohammed Seyam	Virginia Tech, USA
Atsushi Shimada	Kyushu University, Japan
Kazutaka Shimada	Kyushu Institute of Technology, Japan
Jinnie Shin	University of Florida, USA
Aditi Singh	Cleveland State University, USA
Anuradha Singh	Banaras Hindu University, India
Jasbir Singh	University of Otago, New Zealand
Kelly Smalenberger	Belmont Abbey College, USA
Michael Smalenberger	University of North Carolina at Charlotte, USA

Álvaro Sobrinho	Federal University of Pernambuco, Brazil
Aakash Soni	ECE Paris, France
Victor Sotelo	UNICAMP, Brazil
Zhongtian Sun	University of Cambridge, UK
Rushil Thareja	IIIT Delhi, India
Danielle R. Thomas	Carnegie Mellon University, USA
Liu Tianrui	University of Sydney, Australia
Anne Trumbore	University of Virginia, USA
Maomi Ueno	University of Electro-Communications, Japan
Jim Wagstaff	Noodle Factory Learning, Singapore
Deliang Wang	Beijing Normal University, China
Jiahui Wu	Beijing University of Posts and Telecommunication, China
Yuhang Wu	University of Hong Kong, China
Yu Xiong	Chongqing University of Posts and Telecommunications, China
Mingrui Xu	Beijing University of Posts and Telecommunications, China
R. Yamamoto Ravenor	Ochanomizu University, Japan
Koichi Yamashita	Tokoha University, Japan
Andres Felipe Zambrano	University of Pennsylvania, USA
Kamyar Zeinalipour	University of Siena, Italy
Ivy Chenjia Zhu	University of Hong Kong, China

Industry, Innovation, and Practitioner Track Program Committee Members

Cristian Cechinel	Federal University of Santa Catarina, Brazil
Geiser Challco	Federal University of the Semi-arid Region, Brazil
Rafael Ferreira Mello	Universidade Federal Rural de Pernambuco, Brazil
Elyda Freitas	University of Pernambuco, Brazil
Huy Nguyen	Carnegie Mellon University, USA
Ranilson Paiva	Federal University of Alagoas, Brazil
Cleon Xavier Pereira Júnior	Federal Institute of Goiás, Brazil
Faisal Rachid	German Research Center for Artificial Intelligence, Germany
Luiz Rodrigues	Federal University of Alagoas, Brazil
Shashi Kant Shankar	Amrita University, India
Tiago Thompsen Primo	Federal University of Pelotas, Brazil

International Artificial Intelligence in Education Society

Management Board

President

Olga C. Santos   UNED, Spain

Secretary/Treasurer

Vania Dimitrova   University of Leeds, UK

Journal Editors

Vincent Aleven   Carnegie Mellon University, USA
Judy Kay   University of Sydney, Australia

Finance Chair

Benedict du Boulay   University of Sussex, UK

Membership Chair

Benjamin D. Nye   University of Southern California, USA

IAIED Officers

Yancy Vance Paredes   North Carolina State University, USA
Son T. H. Pham   Nha Viet Institute, USA
Miguel Portaz   UNED, Spain

Executive Committee

Akihiro Kashihara   University of Electro-Communications, Japan
Amruth Kumar   Ramapo College of New Jersey, USA
Christothea Herodotou   Open University, UK
Jeanine A. Defalco   CCDC-STTC, USA
Judith Masthoff   Utrecht University, Netherlands
Maria Mercedes T. Rodrigo   Ateneo de Manila University, Philippines
Ning Wang   University of Southern California, USA

AIED 2024 Panel

Do We Need AIED If We Have Powerful Generative AI?

Julita Vassileva[1] (iD) and Alexandra I. Cristea [2] (iD)

[1]University of Saskatchewan, Saskatoon, SK, Canada
jiv@cs.usask.ca
[2]University of Durham, Durham, UK
alexandra.i.cristea@durham.ac.uk

Abstract. The advent of advanced AI systems, particularly generative pre-trained transformers (GPTs) and large language models (LLMs) like ChatGPT, Gemini, Claude, has significantly changed the landscape of artificial intelligence and its applications across various domains. GPTs have demonstrated impressive capabilities in generating human-like text and solving complex problems. Educators are finding creative ways to deploy GPTs systems in their practice [1]. Many schools are struggling with the question of whether to allow or ban the use of AI, because of the potential risks (overreliance, academic integrity). The anticipation of change in the education system is palpable. What does the future hold? This raises important questions about the role of artificial intelligence in education (AIED) research. This panel will discuss the reasons why and ways in which AIED research should or should not remain essential in these transformative times for education, as well as ways in which it should change.

Keywords: AI systems · generative AI · LLMs · AIED

1 Panel Description

Generative AI has entered Education. Many teachers are finding creative ways to deploy the myriads of GPT applications in their classes; schools are debating whether to ban or allow the use of generative AI tools. Is there a place for AIED research in this new world? There are a lot of reasons for optimism and many open research directions. Education is a deeply individualized process. AIED focuses not just on delivering information but on understanding and adapting to individual learning styles, needs, and learners' emotional states. While GPTs can provide answers and explanations, AIED systems are designed to engage learners more interactively and adaptively, aiming to mimic the nuanced responses of a human teacher. While the deployment of generative AI in educational settings raises concerns about pedagogical effectiveness, AIED systems developed with specific educational theories as underpinning can be tailored to foster not just cognitive skills but also critical thinking, creativity, and emotional intelligence. On the other hand, several papers in the AIED'2024 conference program demonstrate that the research community already incorporates GPTs to power systems that accomplish these aspirations.

However, the AIED community needs to be careful to avoid raising too high expectations. Many of us remember the "AI winter" (a term coined at the AAAI conference in 1984 [2]), which lasted nearly 20 years. According to some [3], Generative AI has reached the 'Peak of Inflated Expectations' in the Gartner Hype Cycle curve in September 2023, where the initial excitement and hype are at their highest. However, the critical concern is whether the generative AI technologies will descend into the 'Trough of Disillusionment,' as past AI technologies have, potentially leading to another AI winter. Many factors can lead to this, including concerns about privacy, misinformation, job displacement, and environmental concerns, leading to public and regulatory backlash. Specific areas of concern in applying generative AI applications in education are the current limitations of GPTs in understanding context, reasoning, and providing explanations based on correlations rather than causal relationships [4]—also, their computational resource intensity. If access to Generative AI systems is limited to populations in affluent regions, educational disparities, inherent biases, digital divides, and inequities will be amplified, and vulnerable workers in specific sectors will be disproportionately displaced. Moreover, disillusionment with deep models may lead to general disillusionment with AI, as society tends to use these terms almost interchangeably [5]. Preventing a new AI winter depends on new research to minimize the algorithmic bias and the computational resources needed to run GPTs, as well as on the global society to address these challenges through improved access, developing regulatory frameworks, and effective methods for public awareness and education.

From an AI in education perspective, there seems to be a need for new research about how good pedagogical practices [6] and clearer pathways to psychological methodological underpinning, such as specific, measurable ways for embedding, extracting and evaluating motivational theories [20] for educational contexts [7] can be applied within GPTs, or implemented alongside GPTs. Do other classic AIED topics, such as learner models [8, 9, 10], semantic web and concept models [11, 12], personalized game-based learning [13] and gamification [14, 15, 16], affect [17] and engagement [7, 8], equity, inclusion and ethics [14, 18], scrutable AIED [19], and low-cost solutions for developing societies [14] still have a place, and if so, how?

The AIED community will need to address these issues. We hope the panel will bring new insights into how to approach them and raise new questions that may lead to future avenues for the community to explore.

References

1. Johnson, M.: Generative AI in Computer Science Education. Commun. ACM (2024)
2. Russell, S.J., Norvig, P.: Artificial Intelligence: A Modern Approach, 2nd edn. Prentice Hall, Upper Saddle River (2023). ISBN 0-13-790395-2

3. Gartner (2023). https://www.gartner.com/en/articles/what-s-new-in-the-2023-gartner-hype-cycle-for-emerging-technologies

4. Feder, A., et al.: CausaLM: causal model explanation through counterfactual language models. Comput. Linguist. **47**(2), 333–386 (2021). https://doi.org/10.1162/coli_a_00404

5. IBM Data and AI team, AI vs. Machine Learning vs. Deep Learning vs. Neural Networks: What's the difference? (2023). https://www.ibm.com/blog/ai-vs-machine-learning-vs-deep-learning-vs-neural-networks/

6. Al-Khresheh, M.H.: Bridging technology and pedagogy from a global lens: Teachers' perspectives on integrating ChatGPT in English language teaching. Comput. Educ. Artif. Intell. **6**, 100218 (2024). https://doi.org/10.1016/j.caeai.2024.100218

7. Cristea, A.I., et al.: The engage taxonomy: SDT-based measurable engagement indicators for MOOCs and their evaluation. User Model. User-Adap. Inter. (2023). https://doi.org/10.1007/s11257-023-09374-x

8. Orji, F. A., and Vassileva, J.: Modeling the impact of motivation factors on students' study strategies and performance using machine learning. J. Educ. Technol. Syst. **52**(2), 274–296 (2023). https://doi.org/10.1177/00472395231191139

9. Li, Z., Shi, L., et al.: Sim-GAIL: a generative adversarial imitation learning approach of student modelling for intelligent tutoring systems. Neural Comput. Appl. **35**, 24369–24388 (2023). https://doi.org/10.1007/s00521-023-08989-w

10. Alrajhi, L., et al.: Solving the imbalanced data issue: automatic urgency detection for instructor assistance in MOOC discussion forums. User Model. User-Adap. Inter. (2023). https://doi.org/10.1007/s11257-023-09381-y

11. Gkiokas, A., Cristea A.I.: Cognitive agents and machine learning by example: representation with conceptual graphs. Comput. Intell. **34**, 603–634. (2018). https://doi.org/10.1111/coin.12167

12. Aljohani, T., Cristea, A.I.: Learners demographics classification on MOOCs during the COVID-19: author profiling via deep learning based on semantic and syntactic representations. Front. Res. Metrics Anal. **6**, 673928–673928 (2021)

13. Kiron, N., Omar, M.T., Vassileva, J.: Evaluating the impact of serious games on study skills and habits. In: European Conference on Games Based Learning, vol. 17, issue number 1, pp. 326–335 (2023)

14. Toda, A., et al.: Gamification through the looking glass - perceived biases and ethical concerns of brazilian teachers. In: Rodrigo, M.M., Matsuda, N., Cristea, A.I., Dimitrova, V. (eds.) AIED 2022. LNCS, vol. 13356, pp. 259–262. Springer, Cham (2022). https://doi.org/10.1007/978-3-031-11647-6_47

15. Toda, A.M., et al.: Analysing gamification elements in educational environments using an existing gamification taxonomy. Smart Learn. Environ. **6**, 16 (2019). https://doi.org/10.1186/s40561-019-0106-1

16. Toda, A., et al.: Gamification Design for Educational Contexts: Theoretical and Practical Contributions. Springer, Cham (2023). https://link.springer.com/book/10.1007/978-3-031-31949-5

17. Sümer, Ö., et al.: Multimodal engagement analysis from facial videos in the classroom. IEEE Trans. Affect. Comput. **14**(2), 1012–1027 (2023). https://doi.org/10.1109/TAFFC.2021.3127692

18. Jafari, E., Vassileva, J.: Ethical issues in explanations of personalized recommender systems. In: ACM Conference on User Modeling Adaptation and Personalization, UMAP 2023 (2023). https://dl.acm.org/doi/abs/10.1145/3563359.3597383. ISBN 978-1-4503-9891-6
19. Kay, J., et al.: Scrutable AIED. In: Handbook of Artificial Intelligence in Education, pp. 101–125. Edward Elgar Publishing (2023)
20. Orji, F.A., Gutierrez, F.J., Vassileva, J.: Exploring the influence of persuasive strategies on student motivation: self-determination theory perspective. In: Baghaei, N., Ali, R., Win, K., Oyibo, K. (eds.) PERSUASIVE 2024. LNCS, vol. 14636, pp. 222–236. Springer, Cham (2024). https://doi.org/10.1007/978-3-031-58226-4_17

AIED 2024 Keynotes

Navigating Strategic Challenges in Education in the Post-pandemic AI Era

Blaženka Divjak⊚

Faculty of Organization and Informatics, University of Zagreb, Croatia
`blazenka.divjak@foi.unizg.hr`

Abstract. Navigating strategic challenges in education in the post-pandemic AI era has the characteristics of a *complex* decision-making context. The current challenges and possible approaches, supporting decisions about future-oriented education, are discussed.

Keywords: strategic decision-making · education · pandemic · artificial intelligence

1 Introduction

Strategic planning in education needs to consider the contemporary challenges and the type of a decision-making context. Today's education has been significantly impacted by two major unprecedented events which have had a strong transformative effect: the COVID-19 pandemic and the rapid spreading of AI.

To better understand the decision-making in such contexts, it is useful to look at them through the lens of the Cynefin framework. As explained by Snowden [1], "Cynefin creates four open spaces or domains of knowledge all of which have validity within different contexts". In Cynefin, contexts are "defined by the nature of the relationship between cause and effect" [2]. In the four contexts, simple, complicated, complex, and chaotic, leaders need to recognize the situation and act appropriately.

The COVID-19 context was a chaotic one, in which it was not possible to identify the relationships between cause and effect and there were no patterns, so leaders had to act immediately and work to shift from *chaos* to *complexity*. While in chaos *trust* in leaders is necessary, it also presents a danger for democratic processes if it is blind.

Once the world shifted to a *complex* context, another challenge emerged: the main-streaming of AI. Both challenges have a significant impact on education today, on the global, national, institutional, and on the level of individual learners and educators.

Strategic decision-making today is therefore double-burdened. On the one hand, it has to consider the possibility of new major global threats, like a pandemic or a war, and their possible tremendous effect on education. On the other hand, careful thought should be given to AI, which can be a powerful assistant to deal with threats like this, but also a hidden enemy to ethical and meaningful teaching and learning.

2 Decision Making in the *Chaotic* Domain

Research on educational decision-making in the chaotic domain, characterized by the imperative to master unfamiliar threats, has been scarce, at least until the unprecedented global challenge posed by the pandemic. The pandemic presented us with a rich research environment, enabling us to investigate and draw conclusions which can be generalized and applied to other chaotic contexts.

Educational decision-making during the pandemic was done at different levels: global, (multi)national, institution, individual (teacher, student). So, when analysing the decisions and decision-making processes, it makes sense to consider the said levels.

At the onset of the pandemic, I was the chair of the Council of the European Union (EU), gathering ministers in charge of education of the 27 EU Member States. So, looking at the multinational perspective, I personally witnessed an increase in the inclination of educational ministers to work together closely in making decisions in this *chaotic* context. Without delay, new formal and informal instruments for fast data collection, mutual support, learning and exchange of important decisions and processes were developed, as well as the sharing of good and bad practices. We collected primarily qualitative data, based on the responses of the Member States' educational authorities, that were aimed to support the understanding of the situation and finding/recognizing common denominators (patterns). There was no time to collect structured quantitative data, as decision-making had to be instant, making sure that lives and health were safeguarded, and educational processes continued in alternative formats. The public was often not aware of the efforts done at decision-making levels, as in this *chaotic* context [2], it was essential to trust the leaders and their dedication to ensuring the wellbeing of those related to educational institutions. Similar processes striving for fast and open collaboration were also recognized on institutional [3] and global levels (e.g., [4]).

Regardless of the global and regional efforts and collaboration, there were still differences in the success of educational systems' responses to the pandemic. Importantly, previous research has shown that the agility and fitness-for-change of educational systems and their main actors is essential in responding to and coping with major challenges [5].

3 Decision Making in the *Complex* and *Complicated* Domain

One of the most important goals in the *chaotic* domain is to lower the level of chaos and push towards the *complex* context, where we have a clearer picture of what we do and do not know, and have a toolbox of decision-making theories and instruments that can be used to identify at least some patterns for decision-making. In this context, we still need to be agile in decision-making, exploiting the benefits of concepts and tools already at hand, as well as the emerging technologies and approaches.

Currently, we are working in a *complex* context, significantly challenged by the rapid development and increasing accessibility of AI, intertwined with the consequences of the pandemic. To respond to the contemporary requirements, educators should be continuously strengthened and motivated to harness the potential of learning design [6], as a possible universal language, and innovative pedagogies. Meaningful learning design and implementation of future- and learner-oriented teaching and learning practices can be enhanced by the ethical and creative use of AI. Sensible use of AI and minimising the black-box effect require changes in curricula, as well as initial education and continuous professional development of educators [7].

Besides being used as a teaching and learning tool, AI can provide important assistance in decision-making, particularly as an indispensable source of technology, methods and tools for learning analytics. We strive for evidence-based decision-making supported by explainable AI, but this opens up a range of questions. How can we make and implement decisions about learners? How do regulatory frameworks streamline the process of decision-making supported by AI? For example, on the EU level, the new overarching AI regulation places AI systems supporting some kinds of decisions about learners among high-risk AI systems. Such frameworks require adjustments at both national and institutional level, to enable lawful, ethical and meaningful use of AI and learning analytics.

An important area of learning analytics, strongly supported by AI, refers to predictive learning analytics, relying on the development of predictive models. Although predictive modelling has been increasingly used and gaining significance, there are concerns regarding its adequacy in decision-making without human supervision [8]. A need for stronger governance has been identified, as algorithms no longer serve only for informing, but also provide and steer decision-making. As such, it is essential that algorithms are explainable [9], and learning analytics trustworthy. To achieve this trustworthiness, we should take a look through the human-centred lens and consider the perspectives of potential beneficiaries, primarily learners [10]. But above all, we should not trust machines to autonomously make decisions about humans: human-related decisions should be human-made.

Besides being evidence-based and oriented towards hard goals, decision-making should be balanced and fair. This calls for participatory decision-making, which is easily achievable in *complex* and *complicated* domains, because of well-established group decision-making approaches. However, we should consider different perspectives, especially at the higher decision-making (managerial) levels. How can we answer the common situation in education in which men are more involved in managerial decision-making, while a majority of the teaching workforce are women [11]?

Finally, it is crucial that decision-makers are ready to adequately respond to the given *complex* context, being agile, collaborative and relying on relevant expertise and evidence provided by learning analytics [2]. Both educators and decision-makers need future-literacy to be able to imagine possible futures, as well as practical skills to streamline education towards the desired future.

4 Conclusion

The recent developments in AI, with the lessons learnt from the pandemic, and the EdTech legacy, leave us at a new beginning. Besides challenges, we have a valuable opportunity to rethink and streamline education to be more future-looking. It is crucial to innovate decision-making processes as well. We should be aware of the level of complexity of the context and mindful that opening up and widening participation contributes to the relevance of our decisions and the quality of education. Finally, we should make sure that human-related decisions are made by humans, and not machines.

Acknowledgments. This work was partially supported by the *Trustworthy Learning Analytics and Artificial Intelligence for Sound Learning Design* (TRUELA) research project financed by the Croatian Science Foundation (IP-2022-10-2854).

References

1. Snowden, D.: Complex acts of knowing: paradox and descriptive self-awareness. J. Knowl. Manag. **6**(2), 100–111 (2002). https://doi.org/10.1108/13673270210424639
2. Snowden, D.J., Boone, M.E.: A leader's framework for decision making. Harv. Bus. Rev. **85(11), 68–76 (2007)**
3. Beauvais, A., Kazer, M., Rebeschi, L.M., Baker, R., Lupinacci, J.H.: Educating nursing students through the pandemic: the essentials of collaboration. SAGE Open Nurs. **7**, 237796082110626 (2021). https://doi.org/10.1177/23779608211062678
4. OECD. The state of higher education: One year into the COVID-19 pandemic, June 2021. Accessed 21 Sept. 2021. https://www.oecd-ilibrary.org/docserver/83c41957-en.pdf?expires=1632215384&id=id&accname=guest&checksum=CA1E82861 929056554205870BB8DDCEB
5. Svetec, B., Divjak, B.: Emergency responses to the COVID-19 crisis in education: a shift from Chaos to complexity. In: EDEN Conference Proceedings, no. 1, pp. 513–523, September 2021. https://doi.org/10.38069/edenconf-2021-ac0051
6. Grabar, D., Svetec, B., Vondra, P., Divjak, B.: Balanced learning design planning. J. Inf. Organ. Sci. **46**(2), 361–375 (2022). https://doi.org/10.31341/jios.46.2.6
7. Rienties, B., et al.: Online professional development across institutions and borders. Int. J. Educ. Technol. High. Educ. **20**(1), 30 (2023). https://doi.org/10.1186/s41239-023-00399-1
8. Khosravi, H., et al.: Intelligent learning analytics dashboards: automated drill-down recommendations to support teacher data exploration. J. Learn. Anal. **8**(3), 133–154 (2021). https://doi.org/10.18608/jla.2021.7279
9. McConvey, K., Guha, S., Kuzminykh, A.: A human-centered review of algorithms in decision-making in higher education. In: Proceedings of the 2023 CHI Conference on Human Factors in Computing Systems, pp. 1–15, April 2023. https://doi.org/10. 1145/3544548.3580658
10. Divjak, B., Svetec, B., Horvat, D.: Learning analytics dashboards: what do students actually ask for? In: LAK23: 13th International Learning Analytics and Knowledge Conference, pp. 44–56, March 2023. https://doi.org/10.1145/3576050.3576141
11. Grinshtain, Y., Addi-Raccah, A.: Domains of decision-making and forms of capital among men and women teachers. Int. J. Educ. Manag. **34**(6), 1021–1034 (2020). https://doi.org/10.1108/IJEM-03-2019-0108
12. Divjak, B., Svetec, B., Horvat, D., Kadoić, N.: Assessment validity and learning analytics as prerequisites for ensuring student-centred learning design. Br. J. Educ. Technol. **54**(1), 313–334 (2023). https://doi.org/10.1111/bjet.13290

Navigating the Evolution: The Rising Tide of Large Language Models for AI and Education

Peter Clark ⓘ

Allen Institute for AI, Seattle, WA, USA
petec@allenai.org

1 Introduction

AI has labored for 60 years trying to build systems with some degree of "understanding" of the world, with the promise of many potential benefits including in the world of Education and Intelligent Tutoring Systems. After all, a system that can answer questions and explain its answers should surely be a boon for education? Now, with the emergence of language models (LMs), arguably such systems have suddenly appeared seemingly from nowhere. Yet we don't really understand how they work, they make mistakes, and can be inconsistent. What should we make of this new world? To answer this, I'll briefly retrace my journey through this rapidly changing space, and then look to the future.

2 The Rapidly Changing World

2.1 Project Halo

The late Paul Allen, co-founder of Microsoft, had a long-standing dream to build a "Digital Aristotle" - an AI system that contained and understood much of the world's knowledge, so as to help humans in their endeavors, with particular emphasis on science and education. Project Halo, which ran from 2003-2013, was an embodiment of this vision, aiming to build a knowledgeable AI system with a deep understanding of a specific subject - college-level biology - that could be used for educational purposes [4]. The project constructed a large, formal knowledge-base that could answer novel biology questions and explain them by drawing on its internal rules, allowing users to understand the underlying biology behind its answers. This capability was deployed in an educational setting, as an interactive eTextbook called Inquire, where students could not only browse and explore the book but also ask it questions and get reasoned answers back [1]. The application was exciting, and extensive evaluations showed substantial educational benefit. However, the cost of manually building the knowledge base was prohibitive, and ended up as a show-stopper for the work.

2.2 Project Aristo

In 2014, this quest for a knowledgeable AI system was restarted at the then-new Allen Institute for AI (AI2), with machine learning and natural language processing (NLP) at the fore. Specifically, rather than manually build a knowledge-base, the new system - called Aristo - acquired its knowledge through reading and processing of text. With the addition of early language models, specifically BERT and RoBERTa, Aristo's ability accelerated, and in 2019 was able to score over 90% on the non-diagram, multiple choice part of the Regents 8th Grade Science Exams, reflecting the breakthrough that language models (LMs) was bringing to AI [2, 6].

However, illustrating the flip side of LMs, this success came at a cost, namely (unlike in Project Halo) Aristo was unable to systematically explain its answers, a problem which still plagues LM technology today. In particular, if the target application is in an educational setting, offering systematic, reliable explanations for answers is a key requirement, and one that Aristo did not meet. Much of our work since then has been devoted to identifying and extracting the "rational" domain models that underly a LM's normally typically opaque answers.

2.3 Human and Machine Mental Models

Part of teaching involves conveying an accurate, *mental model* of the world from the teacher to a student, allowing the student to understand observations and make good predictions. Do LMs even have "mental models"? Arguably, early LMs did not – their answers were highly correlated with superficial linguistic patterns rather than reflecting a deep understanding of a topic [7]. However, modern LMs perform amazingly well on a variety of tasks, including answering complex, novel questions, suggesting that they form *some* kind of internal world model when answering [5]. If we can expose these internal models that LMs have, many opportunities open up for learning and understanding.

Of course, we humans do not directly communicate our neural, mental models directly either (indeed, we are not fully aware of what all the neurons in our brain are doing). Rather, we express our thoughts in a symbolic formalism – natural language - that *can* be communicated easily, reflecting our understandings of the world that we have been taught, and that we have formed ourselves. Expressing our knowledge in natural language can be viewed as a *constructive* process, in which we generate (statements of) facts and rules that reflect our beliefs, and that chain together coherently to produce useful new knowledge. Indeed, a good textbook will convey concepts, facts, and rules about the world that have been carefully designed to reflect reality and allow new knowledge to be inferred.

2.4 Entailer

Can we evoke this process in LMs, namely have them articulate their "mental models" (coherent set of beliefs) also, and hence have them "teach" a user what they (the LMs) know? It turns out that we can, through a combination of (a) asking a model to generate systematic explanations for its answers and (b) checking the model actually believes its generations (i.e., it "believes what it says"). In a system called Entailer, we trained a LM to explain its answers in a *systematic* way, by producing a chain of reasoning (facts and rules expressed in NL) that objectively concludes the answer (rather than just an informal textual paragraph) [8]. This opens a window into the LM's model of the world, and - to the extent that window is accurate - provides a mechanism for conveying its underlying knowledge to a user.

2.5 TeachMe

Of course, LMs are not perfect, and have erroneous "beliefs" about the world. Systems like Entailer, above, can help users identify those machine misconceptions. Specifically, if the LM produces a wrong answer justified by a model-believed chain of reasoning, the user can then isolate the source of the LM's mistake by tracing down that chain to find the problem. For example, in simple reasoning about physics, Entailer incorrectly concluded that *a penny is magnetic*. By examining the LM's explanation, the user could trace the error to the (bad) LM belief in the reasoning chain that *metals are magnetic* (while in reality, copper is not magnetic). Furthermore, if the user provides the corrected knowledge to the LM (*copper is not magnetic*), they can correct the system's reasoning. This gave rise to a "flipped" educational tool that we developed, called TeachMe, in which the student learns by "teaching" the AI system [3]. Given an AI system that is making mistakes on a test, the student is tasked with identifying the system's mistakes and correct them (including using auxiliary material, if necessary), and in so doing, develops and reinforces their own understanding of the topic. This follows Feynman's maxim that: "If you want to master something, teach it."

2.6 A Never-Ending Research Assistant

Even in the last year, LMs have continued to rapidly advance. Today's LMs not only chat and answer questions, but can search document repositories, read technical documents, generate and execute software code, run (software) experiments, and even generate research hypotheses that might be worthy of exploration. Our current work seeks to integrate these capabilities together into a smart research assistant, able to assist sci-entists in their work in a variety of ways. The AI assistant can help manage the vast information overload that scientists experience today, discuss specific technical papers with the scientist, and semi-autonomously execute complex workflows (experiments, dataset analyses). Similarly, the human scientist can provide top-level research direc-tions and hypotheses to expore, and guide and direct the AI assistant. In this way, both the human scientist and AI Assistant can learn from each others' strengths. And, following the earlier theme, a key task for the AI assistant is maintaining a rational understand-ing ("mental model") of the research topic: namely a graph of the claims, hypotheses, evidence, experiments, and the inferential relationships between them all, and ensuring that this understanding remains consistent. And, unlike the earlier systems, there is one key difference: this understanding is dynamic, constantly changing as new papers are published and new research results obtained - and, with a bit of luck, perhaps help the scientist discover breakthroughs that would otherwise have been missed.

3 Conclusion

The rapidly evolving capabilities of LMs is perhaps one of the most surprising results that the field of AI has had, and is one that is likely to touch all aspects of life, including in education. I've traced the rapid evolution of my group's quest to build knowledgeable systems, from hand-built knowledge bases to a modern, LM-based research assistant agent. While the possibilities are endless, let me also offer three concluding thoughts: embrace the technology; be cautious (the technology is still highly fallible); and, to the extent possible, don't only look at the LM's answers but also at the knowledge and reasoning used arrive at those answers – I've outlined some of the techniques we have been using to expose that information, and in the end it is that which will help us have confidence in the information that LMs are providing, and help us learn from and interact with these new systems appropriately.

References

1. Chaudhri, V.K., et al.: Inquire biology: a textbook that answers questions. AI Mag. **34**, 55–72 (2013). https://api.semanticscholar.org/CorpusID:9127231
2. Clark, P., et al.: From 'f' to 'a' on the n.y. regents science exams: an overview of the aristo project. AI Mag. **41**, 39–53 (2019). https://api.semanticscholar.org/CorpusID: 202539605
3. Dalvi, B., Tafjord, O., Clark, P.: Towards teachable reasoning systems: Using a dynamic memory of user feedback for continual system improvement. In: EMNLP (2022)
4. Gunning, D., et al.: Project halo update – progress toward digital aristotle. AI Mag. **31** (2010)
5. Hinton, G.: Two paths to intelligence. ACL (2023). https://doi.org/10.48448/hf7y-x909. (Keynote Talk)
6. Metz, C.: A breakthrough for AI technology: passing an 8th-grade science test. NY Times (2019). https://www.nytimes.com/2019/09/04/technology/artificial-intelligence-aristo-passed-test.html
7. Ribeiro, M.T., Guestrin, C., Singh, S.: Are red roses red? Evaluating consistency of question-answering models. In: ACL (2019)
8. Tafjord, O., Dalvi, B., Clark, P.: Entailer: Answering questions with faithful and truthful chains of reasoning. In: EMNLP (2022)

Contents – Part I

WideAIED

Late-Breaking Results

Contents – Part II

Doctoral Consortium

Workshops and Tutorials

Blue Sky

Aligning AIED Systems to Embodied Cognition and Learning Theories

Ivon Arroyo[1]([✉])[ID], Injila Rasul[1], Danielle Crabtree[1], Francisco Castro[2][ID], Allison Poh[1][ID], Sai Gattupalli[1], William Lee[1], Hannah Smith[3][ID], and Matthew Micciolo[4]

[1] University of Massachusetts Amherst, Amherst, MA 01003, USA
ivon@cs.umass.edu
[2] New York University, New York, NY 10012, USA
[3] Assumption University, Worcester, MA 01609, USA
[4] Worcester Polytechnic Institute, Worcester, MA 01609, USA

Abstract. Embodied Learning involves the development of perceptuo-motor schemas as students manipulate physical objects that embody a concept, or when students move in domain-relevant ways, physically enacting solutions to problems, as well as gesturing as they interact with peers face-to-face. Teaching standards (e.g., for K-12 math education) align with such theories of learning. For instance, teachers are encouraged to have students "talk mathematics", construct viable arguments and critique the reasoning of others, model with mathematics, and use tools with precision, among others. Yet most learning technologies and tutoring systems are not designed to facilitate embodied learning; instead, they tend to align with more traditional views of student learning. This paper explores possible ways AIED systems can reflect embodied learning theories, within K-12 STEM education. We present the *WearableLearning* platform as an example that intends to bring embodied learning to mathematics and computing education for students at the K-12 level, with implications and suggestions on how to use Artificial Intelligence to make embodied learning environments more effective.

Keywords: Embodied Learning · Math Games · Mobile Devices · Learning in Motion

1 Introduction

Entering a K-12 classroom and witnessing an experienced math instructor in action is a delightful experience. Within these classrooms, you might see students using tangible tools, such as measurement instruments, number lines, fraction tiles, volumetric objects, and base-10 blocks. These tools are often utilized to enhance conceptual understanding, taking advantage of the physical classroom environment where students engage in activities on tables or on the floor, and participate in rich discussions with their peers and teachers.

© The Author(s), under exclusive license to Springer Nature Switzerland AG 2024
A. M. Olney et al. (Eds.): AIED 2024 Workshops, CCIS 2150, pp. 3–17, 2024.
https://doi.org/10.1007/978-3-031-64315-6_1

Mathematical objects serve important roles, acting as bridges between familiar, intuitive prior knowledge and new concepts, supporting students to develop abstract mathematical thinking [10, 13]. The main role of teachers is to facilitate the bridge from concrete representations to abstract concepts, and this practice continues in good math classrooms throughout secondary school. Alternative methods involve students expressing solutions to math problems through their entire bodies. This may include using the body to measure objects, people, or spaces, and even physically walking out numerical mathematical solutions.

These math teaching practices align with theories about how people learn: Embodied Cognition and Embodied Learning posit that sensory-motor action, including gestures, are essential for teaching and learning, as ideas are distributed among the mind, the world, and the social context, and that hand gestures may support essential communication during learning [19]. Students explore mathematical relationships and concepts by manipulating real objects in the environment and communicating face-to-face with peers and teachers.

Yet, there is a significant disparity between the math learning technologies employed in schools and the effective practices of mathematics educators. Conventional K-12 mathematics educational technologies and tutoring systems are typically not designed to facilitate embodied cognition; instead, they tend to align with more traditional perspectives on student learning. Teaching standards for mathematics have evolved, influencing how math educators are expected to instruct and engage with students. Most traditional learning technologies, however, still resemble digital worksheets. Embracing embodied learning implies moving the Human Computer Interaction (HCI) "off the keyboard", to experience concepts, engage in hands-on and socially-rich activities, and obtain support during moments of struggle and cognitive conflict in real-time.

This article poses how learning technologies might address embodied learning, a shift that entails a focus on refining and deepening interactions with learning technology. For instance, teachers should implement tasks that are hands-on and cognitively demanding, introduce discussion, and require strategic thought and rich problem solving by students [12]. We propose how learning technologies could support learning via the best of two worlds: a) intelligent tutoring systems that assess and trace students' knowledge and affect, and can personalize instruction based on moment-by-moment tracking of students' mental states (i.e., knowledge and affective tracing); and b) Embodied Cognition and multi-modal interaction, encouraging students to physically act on the environment in meaningful ways, for sense making and understanding of ideas.

2   Background Research

Teaching Mathematics. K-12 Teachers are encouraged to employ various representations and tools in the exploration of math ideas, and incorporate the manipulation of various tools, shapes, models, or technology. In the United States, The Math Practice Common Core Standards (MPCCS) encompass all grade levels and emphasize discourse and communication among students in the classroom [11]. Teachers should facilitate discussions, ask appropriate questions, and solicit student ideas to support students' mathematical thinking [24], offering ample opportunities for students to "talk mathematics" and share ideas in groups [28]. Discourse involves an explanation of mathematical strategies and students' own justification of their thinking and reasoning [13]. Explanations focus not only on how students obtain solutions but why certain strategies are appropriate.

Traditional Learning Technologies & AIED Systems. The affordances of tutoring systems and cognitive tutors as supports to student learning and teacher's pedagogical goals, are fairly well known: a) they provide immediate feedback when students solve problems and enter correct/incorrect answers [31]; b) they provide detailed personalized support when students make mistakes (help provision) or when they seek help during cognitive conflict [3]; c) they enable the use of detailed student logs from student interactions with the computer to provide "stealth assessments" even within educational games, without stopping to test [26]; and d) they allow personal pacing of problem solving work, with students working at their own pace with a "more knowledgeable other" (computer versus human) enabling, to some extent, replication of a reasonable form of a teacher/tutor into as many students in the class, albeit a digital tutor that supports students even emotionally as they learn [8]. Yet, underlying most learning technologies is an information-processing view of learning based on memory, problem solving and practice, where knowledge is considered an internal process in the mind.

Embodied Cognition. The exploration of cognition in research is expanding to encompass a wider viewpoint, with the recognition that it is distributed across the mind, senses, motor capabilities, and social interactions [29]. This theory of embodied cognition assumes that sensory perceptions, motor functions, and sociocultural contexts shape the structure and development of thinking skills, including mathematical thinking. Learning involves the creation, manipulation, and sharing of meaning through bodily interactions [15]. Students may be guided to encounter, discover, and rehearse perceptuo-motor schemas in relation to mathematical concepts and relationships. Research on hand motion gestures while teaching/learning suggests that a motor encoding of our mathematical ideas exists, and thus, that mathematics teaching should use bodily motion, action and hand gesturing [1]. Yet, one problem in embodiment research is the separation between embodied cognition/learning research, and teachers and education practitioners [18].

Frameworks of Embodied Educational Media. Abrahamson et al. [2] proposed a theoretical framework that encapsulates the elements mentioned, offering an ecological-dynamics analysis of learning mathematics through interaction with educational media designed for embodied experiences. This framework includes investigating and informing social practices in teaching and learning of physical movement, which believes in instructional affordances of instructional materials. We naturally ground new concepts in perceived affordances for enacting movement forms that solve motor-control problems. Similarly, the theory of 4E Cognition (embodied, embedded, enacted, and extended) [25], suggests that mathematical ideas are grounded in perceived dynamic images that develop through sensorimotor interaction, and then rise to consciousness through languaging. These theories bear practical implications for how we design learning environments. Students can develop new mathematical knowledge by first learning to move in new ways and only then making sense of this emergent capacity within a discipline --such as mathematics.

Importantly, we are not claiming that every movement will lead to learning. Certain kinds of movement (e.g. running to the end of the room) might contribute to the engagement/enjoyment, but it is unrelated to learning and not what we are referring to in this paper as "embodiment". Instead, we are referring to activities that involve action, movement and gestures that match, signify, and externalize mathematical ideas. Past research has shown that random movement does not lead to math learning, rather movement encouraged by math-embodied activities has to match, capture, and/or express mathematical concepts to be learned [18].

Embodied Design Frameworks. Principles of embodied design can be used as a critical lens on how digital resources are created and brought to life into mathematics learning tasks. *WearableLearning* facilitates the creation of learning activities that implement themes of embodied interaction design [14]. Embodied games and simulations have utilized a large breadth of design approaches; thus, we rely on a unified design framework [14] that aggregates different conceptual design approaches for embodied learning systems, which is based on: 1) kinds of physical interaction, classified into embodied, enacted, manipulated, surrogate, or augmented; 2) kinds of social interactions that the activity requires (e.g., collaborative, competitive and the role of Non-Player-Characters such as supportive teachers); and 3) role of the environment, and the degree of physical/virtual/mixed world that it involves. We expect teams of teachers and researchers to create activities spanning different areas of this framework, as our existing math game-like activities already cover different areas.

3   A Case Study: WearableLearning

We present a case study on *WearableLearning* as an example of a learning technology that employs embodied learning theories, both for Mathematics learning and early Computer Science, designed for children in grades 3–8. *WearableLearning* enables K-12 mathematics thinking and learning via physically active and multiplayer math games, using mobile devices (e.g., Wi-Fi-enabled phones or tablets); these are embodied mathematics experiences, quests, measuring objects and space, and even walking out math solutions in physical spaces [4, 5]. Students participate in active and collaborative math activities, finding mathematics in their school environments, and using manipulatives and other typical materials in a math classroom (e.g., geometric shapes, measurement tools, fraction tiles). Mobile devices become "virtual assistants," digital tutors that provide instructions and support. Assistants are coordinated across students (for team-based collaboration or cooperation), and multiplex the teacher in as many parts as students, orchestrating activities, providing just-in time help/hints and feedback as they move and "play" by exploring the actual space/objects in real-time.

Mobile devices provide great potential to augment educational contexts and promote tangible, mobile, and ubiquitous learning, as technological devices have become more portable, less cumbersome, and even wearable. Mobile technologies and tools are light, easy to wear, and offer the potential of tracking large amounts of physical activity and individual information. The promise is a new genre of learning technologies that blends with classroom culture, hands-on manipulatives, encouraging sense-making, while still retaining the benefits of intelligent tutoring systems (ITS). *WearableLearning* both promotes human-to-human interaction, and leverages technology for what it affords best: self-pacing, just-in-time support, coordination, cooperation.

Fig. 1. Children playing *EstimateIT!* in schools and outside, via *WearableLearning*

Game-like activities created by teachers and researchers using the *WearableLearning* platform help grade 5–8 students learn mathematics [4]. These embodied math activities were created for a range of grades and standards, from numbers and operations to algebra to geometry and measurement, for children grades 3–8. The *WearableLearning* Platform and Curriculum have been used to support multi-student math active learning, using mobile devices in real K-12 school settings and afterschool programs. The platform is web-based, supporting embodied developmental mathematics game play for multiple students, blending hands-on activities and collaborative games within the classroom culture. *WearableLearning* supports collaborative, game-based learning that enables students to move about and explore their environment, while guided by mobile devices. It incorporates embodied activities into the realities of classrooms, curricula, and standards [4] and enables students to play social and active learning games that stimulate mathematical thought. It supports multiplayer, interactive, gamified experiences that use smartphones in full classes of K-12 students. While using *WearableLearning*, students can manipulate objects available in the physical environment to take measurements, compute, estimate, and gesture while interacting with their peers and teachers in rich face-to-face discourse. As game players, students use *WearableLearning* through smartphones to solve quests and math problems. As game creators, teachers deepen their math knowledge by designing math games that target the curriculum their students are learning.

Prior research has analyzed the feasibility of teachers, researchers, and students creating physically-active multiplayer math games with *WearableLearning*, and identifies challenges they might encounter [27]. Math game playing and creation with *WearableLearning* has so far been implemented with roughly 500 students, as well as 20 math teachers, who also have created a variety of games for students to play as part of an eight-hour teacher professional development program [28].

WearableLearning activities apply principles of embodied interaction design [14], in that they involve *thinking through doing,* as mathematical reasoning uses bodily capacities; it implements the *bodies matter in performance* principle, as motion/action enable better performance; it implements *high visibility for sharing,* as visible mathematical objects/artifacts support synchronous collaboration. Embodied activities also involve *risk-taking,* as a physical action is characterized by risk; choosing an action requires commitment, personal responsibility, and accountability to the team. This results in high levels of affective engagement with tasks, as opposed to technologies that often strive to minimize or eliminate risk.

The promise of the *WearableLearning* technology is a new form of learning environment, augmented by technology, that naturally blends into the classroom to facilitate interactions between teachers, students, and the curriculum, and where collaboration and face-to-face discourse and gesturing between students and teachers is a natural ecological part of the problem-solving and learning process (see Fig. 1). A challenge of embodied technologies is maintaining the affordances of embodied interaction while deploying these platforms with full classes of students, at scale.

3.1   WearableLearning and Its Math Games Library

The *WearableLearning* Platform enables teachers to play, manage, and create educational game experiences that involve several students, physical movement, and objects in the environment. Mobile devices are worn on the lower arm (as many games require free hands for hand actions, e.g. Fig. 3), yet for other games students may carry them (when the math games involve mostly their legs, e.g. stepping, Fig. 2). Anybody can browse public *WearableLearning* games from its library, to play with students as class-wide activities. WearableLearning users can take on different roles within the platform: Game Players, Managers or Creators.

A math games library is accessible from WearableLearning.org, with their designated Math Common Core Standards (MCCS), which US teachers understand. These games were designed and implemented in *WearableLearning's* Game Editor, and are deployed with its Game Player, fit a variety of Math Common Core standards [11] and student ages/grade-levels, concretely for MCCS areas 3.G, 3.NF, 4.MD, 4.NF, 5.NF, 6.NS, 7.NS., 7.RP, 8.G and 8.EE. These games were created either by researchers and/or by a team of math and STEM teachers during an 8-session workshop. The process of creating games is well understood by now, and an 8-session Professional Development (Smith et al., 2023) already exists to prepare teachers to create math games in *WearableLearning*, via the Game Editor (see Fig. 4). The following are examples of the games available in the *WearableLearning* games' library.

Tangrams Race is a relay-race game designed to teach geometric concepts. Each player within a team wears a smartphone and takes turns to run to an end-line to retrieve geometric shapes whose description is specified in the smartphone. Once students retrieve a geometric piece, they enter a color code that identifies the shape (a sticker on the shape) and receive immediate feedback: if they picked the correct shape, the player runs back with the shape to tag the next teammate. Otherwise, they can ask for hints until they retrieve the correct piece. Once all correct pieces have been retrieved, students work as a team to create a final tangram with their shapes [34].

Estimate IT! is a measurement and estimation scavenger hunt-style game for elementary school children where teams of students search for volumes described in their smartphones in mathematical terms (e.g., "Find a cube with a 6″ side"). Players are given an unmarked 12-inch stick to support measurement estimations. Players collaborate and discuss geometry concepts present in the game (Fig. 1). The game provides hints about how to measure and/or estimate to find the correct volume (e.g., "Once you find a cube, use the dowel to estimate its height, which should be 6 inches tall"). Once students find the correct volume, they enter an identifier (colored squares sequence, sticker on the volume) into the phone and receive immediate feedback [35].

Integer Hopscotch gives students word problems on addition/subtraction of integers, as players walk along a 20-foot number line to find the answer. Each integer on the number line has a color code, which students input into a mobile device to receive immediate feedback and can request a hint. e.g. about whether they should move right or left. The goal of the game is to work in pairs to complete all the integer problems by "walking along the number line".

What's My Line? is a slope-equation and graph-matching game that involves slope-intercept form of equations to ensure that students can read and interpret equations. Students move through different stations around the classroom and are asked to match equations presented on a smartphone with the corresponding line on a large paper graph where several lines are drawn. Students recreate the correct lines on their own paper pads as they move between stations, before other teams can finish the game. The drawn lines on the team's paper-pad reveal a secret shape.

Flip, Slide, Turn is a game about geometry on a giant coordinate-plane tarp on the floor of a classroom, gym, or a square-tiled patio (Fig. 2). Pairs of students share a device—one of the students reads the instructions aloud while the other student becomes a point on the coordinate-plane tarp, and is reflected, translated, or rotated around the origin by walking on the tarp. One student enacts the solution, the other reads aloud and commands (surrogates); students think, argue, celebrate successes [22].

Teachers and researchers have been capable of creating highly innovative math activities with game-like qualities, incorporating various forms of embodiment (enacted, manipulated, surrogate, augmented). The activities also constitute effective tools for math learning, as evidenced by students' consistent improvement in test performance from before to after exposure to these math game-like activities, which encompass diverse forms of embodiment, including the manipulation of objects, enactment of solutions (such as measuring or performing addition/subtraction on a number line), and embodiment through actions like Flip, Slide, Turn, where students represent x-y points.

These engaging embodied game-like activities, supported by *WearableLearning*, are the beginning of new research at the intersection of problem solving, embodied action and gesture, and mobile learning technologies. The games that teachers and researchers created were implemented in real math classroom settings and showed improvement in math problem solving performance across different areas of math and grade levels, even after 30–40-min exposures. All games have led to significant learning gains so far, comparing pretests to posttests, and while the results presented don't explicitly reveal it, students were observed to be incredibly engaged while playing these games.

Fig. 2. *WearableLearning* utilizes mobile devices to provide teams of students with problem-solving challenges. In the *Flip, Slide, Turn* game, students collaborate and enact the solutions with their full body on a coordinate plane.

The kinds and degree of embodiment varied with each math game, suggesting that future research and development should focus on understanding how physical actions in each game mediates student learning and the affordances of the materials, objects, space, and human interactions, as mediators of math learning. Teacher-created games appeared to be more effective than researcher-only created games in general, probably because teachers have more pedagogical and classroom knowledge, which allowed them to target specific areas where students needed improvement, and created tests that better assessed the material taught by each game.

4   Research Line #1: Analyzing the Benefit of Embodiment in LTs

Similar to much research on effective instructional strategies within Learning Technologies [3, 24], an important line of research relates to understanding the potential of embodiment (domain-relevant motor action and gesturing) compared to more traditional stationary learning technologies, and compared to business as usual in the classroom (e.g. math class using manipulatives but without technology). For instance, initial research studies by the authors have attempted to understand the potential of the embodied *WearableLearning* concept, in comparison to control groups. One of the studies compared a classroom lecture to the *Tangrams Race*, implemented via a prototype of *WearableLearning*, and yielded a medium effect size for math improvement, where an experimental *Tangrams Race* condition outperformed a control business as usual condition [34].

A second study analyzed the potential contribution of the affordances of the physically active *TangramsRace*, compared to a stagnant one. The experimental condition was *TangramsRace* in *WearableLearning*, as shown in Fig. 3. The control condition was a digital platform that provided the same questions and the same hints, but the it required students to sit down and solve questions on a traditional math practice technology, with immediate feedback, and with hints available upon request, similarly to the Tangrams Race in *WearableLearning* (experimental condition). Participants were N = 54 students from grades 3–5 in afterschool programs in the United States, who were randomly assigned to the experimental (*Tangrams Race*) or sitting down (control) condition. The study yielded no significant results (p < 0.17), yet, the effect size corresponded to a medium effect size, Cohen's d = 0.36, favoring the Tangrams Race condition.

Fig. 3. A student playing the *Tangrams Race game*. From left to right: a) Student reads the mathematical constraints of the sought object; b) student rotates triangle and compares sizes using hands; c) student enters color code for identifying object; d) after receiving feedback, the student picks the object and runs back; e) at the start line, students collaborate to build the requested Tangram.

Embodied Hints and Support. Researchers have analyzed hand gestures as part of STEM lectures to improve concept understanding [9]. Given the embodied nature of our learning tasks, i.e., actions to succeed in the math games are highly physical, gesture-based hints in embodied tutoring systems should allow struggling students to receive improved support while playing games. In one research study [7] we performed an *embodied cognitive task analysis* of children's gestures/actions during measurement tasks. Using data from that study, the authors created hints that conveyed the most common motor strategies observed from college students, which acknowledged the most common mistakes of elementary students. A controlled study analyzed the benefit of having "embodied hints" on the phones [35], which tapped on those typical estimation/measurement errors. This study yielded a medium effect size of Cohen's $d = 0.37$, in favor of the "embodied hints" condition. Given the results we consider that the comparison of carefully designed embodied games, deployed via *WearableLearning*, and the creation of digital embodied support, has good promise of yielding a significant benefit compared to business as usual in future studies.

Authoring Embodied Games w/Narratives. Embodied tutoring systems need an authoring environment to create content for it. In *WearableLearning*, the Game Editor environment is designed to create/program unique games with a wide variety of content, levels, and modes of collaborative or competitive play, for players or teams to engage in. As Game Editors, students define games as Finite State Machines (FSMs) that describe the behavior of players' mobile devices (see Fig. 4). FSMs are computational models that can be used to simulate phenomena, sequential logic, and the behavior of computer programs. They include a finite number of states and directional transitions between states (including loops). Game Editors program their games using *Wearable-Learning's* drag-and-drop FSM-based programming language, where they specify output states to indicate media to display on mobile screens (e.g., text, images, videos) for the purpose of providing game instructions, questions, or hints, that involve interacting with objects and the environment. These instructions in the phone follow a logic towards a narrative in the game, which can vary per game. For instance, in the *Tangrams Race*, the

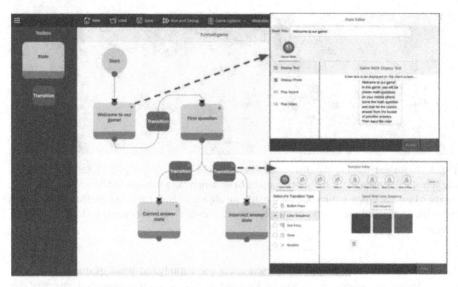

Fig. 4. Authoring Games as Finite-State Automata: the *Game Editor* in WearableLearning

game needs to guarantee that all seven pieces required for the final puzzle are collected by different students at the end line, for them to succeed at putting the puzzle together in the end. Thus, there are constraints about objects that need to be retrieved by children, that guide them to find, manipulate and retrieve correct objects.

States correspond to *screen outputs* and can be configured to display different media to make interesting prompts such as displaying a math problem with pictures related to the problem, combined with an audio recording. Game Editors also specify transitions between states to indicate a possible input that players can submit (e.g., button presses, text input, color sequences that identify objects) to move from one state to the next (see Fig. 4). Transition input specification enables different game response interactions; for example, game creators can ask players to respond to problems with a text entry (e.g., "How many sides does a triangle have?), or use buttons to enter choices (e.g., "Push the black button if you want a hint"), or type color sequences that identify objects (e.g. red, blue, green may identify a certain kind of object to be sought for). Game Editors can test-run and debug their games in the built-in debugger, which simulates gameplay by launching a phone simulator to play-test each game and player's actions.

Students Creating Physically Active Games Develop Computational Thinking. WearableLearning allows teachers and students to visually program and debug games via a drag-and-drop programming interface to specify the behavior of game players' mobile devices that support players throughout the games. The WearableLearning curriculum [8] was created to entice teachers (PD) and their students to engage in game design and creation. Designed for programming novices, 5th-10th grade students have created games by designing them first on paper and later within the Game Editor. Research studies have suggested that teachers develop CT as they create math games [27]. Yet, it is not only teachers that have created games, but also several research studies have explored how students creating games helps them develop STEM skills by creating the games themselves. The *WearableLearning* math game creation curriculum has been used by ~ 25 teachers and 100 students, and has suggested that this 7 h activity enables them to gain computational thinking skills, particularly algorithmic thinking and knowledge about models and simulations [21].

5   Research Line #2. Enhancing Embodied Technologies with AI

The *WearableLearning* game-based environment operates as a traditional tutoring system, however, this also incorporates bodily movement and interaction with objects, peers, and the physical space. In *WearableLearning*, there are also correct and incorrect answers, for chosen objects in the real world (or for configurations of objects). There are also hints that can be requested. Similarly to game-based tutoring systems, there is a narrative that needs to be followed temporally, and occasionally there are chance actions carried out (player choices).

Assessment of Student Knowledge via Multimodal Learning Analytics. The goal of Learning Analytics is the study of the low-level traces left by the learning process, in order to better understand and estimate one or more learning constructs that are part of the process and, through carefully designed tools [32]. Most AIED systems, including *WearableLearning*, collect detailed log data from students, math-task performance in real-time as students carry on tasks and interact with the technology. A new line of research involves the possibility to feed formative assessments back to teachers after students work with physical objects, or with their full bodies, using sensors for assessment of students' motion and gesture. Teachers spend too much time on assessment (usually teaching, pausing to give tests, and then re-teaching); this takes time away from learning opportunities. *WearableLearning* gathers large amounts of data from students while they work without pause. This enables the possibility of providing motor "stealth assessments" [26] as students act/interact with mobile devices as part of participating in mobile game-like activities.

Like any game-based AIED system, it is possible to estimate students' knowledge while playing a game-like activity. In *WearableLearning*, each "state" can contain metadata, such as the fact that it is a "math question", and also the mathematics "skill (standard)" that is associated with. Other states are "answer" states and may be categorized as "correct" or "incorrect" states. Succeeding or failing at the math question (by arriving at the expected correct answer state, that may correspond to entering a code that indicates a correct/incorrect object in the environment) is evidence of a student's knowledge. This means that knowledge can be modeled with a traditional Bayesian Estimation of students' knowledge based on correctness, errors, and hint requests [26,33], even in these mobile game-based embodied environments. While *WearableLearning* does not yet make such assessments for each student), that is the long-term vision for this technology, especially the possibility of sensing specific motor actions/gestures automatically, when motion sensors are worn.

Figure 5 shows new possibilities of AIED research where lower-level sensor data can help support higher levels behaviors and motor actions first, to later infer physical gestures and actions, which should help assess higher level cognitive/affective constructs, as well as domain-level practices (e.g. mathematics common core practices that are encouraged in a mathematics classroom). The example in Fig. 5 shows that a variety of hand movements, as well as full-body movements, can be inferred from sensors that are worn or carried, either on mobile phones/watches, or via ring/bracelet sensors on the hand. These behaviors, in combination with typical HCI behaviors, can be used to recreate motor gestures and actions that are domain relevant. These, in turn, can be used to infer cognitive/affective states, and even Mathematics Common Core Practices. Creating models that can make these inferences involves collecting data from real students as they engage in these embodied activities in schools, gathering sensor data using sensors, and also videotaping students, to later annotate a variety of motor actions/gestures in videos, as well as cognitive/affective student states. By fusing, triangulating and integrating these sources of data we can obtain an invaluable datasource that captures individual and collaborative behaviors, affective/cognitive states, gestures, actions and interactions with technology and real objects, math strategies, as well as kinematic descriptors and motor data, at a high level of detail. Educational Data Mining and Machine Learning can be used to make these higher-level inferences from lower-level sensorial data streams. Once created, these models can be used to make assessments, which can support not only basic research on how people learn, but also support teachers through Dashboards (under development) to track students' progress and provide formative assessments [20], without stopping to test.

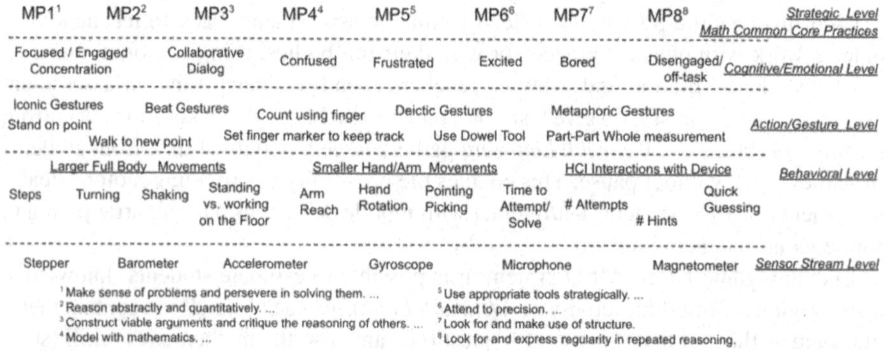

MP1[1]	MP2[2]	MP3[3]	MP4[4]	MP5[5]	MP6[6]	MP7[7]	MP8[8]	_Strategic Level_ _Math Common Core Practices_
Focused / Engaged Concentration	Collaborative Dialog	Confused	Frustrated	Excited	Bored	Disengaged/ off-task		_Cognitive/Emotional Level_
Iconic Gestures Stand on point	Beat Gestures Walk to new point	Count using finger Set finger marker to keep track	Deictic Gestures Use Dowel Tool	Metaphoric Gestures Part-Part Whole measurement				_Action/Gesture Level_

Larger Full Body Movements				Smaller Hand/Arm Movements			HCI Interactions with Device		_Behavioral Level_
Steps	Turning	Shaking	Standing vs. working on the Floor	Arm Reach	Hand Rotation	Pointing Picking	Time to Attempt/ Solve # Attempts # Hints	Quick Guessing	

Stepper	Barometer	Accelerometer	Gyroscope	Microphone	Magnetometer	_Sensor Stream Level_

[1] Make sense of problems and persevere in solving them. ...
[2] Reason abstractly and quantitatively. ...
[3] Construct viable arguments and critique the reasoning of others. ...
[4] Model with mathematics. ...
[5] Use appropriate tools strategically. ...
[6] Attend to precision. ...
[7] Look for and make use of structure.
[8] Look for and express regularity in repeated reasoning.

Fig. 5. Inferring Mathematics Common Core Practices [11] from sensor data

6 Discussion

During the COVID pandemic, teachers and parents felt children spent "too much time" on the computer, resulting in a negative impact on social skills development and human communication [17]. While some tutoring systems may allow collaborative learning via chat, there is something clearly missing regarding the benefits of face-to-face discourse and gesturing in education. Embodied Learning involves collaboration and gesturing with other humans. There is much to study about the benefits of face-to-face student interaction and sharing of spaces, mediated by physical and digital objects, in comparison to chat-based communication, or traditional digital learning environments.

The main benefits of Intelligent Tutoring Systems regard smart pedagogical decisions based on a moment-to-moment automatic assessment and understanding of the student at various levels. In this BlueSky paper, we have argued that it is possible and feasible to create AIED systems that better fit the reality and needs of K-12 classrooms, by aligning to Embodied Learning principles. While embodied learning research is pinning down the precise cognitive role of motor action in conceptual learning [1], the objective of digital resources should be to foster opportunities for students to develop new situated perceptions, mobilizing the prospective enactment of mathematical ideas and mathematical transformations [2]. This implies the need to create novel technologies that enable the enactment of mathematical practices, while at the same time automatically assessing how students move, and the impact these actions/gestures have on cognition, affect and learning.

The idea of using sensors for assessments of affective/cognitive constructs is not new [6]. The main difference with prior research is two-fold: a) the fidelity that embodied learning technologies can bring to the reality of teaching/learning experiences of K-12 classrooms; b) the growing importance of motion, action and gestures in human learning; c) the opportunity, given abundance of sensors in mobile (and wearable) digital devices. Yet, for a meaningful impact, it is essential that embodied digital learning technologies efficiently scale, to meet the demands of real-world settings, including schools and classes with a ratio of 20 students per teacher.

Novel multimodal learning and assessment systems cannot assume that students will always be ready and "in the mood" for learning, nor have all the prerequisite knowledge available to them. Yet this is the strong point of intelligent tutoring systems and other AIED systems, which can respond to students in a variety of ways depending on a variety of cognitive/affective states. There are many opportunities ahead at the intersection of AIED and embodied learning, in the K-12 education arena.

References

1. Abrahamson, D., et al.: The future of embodied design for mathematics teaching and learning. In Frontiers in Education, vol. 5, p. 147. Frontiers Media SA, August 2020
2. Abrahamson, D., Ryokai, K., Dimmel, J.K.: Learning mathematics with digital resources: Reclaiming the cognitive role of physical movement. In: Handbook of Digital (Curriculum) Resources in Mathematics Education. Springer, New York (2023). https://doi.org/10.1007/978-3-030-95060-6_22-1
3. Aleven, V.: Help seeking and intelligent tutoring systems: theoretical perspectives and a step towards theoretical integration. In: International Handbook of Metacognition and Learning Technologies, pp. 311–335. Springer, New York (2013)
4. Arroyo, I., Castro, F., Smith, H., Harrison, A., Ottmar, E.: Augmenting Embodied Mathematics Classrooms with Mobile Tutors. AERA 2021 (2021)
5. Arroyo, I., Closser, A.H., Castro, F., Smith, H., Ottmar, E., Micciolo, M.: The Wearable-Learning Platform: A Computational Thinking Tool Supporting Game Design and Active Play. Technology, Knowledge and Learning, 1–10 (2022)

6. Arroyo, I., Cooper, D.G., Burleson, W., Woolf, B.P., Muldner, K., Christopherson, R.: Emotion sensors go to school. In: Artificial intelligence in Education, pp. 17–24. IOS Press (2009). https://doi.org/10.3233/978-1-60750-028-5-17

7. Harrison, A., Smith, H., Botelho, A., Ottmar, E., Arroyo, I.: For good measure: identifying student measurement estimation strategies through actions, language, and gesture. In: Gresalfi, M., Horn, I.S. (eds.) The Interdisciplinarity of the Learning Sciences, pp. 869–870, June 2020, ISSN: 1573–4552

8. Arroyo, I., Woolf, B.P., Burelson, W., Muldner, K., Rai, D., Tai, M.: A multimedia adaptive tutoring system for mathematics that addresses cognition, metacognition and affect. Int. J. AI Educ. **24**(4), 387–426 (2014)

9. Bentley, B., Walters, K., Yates, G.C.: Using iconic hand gestures in teaching a year 8 science lesson. Appl. Cogn. Psychol. **37**(3), 496–506 (2023)

10. Boggan, M., Harper, S., Whitmire, A.: Using manipulatives to teach elementary mathematics. J. Inst. Pedagogies **3** (2010)

11. National Governors Association Center for Best Practices & Council of Chief State School Officers. *Common Core State Standards for Mathematics*. Washington, DC: Authors (2010)

12. Clarke, D., Roche, A.: Using contextualized tasks to engage students in meaningful and worthwhile mathematics learning. J. Math. Behav. **51**, 95–108 (2018)

13. Cobb, P., Wood, T., Yackel, E., McNeal, B.: Characteristics of classroom mathematics traditions: an interactional analysis. Am. Educ. Res. J. **29**(3), 573–604 (1992). https://doi.org/10.2307/1163258

14. Klemmer, S.R., Hartmann, B., Takayama, L.: How bodies matter: five themes for interaction design. In: Proceedings of the 6th Conference on Designing Interactive Systems, pp. 140–149, June 2006

15. Lindgren, R., Johnson-Glenberg, M.: Emboldened by embodiment: Six precepts for research on embodied learning and mixed reality. Educ. Res. **42**(8), 445–452 (2013)

16. Melcer, E.F., Isbister, K.: Learning with the body: a design framework for embodied learning games and simulations. In: Software Engineering Perspectives in Computer Game Development, pp. 161–195. Chapman & Hall/CRC (2021)

17. Misirli, O., Ergulec, F.: Emergency remote teaching during the COVID-19 pandemic: parents experiences and perspectives. Educ. Inf. Technol. **26**(6), 6699–6718 (2021)

18. Nathan, M.: Foundations of Embodied Learning: A Paradigm for Education. Routledge (2021)

19. Novack, M., Goldin-Meadow, S.: Learning from gesture: how our hands change our minds. Educ. Psychol. Rev. **27**, 405–412 (2015)

20. Poh, A., Castro, F. E.V., Arroyo, I.: Design principles for teacher dashboards to support in-class learning. In: Proceedings of the International Conference of the Learning Sciences. International Society of the Learning Sciences (2023)

21. Rasul. I., Crabtree, D., Castro, F., Poh, A., Gattupalli, S., Kathala, K., Arroyo, I.: WearableLearning: developing computational thinking through modeling, simulation and computational problem solving. In: Proceedings of the International Conference of the Learning Sciences. Int. Soc. of Learning Sciences (2023)

22. Crabtree, D., Rasul, I., Arroyo, I.: Implementing WearableLearning in the math classroom: an exploration into the affordances of embodied learning through game-play compared to traditional learning technologies. In: Proceedings of the International Conference of the Learning Sciences. Int. Soc. of Learning Sciences (2024)

23. Schoenfeld, A.H.: Learning to think mathematically: problem solving, metacognition, and sense making in mathematics. J. Educ. **196**(2), 1–38 (2016)

24. Schworm, S., Renkl, A.: Computer-supported example-based learning: When instructional explanations reduce self-explanations. Comput. Educ. **46**(4) (2006)

25. Newen, A., De Bruin, L., Gallagher, S. (eds.): The Oxford Handbook of 4E Cognition. Oxford University Press (2018)

26. Shute, V., Wang, L.: Assessing and supporting hard-to-measure constructs in video games. In: The Wiley Handbook of Cognition and Assessment: Frameworks, Methodologies, and Applications, pp. 535–562 (2016)
27. Smith, H., Harrison, A., Ottmar, E., Arroyo, I.: Developing Math Knowledge and Computational Thinking Through Game Play and Design: A Professional Development Program. Contemporary Issues in Technology and Teacher Education 20(4) (2020). https://www.learnt echlib.org/p/215216/
28. Smith, H., Rushton, N. Reis, S., Behning, C., Ottmar, E., Smith, G.: Wearable Learning Teacher Curriculum, 17 February 2023. Retrieved from osf.io/d7uhv
29. Wilson, R.A., Foglia, L.: Embodied cognition. Stanford Encyclopedia of Philosophy (2011)
30. Yackel, E., Cobb, P., Wood, T.: Small-group interactions as a source of learning opportunities in second-grade mathematics. J. Res. Math. Educ. 22(5), 390–408 (1991)
31. Razzaq, R., Ostrow, K.S., Heffernan, N.T.: Effect of immediate feedback on math achievement at the high school level. In: Proc. of the International Conference on AI in Education, pp. 263–267. Springer, Cham, June 2020. https://doi.org/10.1007/978-3-030-52240-7_48
32. Ochoa, X., Lang, C., Siemens, G., Wise, A., Gasevic, D., Merceron, A.: Multimodal learning analytics-Rationale, process, examples, and direction. In: The Handbook of Learning Analytics, pp. 54–65 (2022)
33. van De Sande, B.: Properties of the Bayesian knowledge tracing model. J. Educ. Data Mining 5(2), 1–10 (2013)
34. Liu, Y.: Tangram Race Mathematical Game--Combining Wearable Technology and Traditional Games for Enhancing Mathematics Learning, Masters Thesis, Worcester Polytechnic Institute (2014)
35. Valente, R.C.: Teaching students mathematical embodiment techniques using online learning game platform (WLCP). Interactive Qualifying Project. Worcester Polytechnic Institute E-Projects Library (2019)

Ethical AIED and AIED Ethics: Toward Synergy Between AIED Research and Ethical Frameworks

Conrad Borchers[1]([⊠]) [iD], Xinman Liu[2]([⊠]) [iD], Hakeoung Hannah Lee[3]([⊠]) [iD],
and Jiayi Zhang[4]([⊠]) [iD]

[1] Carnegie Mellon University, Pittsburgh, PA 15213, USA
cborcher@cs.cmu.edu
[2] University of Cambridge, Cambridge, UK
xl505@cam.ac.uk
[3] The University of Texas at Austin, Austin, TX 78712, USA
hklee@utexas.edu
[4] University of Pennsylvania, Philadelphia, PA 19104, USA
joycez@upenn.edu

Abstract. Ethical issues matter for artificial intelligence in education (AIED). Simultaneously, there is a gap between fundamental ethical critiques of AIED research goals and research practices doing ethical good. This article discusses the divide between AIED ethics (i.e., critical social science lenses) and ethical AIED (i.e., methodologies to achieve ethical goals). This discussion contributes paths toward informing AIED research through its fundamental critiques, including improving researcher reflexivity in developing AIED tools, describing desirable futures for AIED through co-design with marginalized voices, and evaluation methods that merge quantitative measurement of ethical soundness with co-design methods. Prioritizing a synthesis between AIED ethics and ethical AIED could make our research community more resilient in the face of rapidly advancing technology and artificial intelligence, threatening public interest and trust in AIED systems. Overall, the discussion concludes that prioritizing collaboration with marginalized stakeholders for designing AIED systems while critically examining our definitions of representation and fairness will likely strengthen our research community.

Keywords: ethics · bias · fairness · equity · representation · design · justice · reflexivity

1 Situating AIED Ethics and Ethical AIED

The field of artificial intelligence in education (AIED) increasingly embraces emergent ethical issues of artificial intelligence (AI). In a landmark paper by Holmes et al. [1], the field recognized the need to consider unintended implications of its technology regarding fairness, bias, equity, and representation. The authors delineate a gap between *doing ethical things* and *doing things ethically*, wherein there is a potential mismatch between the ethical implications of research goals (ethical things) and ethical practices within potentially unethical research goals (ethical doing). Specifically, current AIED research practices are predominantly concerned about "doing things ethically"—the technical and procedural validity (e.g., minimizing bias, improving efficiency, prioritizing privacy) of deploying AIED systems, sidestepping the matter of "doing ethical things"—the normative validity concerning the purpose of an AIED system and whether it is in itself an ethical pursuit [1]. This mismatch begs the question: how can AIED productively integrate more fundamental critiques of its research with its practices? For the purposes of this study, we contrast critical and theoretical work on ethics in AIED (*AIED ethics*) and with AIED research practices aimed at promoting ethical good or avoiding ethical bad (*ethical AIED*). Prior work on general AI ethics has argued that ethical principles and frameworks are "toothless and useless" as they are isolated from practices and are not consequential [2]. However, counterexamples exist, where critical policy research collaborates with technical AI research to develop guidelines for auditing AI-based models, such as emerging large-language models, with implications for regulation and technical model audits at the core of research practice [3]. How may AIED achieve a similar synthesis to strengthen its ethical research practices? The present study contributes paths forward to synergistically integrate these differences in future AIED research.

AIED ethics and ethical AIED follow different definitions of ethics, with AIED ethics being *justice-oriented* and ethical AIED being *measurement-oriented*. AIED ethics stems from ethnographic and critical traditions in the humanities. It offers various *frameworks* that research and policy can learn from to enhance ethical AIED, with many calling for a more fundamental shift in research goals in AIED toward justice (e.g., prioritizing equity-enhancing design over technological advancements [4]). In contrast, ethical AIED research employs quantitative methods, increasingly aiming to *measure* issues such as bias and fairness in AIED systems [5]. While gaps between both approaches have been noted in critical algorithm studies [6], the present study describes how both discourses contrast in terms of specific ethical topics of interest in current AIED research: personalization, equity, representation, and bias. For the former, we describe contemporary methodologies to approach, measure, and improve each issue. For the latter, we summarize common definitions and critiques of each issue regarding AIED research and its ethical implications. Through this contrast, we synthesize *paths forward* of how AIED as a research field can produce practice and output that is not only ethical doing (ethical AIED) but also accomplishes ethical research goals (AIED ethics). We do not aim to draw a pessimistic stance that argues that both lenses are mutually exclusive. On the contrary, we describe recent examples of research that move the field toward achieving such synthesis and discuss what might be on the horizon for AIED researchers when such synthesis is achieved. Prioritizing such synthesis could make the AIED research community more resilient in the face of rapidly advancing technology and AI, which may threaten public interest and trust in AIED systems.

2 Evidence for a Disconnect Between AIED Ethics and Ethical AIED

In discussing AIED ethics, we acknowledge that there are multiple traditions of ethical thinking related to AIED. For instance, Fox [7] delineates four traditions of consequential, ecological, relational, and deontological ethics in AIED research. Holmes and colleagues [8] have also advocated for a rights-based approach to AIED ethics. In recognizing ethical pluralism, the present paper broadly takes on a justice-oriented lens as encompassing the recent discourse in critical studies of AIED ethics. This choice does not imply other ethical lenses on AIED are irrelevant; rather, it favors context-specific approaches to ethical justification that acknowledge the legitimacy of multi-stakeholder grounded realities [9]. As we will argue, a justice-oriented lens to AIED ethics, as opposed to incoherent sets of "toothless and useless" ethical principles detached from real practices, promises more synergies with ethical AIED and paths forward to improving AIED research practices [2].

Past positions of justice-oriented AIED ethics on AIED research practice can be broadly characterized as ones criticizing *measurable ethics*. Biesta [10] contests that contemporary education systems neglect the normative validity of educational measurements (i.e., what *should* be valued and thereby measured). In other words, focusing on the efficiency and effectiveness of learning processes sidesteps the normative questions of what defines a good education in the first place—the aims and ends for which these processes are directed. Commonly voiced concerns over AIED research goals today still echo this critique of measurement and effectiveness: intelligent tutoring systems prioritizing learning efficiency at the expense of collaborative and social interactions [11] or the datafication of student and teacher subjects potentially disintegrating into surveillance [12].

Taking ethical concerns seriously has been increasingly central in AIED, alongside the recognition that addressing ethical concerns upfront makes our community more resilient in the age of rapid technological progress. Still, awareness and debates about the ethics of AIED as a research field were not spotlighted until relatively recently [1]. This nascent discourse centers on computational approaches to boosting fairness and doing ethical good without much accounting for the *ethics of education*—that is, the purpose of learning, choice of pedagogy, human-computer relations, and access to education [1]. Contrasting these computational approaches, AIED ethics recognizes that data and algorithmic systems do not pre-exist the social actors, techno-scientific practices, institutional applications, and power struggles that bring them into being [13]. As such, AIED innovation should be understood as "a knot of social, political, economic and cultural agendas that is riddled with complications, contradictions and conflicts" [14], p. 6]. Can both lenses be integrated? In the following, we discuss how both AIED ethics and ethical AIED approach emerging areas of interest in AIED. After summarizing emerging AIED methodologies and discourse around each issue, we describe their discourse through AIED ethics, as signified by headers beginning with "beyond."

2.1 Four Issues of Ethical Discourse In and Around AIED

2.1.1 Issue 1: Personalization

Through the use of AI, personalization adopted in AIED systems presents an opportunity to provide high-quality and equitable learning access to students at scale: an ethical good. It is often modeled after human tutoring, analyzing students' needs and delivering tailored instructions and adaptive feedback [15]. This approach allows broader access to high-quality instruction, potentially enabling equitable learning opportunities for larger populations. It enables students to learn and progress without being held back or left behind, which can often happen in a traditional classroom where instruction is standardized and delivered based on a fixed schedule [16]. The use of personalized instruction and learning pathways has been found to be beneficial in improving student engagement and learning [15].

Beyond Personalization: AIED ethics can guide thinking beyond the boundaries of typical personalization in AIED learning environments. While personalization seeks to deliver effective educational experiences to learners, it is often limited to micro-level decisions that offer learners individualized contexts, pacing, groupings, and pathways through prearranged materials. These personalized AIED technologies have been criticized for operating on behaviorist and instructionist pedagogies underpinned by a narrow understanding of personalization where "the pathway may be personalized but not the destination" [8, p. 34]. As such, learner agency is pre-determined and constrained within a set of universally standardized outcomes. Real personalization (according to AIED ethics)—or "subjectification" and "individuation" in Biesta's terms—involves cultivating learners' autonomy and capabilities to self-actualize and achieve what they individually want to achieve [10].

2.1.2 Issue 2: Equality and Equity

A central AIED research goal is to create systems that work equally well for different groups of learners. A common concern related to the issue of equity is the so-called "EdTech Matthew Effect," where AIED learning systems specifically benefit learners with high prior knowledge, deepening existing achievement gaps [17]. Equity relates to a constant relationship between effort and learning in AIED learning environments, such as learning opportunities in intelligent tutoring systems and learning gain. Recent research argued that learning rates (i.e., the average improvement in student accuracy per completed problem-solving step with feedback) are highly regular across students and within various learning domains [18]. Equality, or the absence of achievement gaps, could then be achieved if disadvantaged learners receive more learning opportunities in well-designed AIED systems. To evaluate and promote equity, AIED research has called for increasing use of school-level demographic variables to compare how different behaviors in AIED learning environments (e.g., help-seeking) differentially relate to learning outcomes across populations [19].

Beyond Equality and Equity: AIED ethics points attention to the fact that most AIED learning technologies are designed to intervene: they measure when a student struggles or does not reach a certain level of attainment. One risk of this deficit view is that realities of educational experiences could be masked by quantifiable participation and completion rates set by dominant institutions and regimes. AIED ethics cautions about leaving the measurement of what learners are lacking unquestioned. Why might learners be better off if AIED systems are mindful of setting educational goals? Interventions focusing on measurable outcomes overlook the underlying reasons necessitating additional support. For example, struggling to learn might relate to a range of cognitive and motivational factors. While predictive models can discern a subset of different sources of struggle [20], human teachers are often in a better position to diagnose learner needs, in line with a model of human-AI complementary permeating recent AIED successes [21]. However, learner needs might have cultural roots that are left for critical AIED research to examine and design around. For instance, even when help is sought during the usage of intelligent tutoring systems, help-seeking behaviors vary across different sociocultural contexts, influenced by socioeconomic status, religion, power dynamics, and degree of individualism or collectivism from classroom to national levels of culture [22]. This variability underscores that the modeling and assessing these behaviors cannot be universally applicable. Similarly, learner access to AIED systems is rarely considered in tool design, with few notable exceptions [23]. These considerations underscore that more holistic considerations of learner contexts and obstacles to learning could improve learning environments and their ethical potential to reduce inequality [12].

2.1.3 Issue 3: Representation

Representation in ethical AIED focuses on integrating perspectives of users in the design, development, deployment, and evaluation of educational technologies and ensuring diversity encoded in demographic markers in datasets. This includes co-design practices involving educators, students, and other stakeholders in the creation of AIED systems [21]. Collaboration between researchers and stakeholders can potentially increase impact through more effective implementation and use. Research has also called for including diverse voices in the development process to ensure that AIED systems are reflective of a wide range of learning contexts and the needs of different learner populations [19]. By prioritizing representation, developers can mitigate the risk of perpetuating existing biases and ensure that technologies are inclusive and beneficial to a broad spectrum of users. Regarding the deployment and evaluation of AIED systems, past work has identified the need to study AIED systems in diverse cultural contexts and study their efficacy [23]. Recent research calls for the systematic study of AIED tools through the lens of the demographic groups that use them, which can be (among others) inferred from census data of schools AIED studies are run in, as student-level demographic data collection is often not feasible [19].

Beyond Representation: While ethical AIED highlights evaluating the effectiveness of AIED systems across social and cultural contexts [8, 24], AIED ethics emphasizes a critical attitude towards constructing and measuring learner categories. The most immediate approach to improve representation is to add, combine, and overlap identity categories such as class, race, gender, sexuality, ethnicity, ability, nationality, and age. However, AIED ethics argues that this approach leaves institutional forms of racialized, classed, gendered processes perpetuated by the dominant regimes of power unexamined. An additive approach to representation may be reductionist in serving as "a palliative to keep marginalized groups... from rebelling against a system that promotes structural inequality" [24], p. 25]. Accordingly, critically examining how AIED classifies learners (e.g., in terms of race, gender, or class) could improve the benefits of AIED systems for marginalized and underrepresented learners [25, 27]. Past research offers examples of this issue in education: without rethinking gender binaries within classification systems, we risk marginalizing non-binary learners [26]. Similarly, it is worthwhile to reflect on how alternative demarcations of race and ethnicity could serve underrepresented learners beyond North American contexts better. Within AIED, the construction of classification systems can constrain learners by overlooking within-group differences. For instance, one study argued that categorizing Laotian, Cambodian, and Vietnamese American students within the broad "model minority" stereotype associated with East Asian Americans from China, Japan, and Korea ignores crucial within-group differences—such as academic performance and family resources—that are more predictive of educational outcomes than broader racial group distinctions [25].

How can critical lenses on representation inform AIED co-design practices? Design justice and liberatory philosophies that center on community-led and co-constructive practices [27] increasingly allow researchers to adopt participatory approaches and listen to marginalized voices among educational stakeholders [28]. Participatory approaches including diverse stakeholders in the design process) are compatible with current AIED co-design practices. However, through a critical lens, AIED ethics emphasizes a deliberate analysis of power within individuals, institutes, and where they intersect [25]. Consequently, they focus co-design efforts on listening to marginalized voices by establishing design spaces where marginalized groups can easily participate and envision alternative designs to current solutions [27]. Beyond attempts to measure representational fairness through categories [19], this approach can present opportunities for AIED tool design otherwise invisible to researchers.

2.1.4 Issue 4: Bias and Fairness

Student modeling in AIED systems involves using log data to model student behaviors and predict learning outcomes. Bias and fairness in these models are especially investigated from an algorithmic standpoint in emerging ethical AIED research. Algorithmic bias or fairness refers to the collective effort of examining the performance of student models, ensuring that the models are capable of providing unbiased evaluation for all learners regardless of their attributes [5, 29]. Algorithmic bias describes the problem where a data-driven predictive model functions better for some populations than others, producing disparate and poorer impacts for historically underrepresented or protected groups [29]. As predictions are often used to inform decisions and actions, algorithmic bias in a model can cause unfairness in the allocation of resources and misplacement of treatment. An increasing number of works in the field of AIED have dedicated their efforts to evaluating and improving the fairness of student models. Among them, a range of models have been examined across different student populations and intersectional groups [30]. To improve model fairness, studies suggested increasing data collection for minority students [31] and being critical with the decisions on the inclusion of demographic data [19].

Beyond Bias and Fairness: Efforts to improve algorithmic fairness in education assume the general benefit of innovation and aim to distribute these benefits equitably; in other words, no groups of learners should systematically benefit more from technology than others. AIED ethics asks about broader considerations of whether these approaches promote more ethical good or put historically disadvantaged learner groups at risk through second-order effects of technology. For example, while tools to profile learners may help prevent poverty-stricken students from dropping out of school, their intrusiveness can also undermine learners' rights to privacy and be misused by governing entities to distribute and rescind welfare. This is the case for Brazil's Bolsa Familia program (a direct income transfer program aimed at helping families out of poverty), where the use of facial recognition technology in public schools may lead to punitive consequences for families dependent on welfare linked to monitored school attendance [32]. Such a program describes the ethical dilemma in algorithmic attempts to measure and enhance fairness in education requiring demographic data collection. AIED ethics highlights that these processes may heighten the visibility and, thereby, the vulnerability of historically low-income and marginalized groups. Eubanks [4] argues that while the expansion of digital systems in criminal justice, welfare, and education has increased the visibility of working-class women seeking public assistance, these systems also exacerbated their physical and economic vulnerability through behavioral surveillance and discoveries that would have gone unnoticed in the privacy afforded by wealthier families. Similarly, researchers warned against the danger of algorithms in increasing the vulnerability of already marginalized learners through further stigmatization [37].

What does AIED ethics suggest AIED research could do better about historical biases in present-day data? AIED ethics lenses on bias emphasize limitations of the promises of data neutrality and objective calculations in model-based approaches to bias. AI is not an inherently neutral set of technologies but rather embedded in and produced from human-run systems where historical biases are often entangled and untraceable [33]. As such, beyond improving accuracy and eliminating bias, AIED researchers and attempts should foreground their own positionality and reflexivity, including premeditation of how systems and algorithms they develop could be harmful to vulnerable learner populations. Researchers should "start from interrogating the existing inequalities, reflect their own position in the system of these inequalities and actively ask which constituencies will or will not benefit" [34], p. 331] rather than construing their personal involvement (e.g., motivations, beliefs, roles) in data protection as bad or biased practices. In AIED, special attention can be paid to the power disparities between those initiating and those subjected to AIED interventions [12]. We acknowledge that not all potential harm can be preemptively detected and no research to promote equality and support vulnerable populations is "risk-free." Still, AIED systems (including their public perception and likelihood of doing ethical good) could benefit from a holistic assessment of learner contexts and potential intervention risks in these populations.

3   Examples of Research Bridging AIED Ethics and Ethical AIED

Having delineated differences in how AIED ethics (i.e., justice-oriented, framework-heavy critical social science work) and ethical AIED (i.e., measurement-oriented, quantitative research practices) take on emerging challenges in AIED, the present study is not intended to paint a picture of insurmountable divides between both approaches. Rather, to aim at a synthesis of both approaches, we briefly summarize two example directions of successful synergies between both strands: a) research bridging AIED stakeholders and researchers and b) research reflecting on AIED research goals, practices, and conditions.

First, research bridging AIED stakeholders (e.g., students, teachers, caregivers) and researchers could promote ethical good and advancements in AIED. Practices that involve listening and designing around learner and educator needs are not only expected to amplify underrepresented voices but also lead to more effective AIED systems by listening to "weak signals" [35]. Studying AIED systems in the context of closely working with AIED stakeholders and studying the adoption of such systems through observational methods also bears the potential of mitigating some of the concerns summarized around AIED ethics discourse earlier. First, studying where AIED systems break, fail, or do not help learners and why can address more fundamental issues around bias. Second, studying cultural practices beyond monitoring demographic data can reveal limitations in contemporary systems of classifying learners (representation). Third, studying adoption and system use beyond short-term studies could discover the potential harms of advanced technology on vulnerable populations (as in the Bolsa Familia program [32]). Fourth, designing around learner needs and concerns could potentially support learner self-actualization by setting broader learning goals beyond personalization within set learning goals. For research practice, observational methods could be supported by co-designing AIED tools with teachers [21] or specifically listening to underrepresented groups for whom existing AIED tools might not work well [27]. These approaches also focus on improving adoption before technology rollout, aligning the visions and needs of stakeholders with AIED tool development [36]. To make this vision of stakeholder involvement an actuality, AIED as a research community could support the creation of spaces where AIED stakeholders meet and work with researchers. A benefit of spaces for meeting and communicating can be an increase in trust in and adoption of AIED systems [12, 37]. Communication channels are expected to deepen trust between AIED stakeholders and research. To involve minorities in research more, efforts could focus on issues that research found relevant to trust in AI systems, for example, building consensus and best practices around data privacy standards [37].

The second example of bridging ethical AIED and AIED ethics is research reflecting on the positive and negative impacts of AIED research goals, practices, and conditions coming from within the AIED community. For example, recent work has qualitatively studied how the presence of analytics in AIED systems can introduce tensions in student mentoring relationships in higher education [12]. Specifically, the study noted that discrepancies between reported activity and data in activity reports could undermine trusted relationships between mentors and mentees. Therefore, the study calls for increased research on how AIED systems transform existing practices (e.g., in the classroom) and how participants perceive these systems rather than studying outcomes and fairness metrics based on performance alone. Similarly, overview articles reflecting on research paradigms and assumptions in AIED research can steer conversations around methodologies and approaches to measuring how AIED systems and research transform learners' lived experiences and promote ethical good. This includes position articles: a recent article argued that the majority of AIED systems operate on a deficiency-based lens, where intervention is provided to students who are lacking, at-risk, or not learning well, which underappreciates opportunities to design adaptivity around learner strengths and assets [38].

4 Paths Forward

Synthesized from the issues above and their differences when viewed from AIED ethics frameworks and ethical AIED research practice, we suggest paths forward that could promote the synthesis of both lenses.

4.1 Path 1: Researcher Reflexivity

Ethical AIED frameworks suggest that researchers question their research goals, definitions of issues such as bias, and the demographic lines along which learners are represented and studied. How can this lens be productively integrated into AIED research practice? Practically speaking, how does one go from theory to conceptualization to practice? One approach could be to carefully evaluate and communicate the feasible expectations and limitations of AIED systems to those involved and affected (e.g., learners, teachers, caregivers). Practicing ethics of care that involves designing around the concerns and needs of marginalized stakeholders is especially important given the uneven power relations between researchers and those stakeholders [39]. For instance, in discussing the data sources used, AIED researchers could reflect on their positionality concerning the participants and end users. Beyond informed consent as a standard practice in AIED, this includes acknowledging the additional roles and responsibilities implied for those providing the data to ensure they are not merely reduced to data subjects without rights and agency [7, 12]. For example, while surveillance and monitoring mechanisms in learning analytics may serve learners, they may also increase their vulnerability and raise concerns that could make learners less likely to benefit from AIED systems if left unexamined [12, 39]. Further, rather than building tools around feasibility, reflexivity creates a space to be transparent about dynamics during the design process of AIED tools, including how much weight was given to different stakeholders in design decisions. Transparency can then surface more ways in which AIED could incorporate critical theories and justice-driven design into its practices, supported by research community discussion.

In practice, achieving the level of transparency and reflexivity advocated may take much work for AIED researchers. Next to a lack of training resources for researchers or community platforms to engage in reflexivity, it is an open question how AIED researchers should best respond to discrepancies between current research goals and reflexivity that might question them. As research programs operate on medium-term time horizons of a few to several years, reflexivity could not only focus on broader research goals but also on smaller changes in research practices, such as prioritizing working with diverse samples, taking more time to solicit broad feedback from stake-holders during the AIED tool design process, or dedicating more time for needs finding rather than prioritizing rolling out novel capabilities early (e.g., generative AI-based AIED tools). This raises the question: Is the responsibility for ethical AIED research primarily at the individual researcher level, or does it necessitate broader institutional or community-wide commitment? Institutional and community-wide practices could play a critical role in creating the environment and providing the resources necessary for such ethical considerations to be integrated into the research process. To further encourage these practices, the AIED research community could advocate for including positionality statements and ethics requirements as part of the evaluation criteria for paper submissions. For instance, Cambo and Gergle [40] presented concepts of model positionality and computational reflexivity to encourage data scientists to reflect on the sociocultural contexts of model development, along with the backgrounds of annotators and researchers and their position within power dynamics with research subjects.

Establishing norms for conducting discussions that prioritize constructive and crit-ical engagement, alongside fostering a culture that values and rewards reflexivity in research activities, are essential steps toward establishing a more justice-oriented AIED field. Furthermore, AIED researchers could engage with wider education policies. For instance, to what extent is centering AIED research around institutional and national cur-ricula or policies of standardized knowledge and skills desirable or constraining? How is AIED research hindered or enabled by wider cultural or policy factors, and to what extent might AIED researchers be positioned to challenge them? AIED researchers—along with policymakers, educators, learners, and other relevant stakeholders—could actively reflect on their positionalities and ask which constituencies will or will not ben-efit from the development and deployment of AIED systems: Whose perspectives are we looking from? Who benefits from such perspectives and is at a structural disadvantage [34]? Deliberately engaging with reflexivity kickstarts further initiatives to incorporate diverse voices from the wider community, creating more equitable and responsible AIED systems and practices.

4.2 Path 2: Increasing Diverse Stakeholder Collaboration and Advancing Research Methods for Diverse Stakeholder Involvement

As one solution of integrating concerns of AIED ethics into ethical AIED research prac-tice, we have argued for including diverse perspectives and experiences in the design, development, and deployment of AIED systems. How can AIED develop, refine, and promote design research practices that listen to stakeholder needs and voices? Expand-ing methodologies for diverse stakeholder collaboration to envision desirable futures can involve integrating co-design principles and proactive adjustments to system design from the outset. Co-design methodologies emphasize the involvement of various stakeholders throughout the design process, ensuring that their perspectives, needs, and aspirations are incorporated into the final product or outcome [21]. This approach fosters inclusivity and ensures that the resulting solutions are more reflective of the diverse range of voices and experiences involved. Additionally, integrating changes to system design upfront allows for identifying and mitigating potential ethical concerns or unintended conse-quences early in the development process [41]. By actively involving stakeholders and considering ethical implications from the beginning, this approach promotes creating more robust, inclusive, and responsive solutions, fostering trust.

What does a coupling of inclusive co-design with continuous measurement look like in research practice? Related human-computer interaction methodologies emphasize aligned values with AIED stakeholders throughout the design and deployment lifecycle. While research methodologies such as community-based participatory research [27] have emphasized the inclusion of stakeholders in the *design* process of AIED systems, these collaborative efforts must be embedded in and extended through the *adoption* process of AIED systems in specific educational settings to ensure sustainable innovation [42]. Design-based implementation research methodology centers the design and implementation of educational tools around identifying "persistent problems of practice from multiple stakeholders' perspectives" from the very outset and is committed to "developing capacity for sustaining change in systems" [42, p. 142–243]. This is especially important for AIED as discrepancies might exist between what role AI should play in issues according to different stakeholders [36]. AIED research can also take inspiration from participatory research models, such as Research-Practice-Industry Partnerships, aligning the design of AIED systems to practitioners' needs while incorporating a critical research lens [44].

4.3 Path 3: Combining Co-design with Quantitative Measurement

One lesson learned from studying potential synergies between AIED ethics and ethical AIED research practice is that measurement-based approaches to strengthening ethics in AIED research are not necessarily bad but rather *limited*. Prioritizing measurement-based approaches is unlikely to eliminate all ethical issues in AIED systems (e.g., remediating historical underrepresentation of certain demographic groups in AIED system design). We propose that co-design with underrepresented stakeholders could be combined with regular measurements of variation in learning rates and other outcomes of interest in different learner populations to achieve ethical AIED research goals. Further, coupling inclusive co-design practices with quantitative measurement of educational effectiveness could derive more general principles that make AIED systems effective for different learner populations by comparing different design variations of systems.

A research opportunity in AIED exists to study whether community-based design through critical lenses can create more favorable learning outcomes (as measured through established measures of AIED learning environment effectiveness, such as learning gains and learning rates). Inclusive design could facilitate appropriate and sustainable adoption by creating intentional feedback loops that elevate the voices and needs of all involved parties, achieving desirable outcomes for practitioners and learners. Rather than being perceived as a constraint to innovation in AIED, justice-oriented ethics could leverage hidden design opportunities by paying attention to the "weak signals" in social and education systems. Within these systems, individuals from marginalized standpoints are best equipped to identify alternative solutions to systemic flaws, as they are more vulnerable to current risks and cognizant of the fundamental social regularities often invisible to those in more privileged or dominant positions [35]. As such, beyond addressing measurement-oriented improvements to AIED systems, AIED research could simultaneously serve as a justice-oriented dialogic space that proactively bridges the concerns and visions of different stakeholders involved and affected by AIED.

5 Summary and Outlook

The present study discussed the relationship between AIED ethics and ethical AIED, highlighting a gap between critical ethical perspectives on the AIED research goals and the practical methodologies to address ethical concerns. This discussion contributed paths to bridging both lenses. Researcher reflexivity regarding their standpoints and definitions of potential issues in AIED systems (e.g., bias and representation) offer one entry point to bridge the ethical frameworks of AIED with its research practice. Promising avenues for resulting research include advancing methods for co-design with marginalized communities, which could be combined with established learning measurements (e.g., learning rates) in relation to technology design. Further, studying the disciplinary overlap between AIED ethics and ethical AIED, including systematic review papers and quantitative inquiry into topic centers, could guide synergy. Prioritizing a synthesis between AIED ethics and ethical AIED could make our research community more resilient in the face of rapidly advancing technology and AI, threatening public interest and trust in AIED systems. Acknowledging that paths toward ethical AIED are intricate and multifaceted, we hope this discussion fosters ongoing dialogue, collaboration, and reflexivity among researchers, practitioners, and the communities they aim to serve.

Acknowledgments. We extend our heartfelt thanks to the Learnest Ethical AI in Education Fellowship for providing the space and resources that enabled us, the authors, to collaborate.

References

1. Holmes, W., et al.: Ethics of AI in education: towards a community-wide framework. Int. J. Artif. Intell. Educ. **32**, 504–526 (2022)
2. Munn, L.: The uselessness of AI ethics. AI Ethics. **3**, 869–877 (2023)
3. Mökander, J., Schuett, J., Kirk, H.R., Floridi, L.: Auditing large language models: a three-layered approach. AI Ethics (2023)
4. Eubanks, V.: Digital Dead End: Fighting for Social Justice in the Information Age. The MIT Press (2011)
5. Baker, R.S., Hawn, A.: Algorithmic bias in education. Int. J. Artif. Intell. Educ. **32**, 1052–1092 (2022)
6. Moats, D., Seaver, N.: "You Social Scientists Love Mind Games": Experimenting in the "divide" between data science and critical algorithm studies. Big Data Soc. 6 (2019)
7. Fox, A.: Educational research and AIED: Identifying ethical challenges. In: Holmes, W., Porayska-Pomsta, K. (eds.) The Ethics of Artificial Intelligence in Education. Routledge (2022)
8. Holmes, W., Persson, J., Chounta, I.A., Wasson, B., Dimitrova, V.: Artificial Intelligence and Education a Critical View Through the Lens of Human Rights, Democracy and the Rule of Law. The Council of Europe, France (2022)

9. Franzke, A.S., Bechmann, A., Zimmer, M., Ess, C.M.: Internet Research: Ethical Guidelines 3.0. Assoc. Internet Res. (2020)
10. Biesta, G.: Good education in an age of measurement: ethics, politics, democracy. Paradigm Publishers, Boulder, Colo (2010)
11. Holmes, W.: The Unintended Consequences of Artificial Intelligence and Education. Education International: Brussels, Belgium (2023)
12. Lee, H.H., Gargroetzi, E.: "It's like a double-edged sword": Mentor Perspectives on Ethics and Responsibility in a Learning Analytics-Supported Virtual Mentoring Program. J. Learn. Anal. **10**, 85–100 (2023)
13. Williamson, B.: Datafication of Education: A Critical Approach to Emerging Analytics Technologies and Practices. In: Rethinking Pedagogy for a Digital Age. Routledge (2019)
14. Selwyn, N.: Distrusting Educational Technology: Critical Questions for Changing Times. Routledge, Taylor & Francis Group, New York; London (2014)
15. Morgan, B., Hogan, M., Hampton, A.J., Lippert, A., Graesser, A.C.: The need for personalized learning and the potential of intelligent tutoring systems. In: Handbook of Learning from Multiple Representations and Perspectives. Routledge (2020)
16. Hill, J.R., Hannafin, M.J.: Teaching and learning in digital environments: the resurgence of resource-based learning. Educ. Technol. Res. Dev. **49**, 37–52 (2001)
17. Reich, J.: Failure to Disrupt: Why Technology Alone Can't Transform Education. Harvard University Press, Cambridge, Massachusetts (2020)
18. Koedinger, K.R., Carvalho, P.F., Liu, R., McLaughlin, E.A.: An astonishing regularity in student learning rate. Proc. Natl. Acad. Sci. **120** (2023)
19. Karumbaiah, S., Ocumpaugh, J., Baker, R.S.: Context matters: differing implications of motivation and help-seeking in educational technology. Int. J. Artif. Intell. Educ. **32**, 685–724 (2022)
20. Mu, T., Jetten, A., Brunskill, E.: Towards Suggesting Actionable Interventions for Wheel-Spinning Students. International Educational Data Mining Society (2020)
21. Holstein, K., McLaren, B.M., Aleven, V.: Co-designing a real-time classroom orchestration tool to support teacher–AI complementarity. J. Learn. Anal. **6**, 27–52 (2019)
22. Ogan, A., Walker, E., Baker, R., Rodrigo, Ma.M.T., Soriano, J.C., Castro, M.J.: Towards understanding how to assess help-seeking behavior across cultures. Int. J. Artif. Intell. Educ. **25**, 229–248 (2015)
23. Kwon, C., Butler, R., Stamper, J., Ogan, A., Forcier, A., Fitzgerald, E., Wambuzi, S.: Learning analytics for last mile students in Africa. In: Proceedings of the 13th Learning Analytics and Knowledge Conference (2023)
24. Banks, J.A.: The Routledge International Companion to Multicultural Education. Routledge, Taylor & Francis Group, New York; London (2009)
25. Hancock, A.M.: When multiplication doesn't equal quick addition: examining intersectionality as a research paradigm. Perspect. Polit. **5**, 63–79 (2007)
26. D'Ignazio, C., Klein, L.F.: Data Feminism. The MIT Press, Cambridge, Massachusetts (2020)
27. Harrington, C., Erete, S., Piper, A.M.: Deconstructing community-based collaborative design: towards more equitable participatory design engagements. Proc. ACM Hum.-Comput. Interact. **3**, 216:1–216:25 (2019)
28. Brossi, L., Castillo, A.M., Cortesi, S.: Student centered requirements for the ethics of AI in education. In: Holmes, W., Porayska-Pomsta, K. (eds.) The Ethics of Artificial Intelligence in Education: Practices, Challenges, and Debates, pp. 91–112. Routledge, New York (2022)
29. Kizilcec, R.F., Lee, H.: Algorithmic fairness in education. In: Holmes, W., Porayska-Pomsta, K. (eds.) The Ethics of Artificial Intelligence in Education: Practices, Challenges, and Debates, pp. 174–202. Routledge, New York (2022)

30. Zambrano, A.F., Zhang, J., Baker, R.S.: Investigating algorithmic bias on bayesian knowledge tracing and carelessness detectors. In: Proceedings of the 14th Learning Analytics and Knowledge Conference, pp. 349–359 (2024)
31. Bird, K.A., Castleman, B.L., Song, Y.: Are algorithms biased in education? exploring racial bias in predicting community college student success. J. Policy Anal. (2024)
32. Canto, M.: Global Information Society Watch 2019 Artificial intelligence: Human rights, social justice and development: Brazil. Instituto de Pesquisa em Direito e Tecnologia do Recife (2019)
33. Browne, J.: AI and structural injustice: a feminist perspective. In: Browne, J., Cave, S., Drage, E., McInerney, K. (eds.) Feminist AI: Critical Perspectives on Algorithms, Data, and Intelligent Machines (2023)
34. Draude, C., Klumbyte, G., Lücking, P., Treusch, P.: Situated algorithms: a sociotechnical systemic approach to bias. Online Inf. Rev. **44**, 325–342 (2019)
35. Treviranus, J.: Learning to learn differently. In: Holmes, W., Porayska-Pomsta, K. (eds.) The Ethics of Artificial Intelligence in Education: Practices, Challenges, and Debates. Routledge, Taylor & Francis Group, New York, NY (2023)
36. Rahm, L., Rahm-Skågeby, J.: Imaginaries and problematisations: a heuristic lens in the age of artificial intelligence in education. Br. J. Educ. Technol. **54**, 1147–1159 (2023)
37. Prinsloo, P., Slade, S.: Big data, higher education and learning analytics: beyond justice, towards an ethics of care. Big data and learning analytics in higher education: Current theory and practice, pp. 109–124 (2017)
38. Ocumpaugh, J., Roscoe, R.D., Baker, R.S., Hutt, S., Aguilar, S.J.: Toward asset-based instruction and assessment in artificial intelligence in education. Int. J. Artif. Intell. Educ. (2024)
39. Prinsloo, P., Slade, S.: Student vulnerability, agency and learning analytics: an exploration. J. Learn. Anal. 3, (2016)
40. Cambo, S.A., Gergle, D.: Model positionality and computational reflexivity: promoting reflexivity in data science. In: Proceedings of the 2022 CHI Conference on Human Factors in Computing Systems, pp. 1–19. Association for Computing Machinery, New York, NY (2022)
41. Bhimdiwala, A., Neri, R.C., Gomez, L.M.: Advancing the design and implementation of artificial intelligence in education through continuous improvement. Int. J. Artif. Intell. Educ. **32**, 756–782 (2022)
42. Underwood, S.M., Kararo, A.T.: Design-based implementation research (DBIR): an approach to propagate a transformed general chemistry curriculum across multiple institutions. J. Chem. Educ. **98**, 3643–3655 (2021)
43. Fishman, B.J., Penuel, W.R., Allen, A.R., Cheng, B.H., Sabelli, N.: Design-based implementation research: an emerging model for transforming the relationship of research and practice. Teach. Coll. Rec. **115**, 136–156 (2013)
44. Pautz Stephenson, S., Banks, R., Coenraad, M.: Outcomes of Increased Practitioner Engagement in Edtech Development: How Strong, Sustainable Research-Practice-Industry Partnerships will Build a Better Edtech Future. Digital Promise (2022)

Enhancing LLM-Based Feedback: Insights from Intelligent Tutoring Systems and the Learning Sciences

John Stamper[1]([⊠])(iD), Ruiwei Xiao[1]([⊠])(iD), and Xinying Hou[2]([⊠])(iD)

[1] Carnegie Mellon University, Pittsburgh, PA 15213, USA
{jstamper,ruiweix}@cs.cmu.edu
[2] University of Michigan, Ann Arbor, MI 48109, USA
xyhou@umich.edu

Abstract. The field of Artificial Intelligence in Education (AIED) focuses on the intersection of technology, education, and psychology, placing a strong emphasis on supporting learners' needs with compassion and understanding. The growing prominence of Large Language Models (LLMs) has led to the development of scalable solutions within educational settings, including generating different types of feedback in Intelligent Tutoring Systems. However, the approach to utilizing these models often involves directly formulating prompts to solicit specific information, lacking a solid theoretical foundation for prompt construction and empirical assessments of their impact on learning. This work advocates careful and caring AIED research by going through previous research on feedback generation in ITS, with emphasis on the theoretical frameworks they utilized and the efficacy of the corresponding design in empirical evaluations, and then suggesting opportunities to apply these evidence-based principles to the design, experiment, and evaluation phases of LLM-based feedback generation. The main contributions of this paper include: an avocation of applying more cautious, theoretically grounded methods in feedback generation in the era of generative AI; and practical suggestions on theory and evidence-based feedback design for LLM-powered ITS.

Keywords: Intelligent Tutoring System (ITS) · Large Language Models (LLMs) · Generative AI (GenAI) · Hint · Formative Feedback

1 Introduction

While the term Generative AI (GenAI) has become synonymous with LLMs and exploded with the release of ChatGPT in 2022 [6], it is important to note that GenAI has been a part of the AIED community for many years. A long line of work has been seen in areas where LLMs have shown particular usefulness such as content generation [12], including generating formative feedback [51]. The rush to apply LLMs to these areas to improve AIED certainly represents a

A. M. Olney et al. (Eds.): AIED 2024 Workshops, CCIS 2150, pp. 32–43, 2024.
https://doi.org/10.1007/978-3-031-64315-6_3

good opportunity to help in education, but the community should not discount the decades of work that has been done in these areas before the current interest in the latest LLMs. In particular, the AIED community has decades of research on the proper way to implement hints and feedback, and while LLMs seem to be helpful in different phases of feedback generation, the approaches used should be carefully designed and evaluated through theories and empirical evidence.

This work sets forth a balanced discourse on the integration of LLMs with learning science work, focusing on leveraging both previous insights on feedback design in intelligent tutoring systems (ITSs) and contemporary advancements in generative AI. It outlines a strategic blueprint for infusing LLM-based feedback by building on previous contributions around ITSs and the learning sciences, aiming to refine feedback generation processes and achieve more effective learning results. This work underscores the significance of adhering to established educational frameworks and validates the potential of LLMs to improve feedback components in educational systems, thus offering a pathway toward more effective and responsible future educational technologies.

2   The Development of ITS Feedback Generation

In this section, we synthesized existing research on ITS feedback generation based on how they were generated. Three primary methods were identified for generating feedback: the expert-created learner model, the data-driven learner model, and the use of large language models. The first method involves experts manually inputting models of learners' potential behaviors or constraints of the problem. The second one compiles data from learners' interactions with similar problems to automatically build a model of learner behavior. The third method leverages LLMs, focusing on supplying the appropriate context and requirements, striving to generate more adaptive feedback with less human labor.

2.1   Feedback Generated from Expert-Generated Learner Model

To generate feedback for students, traditional ITSs heavily rely on experts' input on learner modeling, with an emphasis on student problem-solving states. There are two main lines of learner modeling methods in these intelligent tutors: the production rules model, which originated from Anderson's ACT-R theory [5], and the constraint-based model (CBM), which is based on Ohlsson's theory of learning from performance errors [35]. Specifically, for a given problem, a production rules model uses a set of if-then rules or example solutions [3] generated by experts to model the knowledge and possible decision-making process to solve this problem. A CBM is composed of a set of expert-written constraints that should not be broken by any potential correct solutions. When generating feedback, the learner's solution will be compared to the rule-based model or constraint-based model, and the point the student gets astray from the right track will be included in the delivered feedback.

The effectiveness of expert-generated models for both approaches has been repeatedly proven by evidence from theories and classroom studies. For example, feedback in CTAT tutors [2] is generated from production rules. Until 2016, there were 18 CTAT tutors distributed in real educational settings, used by approximately 44,000 students. One of the CTAT tutors, Andes Physics Tutor [55] with its immediate feedback, helping students achieve significantly better learning in five years' repeated measurements. CBM has also proven to be extremely effective and efficient on highly structured procedural tasks and open-ended tasks such as programming. For instance, the feedback generated by CBM in SQL-Tutor led to significantly higher learning outcomes in 4 studies during 1998–2000 [32]. The learning curve analysis further grounded the experiment result with sound psychological foundations aligning with the smooth learning curve criterion [29].

2.2   Feedback Generated from Data-Driven Learner Model

Regardless of their effectiveness, building expert models can be extremely taxing, making them hard to scale up. Furthermore, these methods have inherent limitations: experts may overlook common mistakes made by students, and both strategies often yield less-optimal results in ill-defined domains [14]. Therefore, researchers have started to seek scalable solutions by applying data-driven methods to aggregate previous students' solutions and construct the learner's model. The automated feedback generation nature of the learner model positions the data-driven feedback approach as an initial application of generative AI in creating feedback.

The earliest work of generative AI on feedback generation can be traced back to the DIAG system [12]. The system utilized the NLP model to aggregate system messages in various structures and found the feedback aggregated by functions led to higher learning gain in the classroom study. Another stream of data-driven works built on the production rule approach automated the construction of the learner's cognitive model during problem-solving. The foundation work of the data-driven cognitive model, regardless of its non-generative nature, can be found in 1997 [11], where researchers initially applied Bayesian Networks to students' problem-solving data for plan recognition and action prediction. This approach was further developed by the Hint Factory [52], which employed Markov decision processes to analyze Logic Proof Intelligent Tutor submission data [50], generating production rules and comparing student submissions to these rules to tailor hints. This method of production-rule-based feedback generation garnered attention across ITS research in more disciplines [13,14,41]. The subsequent classroom studies and learning factor analyses [7] confirmed this approach's efficacy and versatility across various domains, demonstrating its potential with minimal data requirements [46].

2.3   Feedback Generated from LLMs

Some major shortcomings of the data-driven approach are: 1) it relies on the quantity and quality of the training data, and 2) the feedback in these systems is in fixed templates with limited adaptation. The recent prevalence of LLMs provides opportunities to advance the field of feedback generation by generating adaptive, human-like feedback without training data. For the feedback generation pipeline, most LLM-based work takes the student's current state together with certain prompts asking for feedback as input, treats the LLMs as a black box to process the prompt, and directly uses the output as the personalized feedback.

Indeed, LLMs enhance the scalability of adaptive feedback across various domains [20,25,31,34,48], and help in bypassing the expert model or cognitive model building process from scratch. However, there remains a significant amount of work overlooked by many researchers in the pedagogical design of feedback and the evaluation of its impact on learning. Few works are backed by learning sciences principles such as learning-by-teaching [48] and self-reflection [24], while others did not elaborate learning design considerations in their design rationale. Moreover, most of the evaluations on the LLM-based feedback systems only reported classroom usage data, and there is a lack of theoretical support or evidence on learning for these emerging systems. To guide better LLM-based feedback design and evaluation, in the next section, we highlight theoretical and empirical evidence from previous ITS and learning sciences work, hence suggesting implications for LLM-based feedback accordingly.

3   Implications for LLM-Based Feedback

Prior research in ITSs has built strong groundwork for LLM-based feedback generation. For instance, the left figure in Fig. 1 illustrates the key functions of a traditional intelligent tutoring system, showing how feedback is triggered, generated, and delivered in an ITS [21]. However, methods proven effective in traditional ITS have not yet been fully leveraged in LLM-based feedback research. Additionally, technological advancements in GenAI now enable scaling approaches previously limited by technological constraints.

This section discusses recommendations grounded in prior ITS research and new scalable opportunities from GenAI advancements in the feedback design, generation, delivery, and evaluation phases. The right figure in Fig. 1 shows how new opportunities in LLM and GenAI could fit into and extend the traditional key functions in ITS (the left one in Fig. 1), particularly around the feedback.

3.1   Trigger to Deliver the Feedback

Feedback in ITSs is often designed to be triggered either by the system or initiated by the students [3,33]. One special type of feedback is suggestions or supportive materials to help students move forward. It is common to provide

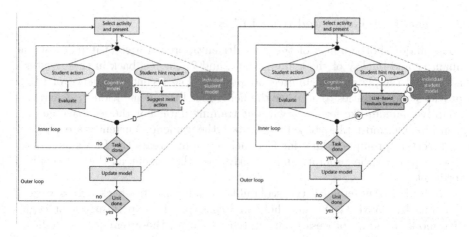

Fig. 1. Key functions of Intelligent Tutoring Systems from Koedinger et al. [21] (left) and Opportunities to Optimize LLM-based Feedback in ITS (right, adapted from [21]): I - Trigger to deliver the feedback; II - Information needed to communicate with LLMs as input; III - Content requested in the generated LLM feedback output; IV - Modality for delivering feedback to students.

such feedback based on students' request actions [17,18,30,46,57], as shown in A of Fig. 1-left. However, in some other work, concerning that novices might lack the metacognitive skills to identify when and what to ask for help, researchers believed that the model-guided feedback delivery could scaffold learners better. Therefore, Help Tutor was designed to provide metacognitive feedback automatically after the system identifies learners' misbehaviors [1]. While researchers found that having this did not lead to higher learning gains, the concerns about abusing feedback can be more serious when LLM is widely applied in educational settings. On the one hand, students nowadays can access commercial LLM products easily for timely formative feedback on their progress or even direct answers, so they are more likely to abuse them by asking for feedback too frequently. On the other hand, to avoid over-reliance on these LLM tools, some self-regulated learners might avoid asking for formative feedback, even when they are struggling or unsure about their answers. With those two new phenomena in mind, future research needs to find a balanced timing and optimized trigger to deliver feedback and establish a more effective learning experience (I in Fig. 1-right).

3.2 Information Needed to Communicate with LLM as Input

Prompt engineering is an important step when using LLMs to generate feedback in the existing learning systems. Prompt engineering refers to the practice of creating and optimizing prompts to communicate with large language models effectively. There are guidelines about how to generate effective prompts for LLMs. For example, the CLEAR framework comprises five key principles to achieve prompts for more effective AI-generated content creation [26]. Besides these general guidelines, the required information in prompts can vary based on the specific roles AI is expected to take. When AI takes on the role of providing feedback to students' answer inputs [27,34], information resources like the grading rubric, the student answers, and the task description, are commonly included in the prompt. While current implications have already integrated such basic related information to generate prompts, one future direction could involve incorporating various types of information into prompt engineering. For example, many traditional ITSs represent students' current understanding of the subject as a model and adapt instructions to students' needs based on it [54] (B in Fig. 1-left). Such student models play an important role in understanding and identifying students' needs before offering appropriate adaptation [9]. Therefore, to enhance the quality of LLM outputs in meeting individual student's needs, different layers of information from a student model, such as the estimations of the student's knowledge level, cognitive state, strengths or weaknesses, affective state, and meta-cognitive skills, can be integrated when conducting prompt engineering in the context of education [9] (II in Fig. 1-right).

3.3   Content Requested in the Generated LLM Feedback Output

One type of ITS feedback output is a suggestion for the next action (C in Fig. 1-left). When integrating LLM into learning systems, such output can also be optimized using learning science theories (III in Fig. 1-right). Particularly, the prevalent form of feedback in recent LLM-based learning systems is a high-level natural language explanation without revealing the answer. In certain systems, persistent student inquiries for the answer lead the feedback to progressively reveal more comprehensive hints, culminating in responses that closely resemble complete, bottom-out solutions. Although such feedback can assist students in solving their problems, its effectiveness in promoting long-term learning is uncertain. Should this feedback fail to facilitate learning, the LLM-based Intelligent Tutoring System (ITS) might merely become an ineffective variant or, in the best-case scenario, a duplicate of ChatGPT. To better trigger learning, this section aims to encourage future feedback implementation guided by learning sciences frameworks such as Bloom's Taxonomy [23] and Knowledge-Learning-Instruction (KLI) framework [22], and we employ the KLI framework as a guiding example to illustrate how to select feedback content based on the knowledge type to enhance learner outcomes. The KLI framework allocates learning processes into three types: memory and fluency, induction and refinement, understanding, and sense-making, and lists seven instructional principles that are effectively robust within each category, all of which are backed by substantial experimental validation.

For enhancing memory and fluency, instructional principles such as spacing, testing, and optimized scheduling are pivotal. They focus on the timing and frequency of practice repetitions. Feedback mechanisms like flashcards, visual cues [25], or multiple-choice questions, inspired by the testing principle, offer repeated practice opportunities at short intervals, reinforcing learning. While spacing and optimization extend beyond feedback's scope, their implementation in ITS is beneficial to effective instructional strategies in memory and fluency processes.

For tasks involving induction and refinement processes, such as solving programming or math problems, faded worked examples are effective in demonstrating the desired process with the appropriate level of assistance needed by fading the demonstration to ensure the scaffolding is within Vygotsky's Zone of Proximal Development [28], tailored to the learner's existing knowledge and areas for growth. The implementation of faded worked examples into intelligent tutors also resulted in higher learning efficiency and a deeper conceptual understanding of the problems in existing ITS works [49].

For understanding and sense-making processes, self-explanation questions are one of the most effective forms of feedback that elicit active learning and deeper thinking [15]. Previous ITS works also support its effects on consolidating learners' understanding of fractions [44].

3.4   Modality for Delivering Feedback to Students

Due to technology limitations, traditional ITS systems mainly deliver feedback to students as text (D in Fig. 1-left). Recent advancements in GenAI techniques provide opportunities to expand the spectrum of feedback modalities (IV in Fig. 1-right) from text to images, audios, videos, and combinations of these modalities. As the main focus of this paper is LLM-based feedback, we look into how GenAI techniques could deliver such feedback to students in diverse modalities. We first referred to classic multimedia learning principles [10] and chose related principles that provide recommendations on how feedback might be delivered. Then, we provided examples of recent GenAI techniques and corresponding commercial applications that could be used to achieve such feedback modalities. Finally, for each principle-based feedback modality, we provided one example use scenario. We summarized these in Table 1. As new GenAI technologies are continuously growing, we believe it is important for future work to look at both the principle and technology sides to decide what feedback modalities should and could be provided in different learning contexts.

3.5   Evaluate the Quality of the Generated Feedback

Existing studies has tested the scope of LLMs' feedback generation capabilities with data-driven system evaluation (whether the system can perform well on a comprehensive testing dataset) [16] and expert evaluation (given an evaluation matrix, the expert(s) would rate the performance of each feedback) [34,47,57] on mainly precision and coverage [39,40]. Moreover, it is important to recognize the contributions of GenAI-generated feedback as it expands the array of metrics for assessing feedback quality beyond the traditional ITS evaluation scope, with new metrics like appropriateness, conciseness, and comprehensiveness. Compared to traditional feedback generation approaches, LLMs' advantages in natural language tasks enable the generation of personalized feedback with an encouraging tone [47,57]. Taking a step further from examining the tone, studies have

Table 1. Example feedback modalities supported by Generative AI techniques

Principle (Selected from [10])	Feedback modality	Example GenAI Techniques Involved (Example GenAI Applications)	Example Feedback Scenario
Multimedia	Images	Text-to-image (OpenAI DALL.E [36]; Stable Diffusion[a]; Runway[b])	Content-related images as visual cue for vocabulary memorization
Personalization	Text	Text-to-text (OpenAI GPT[c]; Anthropic Claude[d]; Meta Llama[e])	Text feedback in more informal and conversational styles
Embodiment	Human-like agent	Text-to-image; Text-to-video (Runway[f]; OpenAI Sora [37]) Text-to-speech (Deepgram[g]; OpenAI TTS models[h]; WellSaid Labs[i])	Virtual teaching assistant avatar talking about the next-step hints
Modality	Audio	Text-to-speech	Add audio feedback as a new option to existing systems
Segmentation	Segmented Text	Text-to-text	Apply LLMs to segment one feedback into multi-levels of feedback.
Redundancy	Audio, Image, Video	Text-to-speech; Text-to-image; Text-to-video	When both audio and visual feedback is generated, allow users to turn off the text on screen
Temporal Contiguity	Audio, Image, Video	Text-to-speech; Text-to-image; Text-to-video	When both audio and visual feedback are generated, present them simultaneously
Spatial Contiguity	Text, Video	Text-to-text; Text-to-video	Place feedback close to the part that it elaborates on
Coherence	Text, Image	Text-to-text; Text-to-image	Evaluate whether existing content are on-topic or not
Signaling	Text	Text-to-text	Highlight the errors on the screen

[a] https://stability.ai/stable-image
[b] https://runwayml.com/ai-tools/text-to-image/
[c] https://openai.com/gpt-4
[d] https://www.anthropic.com/news/claude-3-family
[e] https://llama.meta.com/
[f] https://runwayml.com/ai-tools/gen-2-text-to-video/
[g] https://deepgram.com/
[h] https://platform.openai.com/docs/models/tts
[i] https://wellsaidlabs.com/features/api/

also assessed the capability of GenAI to play specific roles (e.g., instructor [53], student[48]) in the context of feedback generation. These initial assessments illuminate the potential to scale both the implementation and evaluation of specific agents and their behaviors that are crafted according to learning sciences principles, such as facilitating a growth mindset as an instructor or enhancing the dynamics of a collaborative learning environment as a helpful peer.

In addition to system and expert evaluation, classroom deployment allows researchers to gather feedback from real learners in authentic educational settings. Under this context, student usage log data [19, 20, 24, 43] and student self-reported survey data [20, 24, 38] are two commonly gathered data types for LLM-generated material evaluation. With few existing LLM-generated feedback works applying these approaches, the next steps for evaluating this feedback could involve: 1) designing controlled experiments with pre-post tests to assess learning outcomes [4, 42, 55], and 2) employing learning analytics on log data to explore feedback's effectiveness on help-seeking [57] and learning [38] with greater detail and nuances. For instance, the impacts of feedback on learning outcomes can be elucidated by performing a learning curve analysis to compare the slopes, which represent varying learning rates under different feedback conditions [45]. Finally, ethical concerns such as biases [56] and hallucination [8] associated with LLMs could jeopardize an effective and equitable learning environment. Therefore, it is crucial to incorporate relevant metrics into the evaluation of feedback quality and mitigate these issues with automated approaches or human intervention.

4   Conclusion

To fully utilize the potential of generative AI in the educational context, it is essential to approach its integration with pedagogical design. This paper synthesized the progression of AIED focusing on feedback generation, emphasized the importance of grounding the current LLM-based feedback generation in theoretical frameworks and evidence-based approaches to prompt its learning effectiveness, and suggested corresponding evidence-grounded implications through four feedback generation stages. We aim to evoke the awareness of AIED researchers and practitioners on the legacy of pre-LLM feedback generation efforts, and offer a toolkit as a foundational resource for designing pedagogical LLM-based feedback generation systems that foster the advancement of the AIED field in the era of generative AI.

References

1. Aleven, V., Mclaren, B., Roll, I., Koedinger, K.: Toward meta-cognitive tutoring: a model of help seeking with a cognitive tutor. Int. J. Artif. Intell. Educ. **16**(2), 101–128 (2006)
2. Aleven, V., McLaren, B.M., Sewall, J., Koedinger, K.R.: The cognitive tutor authoring tools (CTAT): preliminary evaluation of efficiency gains. In: Ikeda, M., Ashley, K.D., Chan, T.-W. (eds.) ITS 2006. LNCS, vol. 4053, pp. 61–70. Springer, Heidelberg (2006). https://doi.org/10.1007/11774303_7
3. Aleven, V., Mclaren, B.M., Sewall, J., Koedinger, K.R.: A new paradigm for intelligent tutoring systems: example-tracing tutors. Int. J. Artif. Intell. Educ. **19**(2), 105–154 (2009)
4. Alhazmi, S., Thevathayan, C., Hamilton, M.: Interactive pedagogical agents for learning sequence diagrams. In: Bittencourt, I.I., Cukurova, M., Muldner, K., Luckin, R., Millán, E. (eds.) AIED 2020. LNCS (LNAI), vol. 12164, pp. 10–14. Springer, Cham (2020). https://doi.org/10.1007/978-3-030-52240-7_2
5. Anderson, J.R.: Act: a simple theory of complex cognition. Am. Psychol. **51**(4), 355 (1996)
6. Baidoo-Anu, D., Ansah, L.O.: Education in the era of generative artificial intelligence (AI): Understanding the potential benefits of ChatGPT in promoting teaching and learning. J. AI **7**(1), 52–62 (2023)
7. Cen, H., Koedinger, K., Junker, B.: Learning factors analysis – a general method for cognitive model evaluation and improvement. In: Ikeda, M., Ashley, K.D., Chan, T.-W. (eds.) ITS 2006. LNCS, vol. 4053, pp. 164–175. Springer, Heidelberg (2006). https://doi.org/10.1007/11774303_17

8. Chen, Y., et al.: Hallucination detection: robustly discerning reliable answers in large language models. In: Proceedings of the 32nd ACM International Conference on Information and Knowledge Management, pp. 245–255 (2023)
9. Chrysafiadi, K., Virvou, M.: Student modeling approaches: a literature review for the last decade. Expert Syst. Appl. **40**(11), 4715–4729 (2013)
10. Clark, R.C., Mayer, R.E.: E-learning and the Science of Instruction: Proven Guidelines for Consumers and Designers of Multimedia Learning. Wiley, Hoboken (2023)
11. Conati, C., Gertner, A.S., VanLehn, K., Druzdzel, M.J.: On-line student modeling for coached problem solving using Bayesian networks. In: Jameson, A., Paris, C., Tasso, C. (eds.) User Modeling. ICMS, vol. 383, pp. 231–242. Springer, Vienna (1997). https://doi.org/10.1007/978-3-7091-2670-7_24
12. Di Eugenio, B., Fossati, D., Yu, D., Haller, S.M., Glass, M.: Natural language generation for intelligent tutoring systems: a case study. In: AIED, pp. 217–224 (2005)
13. Fossati, D., Di Eugenio, B., Ohlsson, S., Brown, C., Chen, L.: Data driven automatic feedback generation in the ilist intelligent tutoring system. Technol. Instr. Cogn. Learn. **10**(1), 5–26 (2015)
14. Fournier-Viger, P., Nkambou, R., Nguifo, E.M.: Learning procedural knowledge from user solutions to ill-defined tasks in a simulated robotic manipulator. In: Romero, et al. (eds.) Handbook of Educational Data Mining, pp. 451–465 (2010)
15. Hausmann, R.G., VanLehn, K.: Explaining self-explaining: a contrast between content and generation. Front. Artif. Intell. Appl. **158**, 417 (2007)
16. Hellas, A., Leinonen, J., Sarsa, S., Koutcheme, C., Kujanpää, L., Sorva, J.: Exploring the responses of large language models to beginner programmers' help requests. In: Proceedings of the 2023 ACM Conference on International Computing Education Research-Volume 1, pp. 93–105 (2023)
17. Hou, X., Ericson, B.J., Wang, X.: Using adaptive parsons problems to scaffold write-code problems. In: Proceedings of the 2022 ACM Conference on International Computing Education Research-Volume 1, pp. 15–26 (2022)
18. Hou, X., Ericson, B.J., Wang, X.: Understanding the effects of using parsons problems to scaffold code writing for students with varying CS self-efficacy levels. In: Proceedings of the 23rd Koli Calling International Conference on Computing Education Research, pp. 1–12 (2023)
19. Kazemitabaar, M., Hou, X., Henley, A., Ericson, B.J., Weintrop, D., Grossman, T.: How novices use LLM-based code generators to solve cs1 coding tasks in a self-paced learning environment. In: Proceedings of the 23rd Koli Calling International Conference on Computing Education Research, pp. 1–12 (2023)
20. Kazemitabaar, M., et al.: Codeaid: evaluating a classroom deployment of an llm-based programming assistant that balances student and educator needs. arXiv preprint arXiv:2401.11314 (2024)
21. Koedinger, K., Brunskill, E., Baker, R., Mclaughlin, E., Stamper, J.C.: New potentials for data-driven intelligent tutoring system development and optimization. AI Mag. **34**, 27–41 (2013). https://api.semanticscholar.org/CorpusID:13189100
22. Koedinger, K.R., Corbett, A.T., Perfetti, C.: The knowledge-learning-instruction framework: bridging the science-practice chasm to enhance robust student learning. Cogn. Sci. **36**(5), 757–798 (2012)
23. Krathwohl, D.R.: A revision of bloom's taxonomy: an overview. Theory Pract. **41**(4), 212–218 (2002)

24. Kumar, H., Musabirov, I., Williams, J.J., Liut, M.: Quickta: exploring the design space of using large language models to provide support to students. In: Learning Analytics and Knowledge Conference. Learning Analytics and Knowledge Conference 2023 (LAK'23). ACM, Arlington, Texas (2023)

25. Lee, J., Lan, A.: Smartphone: exploring keyword mnemonic with auto-generated verbal and visual cues. In: Wang, N., Rebolledo-Mendez, G., Matsuda, N., Santos, O.C., Dimitrova, V. (eds.) AIED 2023. LNCS, vol. 13916, pp. 16–27. Springer, Cham (2023). https://doi.org/10.1007/978-3-031-36272-9_2

26. Lo, L.S.: The clear path: a framework for enhancing information literacy through prompt engineering. J. Acad. Librariansh. **49**(4), 102720 (2023)

27. Lu, C., Cutumisu, M.: Integrating deep learning into an automated feedback generation system for automated essay scoring. International Educational Data Mining Society (2021)

28. Malik, S.A.: Revisiting and re-representing scaffolding: the two gradient model. Cogent Educ. **4**(1), 1331533 (2017)

29. Martin, B., Koedinger, K.R., Mitrovic, A., Mathan, S.: On using learning curves to evaluate its. In: AIED, pp. 419–426 (2005)

30. McLaren, B.M., Richey, J.E., Nguyen, H., Hou, X.: How instructional context can impact learning with educational technology: lessons from a study with a digital learning game. Comput. Educ. **178**, 104366 (2022)

31. McNichols, H., Zhang, M., Lan, A.: Algebra error classification with large language models. In: Wang, N., Rebolledo-Mendez, G., Matsuda, N., Santos, O.C., Dimitrova, V. (eds.) AIED 2023. LNCS, vol. 13916, pp. 365–376. Springer, Cham (2023). https://doi.org/10.1007/978-3-031-36272-9_30

32. Mitrovic, A., Mayo, M., Suraweera, P., Martin, B.: Constraint-based tutors: a success story. In: Monostori, L., Váncza, J., Ali, M. (eds.) IEA/AIE 2001. LNCS (LNAI), vol. 2070, pp. 931–940. Springer, Heidelberg (2001). https://doi.org/10.1007/3-540-45517-5_103

33. Mitrovic, A., Ohlsson, S., Barrow, D.K.: The effect of positive feedback in a constraint-based intelligent tutoring system. Comput. Educ. **60**(1), 264–272 (2013)

34. Nguyen, H.A., Stec, H., Hou, X., Di, S., McLaren, B.M.: Evaluating ChatGPT's decimal skills and feedback generation in a digital learning game. In: Viberg, O., Jivet, I., Muñoz-Merino, P., Perifanou, M., Papathoma, T. (eds.) EC-TEL 2023. LNCS, vol. 14200, pp. 278–293. Springer, Cham (2023). https://doi.org/10.1007/978-3-031-42682-7_19

35. Ohlsson, S.: Learning from performance errors. Psychol. Rev. **103**(2), 241 (1996)

36. OpenAI: Dall.e 2 (2023). https://openai.com/dall-e-2/. Accessed 12 Mar 2024

37. OpenAI: Sora (2024). https://openai.com. Accessed 12 Mar 2024

38. Pankiewicz, M., Baker, R.S.: Navigating compiler errors with AI assistance– a study of GPT hints in an introductory programming course. arXiv preprint arXiv:2403.12737 (2024)

39. Phung, T., et al.: Generating high-precision feedback for programming syntax errors using large language models. arXiv preprint arXiv:2302.04662 (2023)

40. Phung, T., et al.: Automating human tutor-style programming feedback: leveraging GPT-4 tutor model for hint generation and GPT-3.5 student model for hint validation. In: Proceedings of the 14th Learning Analytics and Knowledge Conference, pp. 12–23 (2024)

41. Price, T.W., Dong, Y., Barnes, T.: Generating data-driven hints for open-ended programming. International Educational Data Mining Society (2016)

42. Price, T.W., Zhi, R., Barnes, T.: Hint generation under uncertainty: the effect of hint quality on help-seeking behavior. In: André, E., Baker, R., Hu, X., Rodrigo, M.M.T., du Boulay, B. (eds.) AIED 2017. LNCS (LNAI), vol. 10331, pp. 311–322. Springer, Cham (2017). https://doi.org/10.1007/978-3-319-61425-0_26

43. Qi, J.Z.P.L., Hartmann, B., Norouzi, J.D.N.: Conversational programming with LLM-powered interactive support in an introductory computer science course. In: NeurIPS'23 Workshop on Generative AI for Education (GAIED) (2023)

44. Rau, M.A., Aleven, V., Rummel, N.: Intelligent tutoring systems with multiple representations and self-explanation prompts support learning of fractions. In: AIED, pp. 441–448 (2009)

45. Rivers, K., Harpstead, E., Koedinger, K.R.: Learning curve analysis for programming: which concepts do students struggle with? In: ICER, vol. 16, pp. 143–151. ACM (2016)

46. Rivers, K., Koedinger, K.R.: Data-driven hint generation in vast solution spaces: a self-improving python programming tutor. Int. J. Artif. Intell. Educ. **27**, 37–64 (2017)

47. Roest, L., Keuning, H., Jeuring, J.: Next-step hint generation for introductory programming using large language models. In: Proceedings of the 26th Australasian Computing Education Conference, pp. 144–153 (2024)

48. Schmucker, R., Xia, M., Azaria, A., Mitchell, T.: Ruffle&Riley: towards the automated induction of conversational tutoring systems. arXiv preprint arXiv:2310.01420 (2023)

49. Schwonke, R., Wittwer, J., Aleven, V., Salden, R., Krieg, C., Renkl, A.: Can tutored problem solving benefit from faded worked-out examples. In: European Cognitive Science Conference, pp. 23–27 (2007)

50. Stamper, J.: Automating the generation of production rules for intelligent tutoring systems. In: Proceedings of 9th International Conference on Interactive Computer Aided Learning (2006)

51. Stamper, J., Barnes, T., Croy, M.: Enhancing the automatic generation of hints with expert seeding. In: Aleven, V., Kay, J., Mostow, J. (eds.) ITS 2010. LNCS, vol. 6095, pp. 31–40. Springer, Heidelberg (2010). https://doi.org/10.1007/978-3-642-13437-1_4

52. Stamper, J., Barnes, T., Lehmann, L., Croy, M.: The hint factory: automatic generation of contextualized help for existing computer aided instruction. In: Proceedings of the 9th International Conference on Intelligent Tutoring Systems Young Researchers Track, pp. 71–78 (2008)

53. Tack, A., Piech, C.: The AI teacher test: measuring the pedagogical ability of blender and GPT-3 in educational dialogues. arXiv preprint arXiv:2205.07540 (2022)

54. VanLehn, K.: Student modeling. In: Foundations of Intelligent Tutoring Systems, pp. 55–78 (2013)

55. VanLehn, K., et al.: The Andes physics tutoring system: five years of evaluations. In: AIED, pp. 678–685 (2005)

56. Wei, Y., Carvalho, P.F., Stamper, J.: Uncovering name-based biases in large language models through simulated trust game. arXiv preprint arXiv:2404.14682 (2024)

57. Xiao, R., Hou, X., Stamper, J.: Exploring how multiple levels of GPT-generated programming hints support or disappoint novices. arXiv preprint arXiv:2404.02213 (2024)

Flowing Through Virtual Realms: Leveraging Artificial Intelligence for Immersive Educational Environments

Grzegorz Zwoliński[(✉)] and Dorota Kamińska

Lodz University of Technology, 116 Zeromskiego Street, 90-924 Lodz, Poland
{grzegorz.zwolinski,dorota.kaminska}@p.lodz.pl

Abstract. Virtual Reality presents a promising frontier for educational immersion, offering unparalleled opportunities for engaging learning experiences. However, the integration of artificial intelligence algorithms will be critical to realizing its full potential. This paper explores the fusion of VR and AI through the lens of Csikszentmihalyi's flow theory, with the aim of optimizing educational experiences for learners. AI emerges as a critical facilitator by aligning VR environments with the principles of flow, characterized by deep concentration, intrinsic motivation, and a balance between skill and challenge. Through adaptive learning pathways, personalized feedback mechanisms and dynamic content generation, AI-driven VR environments can tailor educational journeys to individual learners, fostering optimal engagement and learning outcomes. This paper discusses the conceptual framework for integrating AI into VR-based educational environments, highlighting the potential synergy between these technologies in creating immersive learning experiences conducive to flow states. At the same time, we propose a subjective, multidimensional assessment tool for evaluating VR educational applications, which assesses perceived flow based on feedback from students using VR during their educational journey.

Keywords: VR · AI · Education · Flow theory · Flow measure

1 Introduction

Irrespective of the level of the educational system, professionals in the area report similar difficulties related to learning and behavioral disorders among their students [36]. Lack of interest and decreased motivation, concentration disorders, deterioration in communication and interpersonal relationships, or low levels of students' emotional intelligence make it difficult for many educators to reach their students effectively [23]. All these symptoms may be an indication of the need for a change in the approach to the educational process [4].

Virtual reality (VR) is not the solution that initially comes to mind when discussing adaptation and recreation of educational processes. However, research

A. M. Olney et al. (Eds.): AIED 2024 Workshops, CCIS 2150, pp. 44–57, 2024.
https://doi.org/10.1007/978-3-031-64315-6_4

increasingly points to various benefits stemming from VR in educational utilization [18]. For clarification purposes, we decided to coin the term EdVR, and this is how VR-based education will be referred to within the article. The effectiveness of visualizing complex concepts drives the use of VR technologies in education. These technologies provide immersive and interactive learning experiences that engage students and facilitate a deeper understanding of abstract ideas. VR offers immersive journeys into alternate realities that closely resemble the real world. In education, VR is a tool for understanding complex subjects and promoting cultural comprehension among students. Through VR, educators can teach the importance of embracing and respecting cultural diversity and foster an environment where students value differences. The interactive nature of the technology allows students to interact with different environments and scenarios, promoting inclusivity and empathy within the learning environment. For instance, the application of VR in soft skills training can lead, among other things, to changes in social behaviors [19]. The results of an experiment conducted at Stanford University demonstrate that individuals who were required to engage in altruistic behaviors while participating in VR simulations exhibited an increased number of such reactions in real life compared to individuals who had similar VR experiences but were not required to engage in altruistic behavior simulations [37]. Scholars from the University of Maryland unveiled that individuals tend to retain information more effectively when presented in VR as opposed to being conveyed through two-dimensional imagery displayed on personal computer screens, smartphones, or tablets [21]. Furthermore, in [1], the authors demonstrated that lessons deliberately employing VR fostered curiosity among students, heightened interest in the subject matter, and facilitated sustained engagement of children in activities at a significantly higher level compared to sessions conducted solely through conventional methods. Within the biology field, VR offers students the opportunity to delve into the human body in three dimensions, providing a visual representation of intricate systems and organs beyond what traditional textbooks can offer [12]. Similarly, in physics, VR simulations allow students to visualize and interact with abstract concepts like electromagnetic fields [11] or quantum mechanics, aiding their comprehension of these challenging topics [15,42]. Additionally, VR can transport students to historical epochs, such as ancient civilizations [45] or pivotal events [14], enabling them to immerse themselves in the historical context and gain a tangible understanding of the subject matter. By visualizing complex concepts, VR technologies create a dynamic and captivating learning environment that enhances comprehension and retention, rendering education more interactive and impactful. In the landscape of educational applications, achieving optimal learning outcomes requires a strategic blend of effective teaching methods and sustained user engagement. This quest is further complicated by the need to strike a delicate balance between challenging users sufficiently to keep them at the edge of their cognitive capabilities, while minimising the sensor intrusion required for such engagement. A cornerstone concept in this pursuit is Csikszentmihalyi's Flow Theory [29], which posits that learning reaches its peak when individuals are immersed in tasks that precisely match their skill level, creating a harmonious balance between challenge and mastery. Users experiencing this flow state exhibit deep concentration and absorption in their activities. Hence, embedding elements that evoke flow-like experiences within educational applications can markedly heighten learning efficiency [13,17,43]. Parallel to this theory is Jerome Bruner's Comfort Zone Concept, akin to the Zone of Proximal Development, suggesting that optimal learning transpires when individuals operate slightly beyond their current level of ability, but within reach with appropriate support. Educational applications should delicately nudge users beyond their comfort zones while providing scaffolded assistance to facilitate learning [40]. The practical implementation of promoting a flow state within educational applications involves crafting tasks that challenge users without overwhelming them. By integrating real-time feedback mechanisms, such as monitoring physiological responses like heart rate variability, skin conductance, and blink frequency, a promising avenue emerges. These physiological metrics serve as proxies for user engagement and cognitive load, enabling dynamic adjustments within the application to sustain an optimal learning environment [31]. The optimization of learning effects in educational applications necessitates a nuanced approach that melds effective teaching methodologies with strategies to maintain user engagement. Leveraging theories such as the Flow Theory and the Comfort Zone Concept furnishes invaluable insights into designing pedagogically sound methods and nurturing sustained user engagement for superior learning outcomes. Through the judicious use of real-time feedback mechanisms and the prudent minimisation of sensor intrusion, educational applications can pave the way for the promotion of optimal learning experiences for users.

Keeping the user in a state of flow is a significant challenge, leading us to recognise an opportunity to utilize artificial intelligence solutions. By harnessing discernible behavioral cues, including movement patterns and emotional states extrapolated from accessible biomedical indicators, these AI-driven methods will carefully tailor the progression of the educational narrative within VR framework. Such dynamic adjustments to the learning environment, dependent on the user's cognitive and emotional fluctuations, hold great promise for enhancing the effectiveness of educational endeavors and optimizing learning outcomes. This manuscript explores the conceptual framework for integrating artificial intelligence into VR-based educational environments to maintain flow. It highlights the prospective harmony between these technologies in creating immersive learning contexts conducive to achieving flow states.

The rest of the paper is organized as follows: Sect. 2 provides the most important theoretical background. In Sect. 3, we discuss methods for keeping the user in the flow state during the EdVR experience and propose a metric to evaluate the perceived flow in EdVR subjectively. The conclusions are presented in Sect. 5.

2   Background

2.1   Theoretical Foundations of Flow

Csikszentmihalyi's groundbreaking research led him to a profound realization: happiness is not merely a byproduct of external circumstances but an internal state of being. His seminal work, [5], suggests that happiness can be cultivated through the experience of flow. He posits that happiness is not static, but requires active engagement and effort. Beyond each individual's baseline level of happiness, Csikszentmihalyi identifies a realm of happiness over which individuals have agency. Through his studies, he discovered that individuals experience their highest levels of creativity, productivity and joy when they are in a state of flow.

Csikszentmihalyi's inquiries extended to athletes, musicians, and artists, seeking to understand the moments when they achieve peak performance and how they perceive these experiences. He coined the term "flow state" and observed that people often describe their peak performance moments as instances when work flows effortlessly. His aim was to uncover the factors that stimulate creativity, particularly in professional settings, and how this creativity translates into productivity. He emphasised that flow is essential not only for productivity but also for overall job satisfaction. According to Csikszentmihalyi, flow is characterised by complete immersion in an activity where nothing else seems to matter. The experience is so pleasurable that individuals are willing to pursue it at great personal cost, purely for the intrinsic pleasure it brings. Let us delve deeper into the theoretical underpinnings of flow state identification as elucidated by Csíkszentmihályi [7,9,10]:

- **Q1 Having a clear understanding of what you want to achieve:** This involves setting precise goals or objectives that are relevant to the task at hand. When individuals have a clear understanding of their goals, it provides them with direction and purpose, thereby increasing the likelihood of achieving a state of flow.
- **Q2 Being able to concentrate for extended periods:** Concentration plays a key role in achieving flow as it allows individuals to fully immerse themselves in the task without interruption. By minimising distractions, individuals can maintain their focus and engage deeply in the activity at hand.
- **Q3 Losing awareness of oneself:** During the flow state, the individual experiences a sense of selflessness in which they are completely absorbed in the task at hand. This lack of self-awareness allows the individual to transcend their ego and focus solely on the activity, thereby enhancing their overall experience.
- **Q4 Discovering that time passes quickly:** The state of flow often involves a distortion of the perception of time, resulting in the individual losing track of time and becoming completely absorbed in the present moment. This sense of timelessness can enhance the immersive quality of the experience and serve as an indicator of the presence of flow.

- **Q5 Receiving immediate and direct feedback:** Feedback is paramount in guiding behavior and providing individuals with insight into their progress towards their goals. Immediate and direct feedback enables individuals to adjust their actions in real time, increasing their sense of control and mastery over the task at hand.
- **Q6 Experiencing a balance between your skills and the challenges presented:** Flow occurs when individuals perceive a balance between the challenge level of the task and their own skills or abilities. When the level of challenge matches or slightly exceeds their skill level, individuals are incentivised to engage wholeheartedly with the task and experience flow.
- **Q7 Feeling a sense of personal control over the situation:** The flow state is characterised by a sense of agency and control, where the individual feels empowered to overcome the challenges of the task. This sense of control fosters feelings of confidence and competence, thereby enhancing the flow experience.
- **Q8 Sensing that the activity itself is inherently satisfying:** Flow occurs when individuals discover that the activity is inherently rewarding and enjoyable. They are intrinsically motivated by the task and derive satisfaction from the process of engaging in it, prioritising intrinsic enjoyment over external rewards or outcomes.
- **Q9 Lacking awareness of bodily needs:** During the flow state, individuals may temporarily lose awareness of bodily sensations such as hunger, fatigue or discomfort. This increased focus on the task at hand allows individuals to prioritize their cognitive and emotional engagement, fostering a more immersive experience.
- **Q10 Being completely absorbed in the activity itself:** The flow state is characterised by a state of total engagement, where the individual is completely immersed in the task and lost in the experience. This deep absorption allows the individual to enter a state of heightened focus, creativity and productivity.

In addition Csíkszentmihályi identified three essential conditions necessary to enter a state of flow [7,9,10]: goal, balance and feedback.

2.2 Flow in Educational Context

Although initially conceived as a model relating to engagement in a broader sense, the flow model is of considerable relevance and applicability to academic engagement [8]. Research substantiates the existence of conditions conducive to promoting flow in educational environments [16]. However, while replicating states of fun and joy may keep students in a state of deep satisfaction and contentment, it is not as effective in terms of acquiring new knowledge and skills. Flow in an educational context represents an optimal psychological state where individuals are fully immersed and engaged in an activity, experiencing a perfect balance between the challenge of the task and their skills (see Fig. 1).

For the student, it is akin to walking on the edge of a cognitive abyss, where they are neither overwhelmed nor bored but fully absorbed and focused [34,35].

Transitioning beyond this state can lead to entering zones schematically labeled energized and even growth. In these zones, individuals may experience heightened arousal, excitement, or personal growth. However, while these states may have their own merits, they are not conducive to effective learning. In the energized state, individuals may feel stimulated and enthusiastic, but this may be at the expense of losing focus and concentration on the learning task at hand. Similarly, in the GROWTH state, individuals may be challenged to expand their skills and abilities, but this may overwhelm them and hinder effective learning.

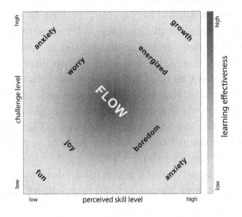

Fig. 1. The flow state within an educational framework - the pinnacle of the human learning experience, wherein individuals become fully immersed in an activity.

Therefore, in order to teach effectively and facilitate optimal learning outcomes, educators should strive to keep students in the flow. This involves carefully managing the level of challenge presented to students to ensure that it is appropriate to their current skill level, thereby fostering a sense of mastery and achievement. In addition, educators should provide timely feedback and support to help students overcome challenges and maintain focus and engagement. By cultivating and sustaining the flow state in the learning environment, educators can maximize students' learning potential and promote meaningful educational experiences [34]. In light of the above, we believe that an educational VR (EduVR) application should incorporate the three Csíkszentmihályi's essential conditions, namely goal, balance, and feedback, to facilitate users' transition into the flow state and maximize educational outcomes. By ensuring that the application establishes clear goals, provides a balanced level of challenge and skill, and offers immediate feedback, designers can create an environment conducive to flow. This immersive experience increases engagement, concentration, and enjoyment, ultimately leading to a more effective learning experience for users.

3 Keeping the User in a State of Flow in EduVR

3.1 Goal

One of the fundamental aspects of setting goals for users in educational tasks is the scenario that clearly and concisely defines these goals. The implementation of the VR application must be aligned with this premise by providing a well-selected set of affordances. These affordances will play a crucial role in providing users with a sense of high control and effectiveness over the application without the need for assistive systems. Ideally, the educational application should be designed to allow users to complete tasks without additional assistance.

In VR educational environments, scenarios serve as blueprints for defining user goals. These scenarios should be carefully crafted to outline specific learning objectives, challenges and paths to success. By presenting clear goals in the context of engaging scenarios, users are motivated to actively participate and pursue those goals. In addition, the design of VR applications should prioritize the selection and implementation of affordances that enable users to navigate the educational content autonomously. Affordances such as interactive elements, intuitive controls and contextual cues should be strategically integrated to facilitate seamless user interaction and task completion. The ultimate goal of VR educational applications should be to enable users to complete tasks independently, relying solely on the affordances provided within VR environment. By designing applications with self-sufficiency in mind, developers can create immersive experiences that empower users to effectively achieve their learning goals.

3.2 Balance

Flow occurs when a delicate equilibrium between the perceived challenges associated with a task and the individual's perceived skills or abilities is achieved. When the level of challenge matches or slightly exceeds a person's skill level, an optimal level of arousal and engagement is attained, facilitating the flow experience. Sustaining engagement and maintaining flow can be achieved by using AI algorithms, drawing on past experience, and leveraging engagement detection through measurements of movement activity [2], relevant biomedical signals [6], and more. The integration of such systems should be aligned with the developed scenario of the VR educational experience to keep the student in a state of flow (individualizing the level of instruction). It is the most promising solution to provide educators with a balanced, student-tailored teaching process [41]. Contrary to appearances, such a solution should not lead to increased stratification in the educational level of students. In traditional shared teaching processes, students with low aptitude often struggle to keep up with the rest, widening the gap in knowledge and skills relative to others. Students with high aptitudes, on the other hand, are hindered in their development. Even a small degree of individualisation of the teaching process, using intelligent, highly immersive educational tools, should bring tangible benefits to society by improving the knowledge, competencies and skills of successive generations [24].

3.3 Feedback

In EdVR, the provision of immediate and clear feedback is a cornerstone for guiding individual behavior and informing users of their progress towards their goals. Feedback is critical in enabling individuals to fine-tune their actions, rectify their trajectory, and improve their outcomes, bolstering their sense of autonomy and proficiency in the task. AI-based systems are emerging as adept solutions for such applications in assessing tasks, providing faster and more accurate feedback, and allowing teachers to devote more time to other teaching duties [33]. Engaging with AI systems to manage feedback guides users through their flow and facilitates seamless navigation and the exploration of new knowledge and skill resources. Immediate feedback ranks among Generation Z's most eagerly anticipated functionalities and will likely remain so for subsequent generations [27]. Considering the current advancements in immersive technologies, feedback mechanisms for user experiences in VR can be conveyed through various communication channels, each catering to different sensory modalities.

It is self-evident that the most natural method of conveying information through VR is visualization. Feedback can be presented as changes in the environment, objects, virtual avatars or other users to provide immediate information to the user about their actions or the state of the virtual world [28] [3]. Effective auditory feedback, a vital component of VR, encompasses a spectrum of elements such as music, sound effects, dialogue, and ambient sounds. This feedback serves to convey emotions, moods, themes, and atmospheres, thereby enriching the immersion and narrative of the VR environment [26]. Analogously works tactile sensations delivered through haptic feedback devices, such as gloves, vests, or controllers, allow users to feel virtual objects or interactions, adding a sense of realism and enhancing user engagement [44].

3.4 Monitoring Flow State in EdVR

In EdVR, managing the user's flow is paramount for optimizing learning outcomes. Quantifying the flow within immersive settings presents a unique challenge, necessitating measurement and analysis systems. Fortunately, in recent years, the rapid development of sensors and AI has made it possible for machines to recognize and analyze human mental states. Humans express their feelings through different signals, thus recognition can be realized by analyzing facial expressions [22], speech [25], behavior [39], or physiological signals [32]. These signals are collected by various sensors (see Fig. 2). These may include:

- **Physiological sensors:** devices measuring physiological responses such as heart rate variability (HRV), skin conductance response (SCR), electromyography (EMG), and electroencephalography (EEG) can offer insights into the user's arousal and cognitive load.
- **Eye-tracking technology:** Eye-tracking systems can monitor gaze patterns and pupil dilation, providing indications of visual attention and cognitive processing.

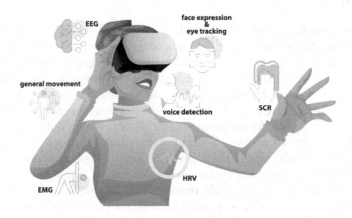

Fig. 2. Bio-signal type that can be used for monitoring flow state in EdVR.

- **Motion capture:** Utilizing motion capture technology, such as inertial measurement units (IMUs) or optical tracking systems, can track users' movements and gestures, reflecting their physical engagement and interaction within the VR environment.
- **Facial expression analysis:** Facial recognition software can analyze facial expressions to gauge emotional states, indicating the user's affective response to the VR experience.
- **Voice analysis:** Voice analysis tools can assess speech patterns, offering insights into the user's emotional state and level of engagement.
- **Behavioral metrics:** Tracking user interactions, such as navigation paths, click rates, and response times, can provide quantitative data on engagement levels and task performance.

By integrating these diverse measurement modalities, AI systems can gain comprehensive information on user engagement, which enables dynamic adjusting of the scenario to the user's current condition, not to break out of the state of flow. It is important to remember that using a VR headset can introduce certain discomforts for users and relying on additional external sensors may disrupt the user's ability to smoothly enter a state of flow, thus impeding full concentration on the VR experience. To mitigate discomfort, it is crucial to minimize the number of external sensors or utilize integrated sensor technology. Therefore, prioritizing integrated sensors directly within VR glasses can significantly enhance user comfort and the intensity of experiences. Fortunately, VR goggle manufacturers understand this issue which can be observed in an increasing number of contemporary HMDs offering highly advanced integrated measurement systems. For example, Meta incorporated built-in sensors and cameras in the Meta Quest Pro model, which provides capabilities for tracking head movements, hand gestures, facial expressions, and eye tracking, enabling real-time assessment of user engagement [30, 47]. HTC Focus 3 and Elite are equipped with motion controllers, which offer precise room-scale tracking and intuitive inter-action mechanisms. Vive also provides facial and eye trackers for those models. Varjo XR-4 has similar features [46]. These advanced HMDs not only enhance the immersion and realism of VR experiences, but also provide rich data streams for monitoring the psychological state of the user. By leveraging the capabilities of these integrated measurement systems, researchers can gain valuable insight into the cognitive and affective responses of the user, driving advancements in educational VR applications. Other supporting measurement systems are avail-able as additional accessories for HMD kits. However, it should be noted that their manufacturers may need to catch up with the latest HMD products. Usu-ally the available products are tailored to work with aging HMD models (e.g. LOOXID Link - user-friendly EEG sensor) [20]. Systems for measuring heart rate variability or skin conductance do not require such deep integration with the HMD. Different solutions are possible for their implementation. It is always important to consider their placement to avoid signal interference during user task performance.

4   Assessment Tool that Rates Perceived Flow in EdVR

In light of our interest surrounding students' perceptions regarding the signif-icance of Csikszentmihalyi's identified properties, we undertook a systematic investigation to describe the alignment between these perceptions and the basic tenets of flow theory. Our overarching objective was twofold: firstly, to establish whether there exists a congruence between students' perspectives and Csikszent-mihalyi's theoretical framework; and secondly, to conceptualize and propose a standardized metric capable of comprehensively assessing application quality in terms of its efficacy in sustaining users in the flow state.

The introduced metric for evaluating user retention in a state of flow, aimed at optimizing learning efficiency, is derived from the framework of the NASA-TLX index [38]. This index, rooted in cognitive psychology, offers a structured approach to assessing the cognitive workload experienced by individuals during task performance. Our adaptation (see Eq. 1) involves integrating principles from Csikszentmihalyi's concept of flow, which emphasizes the optimal psychological state of knowledge by deep engagement and immersion in an activity.

$$\sigma = \sum_{n=Q1}^{Q10} a_n * q_n \tag{1}$$

where: σ - the subjective flow index [from 0 (low) to 100 (high)]; a_n - the weight of the n element and q_n the value of the n element.

To construct our quality assessment index, we synthesized 10 key indicators identified by Csikszentmihalyi that are indicative of flow experiences within edu-cational contexts. These indicators encompass various aspects such as perceived challenge, skill level, clear goals, immediate feedback, and the feeling of control.

Fig. 3. Experimentally determined weighting factors a_n for parameters Q1–Q10.

The fine-tuning of this index was meticulously conducted through comprehensive survey research involving a substantial sample of 254 students with previous exposure to EdVR and acquainted with the theory of flow, aged between 18 and 25 years. By gathering and analyzing their feedback, preferences, and experiences with EdVR applications, we refined the index to accurately capture the nuanced aspects of user engagement and retention conducive to optimal learning outcomes. The resulting form of the quality indicator, as illustrated in Fig. 3, encapsulates the culmination of our empirical investigations and theoretical frameworks, offering a tool for evaluating and enhancing the user experience. It is recommended to interpret the results of perceived subjective flow experiences using the proposed metric σ analogously to the widely recognized NASA-TLX index.

5 Conclusions

The integration of VR into education offers a myriad of new opportunities and possibilities for learners. Traditional training programs often face constraints such as limited real-world scenarios, time, space, and safety concerns. However, virtual environments offer a solution to these challenges by providing an immersive and expansive environment where learners can engage in experiences that may not be feasible or safe in the physical world. Researchers are increasingly emphasizing the potential of the VR to address these barriers. By leveraging the capabilities of the VR, educational programs can provide realistic simulations, hands-on experiences, and interactive learning environments that enhance comprehension and skill acquisition. In the coming decade, a surge in research and applications exploring the use of EdVR is expected. This trend will likely lead to the development of innovative learning platforms, immersive training simulations, and personalized learning experiences catering to diverse educational needs and preferences. Overall, integrating VR into education promises to revolutionize learning experiences and expand access to high-quality training programs. As technology continues to evolve, educators and researchers must continue to explore and harness VR's potential, especially in the context of integrating with AI.

In this paper, we propose a visionary approach to be employed in VR educational applications by utilizing intelligent affordances controlled by AI agents, aiming to keep users in a state of flow. The detection of flow level should be based on various sensors integrated into HMD headsets, as well as others that do not negatively impact user comfort during educational experiences. Maintaining users in a state of flow is intended to ensure the most effective learning path while facilitating the concept of asynchronous learning in student groups. Additionally, the article proposes using a subjective flow index to verify the effectiveness of adaptive sustaining the user in the state of flow, providing a means to assess the efficacy of the proposed approach comprehensively. The presented approach can potentially transform how educational experiences are delivered, enabling personalized learning based on user engagement levels and comfort. Integrating AI with VR applications can create new education opportunities, focusing on information delivery and effectively engaging and maintaining participants' attention in the educational process. The added value of this approach lies in its potential for use in asynchronous learning, where users can access educational applications at any time and place. Thus, VR applications with intelligent affordances can improve the accessibility and effectiveness of the teaching process, leading to better educational outcomes and increased participant engagement.

References

1. Bailey, J.O., Bailenson, J.N., Obradović, J., Aguiar, N.R.: Virtual reality's effect on children's inhibitory control, social compliance, and sharing. J. Appl. Dev. Psychol. **64**, 101052 (2019)
2. Bianchi-Berthouze, N.: Understanding the role of body movement in player engagement. Hum.-Comput. Interact. **28**(1), 40–75 (2013)
3. Brummelman, E., Grapsas, S., van der Kooij, K.: Parental praise and children's exploration: a virtual reality experiment. Sci. Rep. **12**(1), 4967 (2022)
4. Burbules, N.C., Fan, G., Repp, P.: Five trends of education and technology in a sustainable future. Geogr. Sustain. **1**(2), 93–97 (2020)
5. Chikszentmihalyi, M.: Flow: The psychology of optimal experience. New York (1990)
6. Cohen, S.S., et al.: Neural engagement with online educational videos predicts learning performance for individual students. Neurobiol. Learn. Mem. **155**, 60–64 (2018)
7. Csikszentmihalyi, M.: Beyond Boredom and Anxiety. Jossey-bass (2000)
8. Csikszentmihalyi, M., Csikszentmihalyi, M.: Flow and education. In: Applications of Flow in Human Development and Education: The Collected Works of Mihaly Csikszentmihalyi, pp. 129–151 (2014)
9. Csikszentmihalyi, M., Csikszentmihalyi, M.: Intrinsic motivation and effective teaching. In: Applications of Flow in Human Development and Education: The Collected Works of Mihaly Csikszentmihalyi, pp. 173–187 (2014)
10. Csikszentmihalyi, M., Csikszentmihalyi, M.: Toward a psychology of optimal experience. In: Flow and the Foundations of Positive Psychology: The Collected Works of Mihaly Csikszentmihalyi, pp. 209–226 (2014)
11. Cvetkovski, G., et al.: Vimela project: an innovative concept for teaching mechatronics using virtual reality. Przeglad Elektrotechniczny **95**(5), 18–21 (2019)
12. Duarte, M., Santos, L., Júnior, J.G., Peccin, M.: Learning anatomy by virtual reality and augmented reality. A scope review. Morphologie **104**(347), 254–266 (2020)

13. Finneran, C.M., Zhang, P.: Flow in computer-mediated environments: promises and challenges. Commun. Assoc. Inf. Syst. **15**(1), 4 (2005)
14. Georgiou, G., Anastasovitis, E., Nikolopoulos, S., Kompatsiaris, I.: EPANASTASIS-1821: designing an immersive virtual museum for the revival of historical events of the Greek revolution. In: Ioannides, M., Fink, E., Cantoni, L., Champion, E. (eds.) EuroMed 2020. LNCS, vol. 12642, pp. 334–345. Springer, Cham (2021)
15. Georgiou, Y., Tsivitanidou, O., Ioannou, A.: Learning experience design with immersive virtual reality in physics education. Educ. Technol. Res. Dev. **69**(6), 3051–3080 (2021)
16. Heutte, J., Fenouillet, F., Martin-Krumm, C., Boniwell, I., Csikszentmihalyi, M.: Proposal for a conceptual evolution of the flow in education (eduflow) model. In: 8th European Conference on Positive Psychology (ECPP 2016) (2016)
17. Heutte, J., et al.: Optimal experience in adult learning: conception and validation of the flow in education scale (eduflow-2). Front. Psychol. **12**, 828027 (2021)
18. Kamińska, D., et al.: Virtual reality and its applications in education: Survey. Information **10**(10), 318 (2019)
19. Kamińska, D., Zwolinski, G., Dubiel, A.: Mixed reality as empathy machine. case study in universal design course. In: SIGGRAPH Asia 2023 Educator's Forum, pp. 1–8 (2023)
20. Kamińska, D., Zwoliński, G., Merecz-Kot, D.: How virtual reality therapy affects refugees from Ukraine-acute stress reduction pilot study. IEEE Trans. Affect. Comput. (2024)
21. Krokos, E., Plaisant, C., Varshney, A.: Virtual memory palaces: immersion aids recall. Virtual Reality **23**, 1–15 (2019)
22. Krumhuber, E.G., Skora, L.I., Hill, H.C., Lander, K.: The role of facial movements in emotion recognition. Nat. Rev. Psychol. **2**(5), 283–296 (2023)
23. Kruszewska, A., Nazaruk, S., Szewczyk, K.: Polish teachers of early education in the face of distance learning during the COVID-19 pandemic-the difficulties experienced and suggestions for the future. Education 3-13 **50**(3), 304–315 (2022)
24. Lindner, K.T., Schwab, S.: Differentiation and individualisation in inclusive education: a systematic review and narrative synthesis. Int. J. Inclusive Educ. 1–21 (2020)
25. de Lope, J., Grana, M.: An ongoing review of speech emotion recognition. Neurocomputing **528**, 1–11 (2023)
26. Mahmud, M.R., Stewart, M., Cordova, A., Quarles, J.: Auditory feedback for standing balance improvement in virtual reality. In: 2022 IEEE Conference on Virtual Reality and 3D User Interfaces (VR), pp. 782–791. IEEE (2022)
27. Hernandez-de Menendez, M., Escobar Díaz, C.A., Morales-Menendez, R.: Educational experiences with generation z. Int. J. Interact. Des. Manuf. (IJIDeM) **14**, 847–859 (2020)
28. Mestre, D.R., Ewald, M., Maiano, C.: Virtual reality and exercise: behavioral and psychological effects of visual feedback. Annu. Rev. Cyberther. Telemed. **2011**, 122–127 (2011)
29. Nakamura, J., Csikszentmihalyi, M., et al.: The concept of flow. Handb. Positive Psychol. **89**, 105 (2002)
30. Ortiz Díaz, A.A., Nunes De Oliveira, D., Cleger Tamayo, S.: Study on different methods for recognition of facial expressions from the data generated by modern HMDs. In: Stephanidis, C., Antona, M., Ntoa, S., Salvendy, G. (eds.) HCII 2023. CCIS, vol. 1832, pp. 294–302. Springer, Cham (2023). https://doi.org/10.1007/978-3-031-35989-7_38

31. Peifer, C., Kluge, A., Rummel, N., Kolossa, D.: Fostering flow experience in HCI to enhance and allocate human energy. In: Harris, D., Li, W.-C. (eds.) HCII 2020. LNCS (LNAI), vol. 12186, pp. 204–220. Springer, Cham (2020). https://doi.org/10.1007/978-3-030-49044-7_18

32. Pinto, J., Fred, A., da Silva, H.P.: Biosignal-based multimodal emotion recognition in a valence-arousal affective framework applied to immersive video visualization. In: 2019 41st Annual International Conference of the IEEE Engineering in Medicine and Biology Society (EMBC), pp. 3577–3583. IEEE (2019)

33. Rahiman, H.U., Kodikal, R.: Revolutionizing education: artificial intelligence empowered learning in higher education. Cogent Educ. 11(1), 2293431 (2024)

34. Rock, D.: Quiet Leadership. Harper Collins, New York (2006)

35. Rock, D.: Your brain at work: strategies for overcoming distraction, regaining focus, and working smarter all day long. J. Behav. Optom. 21(5), 130 (2010)

36. Rone, N., Guao, N.A., Jariol, M., Acedillo, N., Balinton, K., Francisco, J.: Students' lack of interest, motivation in learning, and classroom participation: How to motivate them? Psychol. Educ. Multidisc. J. 7(8), 636–646 (2023)

37. Rosenberg, R.S., Baughman, S.L., Bailenson, J.N.: Virtual superheroes: using superpowers in virtual reality to encourage prosocial behavior. PLoS ONE 8(1), e55003 (2013)

38. Rubio, S., Díaz, E., Martín, J., Puente, J.M.: Evaluation of subjective mental workload: a comparison of swat, nasa-tlx, and workload profile methods. Appl. Psychol. 53(1), 61–86 (2004)

39. Sapiński, T., Kamińska, D., Pelikant, A., Anbarjafari, G.: Emotion recognition from skeletal movements. Entropy 21(7), 646 (2019)

40. Takaya, K., et al.: Jerome Bruner: Developing a Sense of the Possible. Springer, New York (2013). https://doi.org/10.1007/978-94-007-6781-2

41. Tapalova, O., Zhiyenbayeva, N.: Artificial intelligence in education: aied for personalised learning pathways. Electron. J. e-Learn. 20(5), 639–653 (2022)

42. Tarng, W., Pei, M.C.: Application of virtual reality in learning quantum mechanics. Appl. Sci. 13(19), 10618 (2023)

43. Triberti, S., Di Natale, A.F., Gaggioli, A.: Flowing technologies: the role of flow and related constructs in human-computer interaction. Adv. Flow Res. 393–416 (2021)

44. Våpenstad, C., Hofstad, E.F., Langø, T., Mårvik, R., Chmarra, M.K.: Perceiving haptic feedback in virtual reality simulators. Surg. Endosc. 27, 2391–2397 (2013)

45. Varelas, T., et al.: Activator: an immersive virtual reality serious game platform for highlighting ancient Greek civilization. KN-J. Cartogr. Geograph. Inf. 73(4), 277–288 (2023)

46. Visconti, A., Calandra, D., Lamberti, F.: Comparing technologies for conveying emotions through realistic avatars in virtual reality-based metaverse experiences. Comput. Animat. Virtual Worlds 34(3–4), e2188 (2023)

47. Wei, S., Bloemers, D., Rovira, A.: A preliminary study of the eye tracker in the meta quest pro. In: Proceedings of the 2023 ACM International Conference on Interactive Media Experiences, pp. 216–221 (2023)

Industry, Innovation and Practitioner

The Impact of Example Selection in Few-Shot Prompting on Automated Essay Scoring Using GPT Models

Lui Yoshida[✉]

The University of Tokyo, Tokyo, Japan
luiyoshida@g.ecc.u-tokyo.ac.jp

Abstract. This study investigates the impact of example selection on the performance of automated essay scoring (AES) using few-shot prompting with GPT models. We evaluate the effects of the choice and order of examples in few-shot prompting on several versions of GPT-3.5 and GPT-4 models. Our experiments involve 119 prompts with different examples, and we calculate the quadratic weighted kappa (QWK) to measure the agreement between GPT and human rater scores. Regression analysis is used to quantitatively assess biases introduced by example selection. The results show that the impact of example selection on QWK varies across models, with GPT-3.5 being more influenced by examples than GPT-4. We also find evidence of majority label bias, which is a tendency to favor the majority label among the examples, and recency bias, which is a tendency to favor the label of the most recent example, in GPT-generated essay scores and QWK, with these biases being more pronounced in GPT-3.5. Notably, careful example selection enables GPT-3.5 models to outperform some GPT-4 models. However, among the GPT models, the June 2023 version of GPT-4, which is not the latest model, exhibits the highest stability and performance. Our findings provide insights into the importance of example selection in few-shot prompting for AES, especially in GPT-3.5 models, and highlight the need for individual performance evaluations of each model, even for minor versions.

Keywords: Automated Essay Scoring · Large Language Models · Few-shot Prompting · Bias · General Pre-trained Transformer

1 Introduction

Large Language Models (LLMs) can perform tasks in various domains through user interaction [1, 2] and are expected to impact education significantly, and their potential use in education has already been explored [3–5]. One promising application is automated essay scoring (AES), and initial studies have begun to investigate this area [6–8]. For instance, Yancey et al. [7] validated the performance of GPT-3.5 and GPT-4 on AES using their own dataset and confirmed that including scoring examples in the prompt improves performance. Notably, GPT-4 outperformed GPT-3.5 and achieved expert-level performance.

© The Author(s), under exclusive license to Springer Nature Switzerland AG 2024
A. M. Olney et al. (Eds.): AIED 2024 Workshops, CCIS 2150, pp. 61–73, 2024.
https://doi.org/10.1007/978-3-031-64315-6_5

Several issues need to be addressed when using LLMs for AES. Prompt design is crucial when utilizing LLMs, and few-shot learning, which includes examples of output, is a powerful method for improving general performance [9, 10]. However, research in sentiment analysis [11] has reported that the choice of examples can bias the results. Nevertheless, previous studies on AES have not provided insights into the selection of examples. Therefore, there is a possibility that performance could be further improved by carefully selecting examples, but this has not been verified.

Furthermore, investigating different versions of LLMs is also important. If the performance of GPT-3.5, which is approximately 1/10th the cost of GPT-4, can be improved by selecting appropriate examples, it would be useful from an educational practice perspective. However, this has not been clarified. Moreover, it is unclear whether the impact of example selection is consistent across models, including minor versions. If it is consistent, the insights obtained from one model can be applied to other models; otherwise, each model will need to be validated individually in the future.

Therefore, in this study, we evaluate the effects of the choice and order of examples in few-shot prompting, using several GPT models. First, we perform AES using 119 prompts with different examples and calculate quadratic weighted kappa (QWK), the degree of agreement between GPT and human rater scores. Then, we use regression analysis to assess biases of example selection quantitatively.

2 Background

LLMs and Prompting. Since the output of LLMs is highly dependent on the prompt, many prompting techniques have been developed. For example, few-shot prompting, which includes multiple examples in the prompt, enabled the best performance in various tasks at the time [9, 10]. Research has also demonstrated that employing a Chain of Thought (CoT) prompting, which describes the process of solving a task, can enhance performance [12, 13]. Furthermore, even zero-shot prompts that do not provide examples can improve the performance of LLMs; for example, simply adding "Let's think step by step" can realize CoT and significantly improve task execution performance [14, 15].

Few-shot prompting, which is the focus of this study, is important because it significantly impacts performance. However, bias has been reported where the choice and order of the examples presented affect the results [11, 16, 17]. For example, Zhao et al. [11] pointed out that the accuracy of few-shot prompting is unstable and that the order and type of concrete examples can cause bias and significantly change the results. Specifically, in sentiment analysis, it was confirmed that there is a majority label bias, where the more labels (negative or positive) given to examples, the more the results are given that label, and a recency bias, where the label of the last example given appears more frequently in the results. However, it is unclear whether these representative biases exist in AES, and it is also unknown whether countermeasures are necessary.

AES. AES has been studied for more than 50 years, beginning with Project Essay Grade by Page [18]. In the initial paradigm, researchers manually designed features and created labeled data which were utilized in multiple regression analysis, Latent Semantic Analysis, and machine learning [18–20] in 1960s to 2000s. From the 2000s, neural networks, which do not require manual feature design, began to emerge, shifting the next paradigm. The focus of design moved towards network architecture, including layer structures and propagation mechanisms [21–23]. Then, the introduction of the Transformer [24] marked the beginning of the third paradigm, with the development of models such as GPT [9] and Bidirectional Encoder Representations from Transformers (BERT) [25], employing a pre-training and fine-tuning approach. Particularly, BERT has shown high performance in AES, achieving state-of-the-art results with methods from this paradigm [26–28].

The emergence of LLMs may initiate a new paradigm in AES. By utilizing LLMs in AES, we could overcome the challenges of conventional AES, such as the lack of evaluation based on the content and domain knowledge and the unclear evaluation criteria in neural network-based AES [19, 29]. However, only a few studies have been accumulated [6–8]. For example, Yancey et al. [7] performed AES using GPT-3.5 and GPT-4 with prompt engineering techniques. The results confirmed that the performance was higher when examples were included in a prompt and showed that the evaluation by GPT-4 was almost as good as the expert level. However, there are few studies on selecting and ordering the examples, and there is no knowledge of bias by examples in AES using LLMs.

Through our study, we could be able to gain insights that can further improve performance. Furthermore, by evaluating the impact on different versions of LLMs, we could obtain practically useful knowledge, such as whether there are commonalities between models and whether lower versions can also achieve high performance.

3 Methods

3.1 Automated Essay Scoring Using GPT Models

Dataset. We used TOEFL11 [30] as the essay dataset. The dataset contains eight essay prompts and their corresponding examinee essays, each with approximately 1,000 to 1,600 essays, for a total of 12,100 essays. The data also includes expert ratings of the essays on a three-point scale of high, medium, and low. These ratings were first evaluated by several experts using a 5-point rubric and finally compressed to a 3-point rating according to a set of rules. The rubric ratings are not included in the dataset.

The number of essay scoring by the GPT models in this study is the number of models multiplied by the number of prompts and the number of essays to be evaluated. Therefore, the number of essay scoring is large when all the essays are used; then we sampled the essay data to be evaluated by GPT models. We selected three essays each from the eight essay prompts with a rating value of high, medium, and low, respectively, for a total of 72 essays for evaluation.

GPT Models. In this study, to evaluate the impact of example selection across models, we utilized three instances each of OpenAI's GPT-3.5 and GPT-4 models. For GPT-3.5, we used models released in June 2023 (gpt-3.5-turbo-0613 denoted as Jun23), November 2023 (gpt-3.5-turbo-1106 denoted as Nov23), and January 2024 (gpt-3.5-turbo-0125 denoted as Jan24). Regarding GPT-4, we employed models released in June 2023 (gpt-4-0613 denoted as Jun23), November 2023 (gpt-4-1106-preview denoted as Nov23), and January 2024 (gpt-4-0125-preview denoted as Jan24). The system prompt was not used. The following prompts were used to obtain the API response. For parameters, temperature was set to 0, and default values were used for all other parameters.

Bias. In this study, we treat essay scores (high, medium, and low) of examples in few-shot prompting as labels. We define majority label bias as the bias introduced in essay evaluation by LLMs due to the number of essays exemplifying a specific score. Additionally, we define recency bias as the bias in essay evaluation attributed to the score of the last exemplified essay.

Prompts. We developed prompts based on those used by Yancey et al. [7]. The prompts comprised several components: Instruction, Essay Prompt, Response, Rubric, Rating Examples, and Output Format (see Fig. 1). The Instruction section described the essay evaluation task, while the Essay Prompt section presented the prompt for the essay. The Response section contained the essay to be evaluated. Lastly, the Output Format section was the template for output.

We prepared prompts categorized into four main groups. The first category involves zero-shot prompts that do not include rating examples, labeled as category N. The second through fourth categories consist of few-shot prompts, which include one to three rating examples. Given that evaluation scores are categorized into three levels, 1-shot, 2-shot, and 3-shot prompts have 3, 9, and 27 variants, respectively. Each category is named by concatenating the capital initials (H: high, M: medium, L: low) of the evaluation scores of the examples in order. For instance, a 1-shot prompt with a high evaluation score is categorized as "H", a 2-shot prompt with evaluation scores in the order of medium and low as "ML", and a 3-shot prompt with scores in the order of low, medium, and high as "LMH". While the rubric is based on a 5-point scale, the dataset only includes three types of scores; therefore, in our experiments, the evaluation scores for examples mentioned in few-shot prompts are substituted with high, medium, and low as 4, 3, and 2, respectively. Prior to this experiment, 200 essays were evaluated with GPT-3.5 (Jun23) using three types of 3-shot prompts with the values for high, medium, and low altered

---

You are a rater for writing responses on a high-stakes English language exam for second language learners. You will be provided with a prompt and the test-taker's response. Your rating should be based on the rubric below, following the specified format. There are rating samples of experts so that you can refer to those when rating.

\# Prompt
"""*Essay prompt*"""

\# Response
"""*Essay to be evaluated*"""

\# Rubric
Rubric

\# Rating samples of experts
Examples

\# Output format:
Rating: [<<<Your rating here.>>>]

Fig. 1. A template of a prompt. The parts where data should be inserted are in *italics*.

to [4.5, 3, 1.5], [2–4], and [1, 3, 5], respectively. The average QWK was highest for the values [2–4], which led to these values for this experiment.

When creating prompts for each category, we prepared three sets of examples to ensure they did not overlap with the essay data being evaluated. For instance, for the HHH category prompts, we extracted three essays with high evaluation scores three times randomly to prepare three prompts for the HHH category. We prepared a total of 117 few-shot prompts, which is the sum of the categories for 1, 2, and 3-shot prompts (39) multiplied by three sets. Since the 0-shot prompt does not include examples, there was only one prompt, making a total of 118 prompts prepared across all categories.

GPT Ratings. We obtained 50,976 GPT ratings by using the API of six GPT models described in **Models** during January and February 2024, targeting 72 essays with 118 prepared prompts. Two responses within the category LM by GPT-3.5 (Nov23) did not yield ratings; therefore, these instances were excluded from our analysis.

Agreement Between Experts and GPT Models. To evaluate the agreement between experts and GPT models, we used the QWK, a widely used metric to assess the concordance between machine and human evaluations [31, 32]. Since GPT evaluations used a rubric out of 5 points, we converted these into three levels: scores above 3 as high, 3 as medium, and below 3 as low, and numerically coded these as 3, 2, 1, respectively, to calculate the QWK.

3.2 Regression Analysis for Bias Evaluation

To assess the impact of majority bias and recency bias from few-shot prompt examples on AES using GPT models, we conducted two regression analyses. One analyzed the influence on GPT-generated essay scores, and the other evaluated the effect on QWK. After outlining the common elements of both analyses, we describe each in detail.

Common Component. Both analyses shared independent variables to assess the impact of majority label bias, using the number of examples rated as high, medium, or low (denoted as H_n, M_n, L_n), and to evaluate the effect of recency bias, employing variables indicating if the last example was high, medium, or low (1) or not (0) (denoted as H_l, M_l, L_l). To compare regression coefficients between models in each analysis, these independent variables were standardized to have a mean of 0 and a variance of 1. Due to the small number of examples in 1-shot and 2-shot prompts, only the results of the 3-shot prompts were used for this analysis.

Bias Influence on Scores. The dependent variable for the regression analysis on essay scores was the GPT-generated essay score (*Score*), which was standardized by dividing by the maximum rating value of 5. The regression model can be described as follows, where the coefficients of each independent variable are denoted by $\beta_{H_n}^s, \beta_{M_n}^s, \beta_{L_n}^s, \beta_{H_l}^s, \beta_{M_l}^s, \beta_{L_l}^s$, and the intercept by α_s:

$$Score = \beta_{H_n}^s H_n + \beta_{M_n}^s M_n + \beta_{L_n}^s L_n + \beta_{H_l}^s H_l + \beta_{M_l}^s M_l + \beta_{L_l}^s L_l + \alpha_s \qquad (1)$$

For each essay, regression analysis was performed for each GPT model, calculating the regression coefficients and the coefficient of determination (R^2). A meta-analysis, typically suitable for integrating multiple regression analysis results by treating regression coefficients as effect sizes, was not possible due to essays where all coefficients were zero. Therefore, to test for the presence of bias, i.e., whether the estimated regression coefficients were non-zero, a one-sample t-test at the significance level of 0.05 was conducted, with Holm's correction applied for multiple comparisons.

Bias Influence on QWK. The dependent variable for the regression analysis on QWK was QWK itself. This regression model can also be described using a similar formula where the coefficients of each independent variable are denoted by $\beta_{H_n}^q, \beta_{M_n}^q, \beta_{L_n}^q, \beta_{H_l}^q, \beta_{M_l}^q, \beta_{L_l}^q$, and the intercept by α_q:

$$QWK = \beta_{H_n}^q H_n + \beta_{M_n}^q M_n + \beta_{L_n}^q L_n + \beta_{H_l}^q H_l + \beta_{M_l}^q M_l + \beta_{L_l}^q L_l + \alpha_q \tag{2}$$

Regression analysis was performed, calculating the regression coefficients and the coefficient of determination (R^2) for each model. A t-test at the significance level of 0.05 was used to assess whether the estimated regression coefficients were non-zero, with Holm's correction for multiple comparisons.

4 Results

4.1 QWK Between Experts and GPT Models

The average and standard deviation of the QWK across each category of all GPT models are shown in Fig. 2. The model with the highest average QWK across all categories was GPT-4 (Jun23), while the lowest was GPT-3.5 (Jun23). It was observed that the GPT-3.5 models were more influenced by the example selection compared to the GPT-4 models.

To examine the trends in evaluations for each model, the distribution of GPT evaluations for zero-shot (N), the category with the lowest average QWK, and the highest average QWK are presented in Fig. 3. Given that the number of expert evaluations for each score level is equal in our data, the ideal scenario is for the number of each score level obtained by GPT to be equal as well. For the GPT-3.5 models, results in N generally showed a tendency to overestimate scores in Jun23, underestimate ones in Nov23, and estimate more scores as medium in Jan24. For the GPT-4 models, Jun23 showed a relatively unbiased evaluation, while Nov23 and Jan24 tended to underestimate scores.

4.2 Regression Analysis for Bias Evaluation

The results of the regression analysis on essay scores for each model are documented in Table 1. Although the coefficients of determination were generally low, significant recency bias and majority label bias were identified in the GPT-3.5 models. In the GPT-4 models, some significant recency bias was observed, but majority label bias was not found. Comparison of the absolute values of the regression coefficients showed that recency bias had a larger impact than majority label bias. Furthermore, the absolute values of significant regression coefficients were larger for GPT-3.5 than for GPT-4, indicating that the influence of bias from examples on essay scores was greater for GPT-3.5 than for GPT-4.

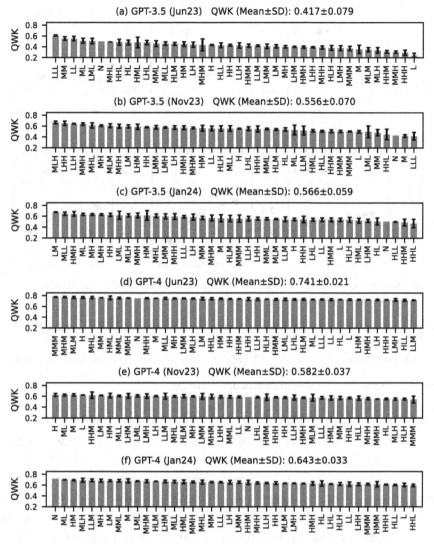

Fig. 2. QWK with variations of few-shot prompts across GPT models

The results of the regression analysis on QWK for each model are documented in Table 2. Similar to previous findings, significant recency bias and majority label bias were confirmed in the GPT-3.5 models. However, for the GPT-4 model, significant majority label bias was observed, but recency bias was not found. The impact of each bias varied between models. For instance, L_l had a positive effect on QWK in GPT-3.5 (Jun23), whereas it had the opposite effect in GPT-3.5 (Nov23, Jan24). Moreover, the absolute values of significant regression coefficients were larger for GPT-3.5 than for GPT-4, indicating again that the influence of bias from examples on QWK was greater for GPT-3.5 than for GPT-4.

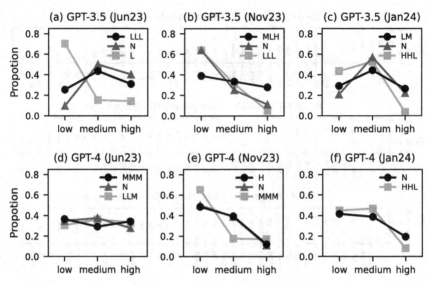

Fig. 3. Distribution of ratings in categories with the highest and lowest average QWK and zero-shot prompt across GPT models.

Table 1. Average regression coefficients and coefficients of determination for essay scores. (*: p < 0.05, **: p < 0.01, ***: p < 0.001). Significant coefficients with the highest absolute value in the models are **bolded**. $\overline{\beta^s_{H_n}}, \overline{\beta^s_{M_n}}, \overline{\beta^s_{L_n}}, \overline{\beta^s_{H_l}}, \overline{\beta^s_{M_l}}, \overline{\beta^s_{L_l}}$ represent the average regression coefficients and $\overline{R^2}$ represents the average coefficient of determination.

Model		$\overline{\beta^s_{H_n}}$	$\overline{\beta^s_{M_n}}$	$\overline{\beta^s_{L_n}}$	$\overline{\beta^s_{H_l}}$	$\overline{\beta^s_{M_l}}$	$\overline{\beta^s_{L_l}}$	$\overline{R^2}$
GPT-3.5	Jun23	−0.0024	0.0060*	−0.0036*	0.0123***	0.0013	**−0.0136*****	0.2010
	Nov23	−0.0005	0.0021	−0.0017	0.0136***	0.0037	**−0.0173*****	0.2094
	Jan24	−0.0049	−0.0007	0.0057**	0.0150***	0.0057**	**−0.0207*****	0.2374
GPT-4	Jun23	0.0014	−0.0013	−0.0001	0.0020*	0.0015	**−0.0035****	0.3040
	Nov23	0.0023	−0.0024	0.0001	−0.0013	**0.0023***	−0.0010	0.1426
	Jan24	−0.0004	−0.0005	0.0009	−0.0013	0.0002	0.0011	0.2597

Table 2. Average regression coefficients and coefficients of determination for QWK (*: p < 0.05, **: p < 0.01, ***: p < 0.001). Significant coefficients with the highest absolute value in the models are **bolded**.

Model		$\beta^q_{H_n}$	$\beta^q_{M_n}$	$\beta^q_{L_n}$	$\beta^q_{H_l}$	$\beta^q_{M_l}$	$\beta^q_{L_l}$	R^2
GPT-3.5	Jun23	−0.0081	−0.0089	0.0171*	−0.0194**	−0.0046	**0.0240*****	0.5165
	Nov23	−0.0108	0.0097	0.0012	**0.0291*****	−0.0103	−0.0189*	0.3173

(continued)

Table 2. (*continued*)

Model	$\beta^q_{H_n}$	$\beta^q_{M_n}$	$\beta^q_{L_n}$	$\beta^q_{H_l}$	$\beta^q_{M_l}$	$\beta^q_{L_l}$	R^2
	Jan24	-0.0221^{***}	0.0187^{**}	0.0034	$\mathbf{0.0234^{***}}$	-0.0196^{**}	-0.0038	0.3321
GPT-4	Jun23	-0.0023	$\mathbf{0.0088^{***}}$	-0.0064^{**}	-0.0008	-0.0034	0.0042	0.3144
	Nov23	0.0014	-0.0041	0.0027	-0.0065	0.0065	-0.0001	0.0764
	Jan24	$\mathbf{-0.0098^{*}}$	0.0056	0.0043	0.0021	0.0001	-0.0022	0.1665

5 Discussion

The impact of example selection on QWK in few-shot prompting varied across models. For instance, GPT-3.5 (Nov23, Jan24) performed better with most few-shot prompting conditions compared to zero-shot prompting (N), with some outperforming GPT-4 (Nov23). Notably, GPT-4 (Jan24) achieved the highest performance with zero-shot prompting. GPT-4 (Jun23) had the highest average QWK among all models and exhibited the lowest standard deviation in QWK, demonstrating robustness to example selection. Overall, comparing the standard deviation of QWK confirmed that GPT-3.5 models were more susceptible to the influence of example selection than GPT-4 models. Regarding practical implications, lower-tier models like GPT-3.5 were significantly influenced by examples, and careful example selection enabled them to outperform some higher-tier models like GPT-4. Therefore, when using lower-tier models for cost-effectiveness, it is recommended to carefully select examples. If cost is not a concern, using GPT-4 (Jun23), the most robust and highest-performing model, is advised. Although recent studies [33, 34] suggested that GPT-4's latest two models, Nov23 and Jan24, would have higher performance, Jun23 exhibited the best results. This demonstrates that the latest models do not always guarantee the highest performance, emphasizing the need for individual performance evaluations of each model, including minor versions. Furthermore, since performance varies across minor versions and previous research papers [7, 8] do not always include minor version information, research papers must include not only major version information but also minor version details when reporting model specifications and results.

Our results demonstrate the presence of biases, such as majority label bias and recency bias, in GPT-generated essay scores and QWK. These biases were more pronounced in GPT-3.5 compared to GPT-4. The impact on essay scores was more influenced by recency bias than majority label bias in GPT-3.5 models, while majority label bias was not significantly detected in GPT-4 models. The impact on QWK was more affected by recency bias in GPT-3.5 models, while GPT-4 only exhibited majority label bias. The impact of these biases varied across models, including minor versions. For example, presenting a lower-scoring essay last (L_l) had a positive effect on QWK in GPT-3.5 (Jun23) but the opposite effect in GPT-3.5 (Nov23, Jan24). The varying influence of examples on QWK across models could be attributed to differences in their inherent scoring tendencies. As depicted in Fig. 3, the evaluation tendencies with zero-shot prompts differed across models, including minor versions. For instance, GPT-3.5

(Jun23) tended to overestimate scores, while GPT-3.5 (Nov23) was inclined to underestimate them. Consequently, L_1 could improve QWK for GPT-3.5 (Jun23) by moderating potentially higher scores but decrease QWK for GPT-3.5 (Nov23) by further lowering scores. However, this explanation does not apply universally to other phenomena, indicating a need for further verification.

The coefficient of determination for QWK in GPT-3.5 (Jun23) exceeded 0.5, but all other coefficients of determination were below 0.4. This suggests that factors other than the ratings of examples are influencing the essay scores and QWK. This research differentiated examples solely based on their scores, thus not conducting an analysis focused on individual examples. Future investigations could enhance our understanding of the impact examples have by incorporating analysis of information beyond scores, such as linguistic characteristics inherent in each example, suggesting a potential for deeper insights into how examples influence outcomes.

Due to the necessity of conducting experiments with multiple models under diverse conditions, we employed a method to extract as diverse a set of essays as possible, albeit with a limited evaluation essay dataset. Nevertheless, the ability to compare models under these conditions yielded valuable insights explained above. Future research can focus on obtaining more detailed findings by increasing the sample size of evaluation essays. Furthermore, while this study focused on GPT models, future research could explore other LLMs like Gemini, Claude, LLaMA, and Vicuna for cross-model insights. Additionally, using different datasets such as the Automated Student Assessment Prize (ASAP) program data [35] or The Cambridge Learner Corpus-First Certificate in English exam (CLC-FCE) [36] could reveal how dataset choice impacts results.

6 Conclusion

In this study, we have explored the impact of example selection on the performance of AES using few-shot prompting with various GPT models. Our findings reveal that the choice and order of examples in few-shot prompts influence the agreement between GPT and human rater scores, as measured by QWK. The impact of example selection varies across models, with GPT-3.5 being more sensitive to the selection of examples compared to GPT-4. Through regression analysis, we have identified the presence of majority label bias and recency bias in GPT-generated essay scores and QWK. These biases are more evident in GPT-3.5 models, while GPT-4 exhibits greater robustness. Interestingly, we find that careful example selection can enable GPT-3.5 models to outperform some GPT-4 models, highlighting the importance of thoughtful prompt engineering. Our study underscores the need for individual performance evaluations of each model, including minor versions, as the impact of example selection varies among models and the latest models do not always guarantee the highest performance.

Acknowledgment. In preparing this manuscript, I used DeepL, Grammarly, ChatGPT, and Claude to improve the language. The tools did not contribute to generating any original ideas. This work was supported by JSPS KAKENHI Grant Number 23K02707 and the research program on "Creation of generative AI learning environment for teachers," conducted at the Tokyo Foundation for Policy Research.

References

1. Min, B., et al.: Recent advances in natural language processing via large pre-trained language models: a survey. ACM Comput. Surv. **56**(2), 30 (2023)
2. Chang, Y., et al.: A Survey on Evaluation of Large Language Models. ACM Trans. Intell. Syst. Technol. (2023)
3. Baidoo-Anu, D., Ansah, L.O.: Education in the Era of Generative Artificial Intelligence (AI): Understanding the Potential Benefits of ChatGPT in Promoting Teaching and Learning. Available at SSRN 4337484 (2023)
4. Kasneci, E., et al.: ChatGPT for good? On opportunities and challenges of large language models for education. Learn. Individ. Differ. **103**, 102274 (2023)
5. Lo, C.K.: What Is the Impact of ChatGPT on Education? A Rapid Review of the Literature. Educ. Sci. **13**(4), 410 (2023)
6. Mizumoto, A., Eguchi, M.: Exploring the Potential of Using an Ai Language Model for Automated Essay Scoring. Res. Methods in Appl. Linguist. **2**(2), 100050 (2023)
7. Yancey, K.P., Laflair, G., Verardi, A., Burstein, J.: Rating Short L2 Essays on the CEFR Scale with GPT-4. In: Proceedings of the 18th Workshop on Innovative Use of NLP for Building Educational Applications, pp. 576–584. Association for Computational Linguistics, Toronto, Canada (2023)
8. Naismith, B., Mulcaire, P., Burstein, J.: Automated evaluation of written discourse coherence using GPT-4. In: Proceedings of the 18th Workshop on Innovative Use of NLP for Building Educational Applications, pp. 394–403. Association for Computational Linguistics, Toronto, Canada (2023)
9. Brown, T., et al.: Language Models are Few-Shot Learners. In: Advances in Neural Information Processing Systems 33 (NeurIPS 2020), pp. 1877–1901. Curran Associates, Inc., Vancouver, Canada (2020)
10. Gu, Y., Han, X., Liu, Z., Huang, M.: PPT: Pre-trained Prompt Tuning for Few-shot Learning. In: Proceedings of the 60th Annual Meeting of the Association for Computational Lin-guistics (Volume 1: Long Papers), pp. 8410–8423. Association for Computational Linguistics, Dublin, Ireland (2022)
11. Zhao, T.Z., Wallace, E., Feng, S., Klein, D., Singh, S.: Calibrate Before Use: Improving Few-Shot Performance of Language Models. In: Proceedings of the 38th International Conference on Machine Learning (ICML 2021), pp. 12697–12706. PMLR, Online (2021)
12. Wei, J., et al.: Chain-of-Thought Prompting Elicits Reasoning in Large Language Models. In: Advances in Neural Information Processing Systems 35 (NeurIPS 2022), pp. 24824–24837. Curran Associates, Inc., New Orleans, USA (2022)
13. Zhang, Z., Zhang, A., Li, M., Smola, A.: Automatic Chain of Thought Prompting in Large Language Models. arXiv preprint arXiv:2210.03493 (2022)
14. Kojima, T., Gu, S.S., Reid, M., Matsuo, Y., Iwasawa, Y.: Large Language Models are Zero-Shot Reasoners. In: Advances in Neural Information Processing Systems 35 (NeurIPS 2022), pp. 22199–22213. Curran Associates, Inc., New Orleans, USA (2023)
15. Sanh, V., et al.: Multitask Prompted Training Enables Zero-Shot Task Generalization. arXiv preprint arXiv:2110.08207 (2022)
16. Gupta, K., et al.: How Robust are LLMs to In-Context Majority Label Bias?. arXiv preprint arXiv:2312.16549 (2023)
17. Nguyen, T., Wong, E.: In-context Example Selection with Influences. arXiv preprint arXiv: 2302.11042 (2023)

18. Page, E.B.: The Imminence of... Grading Essays by Computer. The Phi Delta Kappan. **47**(5), 238–243 (1966)
19. Landauer, T.K.: Automatic Essay Assessment. Assessment in Education: Principles, Policy Pract. **10**(3), 295–308 (2003)
20. Attali, Y., Burstein, J.: Automated Essay Scoring With e-rater® V.2. J. Technol. Learn. Assess. **4**(3) (2006)
21. Dong, F., Zhang, Y.: Automatic Features for Essay Scoring - An Empirical Study. In: Proceedings of the 2016 Conference on Empirical Methods in Natural Language Processing. pp. 1072–1077. Association for Computational Linguistics, Austin, Texas (2016)
22. Alikaniotis, D., Yannakoudakis, H., Rei, M.: Automatic Text Scoring Using Neural Networks. In: Proceedings of the 54th Annual Meeting of the Association for Computational Linguistics (Volume 1: Long Papers), pp. 715–725. Association for Computational Linguistics, Berlin, Germany (2016)
23. Taghipour, K., Ng, H.T.: A Neural Approach to Automated Essay Scoring. In: Proceedings of the 2016 Conference on Empirical Methods in Natural Language Processing, pp. 1882–1891. Association for Computational Linguistics, Austin, Texas (2016)
24. Vaswani, A., et al.: Attention is All you Need. In: Advances in Neural Information Processing Systems 30 (NIPS 2017), Curran Associates, Inc., California, USA (2017)
25. Devlin, J., Chang, M.-W., Lee, K., Toutanova, K.: BERT: Pre-training of Deep Bidirectional Transformers for Language Understanding. In: Proceedings of the 2019 Conference of the North American Chapter of the Association for Computational Linguistics: Human Language Technologies, Volume 1 (Long and Short Papers), Association for Computational Linguistics, Minnesota, USA (2019)
26. Uto, M.: A review of deep-neural automated essay scoring models. Behaviormetrika **48**(2), 459–484 (2021)
27. Yang, R., Cao, J., Wen, Z., Wu, Y., He, X.: Enhancing Automated Essay Scoring Performance via Fine-tuning Pre-trained Language Models with Combination of Regression and Ranking. In: Findings of the Association for Computational Linguistics: EMNLP 2020, pp. 1560–1569. Association for Computational Linguistics, Online (2020)
28. Wang, Y., Wang, C., Li, R., Lin, H.: On the Use of BERT for Automated Essay Scoring: Joint Learning of Multi-Scale Essay Representation. In: Proceedings of the 2022 Conference of the North American Chapter of the Association for Computational Linguistics: Human Language Technologies, pp. 3416–3425. Association for Computational Linguistics, Seattle, USA (2022)
29. Ramesh, D., Sanampudi, S.: Kumar: An automated essay scoring systems: a systematic literature review. Artif. Intell. Rev. **55**(3), 2495–2527 (2022)
30. Blanchard, D., Tetreault, J., Higgins, D., Cahill, A., Chodorow, M.: TOEFL11: A Corpus of Non-Native English. ETS Research Report Series. **2013**(2), i–15 (2013)
31. Ke, Z., Ng, V.: Automated Essay Scoring: A Survey of the State of the Art. In: Proceedings of the Twenty-Eighth International Joint Conference on Artificial Intelligence Survey track, pp. 6300–6308. International Joint Conferences on Artificial Intelligence, Macao, China (2019)
32. Ramnarain-Seetohul, V., Bassoo, V., Rosunally, Y.: Similarity measures in automated essay scoring systems: A ten-year review. Educ. Inf. Technol. **27**(4), 5573–5604 (2022)
33. Gole, M., Nwadiugwu, W.-P., Miranskyy, A.: On Sarcasm Detection with OpenAI GPT-based Models. arXiv:2312.04642 (2023)
34. Large Model Systems Organization: LMSYS Chatbot Arena Leaderboard, https://huggingface.co/spaces/lmsys/chatbot-arena-leaderboard, last accessed 2024/02/05

35. Shermis, M.D.: State-of-the-art automated essay scoring: Competition, results, and future directions from a United States demonstration. Assess. Writ. **20**, 53–76 (2014)
36. Yannakoudakis, H., Briscoe, T., Medlock, B.: A New Dataset and Method for Automatically Grading ESOL Texts. In: Proceedings of the 49th Annual Meeting of the Association for Computational Linguistics: Human Language Technologies, pp. 180–189. Association for Computational Linguistics, Portland, USA (2011)

Bringing AIED, Public Policy and GDPR to Promote Educational Opportunities in Brazil

Carlos Portela[1,2(✉)] , Jorge Rocha[1] , Paula Palomino[1,3] ,
Rodrigo Lisbôa[1,4] , Thiago Cordeiro[1,5] , Alan Silva[1,5] ,
Ig Bittencourt[1,5,6] , Diego Dermeval[1,5,6] , Leonardo Marques[1,5] ,
and Seiji Isotani[1,6,7]

[1] Center for Excellence in Social Technologies (NEES), Maceió, Brazil
jorge.rocha@nees.ufal.br
[2] Federal University of Pará (UFPA), Cametó, Brazil
csp@ufpa.br
[3] São Paulo State College of Technology (FATEC), Matão, Brazil
paula.palomino@fatec.sp.gov.br
[4] Federal Rural University of the Amazon (UFRA), Belém, Brazil
rodrigo.lisboa@ufra.edu.br
[5] Federal University of Alagoas, Maceió, Brazil
{thiago,alanpedro,ig.ibert}@ic.ufal.br, diego.matos@famed.ufal.br,
leonardo.marques@cedu.ufal.br
[6] Harvard Graduate School of Education, Cambridge MA, USA
[7] University of São Paulo, São Paulo, Brazil
sisotani@imc.usp.br

Abstract. This article explores the intersection between Artificial Intelligence in Education (AIED), public policies, and the General Data Protection Regulation (GDPR) in the Brazilian context. We analyze the ethical and legal challenges faced in implementing AIED projects in public schools, focusing on protecting student data. Specifically, we present an overview of the GDPR in Brazil, highlighting the General Data Protection Law (LGPD) and public policies for learning recovery. A specific case study is discussed, showing the application of GDPR in a specific AIED project. Finally, insights, challenges, and recommendations to promote educational opportunities in the Brazilian context are discussed.

Keywords: Artificial Intelligence in Education · Public Policy · Data Protection

1 Introduction

The rapid evolution of Artificial Intelligence (AI) has been transforming various areas of society, and education is no exception [18]. In recent years, Artificial Intelligence in Education (AIED) has emerged as a research and practice area with revolutionary potential for teaching and learning [8]. At the same time, issues related to data protection and privacy have become increasingly prominent [13,21], especially with the implementation of the General Data Protection Regulation (GDPR) [10] in the European Union and its global influence [9,12].

While AI is not explicitly addressed in the GDPR, numerous provisions within the regulation are pertinent to AI applications [9]. Indeed, some aspects are particularly challenged by the novel methods of processing personal data facilitated by AI [12]. A notable tension arises between traditional data protection principles-such as purpose limitation, data minimization, special treatment of 'sensitive data', and limitations on automated decisions-and the expansive utilization of AI and big data. The latter involves gathering extensive datasets concerning individuals and their social interactions, processing them for purposes not fully determined at the time of collection. However, there exist avenues for interpreting, implementing, and evolving data protection principles that align with the beneficial applications of AI and big data [9].

In the Brazilian context, the intersection of AIED, public policies, and GDPR is of particular relevance and complexity. Brazil faces unique challenges related to education and data protection, but it also has a solid foundation of government and non-government initiatives dedicated to promoting the responsible use of technology in education [16,20]. In this scenario, understanding how public policies and data protection regulations can influence and shape the implementation of AIED projects becomes crucial [14,22].

This article explores the intersection between AIED, public policies, and GDPR in the Brazilian context. We intend to examine the ethical and legal challenges that arise when these areas intersect and identify strategies to promote responsible innovation in AIED projects in Brazil. In doing so, we aim to advance knowledge in this area and provide practical insights for researchers, professionals, and policymakers interested in the sustainable development of technology-driven education.

In addition to this introductory section, Sect. 2 provides the necessary foundations to understand the Brazilian scenario regarding GDPR and educational public policies for learning recovery. Then, Sect. 3 details an AIED project to innovate the essay correction process in public schools. Additionally, we present the practical application of GDPR in this AIED project, discussing the guidelines and specific considerations. Section 4 discusses the challenges, opportunities, and ethical and legal implications. Finally, Sect. 5 summarizes the main points discussed throughout the article, highlights important findings, and offers insights for future research and practice in AIED projects associated with public policies that require data protection.

2   Background

2.1   General Data Protection Regulation (GDPR)

According to Wolford [24], the GDPR stands out as the most stringent privacy and security legislation globally. Although it originated from the European Union

(EU), its scope extends to organizations worldwide as long as they handle data about individuals within the EU. Implemented on May 25, 2018, the GDPR [10] reflects Europe's unwavering commitment to safeguarding data privacy and security, especially in an era where cloud services increasingly store personal information and data breaches are alarmingly common. The regulation itself is extensive, comprehensive, and somewhat lacking in specific guidelines, posing a significant challenge for GDPR compliance [12].

The applicability of GDPR guidelines is extraterritorial, covering the processing of data of individuals belonging to the European Union or data located within the EU. This means that regardless of the origin of the controller or processor (whether public or private), if they are conducting operations or providing services under the aforementioned conditions, they must comply with the GDPR [17].

2.2   Brazilian Instantiation of GDPR

In Brazil, data protection comprises an extensive legal framework managed by various agencies operating across the country. Since 2022, privacy has become a fundamental right, as stated in clause X of article 5 of the Federal Constitution of the Federative Republic of Brazil of 1988 (CF/88) [1].

Data governance in the sharing of data within federal public, autonomous, and foundational administration follows the guidelines established in Decree No. 10.046 [3], of October 9, 2019, and must be understood in light of legal restrictions, information and communications security requirements, and the provisions of Law No. 13.709, of August 14, 2018 - General Law for the Protection of Personal Data (LGPD - acronym in Portuguese) [2]. For Brazil, the LGPD is equivalent to the GDPR [19].

There are several highlights regarding the Brazilian data protection structure. Initially, there is the National Data Protection Authority (ANPD - acronym in Portuguese), a public body responsible for legislating on data protection matters in the country and acting as the guardian of LGPD.

For the purposes of this research, we focus on the Data Lifecycle, as proposed in the Best Practices Guide for LGPD Implementation, based on clause X, article 5 of the LGPD [2]. This Data Lifecycle is structured in five phases, according to Fig. 1.

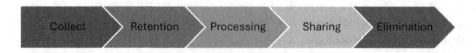

Fig. 1. General Data Lifecycle.

Collection: obtaining, receiving, or producing personal data regardless of the means used (paper document, electronic document, information system, etc.).

Retention: archiving or storing personal data regardless of the means used (paper document, electronic document, database, steel file, etc.).

Processing: any operation involving classification, use, reproduction, processing, evaluation, or control of information, extraction, and modification of personal data.

Sharing: any operation involving personal data transmission, distribution, communication, transfer, dissemination, and sharing.

Elimination: any operation to erase or eliminate personal data. This phase also includes the disposal of organizational assets when necessary for the institution's business.

2.3   Public Policy to Learning Recovery in Brazil

AI has been considered a great ally for various educational stakeholders in recovering students' performance following the significant impact of the COVID-19 pandemic [7]. In May 2019, State ministers, government representatives, academic institutions, members of civil society, and the private sector convened in Beijing, generating the "Beijing Consensus on Artificial Intelligence in Education" [22]. Building upon the Beijing Consensus, UNESCO released a guide on Artificial Intelligence in Education (AIED) for policymakers [23]. In this vein, the Ministry of Education of Brazil and the World Bank collaborated to pass Decree No. 11.079 [5], which established the Brazilian Policy for Learning Recovery in Basic Education, aiming to promote the development and use of AIED Unplugged technologies to help improve current practices ensuring quality and equity in education [16].

In Brazil, text writing is considered the foundation for assessing students' level of comprehension and knowledge articulation [20]. According to the National Literacy Policy [4], writing production consists of the ability to write words and produce texts. Although in early childhood education, children should acquire certain skills and competencies related to writing, it is in primary education that formal literacy begins with essay writing [4]. It is estimated that for these students to improve their texts, they need to practice writing two or more essays per week. This results in overburdening teachers in public schools with the correction of these texts, as it is common to have classrooms with more than 30 students per period.

Given these issues, there is a need to assist teachers in essay evaluation through AI technologies so that students in Brazilian public schools can improve the quality of their writing through teaching practices. These practices are one of the pillars of the Learning Recovery Policy established in Decree No. 11.079 [5]. Thus, a project was proposed to support teachers and students in primary education in generating a pedagogical diagnosis of the texts produced by students, using Unplugged Artificial Intelligence in Education (AIED Unplugged) [16], to adapt to the context of low-income schools, which lack scarce access to the internet and digital devices.

3 Example of Application in Brazil

3.1 An AIED Project in Public Schools

In Brazil, we integrate AI-based educational technologies, requiring no modifications to existing school infrastructure or reliance on stable internet access while accommodating users with limited digital skills. The foremost priority is innovating and improving Foundational Literacy for primary school students by incorporating various AI techniques.

The AI techniques employed are designed to support a national assessment policy evaluating the writing skills of K-9 students in Brazil. By applying the AIED Unplugged concept, we aim to enhance students' writing skills without imposing additional burdens on teachers. This approach takes into account Brazil's social disparities, which encompass schools lacking internet access and digital devices, a high prevalence of students and teachers with low to intermediate levels of digital skills, a shortage of qualified teachers in many vulnerable schools, particularly in the North and Northeast regions of Brazil, and a lack of pedagogical practices focused on providing formative feedback on essay writing.

The core of our solution involves an AI application for mobile devices, enabling the digitization, correction, and pedagogical assessment of handwritten texts in Portuguese. This project was modeled to maintain low transit of personal and/or sensitive personal data, which attributes it to low-risk work processes.

As per the previously specified legal basis, the data flow of the evaluation process was modeled to use personal and/or sensitive personal data minimally. In executing a student's assessment process, the data required are the school where they are enrolled, their school registration number, the class, the school year, and their name.

This data is sourced from students' essays written on sheets of paper with QR codes and markings. The full flow is shown in Fig. 2.

Teachers serve as proxies for students, using the mobile application equipped with a user-friendly menu that automatically populates with students' names. They take photos of the essays, which are stored on the mobile device and uploaded to a server when an internet connection is available. The server employs Computer Vision to segment the image into Portuguese words, achieving an accuracy rate between 92% and 95%. Subsequently, Natural Language Processing (NLP) is utilized to evaluate the essays based on a specific rubric automatically. To support students and teachers, we have developed printable paper-based dashboards, offering feedback based on writing competence, including formal written, thematic coherence, textual typology, and cohesion. This feedback aids students in better understanding the content and producing higher-quality texts.

This solution can ensure cost reduction and alleviate the workload for teachers so that they are free to perform other activities, such as observing other educational needs of the students and providing pedagogical assistance through comments and pointing out errors in students' textual production.

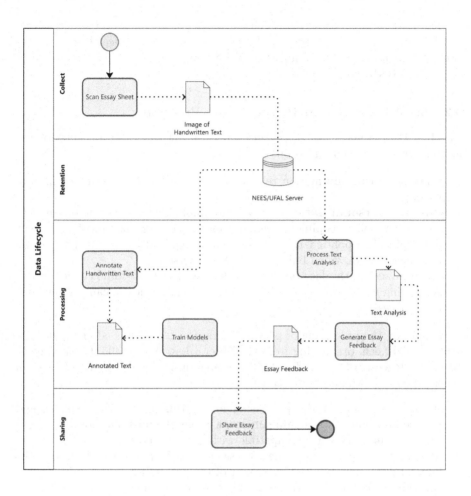

Fig. 2. Project Data Lifecycle.

In our AIED project, the project manager is responsible for implementing best practices, documentation, and work processes involving personal data processing. A schedule consisting of technical and administrative actions was constructed to expand its compliance with the terms and foundations regulated by LGPD.

Since the activities are related to implementing public policies, within the terms and limitations of the LGPD, our AIED project operates under the exception rule, being exempt from requesting consent from the data subjects for processing personal data. It is important to note that, based on the principle of transparency, traceability of the use of public funds, and construction of the evolutionary history of the evaluated students, a significant part of the collected data is not deleted.

The next sections demonstrate the processes of personal and sensitive personal data handling developed by the project and the technical best practices employed with the goal of compliance with the principles and guidelines defined in LGPD (GDPR compliant [19]).

3.2   Legal Principles for Personal Data Processing

The LGPD presents in its scope the structural and relevant concepts for the matter, with special emphasis on:

1. **personal data:** information related to an identified or identifiable natural person;
2. **sensitive personal data:** personal data about racial or ethnic origin, religious belief, political opinion, union membership or organization of a religious, philosophical, or political nature, data concerning health or sexual life, genetic or biometric data, when linked to a natural person;
3. **anonymized data:** data related to a data subject that cannot be identified, considering the use of reasonable and available technical means at the time of its processing.

Regarding the processing of personal data, the LGPD has adopted the concept of good faith, originating from the Federal Law No. 10.406 [6], of January 10, 2002 - Brazilian Civil Code (CCB - acronym in Portuguese), and defines ten other principles as mandatory, namely:

1. **purpose:** conducting the processing for legitimate, specific, explicit, and informed purposes to the data subject, without the possibility of subsequent processing in a manner incompatible with these purposes;
2. **adequacy:** compatibility of the processing with the purposes informed to the data subject, according to the context of the processing;
3. **necessity:** limitation of the processing to the minimum necessary for the achievement of its purposes, encompassing data that are pertinent, proportional, and not excessive concerning the purposes of data processing;
4. **free access:** guarantee to the data subjects of easy and free consultation about the form and duration of the processing, as well as the completeness of their personal data;
5. **data quality:** guarantee to the data subjects of accuracy, clarity, relevance, and updating of the data, according to the need and for the fulfillment of the purpose of their processing;
6. **transparency:** guarantee to the data subjects clear, precise, and easily accessible information about the conduct of the processing and the respective processing agents, observing commercial and industrial secrets;
7. **security:** use of technical and administrative measures capable of protecting personal data from unauthorized accesses and accidental or illegal situations of destruction, loss, alteration, communication, or dissemination;
8. **prevention:** adoption of measures to prevent the occurrence of harm due to the processing of personal data;

9. **non-discrimination:** impossibility of conducting the processing for illicit or abusive discriminatory purposes;
10. **accountability and accountability:** demonstration, by the agent, of the adoption of effective measures capable of proving the observance and compliance with the data protection norms and, including, the effectiveness of these measures.

The AIED project restricts processing to inclusive purposes related to public education policies. It maps the agents involved in the data processing, as defined by the life cycle established by ANPD in Fig. 1. Additionally, we keep the processing of personal and sensitive personal data restricted to the purpose and project goals. This is evident in the project data flow and the use flow of the solution for constructing a textual analysis in Fig. 2.

3.3   Guaranteeing the Rights of Data Holders

In accordance with LGPD, the data holders have been organized according to the following profiles and purposes: user, data subject, press, singular and joint data controllers, control bodies, and the project's Ombudsman.

Such information is essential for us to proceed with protecting your data and, upon authentication and confirmation of your identity and ownership, to facilitate the issuance of a conclusive response to your manifestation.

Under the terms of the LGPD [2], the holders of personal data have specific rights, as provided for in Articles 9, 18, 19, and 20, and these rights must be guaranteed. To this end, various communication channels are made available so that, under the terms of the law, the user can fully exercise their contestation rights and file other requests. To promote these rights, the project communication channels that can be used in this project are publicized and replicated in other internal documents.

3.4   Privacy by Design

This principle is explicit in Article 46 of the LGPD: "The measures referred to in the opening paragraph of this article shall be observed from the design phase of the product or service up to its execution."

As evidenced, it is possible to identify the partial application of the principles defined in the Privacy by Design Framework [15]. This Framework was developed in the 1990s by the Canadian Ph.D. Ann Cavoukian and consists of 7 (seven) structural principles applicable to personal data processing. The application of Privacy by Design ensures that the functionalities of the tools are built with respect to the constitutional right to privacy and guarantees that the subject's personal data are used exclusively for the purposes set out in the Project Work Plan.

Since its conception, our solution has been developed seeking to apply in the best way possible actions to guarantee privacy and the protection of personal data that transit through it.

3.5 Technical and Administrative Measures

All project components that need to access personal data processed in the AIED solutions must be informed about the obligation to adopt measures to protect the data they access. According to LGPD, "Data processing agents or any other person who intervenes in one of the phases of the processing shall ensure the information security provided for in this Law in relation to personal data, even after its conclusion."

As pointed out by the project manager, all components sign a Confidentiality and Secrecy Agreement before starting activities; that is, the project ensures that the data processing agents of the project are guided about their obligations and responsibilities that must be observed during the data processing phase.

3.6 Information Security

In accordance with the legal provision described in Article 47 of the LGPD concerning the relevant projects, information security requirements are applied as per the concept in clause XXIII, of Article 5 of the Federal Decree No. 10.046 [3], of October 9, 2019.

The Article 47 specifies that: "Data processing agents or any other person who intervenes in one of the phases of the processing shall ensure the information security provided for in this Law in relation to personal data, even after its conclusion." The clause XXIII of Article 5 complements that: "information and communication security requirements - actions aimed at enabling and ensuring the availability, integrity, confidentiality, and authenticity of information."

The application of information security in the project was through Non-Functional Requirements: "Check backup routines and policies (daily, weekly, monthly). Ensure data persistence and integrity (LGPD)".

Special emphasis is placed on the data storage process, carried out through Google Cloud Platform [11], a robust cloud computing platform recognized for its excellence in online security. This environment provides an additional layer of protection, ensuring the data is inaccessible to any unauthorized person. The project strictly adheres to data security guidelines and privacy protection established by Google.

4 Discussion

4.1 Insights

The collected data in our AIED project includes information such as the student's name, school, grade, and class, in addition to the handwritten text produced by the student and the evaluation cycle. Regarding data anonymization, this process is carried out logically, restricting access only to the teacher responsible for correcting the students' texts. This AIED support optimizes the teachers' time, helps them, and expands their capacity to build and form a more prepared student to live in society with greater social awareness and critical and collaborative thinking.

To ensure transparency and security, school administrators have exclusive access to monitoring dashboards, which provide information about students' level of proficiency in Portuguese. It is important to note that this access does not include students' personal data, ensuring privacy and compliance with data protection regulations.

This article demonstrates the fulfillment of an AIED project to the legal principles of data processing to guarantee the rights of data subjects in the best way possible, applying privacy resources from the project's design, implementing appropriate technical and administrative measures, and adapting to information security requirements. Transparency and security are emphasized by the exclusive access of school administrators to the monitoring dashboards, which provide information about students' level of fluency in Portuguese without compromising privacy and compliance with LGPD.

4.2 Challenges

The main challenges cited in the literature and identified by the project team are highlighted below:

Equity and Algorithmic Bias: AI algorithms can introduce bias into their decisions, leading to disparities and injustices, especially in educational contexts. For example, if an AI system used to evaluate essays favors certain demographic groups, this can affect equity in education. Ensuring equity in developing and implementing our AIED solutions is a significant ethical challenge [16,23].

Responsibility and Decision-Making: Who is responsible for the essays' feedback based on AI systems in education? Who is held accountable if a student is harmed by an automated decision? Defining the responsibility in our AIED solution is crucial for addressing ethical challenges [13,14].

Transparency and Interpretability: Understanding how AI systems arrive at certain conclusions or recommendations is challenging. This raises concerns about transparency and interpretability, especially regarding decisions that affect students' lives [8,12].

Access and Digital Exclusion: The use of AI technologies in education can exacerbate disparities in digital access and technological literacy among students [7]. This can lead to the exclusion of marginalized or economically disadvantaged groups. Ensuring that all students have equal access to the educational opportunities offered by AIED Unplugged [16] is a significant ethical and legal challenge in our project.

4.3 Recommendations

Some general policy recommendations are outlined below:

For AIED Technologies Developers: It is crucial to conduct longitudinal studies and impact assessments to better understand the unique challenges posed by Brazil's social disparities, limited digital infrastructure, and shortage of qualified teachers in vulnerable regions, while ensuring robust data protection measures. These studies should prioritize the ethical collection, storage, and use of student data to inform the development of tailored AI solutions that address these challenges. By incorporating data protection principles into AI development, we can ensure equitable access and positive outcomes for all students, while safeguarding their privacy.

For Educators and School Administrators: It is important to invest in targeted training programs that address the specific needs of educators in underserved regions. These programs should prioritize building educators' capacity to effectively utilize AI technologies, considering the challenges posed by limited resources and diverse educational contexts in Brazil. It is necessary emphasize ethical considerations and culturally sensitive approaches to AI implementation, ensuring that data privacy and security are central to the training curriculum. This approach not only enhances the effectiveness of AI implementation in diverse educational settings but also fosters trust and confidence among educators, students, and communities.

For Policymakers: The main point of attention is to ensure harmonization between local data protection laws, such as LGPD in Brazil, and international regulations, such as GDPR. This involves ensuring that public policies encourage data protection practices consistent with international standards, promoting transparency, security, and accountability in using AI-based educational technologies. Additionally, develop and implement policies that bridge the digital divide and support equitable access to AI-driven educational tools. This includes initiatives to expand digital infrastructure in underserved areas, increase teacher training opportunities, and establish guidelines for ethically deploying AI technologies in schools with diverse socio-economic backgrounds, as "quilombolas" (descendants of fugitive slave communities) and "ribeirinhas" (near Amazon rivers and islands).

For International Researchers Cooperating with Brazil: It is necessary to understand and respect local data protection laws, especially LGPD. This includes ensuring that research protocols are aligned with legal requirements for privacy and data protection in Brazil, obtaining appropriate consent from participants, and implementing adequate information security measures to protect the collected data. This involves adapting research methodologies to account for social disparities, ensuring that AI solutions are inclusive and beneficial for all students, regardless of their background or geographical location.

By incorporating these tailored strategies into policy recommendations, we can promote responsible and equitable deployment of AIED solutions in Brazil, leveraging AI to overcome educational challenges and improve learning outcomes nationwide while prioritizing the privacy and security of student data. Implementing stringent data protection practices ensures that AI technologies are deployed ethically and transparently, fostering trust among stakeholders and supporting sustainable educational innovation across Brazil.

5 Final Remarks

This article explored the intersection between Artificial Intelligence in Education (AIED), public policies, and the General Data Protection Regulation (GDPR) in the Brazilian context. Significant ethical and legal challenges were identified throughout the analysis, as well as strategies to promote responsible innovation in an AIED project in Brazil.

One of the main highlights was the need to reconcile technological advancement with protecting students' personal data. With the implementation of the LGPD (compliance with GDPR), public policies and AIED projects must ensure transparency, security, and accountability in handling this data, respecting principles of purpose, adequacy, necessity, quality, transparency, security, and prevention.

Additionally, issues related to equity, responsibility, and transparency in decision-making based on AI algorithms were addressed, as well as challenges related to access and digital exclusion. It is essential to ensure that all students have equal access to educational opportunities offered by AI, mitigating existing disparities and promoting inclusion.

In summary, this study contributes to advancing knowledge about the intersection between AIED, public policies, and GDPR in the Brazilian context, including: i) Observance of the legal principles of data processing; ii) Guaranteeing the rights of data subjects; iii) Application of Privacy from the design of the project; iv) Technical and administrative measures applied in the project; and v) Applied Information Security requirements.

The insights and recommendations presented can guide researchers, professionals, and policymakers interested in the sustainable development of technology-driven education. In particular, the results presented in this research are expected to support these agents in promoting innovation in the learning recovery process in Brazil through the use of AIED solutions compliant with the GDPR.

References

1. Brazil: Federal constitution of the federative Republic of Brazil of 1988 (1988). https://www.planalto.gov.br/ccivil_03/constituicao/constituicao.htm
2. Brazil: General law for the protection of personal data (2018). https://www.planalto.gov.br/ccivil_03/_ato2015-2018/2018/lei/l13709.htm
3. Brazil: Decree no. 10.046, of October 9, 2019 (2019). https://www.planalto.gov.br/ccivil_03/_ato2019-2022/2019/decreto/D10046.htm
4. Brazil: National literacy policy (NLP) (2019). http://portal.mec.gov.br/images/banners/caderno_pna_final.pdf
5. Brazil: Decree no. 10.079, of May 23, 2022 (2022). https://www.planalto.gov.br/ccivil_03/_ato2019-2022/2022/Decreto/D11079.htm

6. Brazil: Decree no. 10.406, of January 10, 2022 (2022). https://www.planalto.gov. br/ccivil_03/leis/2002/l10406.htm
7. Chanduvi, J., et al.: Where are we on education recovery? Taking the global pulse of a rapid response (2022). https://www.unicef-irc.org/publications/1391-where-are-we-on-education-recovery-taking-the-global-pulse-of-a-rapid-response.html
8. Chen, X., Zou, D., Xie, H., Cheng, G., Liu, C.: Two decades of artificial intelligence in education. Educ. Technol. Soc. **25**(1), 28–47 (2022)
9. EPRS: The impact of the general data protection regulation (GDPR) on artificial intelligence (2020). https://www.europarl.europa.eu/RegData/etudes/STUD/ 2020/641530/EPRS_STU(2020)641530_EN.pdf
10. European, C.: Rules for the protection of personal data inside and outside the EU (2016). https://commission.europa.eu/law/law-topic/data-protection_en
11. Google: Google cloud platform (2023). https://cloud.google.com/
12. Hijmans, H., Raab, C.: Ethical dimensions of the GDPR, AI regulation, and beyond. Direito Público **18**(100), 63–90 (2022)
13. Holmes, W., et al.: Ethics of AI in education: towards a community-wide framework. Int. J. Artif. Intell. Educ. **32**(3), 504–526 (2022). https://doi.org/10.1007/ s40593-021-00239-1
14. Hong, Y., Nguyen, A., Dang, B., Nguyen, B.P.T.: Data ethics framework for artificial intelligence in education (AIED). In: International Conference on Advanced Learning Technologies, pp. 297–301 (2022). https://doi.org/10.1109/ICALT55010. 2022.00095
15. Hustinx, P.: Privacy by design: delivering the promises. In: IDIS, pp. 253–255 (2020). https://doi.org/10.1007/s12394-010-0061-z
16. Isotani, S., Bittencourt, I.I., Challco, G.C., Dermeval, D., Mello, R.F.: AIED unplugged: leapfrogging the digital divide to reach the underserved. In: Wang, N., Rebolledo-Mendez, G., Dimitrova, V., Matsuda, N., Santos, O.C. (eds.) AIED 2023. CCIS, vol. 1831, pp. 772–779. Springer, Cham (2023). https://doi.org/10. 1007/978-3-031-36336-8_118
17. Lorenzon, L.: Análise comparada entre regulamentações de dados pessoais no brasil e na união europeia (lgpd e gdpr) e seus respectivos instrumentos de enforcement. Revista do Programa de Direito da União Europeia **1**, 39–52 (2021)
18. Ng, A.: Why AI is the new electricity (2017). https://www.gsb.stanford.edu/ insights/andrew-ng-why-ai-new-electricity
19. OneTrust, D., Baptista Luz, A.: Comparing privacy laws: GDPR v. LGPD (2020). https://ec.europa.eu/futurium/en/system/files/ged/dataguidance-gpdr-lgpd-for-print.pdf
20. Portela, C., et al.: A case study on AIED unplugged applied to public policy for learning recovery post-pandemic in Brazil. In: Wang, N., Rebolledo-Mendez, G., Dimitrova, V., Matsuda, N., Santos, O.C. (eds.) AIED 2023. CCIS, vol. 1831, pp. 788–796. Springer, Cham (2023). https://doi.org/10.1007/978-3-031-36336-8_120
21. Subramanian, R.: Emergent AI, social robots and the law: security, privacy and policy issues. J. Int. Technol. Inf. Manag. **26**(3), 81–105 (2017). https://ssrn.com/ abstract=3279236
22. UNESCO: Beijing consensus on artificial intelligence and education (2019). https://unesdoc.unesco.org/ark:/48223/pf0000368303
23. UNESCO: AI and education: guidance for policy-makers (2021). https://unesdoc. unesco.org/ark:/48223/pf0000376709
24. Wolford, B.: What is GDPR, the EU's new data protection law? (2020). https:// gdpr.eu/what-is-gdpr/

Towards Explainable Authorship Verification: An Approach to Minimise Academic Misconduct in Higher Education

Eduardo A. Oliveira[✉], Madhavi Mohoni[✉], and Shannon Rios[✉]

University of Melbourne, Parkville, VIC 3010, Australia
{eduardo.oliveira,shannon.rios}@unimelb.edu.au,
madhavimohoni@gmail.com

Abstract. Academic misconduct poses a growing challenge for higher education institutions worldwide. While AI presents valuable opportunities for learning enhancement, Unauthorized Content Generation (UCG) poses a significant threat to academic integrity. This paper addresses the challenges posed by UCG and explores innovative approaches to detection, focusing on the underutilised concept of authorship verification (AV). Despite the recognition of AV's potential, its application in education has been limited. This study investigates the feasibility of utilising students' academic writing profiles for AV to detect contract cheating and unacknowledged AI usage in academic contexts. Building upon previous research, this study enhances the existing Feature Vector Difference (FVD) AV method by introducing improvements to support better analysis, explainability, and interpretability of the classification process in an educational context. The refined classifier provides probability-based outputs, offering a transparent alternative to traditional "black box" binary outputs, and is able to identify stylometric features suitable for differentiating student's writing profiles. Through this research, we contribute to the advancement of AV technology in education towards explainability, providing educators with a valuable tool to uphold academic integrity and combat the proliferation of UCG in educational environments.

Keywords: generative ai · authorship verification · academic integrity

1 Introduction

While the ethical use of Artificial Intelligence (AI) in educational settings is appreciated and recognised as opportunity [3, 4, 17], Unauthorized Content Generation (UCG) constitutes academic misconduct. Educators are grappling with the integration of AI into education, with responses ranging from outright bans on AI tools like ChatGPT to initiatives promoting AI literacy, new assessment forms and policies among students and teachers [21]. Detection of AI-assisted cheating presents a challenging task, with existing tools exhibiting unreliability and susceptibility to circumvention through paraphrasing [3].

© The Author(s), under exclusive license to Springer Nature Switzerland AG 2024
A. M. Olney et al. (Eds.): AIED 2024 Workshops, CCIS 2150, pp. 87–100, 2024.
https://doi.org/10.1007/978-3-031-64315-6_7

The limitations of current detection methods [7] underscore the need for innovative approaches, such as authorship verification (AV), which examines if two specific documents were written by the same author [12]. Despite its potential, AV remains underutilised in education. In this study we sought to explore the feasibility of utilising student's academic writing profiles for AV to detect contract cheating and unacknowledged genAI usage in academic settings. Additionally, we hope we can leverage discussions and research interest in the area of AV applied in education, which has been relatively dormant in the last few years despite it arguably being a highly promising approach to counter the rise of ghostwriting and AI-generated content.

In this context, this study extends research initiated by [18,21] which evaluated potential automated AV technology to assist educational institutions to validate students' essays and assignments through their writing styles. Here, we expanded and improved the existing Feature Vector Difference (FVD) AV method proposed by [27], which has a relatively simple AV approach, to allow for the creation of more comprehensive academic writing profiles that support better analysis, explainability, and interpretability of the classification process. The output of our new classifier is probability-based, which is preferable in an educational context to a "black box" binary output.

2   Background Literature

2.1   AI in Education and Academic Integrity

The integration of AI technologies, especially Large Language Models (LLMs) such as ChatGPT, into educational systems presents the potential to revolutionise teaching and learning while simultaneously posing significant challenges to academic integrity [17,21]. Due to its potential for misuse, the launch of Chat-GPT received a rapid and controversial response among educators and regulators in general. ChatGPT's rapid adoption has led to its banning in several countries and universities [25,26], highlighting the urgent need for effective strategies to mitigate unauthorised content generation. As educators navigate the complexities of integrating AI into education, there is a growing demand for tools and methods capable of detecting AI-authored content to uphold academic integrity.

The emergence of AI-authored content, however, requires a new approach to combating academic misconduct, with strategies involving policy interventions, detection systems, and advanced detection methods. While AI can significantly contribute to education by acting as a personal tutor, stimulating discussions, offering real-time feedback, and much more [3], its potential for misuse also highlights the critical importance of fostering a culture of ethical and transparent AI use among both students and educators [26]. ChatGPT and other LLMs' ability to generate largely accurate, well-structured and original text makes it extremely easy for students to outsource their essays to these tools and to submit it as their own. It is difficult to distinguish AI-generated text from human-written text [7].

Detecting AI-authored text involves analysing academic submissions for signs of AI authorship. Various approaches have emerged to distinguish AI-generated text from human-authored text. Notably, OpenAI released a RoBERTa-based GPT-2 detector capable of being fine-tuned over newer LLMs [11]. Some other methods employ zero-shot detection to avoid training overhead [14]. "Watermarking" is another approach that incorporates hidden patterns into AI-generated text for detection purposes without affecting text quality [28]. Despite efforts to incorporate AI detectors into academic integrity checkers, their efficacy remains questionable in educational contexts [26]. Stylometric analysis, which examines and focuses on individuals writing styles (instead of on LLM's ones), may provide a more effective alternative for the observed issue.

2.2 Authorship Verification and Academic Students Writing Profiles

AV is the task of determining if a pair of texts were by the same author. Formally, a typical AV problem compares a document of disputed (unknown) authorship d_u, to a set of documents with verified (known) authorship D_{known}, yielding a True (same author) or False (different author) outcome [19]. As a categorisation problem, AV is complex because a single author may intentionally vary their style from text to text for many reasons or may unconsciously drift stylistically over time [12]. Stylometry is one approach to AV that involves the statistical analysis of a writer's work in order to generate a writing style to be used for comparison [20]. The use of stylometry for AV assumes that an author's writing style is consistent and recognisable [13].

The style of a text can be characterised by measuring a vast array of stylistic features that includes lexical (e.g., word, sentence or character-based statistic variation such as vocabulary richness and word-length distributions), syntactic (e.g., function words, punctuation and part-of-speech), structural (e.g., text organisation and layout, fonts, sizes and colors), content-specific (e.g., word n-grams), and idiosyncratic style markers (e.g., misspellings, grammatical mistakes and other usage anomalies) [1]. Stylistic features are the attributes or writing-style markers that are the most effective discriminators of authorship. Over 1000 different style markers have been used in previous research on stylistic analysis, with no consensus on the best set [22].

[21] established and reported an efficient AV mechanism - based on the use of a novel Siamese neural network - capable of distinguishing diverse writing styles of students with reasonable accuracy, while also discerning between students' original work from content generated by an AI model. Their AV method follows the profile-based paradigm, treating all available text samples by one candidate author cumulatively. Text samples are concatenated into a single, often large representative document and then the profile of the author is extracted from that document [19]. This paradigm is essential in the creation and maintenance of students' academic writing profiles across several years. However, their model is not explainable and would pose risks to false positive cases (i.e. student being falsely accused).

3   Methods

Our investigation, an extension from work first presented in [18,21], was organised in four stages: (i) data collection and preprocessing, (ii) adapted FVD AV method development, (iii) adapted FVD AV method evaluation, and (iv) evaluation of academic writing features. Stage 3 was structured in three evaluation phases: (a) evaluation of FVD AV method with PAN-14 dataset, (b) evaluation of FVD AV method with MSR-21 dataset, and (c) evaluation of FVD AV method with MSR-21-GPT dataset.

3.1   Stage 1: Data Collection and Preprocessing

PAN-14. The PAN-14 AV dataset (Table 1) of English essays was derived from the Uppsala Student English (USE) corpus, which is a set of 1,489 essays from 1999–2001, by 440 Swedish students at three distinct academic levels [2]. For PAN-14, the USE corpus was arranged into an AV dataset consisting of 1–5 known texts in each problem, with each author appearing at most twice – once in a same-author and once in a different-author instance.

The PAN-14 dataset emulates academic writing styles and served as a germane testing ground for our AV method aimed at identifying and preventing authorship fraud in academic settings.

Software Engineering Reports 2021 (MSR-21). This data collection was conducted at The University of Melbourne in 2021 [21]. Participants were recruited via e-mail and provided informed consent (Ethics approval #24272). The sample consisted of texts by 20 students from the Master of Software Engineering program.

All participant students were undertaking the yearlong Masters Advanced Software Project (SWEN90013) subject [21]. The aim of this subject is to give students the knowledge and skills required to carry out real life software engineering projects.

This subject included 4 individual writing assessments, including a personal objectives statement, analysis of any digital ethics issue, a report on their reflections through the subject, and an individual contribution statement for their projects. The personal objectives assessment was completed synchronously, in a supervised environment, and submitted to the Canvas Learning Management

Table 1. Statistics of the datasets used in our research.

Dataset	Train size	Test size	Num. problems	Num. docs	Avg. known docs per problem	Avg. words per doc
PAN-14	200	200	400	1447	2.6	840.6
MSR-21	18	18	36	59	2.8	1007.0

System (LMS). The remaining assessments were asynchronous and unsupervised. The collected essays were then cleaned, preprocessed and an AV problem set of known and unknown document pairs were generated similar to PAN-14 (Table 1).

Software Engineering Reports 2021 GPT Data Expansion (MSR-21-GPT). After processing the MSR-21 dataset, we moved into expand this dataset for study 3. The high-level objective of our data expansion process was to simulate a real-world AV scenario in education where students start to make use of generative AI despite potential guidelines for that assessment prohibiting the use of AI writing. Through these expansions, we aimed at investigating if our AV method could distinguish between AI-generated content and students' own works. We used GPT-4 for these expansions, replacing the unknown texts with AI-authored answers to the same questions.

3.2 Stage 2: Adapted FVD AV Method Development

Our final AV method is an adaptation of the [27] submission to PAN20 [8]. FVD extracts a combination of character n-grams and handcrafted features, takes a difference of the two text vectors and models the problem as a binary classification using logistic regression. Despite its relatively simple approach, this method was a top performer in both PAN20 [8] (fan-fiction texts with a closed set of authors across train and test) and PAN21 [9] (fan-fiction texts with unseen authors used in test set).

The FVD method was selected for the current research due to its use of character n-grams, POS n-grams and handcrafted features such as vocabulary richness. Here, we adapted the method to make use of additional features, as shown in Table 2. This allowed us to design a more comprehensive academic writing profile for higher education students. Additionally, we excluded a few FVD original features that were dependent on large texts to allow our method to work in the various academic text types. To handle datasets of limited size, we used k-fold cross-validation during training and testing. This approach divides the dataset into k subsets (folds), taking one fold as a test set and the remaining as the training set. This process is repeated k times, with each subset being used as a test set once. This allows us to use more data for training, preventing overfitting and providing clearer results. We use a variant of this method called Repeated Stratified k-fold cross-validation which repeats the experiment after re-shuffling, and ensures that the distribution of labels in each fold is consistent with that of the entire dataset. Lastly, we also updated the tokenisation and POS tagger used in the original FVD implementation for improved performance.

A logistic regression classifier was used in the original [27] AV method, and although performance could potentially be enhanced with other classifiers such as Neural Networks, the simplicity and interpretability of logistic regression classification made it a preferable choice for our method.

Table 2. Features extracted from text in our adapted FVD method.

Category	Feature	Description
Lexical	Character n-grams	TF-IDF values for character n-grams with $3 \leq n \leq 6$
	Special character frequencies	TF-IDF values of the following special characters: !"#$%&'()*+,-./:;<=>?@[\^-'{|}~
	Function word frequencies	Using 179 stopwords available in the NLTK corpus
	Average characters per word	Average number of characters in a token
	Word length distributions (1–10)	Fraction of tokens of length l, where $1 \leq l \leq 10$
	Vocabulary richness	The ratio of hapax-legomenon (number of words appearing once) to dis-legomenon (number of words appearing twice) divided by the total number of tokens in the text (for scaling)
Syntactic	POS Tag n-grams	TF-IDF values of POS tag tri-grams. In the previous example, this would consider tri-grams over tags ['IN', 'PRP', 'VBP', 'TO', 'VB', 'DT', 'NN', 'NN', 'IN', 'NN', ',', ...]
	POS Tag chunk n-grams	TF-IDF values for POS tag chunk tri-grams (higher level of parse tree). For the sentence above, tri-grams are taken over chunks ['IN', 'NP', 'VP', 'NP', 'IN', 'NP', ',', 'NP', ...]
	POS chunk construction (NP, VP) n-grams	TF-IDF values of each noun phrase and verb phrase expansion. For the sentence above, tri-grams are taken over expansions ['NP[PRP]', 'VP[VB TO VB]', 'NP[DT NN NN]', 'NP[NN]', 'NP[PRP]', ...]

3.3 Stage 3: Adapted FVD AV Method Evaluation

After the development of our adapted AV method, we validated it on the English Essays corpus used in the PAN-14 AV task. To evaluate our adapted FVD method, we applied the identical evaluation framework used in PAN-14, which consists of standard AV metrics of c@1, AUROC and 'final score'. We then compared our results against the top-performing approaches in PAN-14, including [5,15,23]. The goal of this validation was to assure our method showed comparably sound performance to other AV approaches, allowing us to proceed with our experiments to future analyses. We then further validated the method on the MSR-21 and MSR-21-GPT data sets using the same measures of performance for the PAN-14 data set.

3.4 Stage 4: Evaluation of Academic Writing Features

As previously mentioned, the potential for explainability was one of our primary reasons to select logistic regression. For each trained classifier, we analysed patterns on various character n-grams, POS features, function word distributions and special character distributions. Even though logistic regression coefficients alone do not fully capture the relationships between features, the highest coefficient features, combined with an informal analysis of where such stylometric

Table 3. Relative performance of FVD and CNG compared to the 3 top performing methods and baseline in the PAN-14 English Essays task [24].

Method	# methods outperformed	Final score	AUC	c@1	# unanswered
[5]	13	0.513	0.723	0.710	15
[23]	12	0.459	0.699	0.657	2
adapted FVD	12	0.430	0.680	0.633	6
[15]	11	0.372	0.620	0.600	0
PAN-14 Baseline	2	0.288	0.518	0.548	0

fragments tend to occur across the texts, gave us an "under the hood" glimpse of the classifier's decision-making, and pointed us to interesting text fragments that are important in gauging relative dissimilarities across students academic profiles.

4   Findings and Discussions

4.1   Adapted FVD AV Method Evaluation on PAN-14

We began with a validation of our developed adaption of FVD method on the PAN-14 dataset. Table 3 shows the performance of our adapted FVD AV method, the top 3 PAN-14 methods in the English Essays task, and the PAN-14 baseline. Of particular note, our expanded adapted FVD AV method outperforms 12 of the 13 PAN-14 participants.

The PAN-14 English Essays task is known to be particularly challenging [6], and most of the participants - including the overall PAN-14 winner [10] - performed significantly worse on this dataset than on the other PAN-14 AV tasks. Despite this, our method showed relatively strong results, with an AUC of 0.680 and a c@1 of 0.633 (Fig. 1a) and outperformed previous work initiated by [16].

4.2   Adapted FVD AV Method Evaluation on MSR-21

The performance of our adapted FVD on MSR-21 is shown in Table 4 across all folds of RepeatedStratifiedKFold cross-validation (2 folds with 5 repeats). The overall score of FVD on MSR-21 is 0.761, and it achieves an AUC of 0.828. The Average ROC curve also maintains relatively higher distance from the AUC = 0.5 line at all thresholds (Fig. 1b). A stair-like pattern appears due to the size of the dataset, despite testing the dataset across multiple folds. A sharp initial uptick indicates that even at a low decision threshold, our adapted FVD method is able to correctly identify same-author cases. The strong performance on the higher education writing dataset corroborates the use of such a writing profile on technical, specialised and educational writings.

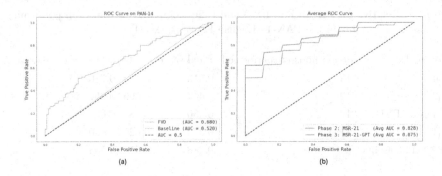

Fig. 1. FVD AV Method Evaluation Phases' Results on Different Datasets: (a) Phase 1 (PAN-14), (b) Phase 2 (MSR-21) and Phase 3 (MSR-21-GPT) ROC curve.

Table 4. Phase 3 evaluation of our adapted FVD on MSR-21.

Method	AUC	c@1	F_1	Overall
MSR-21	0.828	0.723	0.749	0.761
MSR-21-GPT	0.875	0.772	0.787	0.808

4.3   Adapted FVD AV Method Evaluation on MSR-21-GPT

The MSR-21-GPT expansion outperforms MSR-21 on all metrics over the k folds. There were 90 modified test cases (out of 180 total cases after 2 folds with 5 repetitions with test size of 18 in each run). A comparison between FVD performance on MSR-21-GPT and MSR-21 is shown in Table 4. Its ROC is higher than that of MSR-21 at all thresholds, as shown in Fig. 1b. Overall, FVD performance on MSR-21-GPT was greater its performance on MSR-21 in all comparisons conducted.

The strong performance of FVD on GPT expansions (MSR-21-GPT) showed that our developed academic writing profile was robust to GPT authorship. Aside from validating our method, this result is important to show that AV and well-developed student profiles can be a strong option in detecting GPT authorship in education, comparable to detecting human-authored text. Unlike AI detection tools, AV-based detection does not rely on the properties of AI text, which can be easily obfuscated with paraphrasing, and will not have any biases in training, as the comparison is directly against the student's own work. It also has built-in adaptability – since students typically submit written assessments on a regular basis in the course of their education, a writing profile can be continually updated with more known texts. This potentially adds a degree of robustness to change in student writing styles over time, including vocabulary and sentence construction improvements. In future, we aim at making features analysis more explainable to students and educators that are not familiar with NLP terms.

Fig. 2. Academic Writing Analysis - Top 5 most discerning features during classification by our adapted FVD method on (a) PAN-14 and (b) MGE-21 datasets.

4.4 Evaluation of Academic Writing Features

Our FVD AV method now offers significantly improved explainability compared to previous machine learning models. This is evident in our analysis of the most important features in both the PAN-14 and MSR-21 datasets, as depicted in Fig. 2. In PAN-14, our implemented FVD assigns POS tag features with the largest coefficient features in its training on PAN-14. This category was dominated by ('NNS', '.') which represents a plural noun followed by a punctuation, e.g. a sentence that ends with a plural noun. Character n-grams highlighted include variants of 'nur' and 'acters'. In our informal analysis of the appearances of these n-grams in the corpus contents, aside from mentions of 'nurturing' and 'nursing', many students undertook a literature report on a poem with a prominent character who was a nurse, which possibly explain this observed feature pattern.

The most discerning POS chunk feature was (`'NP'`, `'.'`, `'"'`) which refers to a noun phrase followed by a punctuation and a quotation. In an informal analysis, we found that this often captured ends of quotes. Noun phrases constructed as `NP[CD NNP]` were also assigned a high coefficient. This refereed to a cardinal number (counting number) followed by a singular proper noun. This construction appeared in various forms such as dates ("four March"), a year and country ("1994 Sweden") and also a slightly erroneously tagged book title ("*One Flew* Over the Cuckoo's Nest"). The function word 'haven' (as part of 'haven't') is found to be the most important function word, although function words and special characters have relatively lower importance.

In MSR-21, character n-grams were the highest coefficient features while training this classifier. The n-grams involving subwords relating to 'conce' dominated the training, indicating a likely overfit based on a single subword. The subword 'conc' occurs across the texts in various forms ("concern", "concentration", "concepts", etc.). The use of parentheses around noun phrases were an important POS chunk, as well as the use of the POS triplet (`'NNS'`, `'IN'`, `'NNS'`).

4.5   Discussions

Overall, our adapted FVD AV method's analysis of academic writing profiles' patterns, encompassing character n-grams, POS features, function word distributions, and special character distributions, has provided valuable insights into the classification process. These findings not only enhanced the understanding of how the approach works but also validated its effectiveness in AV.

Integrating stylometric analysis tools into subjects could significantly enhance the learning experience of students by providing a framework for advanced writing analysis and feedback. Educators could begin by establishing baseline writing profiles for each student at the start of the subject, using initial assignments (conducted in invigilated environments). As the course progresses, subsequent submissions could be compared against these profiles to monitor development and ensure writing consistency, flagging any deviations for potential academic integrity issues. This ongoing analysis would not only facilitate individualised feedback but also allow for the tailoring of course materials and activities to address common writing deficiencies identified by the tool. These tools can provide objective, quantifiable metrics on various aspects of writing, such as vocabulary diversity, sentence complexity, and punctuation usage.

In incorporating these tools, educators can track longitudinal changes in students' writing styles across subjects, semesters or even entire programs, potentially identifying improvements or areas in need of further instruction. This function is particularly useful in identifying areas where a student is improving or highlighting aspects of their writing that may require further attention and development. With its detailed breakdown of writing stylistic features, educators can offer targeted advice. For instance, a consistently high "Long Count" could indicate a tendency to write run-on sentences, and the tool can guide instructors on how to address this with the student directly. A pattern of low "Unique Count (%)" across student submissions, for example, might suggest the need for more vocabulary-building exercises in the curriculum. The data provided by the tool is also instrumental in constructing individualised learning plans. These plans are tailored to address specific areas where a student may need to enhance their writing skills, offering a customised approach to learning.

However, the deployment of stylometric tools in educational settings is not without challenges. One significant concern is the protection of student privacy and the ethical handling of data collection. Educational institutions must, therefore, ensure that data is securely stored, and that usage complies with relevant privacy laws and institutional policies (as it happens with current use of LMS and tools such as Turnitin). Educational institutions and educators must be transparent with students about how their data will be used. The deployment of such tools should always prioritise the privacy of students. Furthermore, the outputs generated by stylometric tools must be communicated effectively to non-technical students, educators, and administrators. The interpretation of data from these tools requires a certain level of understanding of statistical measures and their implications for writing. To bridge this gap, educators can facilitate workshops or create instructional materials that interpret these metrics in a user-friendly manner. Providing visual aids, such as graphs or colour-coded feedback, can also help convey the significance of the analysis more intuitively. Ultimately, the goal should be to present the data in a manner that reinforces the pedagogical objectives without overwhelming users with overly complex technical details. Human judgment and context should always be part of these review processes.

Building from our findings, our adapted FVD method was recently developed as a stylometric analysis tool (Fig. 3) to allow non-technical educators to create and monitor the development of students' academic writing styles at scale

Fig. 3. Towards explainable AV: Screenshot of our developed tool to allow educators to create and monitor students' writing styles development.

and consistently. Our developed tool was initially designed to analyse students' writing styles, producing a "Final Score" to overall similarities between new textual artifacts and existing ones that are part of students writing profiles, and breaking down the analysis into various stylistic features such as "Long Count (%)," "Word Length," "Unique Count (%)", "Fullstop Count (%)", "Vocabulary Richness", and more. Our initial version of the tool also highlight sections of a written piece in the right pane (indicating areas of potential interest or concern), it allows academics to filter stylistic features based on types (top of screen), and explains to academics how to interpret results ("What does this mean?" link on top right corner of screen).

The tool will be trialed in 2024 semester 2 in the Software Project (COMP90082) subject, which is part of the Master of Information Technology, The University of Melbourne. No integration between the tool and Canvas LMS is anticipated for 2024. The tool will be used as a standalone solution and for research purposes. Integration of this tool and Canvas is planned for 2025. The tool will also be freely available for download.

5   Conclusions and Future Work

This study expanded on the work initiated by [18,21] by exploring the use of a rich set of stylometric features from academic students' writings to build a profile and use it for AV. We found that our approach performed competitively on benchmark AV problem sets and on a contemporary educational dataset, demonstrating its ability to discern authorship on higher education reports and essays. Our method represents a step towards an explainable and evidence-based AV approach that can be effective at representing and comparing student writing profiles, and presents an alternative to current generative AI detection tools.

More experimentation with features can be done to build a stronger underlying profile or AV method for maximising analyses. More formal analysis can also be done on the impact of each feature on AV decision-making. Topic masking, topic modelling or text distortion approaches can be incorporated to reduce dependence on topic. Additionally, a promising avenue of research would be a full-scale experiment on the effects of different types of obfuscation (GPT imitation, paraphrasing, etc.) on AV methods and AI detection methods. This would be beneficial to demonstrate the robustness of AV and writing profiles against obfuscation, which is a critical vulnerability in AI detection tools. Lastly, new investigations could be conducted in other fields (not strictly related to education). It is also not clear yet the full impact on the size and type of writing that is suitable for AV, or the amount of writing samples needed to generate an effective writing profile.

References

1. Abbasi, A., Chen, H.: Writeprints: a stylometric approach to identity-level identification and similarity detection in cyberspace. ACM Trans. Inf. Syst. **26**(2), 1–29 (2008). https://dl.acm.org/doi/10.1145/1344411.1344413
2. Axelsson, M.W.: USE - The Uppsala Student English Corpus: An Instrument for Needs Analysis (2002)
3. Cotton, D.R.E., Cotton, P.A., Shipway, J.R.: Chatting and cheating: ensuring academic integrity in the era of ChatGPT. Innov. Educ. Teach. Int. 1–12 (2023). eprint: https://doi.org/10.1080/14703297.2023.2190148. Publisher: Routledge
4. Foltynek, T., et al.: ENAI recommendations on the ethical use of artificial intelligence in education. Int. J. Educ. Integr. **19**(1), 1–4 (2023). https://doi.org/10.1007/s40979-023-00133-4, https://edintegrity.biomedcentral.com/articles/10.1007/s40979-023-00133-4, number: 1 Publisher: BioMed Central
5. Frery, J., Largeron, C., Juganaru-Mathieu, M.: UJM at CLEF in author identification - Notebook for PAN at CLEF 2014, vol. 1180, pp. 1042–1048 (2014)
6. Jankowska, M.: Author style analysis in text documents based on character and word N-grams, April 2017. https://DalSpace.library.dal.ca//handle/10222/72872. Accepted 27 Apr 2017
7. Kasneci, E., et al.: ChatGPT for good? On opportunities and challenges of large language models for education. Learn. Individ. Differ. **103**, 102274 (2023). https://doi.org/10.1016/j.lindif.2023.102274, https://www.sciencedirect.com/science/article/pii/S1041608023000195
8. Kestemont, M., et al.: Overview of the Cross-Domain Authorship Verification Task at PAN 2020 (2020)
9. Kestemont, M., et al.: Overview of the Cross-Domain Authorship Verification Task at PAN 2021 (2021)
10. Khonji, M., Iraqi, Y.: A Slightly-modified GI-based Author-verifier with Lots of Features (ASGALF) (2014)
11. Kirchner, J.H., Ahmad, L., Aaronson, S., Leike, J.: New AI classifier for indicating AI-written text (2023). https://openai.com/blog/new-ai-classifier-for-indicating-ai-written-text
12. Koppel, M., Schler, J.: Authorship verification as a one-class classification problem (2004). https://doi.org/10.1145/1015330.1015448
13. Laramée, F.D.: Introduction to stylometry with Python. Program. Historian (2018). https://programminghistorian.org/en/lessons/introduction-to-stylometry-with-python
14. Mitchell, E., Lee, Y., Khazatsky, A., Manning, C.D., Finn, C.: Detect-GPT: Zero-Shot Machine-Generated Text Detection using Probability Curvature (2023). https://doi.org/10.48550/arXiv.2301.11305, http://arxiv.org/abs/2301.11305, arXiv:2301.11305 [cs]
15. Moreau, E., Jayapal, A., Vogel, C.: Author Verification: Exploring a Large Set of Parameters using a Genetic Algorithm - Notebook for PAN at CLEF 2014 (2014)
16. Oliveira, E., de Barba, P.G.: The impact of cognitive load on students' academic writing: an authorship verification investigation. In: Annual Conference of the Australasian Society for Computers in Learning in Tertiary Education 2022: Reconnecting relationships through technology, p. e22177. Australasian Society for Computers in Learning in Tertiary Education (2022). https://doi.org/10.14742/apubs.2022.177

17. Oliveira, E., Rios, S., Jiang, Z.: AI-powered peer review process: an approach to enhance computer science students' engagement with code review in industry-based subjects, pp. 184–194. ASCILITE Publications (2023). https://doi.org/10.14742/apubs.2023.482

18. Oliveira, E.A., Conijn, R., De Barba, P., Trezise, K., van Zaanen, M., Kennedy, G.: Writing analytics across essay tasks with different cognitive load demands, pp. 60–70. ASCILITE Publications (2020). https://doi.org/10.14742/ascilite2020.0121

19. Potha, N., Stamatatos, E.: A profile-based method for authorship verification. In: Likas, A., Blekas, K., Kalles, D. (eds.) SETN 2014. LNCS (LNAI), vol. 8445, pp. 313–326. Springer, Cham (2014). https://doi.org/10.1007/978-3-319-07064-3_25

20. Potthast, M., Rosso, P., Stamatatos, E., Stein, B.: A decade of shared tasks in digital text forensics at PAN. In: Azzopardi, L., Stein, B., Fuhr, N., Mayr, P., Hauff, C., Hiemstra, D. (eds.) ECIR 2019. LNCS, vol. 11438, pp. 291–300. Springer, Cham (2019). https://doi.org/10.1007/978-3-030-15719-7_39

21. Rios, S., Zhang, Y., Oliveira, E.: Authorship verification in software engineering education: forget ChatGPT and focus on students' academic writing profiles, pp. 195–204. ASCILITE Publications (2023). https://doi.org/10.14742/apubs.2023.559

22. Rudman, J.: The state of authorship attribution studies: some problems and solutions. Comput. Humanit. **31**, 351–365 (1997)

23. Satyam, A., Dawn, A.K., Saha, S.: A statistical analysis approach to author identification using latent semantic analysis (2014). https://www.semanticscholar.org/paper/A-Statistical-Analysis-Approach-to-Author-Using-Satyam-Anand/3f634c84a27d5ccc152653bdf0c9e86ddfa9d682

24. Stamatatos, E., et al.: Overview of the Author Identification Task at PAN 2014 (2014)

25. UNESCO: ChatGPT, artificial intelligence and higher education: what do higher education institutions need to know? - UNESCO-IESALC (2023)

26. Weber-Wulff, D., et al.: Testing of detection tools for AI-generated text (2023). https://doi.org/10.48550/arXiv.2306.15666, http://arxiv.org/abs/2306.15666, arXiv:2306.15666 [cs]

27. Weerasinghe, J., Greenstadt, R.: Feature Vector Difference based Neural Network and Logistic Regression Models for Authorship Verification (2020)

28. Zhao, X., Wang, Y.X., Li, L.: Protecting language generation models via invisible watermarking (2023). https://doi.org/10.48550/arXiv.2302.03162, http://arxiv.org/abs/2302.03162, arXiv:2302.03162 [cs]

AI in K-12 Social Studies Education: A Critical Examination of Ethical and Practical Challenges

Ilene R. Berson$^{(\boxtimes)}$ ⓘ and Michael J. Berson ⓘ

University of South Florida, Tampa, FL 33620, USA
iberson@usf.edu

Abstract. The integration of Artificial Intelligence (AI) in K-12 social studies education presents a complex blend of opportunities and challenges. This paper explores the transformative potential and ethical dilemmas of employing AI, particularly in engaging with historical primary sources. Utilizing a Delphi methodology, this study captures insights from social studies educators on AI's role in enhancing or complicating the teaching and learning process. The research reveals the transformative potential of AI to deepen historical understanding and empathy through innovative tools, yet it also underscores significant concerns about the accuracy, representation of historical events and figures, and the authenticity of educational content. This exploration presents the paradox inherent in educational AI: its ability to radically transform learning experiences with innovative tools while also threatening to obscure the distinction between historical accuracy and fabrication. The emerging capabilities of generative AI prompt vital questions concerning the legitimacy of information, trustworthiness, and authenticity in historical representation. We consider the implications of AI for pedagogy, emphasizing the need for instructional approaches that engage students in the critical analysis of complex influences shaping historical narratives and visual representations. The paper also addresses the role of AI in perpetuating or challenging historical narratives and biases. It underscores the importance of critical theory in social studies education to analyze the impact of synthetic media and conversational AI agents on students' perceptions of history. We emphasize the need for collaboration between educators, technologists, and historians to ensure that AI tools enrich the educational landscape without compromising the integrity and complexity of historical understanding.

Keywords: Social Studies Education · AI Literacy · Ethical Considerations · Pedagogical Practices · Primary Sources

1 Introduction

1.1 AI in Social Studies Education

The integration of Artificial Intelligence (AI) in K-12 social studies education presents a paradoxical blend of opportunity and challenge [1]. According to the National Council for the Social Studies (NCSS) [2], social studies "is the study of individuals, communities, systems, and their interactions across time and place that prepares students for

A. M. Olney et al. (Eds.): AIED 2024 Workshops, CCIS 2150, pp. 101–112, 2024.
https://doi.org/10.1007/978-3-031-64315-6_8

local, national, and global civic life." This academic discipline utilizes an inquiry-based approach to enable students to explore a vast array of human experiences through generating questions, collecting and analyzing evidence from credible sources, considering multiple perspectives, and applying knowledge and skills pertinent to social studies.

The core objective of social studies in the United States is to prepare learners for effective civic participation by providing them with knowledge of human rights and fostering an understanding of local, national, and global responsibilities. This preparation facilitates the development of skills necessary for a lifelong practice of civil discourse and civic engagement within their communities. The NCSS emphasizes that through examining historical contexts, engaging with current societal issues, and learning to influence future societal conditions, social studies equips students "to work together to create a just world in which they want to live" [2].

The terminology used to describe the components of social studies education can vary significantly among states, districts, and schools, making it challenging to universally categorize the disciplines, fields, and subject titles under this umbrella. However, it generally includes a diverse spectrum of courses aimed at studying human interactions and experiences within various contexts. At the elementary level, social studies is an interdisciplinary subject that encompasses history, geography, economics, and government or civics. It is integrated with language arts, visual and performing arts, and STEM (Science, Technology, Engineering, and Mathematics) education, enriching students' learning experiences and understanding of social contexts. At the secondary level, social studies students further specialize through focused, disciplinary and interdisciplinary studies that enhance their understanding and application of the social sciences. Despite these differences, the core essence of social studies remains the study of the human experience and our interactions within various societal constructs. This definition underlines the fundamental goal of social studies education: to develop informed, engaged citizens capable of contributing thoughtfully to society.

Historical thinking is integral to social studies education across its diverse disciplinary areas, encompassing history, geography, economics, and civics. This mode of thinking engages students in the practices of critical analysis and interpretation of the human past—an essential skill that fosters a comprehensive understanding of the present and informed decision-making for the future.

Incorporating AI tools to facilitate historical thinking through the analysis of primary sources offers innovative avenues for enhancing student learning and engagement. Primary sources, which include documents, artifacts, diary entries, letters, and other materials produced by people who witnessed or participated in historical events, serve as the cornerstone of historical inquiry [3–5]. They offer students a direct window into the past, allowing for a deeper and more nuanced understanding of historical events and figures through the perspectives of those who experienced them firsthand.

This paper explores the transformative potential of employing AI technologies, such as conversational agents and synthetic media, in the engagement with these primary sources. Conversational AI agents can simulate discussions with historical figures or guide students through historical events as if they were there, while synthetic media can recreate or visually represent historical moments with unprecedented realism. These AI-driven approaches provide innovative avenues for students to interact with primary sources, potentially deepening their connection with the past and enhancing their critical thinking skills [1, 6, 7].

However, the use of AI in this context also introduces complex challenges, particularly concerning the authenticity and accuracy of the historical narratives that AI tools help to construct. The ability of AI to alter or generate new primary sources raises ethical questions about the integrity of social studies education and the responsibility of educators to ensure that students can distinguish between authentic historical documents and AI-generated content. This exploration necessitates a critical examination of how AI technologies can be integrated responsibly into social studies curricula, ensuring that they serve as a complement to traditional methods of historical inquiry rather than a replacement.

In this study, we employed a Delphi methodology [8, 9] to engage social studies educators in a dialogue about the opportunities and challenges that AI presents for social studies teaching and learning. This method allowed us to gather diverse insights on how advanced AI tools—ranging from interactive platforms to generative AI capable of creating images, videos, and texts—can enhance student learning experiences. While these technologies hold the promise of making social studies more engaging and interactive, they also introduce significant ethical and practical concerns. Notably, questions arise about the accuracy and authenticity of historical events and figures represented through AI-generated content, highlighting a tension between technological innovation and educational integrity.

The potential of AI to revolutionize social studies teaching is met with mixed reactions within the education community [10–15]. Some educators see AI as a groundbreaking tool that could dramatically transform the field, while others are cautious, voicing concerns over its impact on pedagogical practices and the authenticity of educational content. This paper navigates these divergent viewpoints, aiming to uncover a middle ground that leverages AI's potential to address current challenges in social studies education while developing effective strategies for classroom integration.

In an era marked by the blurring lines between myth and reality—fueled by the proliferation of 'supposition news' and cultural falsehoods [14, 16]—the task of accurately reconstructing historical narratives becomes increasingly complex. This context raises several pivotal questions that guided our research, including: What is the role of social studies in the age of AI? How can educators ethically utilize AI to enhance their teaching? What new skills must be cultivated in students to navigate this landscape? And what cautionary tales should we heed? Through this study, we seek to foster a nuanced discourse that builds on insights and experiences to cultivate an education system that upholds historical accuracy and pedagogical integrity while evolving to meet the needs of students in today's digital society.

2 Theory

2.1 Critical Theory

Building on Henry Giroux's critical pedagogy [17–19], this paper examines AI in social studies education, particularly focusing on how synthetic media and conversational AI agents impersonating historical figures intersect with issues of representation and authenticity in education. Giroux's work provides a framework for analyzing how these AI tools might influence students' understanding of history, potentially distorting their perception of the past. This aligns with Giroux's emphasis on critical engagement with dominant narratives, urging educators and students to scrutinize the power dynamics and ideologies embedded within educational content, especially when advanced technologies like AI are involved in shaping historical narratives Giroux advocates for an education that empowers students to challenge dominant narratives and social injustices. In this context, the use of AI to recreate historical figures raises critical questions about authenticity, bias, and the shaping of historical narratives. This study explores how such representations can either support critical engagement with history or potentially lead to a distorted understanding of the past. The ethical implications are profound, as they relate not only to the accuracy of historical information but also to broader issues of power and ideology in educational content. This exploration aligns with Giroux's call for educators and students to critically examine the influences shaping their educational experiences and the narratives presented to them.

3 Method

3.1 Participants and Procedures

The study utilized a comprehensive Delphi methodology [8, 9], engaging U.S. K-12 social studies educators across three iterative rounds. The goal was to ensure a rich, multi-faceted understanding of the role and implications of AI in social studies education, reflective of diverse educational contexts and experiences.

First Round - Initial Panel Engagement. A panel of 32 K-12 social studies educators, selected for their expertise in both social studies and AI technologies, participated in the initial round. They were presented with a series of open-ended questions aimed at exploring their perspectives on integrating AI into social studies. These questions covered a broad range of topics, including ethical implications, pedagogical challenges, and the potential impact of AI on traditional teaching methods.

Second Round - Expert Group Analysis. In the second round, a separate group of experts, comprising educational theorists, technology specialists, and social studies scholars, was introduced. This group reviewed the synthesized responses from the first round. They provided critical feedback, focusing on refining, expanding, or challenging the initial responses. Their task was to deepen the analysis and contribute their expert insights to the ongoing discussion.

Final Round - Broadening the Consensus. The third round expanded the participant base to include over 200 social studies teachers from various educational backgrounds and regions in the United States. This broader group received a summary of the insights and refined statements from the previous rounds. In small breakout sessions participants were asked to evaluate these statements, offering their agreement or disagreement, and to contribute any additional thoughts or experiences. Notes were taken during these discussions to capture the breadth and depth of the conversations. After the breakout sessions, participants reconvened to share their group's insights and perspectives. This final round aimed to achieve a more comprehensive consensus and to capture a wider range of perspectives on the subject.

The data collected from these discussions were then analyzed through a qualitative lens. This involved identifying common themes, contrasting different viewpoints, and drawing conclusions about the collective stance of social studies educators on the integration of AI in their field. The aim was to glean comprehensive insights into the educators' perceptions, challenges, and opportunities presented by AI in social studies education.

4   Findings

Reflecting insights from the Delphi study, the discussion uncovers nuanced challenges not only in the creation of historical content through AI but also in its utilization within social studies education. This emphasizes the imperative for maintaining accuracy, authenticity, and ethical standards in both the development and deployment of AI-generated materials, ensuring a responsible integration of these technologies in the curriculum.

4.1   Pedagogical Implications

In the analysis of findings from our Delphi study with social studies educators, the pedagogical implications of AI integration emerged as a central theme. This included discussions around the potential of AI to both enhance and complicate teaching methodologies and learning outcomes, which revealed a dual narrative: the transformative potential of AI in enhancing pedagogical practices and the indispensable role of educator expertise in navigating this new terrain. The integration of AI into social studies curricula introduces pedagogical challenges, particularly in balancing AI tools with traditional teaching methods. Educators deliberated on strategies that integrated AI in a manner that enhances rather than replaces traditional educational resources, fostering an environment where AI supplements the established curriculum.

Transformative Pedagogical Practices Through AI. Educators are leveraging AI to make historical content more accessible and engaging. These innovations serve as entry points for deeper exploration to enhance inquiry-based learning.

Discussions on integrating AI into the social studies curriculum highlighted efforts to balance AI tools with traditional teaching methods. Some educators shared examples of integrating AI in ways that supplement rather than supplant the rich, traditional curriculum of social studies. Elementary and middle school educators described the use of AI-driven transcription and translation to extend the accessibility of historical documents, breaking language barriers and broadening the scope of resources available to students. AI's capability to summarize and rewrite text is particularly useful in tailoring historical content to learners of diverse levels. For example, an AI tool can take a dense, complex historical document like the Federalist Papers and generate a summary that captures the main arguments and themes in simpler language. Similarly, AI can rewrite segments of historical texts, such as excerpts from primary source documents or detailed historical analyses, into versions that are more accessible to younger students or those with varying levels of reading proficiency. This application of AI ensures that students of all backgrounds and abilities can engage with and understand crucial historical content without being overwhelmed by the original language's complexity. This facilitates an initial understanding, prompting more detailed discussions and analyses anchored in primary source material.

Another innovative use of AI described by elementary social studies educators involved creating developmentally appropriate definitions of discipline-specific content. AI can analyze the reading level and comprehension abilities of a target student group to tailor definitions and explanations of complex terms and concepts. For instance, a term like "sovereignty" can be explained in simpler terms for younger students, while providing a more nuanced and detailed definition for advanced learners. This customization allowed educators to present disciplinary content in a way that is both accessible and suitable for their students' developmental stages, ensuring a deeper and more meaningful understanding of historical concepts.

In secondary education, teachers described using AI to enhance critical analysis and engagement with primary sources. AI technologies have made significant strides in audio enhancement, particularly beneficial for educational materials where original recordings are garbled or difficult to understand. AI-enabled software can isolate and amplify human speech from a noisy background in archival footage, such as a 1940s newsreel with significant static interference or a speech by a historical figure recorded with old, less sophisticated equipment. By applying advanced algorithms, these tools filter out extraneous noises and enhance the clarity of spoken words, making the content more accessible to students. This not only improves the auditory experience but also aids in the comprehension of historical events, enabling students to engage more deeply with primary sources.

Critical Role of Educator Expertise. The implementation examples detailed by Delphi participants underscore the potential of AI to democratize access to historical resources, making it more comprehensible and engaging for learners. However, educators also reaffirmed the necessity of educator expertise in ensuring that the study of social studies remains rigorous, contextually rich, and ethically responsible. Participants deliberated on the critical balance between embracing the innovative potential of AI in social studies education and the indispensable role of teachers in guiding this integration. A particularly effective model discussed was an integrated learning approach in which AI simulations could act as engaging introductory activities, sparking interest and curiosity among students. This engagement is then deepened through critical discussions and analyses based on primary sources, ensuring that students' understanding remains grounded in authentic historical evidence. This strategy exemplifies how AI can serve as a powerful tool for engagement without compromising the rigor, depth, and integrity of evidence-based historical inquiry.

4.2 Ethical Considerations

Delving into the ethical implications of AI in social studies, each group within our study confronted the challenges linked to the accuracy and authenticity of AI-created historical content and its influence on students' historical understanding. A predominant ethical concern emerged over AI's capacity to perpetuate biases or historical inaccuracies [20–24], leading to a consensus on the necessity for content validation protocols. Social studies educators suggested that these protocols, involving collaboration between historians and cultural experts, should be designed to ensure the diversity and accuracy of AI-generated materials, safeguarding against the oversimplification or misrepresentation of historical narratives.

One notable discussion centered around the use of AI-enabled conversational agents to "revive" historical figures, employing synthetic media to construct speeches or simulate conversations. This blending of historical fact with speculative fiction raised significant ethical questions, blurring the line between educational content and creative storytelling [22, 25–28]. Ensuring content accuracy when these figures never communicated in modern formats, like chatbots, is complex. This raises critical questions about historical authenticity and the potential for anachronistic or misrepresented portrayals. Participants emphasized the importance of transparent disclaimers about AI's involvement in content creation, helping students discern between historical facts and AI-augmented narratives.

A particularly striking concern of social studies educators emerged regarding the use of synthetic media to visually recreate historical events with remarkable realism [12, 14, 22, 24, 25, 29]. While these recreations can vividly bring history to life, enabling students to experience reenacted moments or speeches, they also introduce the risk of inadvertently embedding contemporary biases or misconceptions into these representations that alter historical events or inaccurately portray figures, thus leading to misinformation.

Even accessible applications that some educators had piloted in their classroom instruction, such as the AI-enabled manipulation of historical photographs to alter facial expressions—adding smiles to figures in contexts of serious or somber events—were identified as problematic. This "happy history" [30] manipulation can distort students' understanding of the historical and emotional context of events and mislead students about the gravity of the situation. This application not only challenges students' ability to discern between authentic and AI-generated content but also underscores the need for developing critical thinking skills in students to question and analyze the trustworthiness and accuracy of historical sources. Educators emphasized the importance of teaching students to critically evaluate sources, understanding that AI modifications, while technologically advanced, might not always accurately represent historical events. This approach to critical analysis is crucial in an era where AI-fabricated content is becoming more sophisticated and prevalent.

4.3   Critical Thinking and AI Literacy

The necessity for fostering critical thinking and AI literacy among students emerged as a central theme. At the core of social studies education lies the development of critical thinking and analytical skills. Educators are instrumental in guiding students through the process of questioning, critiquing, and critically analyzing historical sources to assess the credibility of information, uncover biases, and navigate the complexities inherent in historical narratives. Educators emphasized the importance of equipping students with the skills to critically evaluate both authentic historical material and AI-altered content, competencies vital for differentiating between authentic historical representations and imaginative interpretations in today's era of advanced digital manipulations [31].

One highlighted exercise involved students comparing AI-generated narratives with traditional historical accounts to identify inconsistencies and potential biases. This was aimed at developing students' critical thinking skills and proficiency in questioning information, irrespective of its source.

However, Delphi study participants also engaged deeply with a discussion of the educational merits of using AI to introduce alternative narratives. Echoing Saidiya Hartman's concept of critical fabulation [32, 33], some educators advocated for a sophisticated approach to historical narratives that intentionally blurs the boundaries between fiction and history to imaginatively craft counterhistories in response to incomplete or biased official records. AI conversational agents and synthetic media have the potential to bring to life stories that are otherwise lost or underrepresented in traditional historical archives, crafting narratives that imaginatively fill gaps in historical records, particularly those related to marginalized or obscured individuals and events. By employing a method akin to Hartman's critical fabulation [32, 33], these AI tools can create immersive, narrative experiences that speculate on and give context to unrecoverable pasts. This approach can be particularly powerful in addressing the histories of individuals or events that have been left out of mainstream historical discourse.

Others warned that the speculative nature of AI-generated narratives, while valuable in filling historical gaps, risks creating content that distorts historical truth or perpetuates inaccuracies. This approach underscores the necessity for a critical examination of historical content. AI tools, promising as they are, must be applied with discretion to cultivate a comprehensive understanding of history that acknowledges its complexities and the influence of power dynamics.

This pedagogy becomes particularly relevant when exploring the histories of marginalized communities, demanding a cautious use of AI tools to amplify rather than obscure the complexities of history. Hartman's call for a deeply engaged interaction with historical content, irrespective of its origin, champions a more profound and nuanced understanding of the multifaceted nature of history and its power influences. This perspective is imperative for cultivating AI literacy among students, preparing them to traverse the delicate line between fact and fiction and to critically engage with the varied narratives that inform our collective understanding of the past.

Participants highlighted that while all historical archives are inherently incomplete, the endeavor to bridge these voids necessitates a creative yet ethically grounded approach. In an era increasingly marked by the conflation of myth with reality—a trend exacerbated by the proliferation of 'supposition news' and cultural fabrications—a pedagogic approach to filling these voids must be anchored in ethical responsibility that reconstructs history with integrity.

Social studies educators play a pivotal role in navigating the ethical challenges and potential biases inherent in AI-generated content. They can identify and address instances where AI might perpetuate historical inaccuracies or biases. Through their expertise, educators ensure that students not only engage with historical content but also develop an awareness of the ethical implications of how history is presented and interpreted. Students should be encouraged to question who creates these narratives, whose voices are amplified or silenced, and how these narratives shape our understanding of the past.

A significant finding was the need for teacher education to embed AI literacy across the curriculum. Educators voiced the importance of ongoing professional development opportunities to keep pace with evolving AI technologies alongside robust support systems for troubleshooting and pedagogical guidance.

5   Conclusions

The Delphi study conducted with social studies educators has cast a spotlight on the intricate balance needed when integrating AI into social studies education. It has unveiled the potential ethical dilemmas and challenges in historical representation, where AI tools, though enhancing student engagement, risk misrepresenting historical figures and events. This risk introduces ethical dilemmas related to authenticity and accuracy, emphasizing the urgent need for educators to weave critical thinking skills into the curriculum. Such skills are crucial for preparing students to discern between AI-altered narratives and genuine historical material, enabling them to critically navigate the increasingly blurred lines between real and altered narratives.

The impact of AI on student perceptions and the imperative for teacher preparedness further underscore the complexity of implementing AI in educational settings. Educators are tasked with not only incorporating AI tools into their teaching but also addressing their limitations and potential inaccuracies. This dual responsibility highlights the importance of using AI to supplement rather than replace traditional historical inquiry and analysis, reinforcing the commitment to historical accuracy and ethical storytelling.

Similar to the findings of our Delphi study, recent research [34, 35] confirms that social studies teachers frequently utilize self-created or extensively modified instructional materials instead of relying solely on published curricula. These educators are also among the early adopters of AI tools in educational settings, primarily using AI to adapt and enhance instructional content [35]. While these tools offer the potential to introduce diverse perspectives and materials, they also pose risks related to the quality of the modifications, which may not always align with established educational standards. There is, therefore, a pressing need for further research to evaluate the quality of AI-generated content in classroom settings to ensure it supports comprehensive, standards-aligned instruction.

As educators traverse the new landscape shaped by AI technologies, there emerges a clear call for innovative approaches and strategic collaborations. The integration of AI into social studies education prompts a reevaluation of curricula, teaching tactics, and the very framework of social studies education to align with contemporary technological and political realities. This necessitates forging partnerships across various sectors, including technology companies, historians, museums, and government entities, to develop educational strategies that resonate with today's learners. Such collaboration is vital for crafting lessons that not only engage but also ethically inform students about the past.

In weaving together the insights from the Delphi study, it is evident that while AI offers exciting possibilities for enriching social studies education, it requires a cautious and critical approach to ensure these tools enhance rather than compromise the integrity of historical inquiry. The journey ahead calls for educators to navigate these challenges with creativity and critical insight, ensuring that as we step into the future, we do so with a deep and ethically grounded understanding of the past.

References

1. Berson, I.R., Berson, M.J.: The democratization of AI and its transformative potential in social studies education. Soc. Educ. **87**(2), 114–118 (2023)
2. National Council for the Social Studies: New definition of social studies approved. https://www.socialstudies.org/media-information/definition-social-studies-nov2023
3. Berson, I.R., Berson, M.J.: Developing multiple literacies of young learners with digital primary sources. In: Russell, W. (ed.) Digital Social Studies, pp. 45–60. Information Age Publishing (2014)
4. Berson, I.R., Berson, M.J., Dennis, D. V., Powell, R.: Leveraging literacy: research on critical reading in the social studies. In: Manfra, M.M., Bolick, C.M. (eds.) Handbook of Social Studies Research, pp. 414–439. Wiley-Blackwell (2017)
5. Berson, I.R., Berson, M.J., Snow, B.: KidCitizen: designing an app for inquiry with primary sources. Soc. Educ. **81**(2), 105–108 (2017)
6. Luo, W., et al.: Aladdin's genie or Pandora's box for early childhood education? experts chat on the roles, challenges, and developments of ChatGPT. Early Educ. Dev. (2023). https://doi.org/10.1080/10409289.2023.2214181
7. Berson, I.R., Berson, M.J., Luo, W., He, H.: Intelligence augmentation in early childhood education: a multimodal creative inquiry approach. In: Wang, N., Rebolledo-Mendez, G., Dimitrova, V., Matsuda, N., Santos, O.C. (eds.) Communications in Computer and Information Science, vol. 1831, pp. 756–763. Springer, Cham (2023). https://doi.org/10.1007/978-3-031-36336-8_116
8. Martorella, P.H.: Consensus building among social educators: a Delphi study. Theory Res. Soc. Educ. **19**(1), 83–94 (1991). https://doi.org/10.1080/00933104.1991.10505629
9. Fogo, B.: Core practices for teaching history: the results of a Delphi panel survey. Theory Res. Soc. Educ. **42**(2), 151–196 (2014). https://doi.org/10.1080/00933104.2014.902781
10. Danoff, M., Kyung, Y.: Simulating the past: the case for AI in history pedagogy, Teachers College, Columbia University (2022). https://media.journoportfolio.com/users/296141/uploads/67c3735e-b156-4729-93a1-a5c4b8b6bce0.pdf
11. Burgard, K.L.B., Boucher, M.L., Ellsworth, T.M.: Reexamining the classroom simulation: guidelines for making affirming pedagogical choices. Middle Sch. J. **55**(1), 4–12 (2024). https://doi.org/10.1080/00940771.2023.2282600
12. Sheng, X.: The role of artificial intelligence in history education of Chinese high schools. J. Educ. Humanities Soc. Sci. **8** (2023). https://doi.org/10.54097/ehss.v8i.4255
13. Toktamysov, S., Alwaely, S.A., Gallyamova, Z.: Digital technologies in history training: the impact on students' academic performance. Educ. Inf. Technol. **28**, 2173–2186 (2023). https://doi.org/10.1007/s10639-022-11210-5

14. Caporusso, N.: Deepfakes for the good: a beneficial application of contentious artificial intelligence technology. In: Ahram, T. (ed.) Advances in Intelligent Systems and Computing, vol. 1213, pp. 235–241. Springer, Cham (2021). https://doi.org/10.1007/978-3-030-51328-3_33
15. Selwyn, N.: On the limits of artificial intelligence (AI) in education. Nordisk tidsskrift for pedagogikk og kritikk **10**, 3–14 (2024). https://doi.org/10.23865/ntpk.v10.6062
16. Cook, P.: Beyond "fake news": misinformation studies for a postdigital era. In: Parker, L. (ed.) Education in the Age of Misinformation: Philosophical and Pedagogical Explorations, pp. 9–31. Palgrave Macmillan, Cham (2023). https://doi.org/10.1007/978-3-031-25871-8
17. Giroux, H.A.: The crisis of public values in the age of the new media. Crit. Stud. Media Commun. **28**(1), 8–29 (2011). https://doi.org/10.1080/15295036.2011.544618
18. Giroux, H.A.: Critical theory and rationality in citizenship education. Curric. Inq. **10**(4), 329–366 (1980). https://doi.org/10.1080/03626784.1980.11075229
19. Giroux, H.A.: Resisting difference: Cultural studies and the discourse of critical pedagogy. In: Grossberg, L., Nelson, C., Treichler, P. (eds.) Cultural Studies, pp. 199–212. Routledge, New York (1992)
20. Casiraghi, S.: Anything new under the sun? Insights from a history of institutionalized AI ethics. Ethics Inf. Technol. **25** (2023). https://doi.org/10.1007/s10676-023-09702-0
21. The Institute for Ethical AI in Education: The ethical framework for AI in education. https://www.buckingham.ac.uk/wp-content/uploads/2021/03/The-Institute-for-Ethical-AI-in-Education-The-Ethical-Framework-for-AI-in-Education.pdf
22. Rennolds, N., Varanasi, L.: AI chatbots for historical figures are an ethical minefield (2023). https://www.businessinsider.com/ai-chatbots-ethics-dangers-historical-figures-2023-10
23. Adams, C., Pente, P., Lemermeyer, G., Rockwell, G.: Ethical principles for artificial intelligence in K-12 education. Comput. Educ. Artif. Intell. **4** (2023). https://doi.org/10.1016/j.caeai.2023.100131
24. Coburn, C., Williams, K., Stroud, S.R.: Enhanced Realism or A.I.-Generated Illusion? Synthetic Voice in the Documentary Film Roadrunner. J. Media Ethics Explor. Questions Media Morality **37**, 282–284 (2022). https://doi.org/10.1080/23736992.2022.2113883
25. Chan, H.Y., Liu, S.W., Hou, H.T.: Interacting with real-person non-player characters to learn history: development and playing behavior pattern analysis of a remote scaffolding-based situated educational game. Interact. Learn. Environ. (2023). https://doi.org/10.1080/10494820.2023.2192745
26. Pataranutaporn, P., et al.: Living memories: AI-generated characters as digital mementos. In: International Conference on Intelligent User Interfaces, Proceedings IUI, pp. 889–901. Association for Computing Machinery (2023) https://doi.org/10.1145/3581641.3584065
27. Hutson, J., Ratican, J.: Life, death, and AI: Exploring digital necromancy in popular culture—Ethical considerations, technological limitations, and the pet cemetery conundrum. Metaverse **4**(1) (2023). https://doi.org/10.54517/m.v4i1.2166
28. Haller, E., Rebedea, T.: Designing a chat-bot that simulates an historical figure. In: Proceedings - 19th International Conference on Control Systems and Computer Science, CSCS 2013, pp. 582–589 (2013). https://doi.org/10.1109/CSCS.2013.85
29. Noh, Y.G., Hong, J.H.: Designing reenacted chatbots to enhance museum experience. Appl. Sci. **11** (2021). https://doi.org/10.3390/app11167420
30. Jonker, E.: Reflections on history education: easy and difficult histories. J. Educ. Media Memory Soc. **4**(1), 95–110 (2012). https://doi.org/10.3167/jemms.2011.040107
31. Melo-Pfeifer, S., Gertz, H.D.: Learning about disinformation through situated and responsive pedagogy: Bridging the gap between students' digital and school lives. In: Parker, L. (ed.) Education in the Age of Misinformation: Philosophical and Pedagogical Explorations, pp. 225–249. Palgrave Macmillan, Cham (2023). https://doi.org/10.1007/978-3-031-25871-8
32. Hartman, S.: Venus in two acts. Small Axe: A Caribbean J. Criticism **12**(2), 1–14 (2008). https://doi.org/10.1215/-12-2-1

33. Hartman, S.: Wayward lives, beautiful experiments: Intimate histories of social upheaval. W.W. Norton and Company (2019)
34. Diliberti, M.K., Woo, A., Kaufman, J.H.: The missing infrastructure for elementary (K-5) social studies instruction: Findings from the 2022 American instructional resources survey. RAND Corporation, Santa Monica, CA (2023). https://www.rand.org/pubs/research_reports/RRA134-17.html
35. Diliberti, M.K., Schwartz, H.L., Doan, S.Y., Shapiro, A., Rainey, L.R., Lake, R.J.: Using artificial intelligence tools in K-12 classrooms. RAND Corporation, Santa Monica, CA (2024). https://www.rand.org/pubs/research_reports/RRA956-21.html

Enhancing Programming Education with ChatGPT: A Case Study on Student Perceptions and Interactions in a Python Course

Boxuan Ma[1]([✉]) [iD], Li Chen[2] [iD], and Shin'ichi Konomi[1] [iD]

[1] Faculty of Arts and Science, Kyushu University, Fukuoka, Japan
{boxuan,konomi}@artsci.kyushu-u.ac.jp
[2] Faculty of Information Science and Electrical Engineering, Kyushu University, Fukuoka, Japan
chenli@limu.ait.kyushu-u.ac.jp

Abstract. The integration of ChatGPT as a supportive tool in education, notably in programming courses, addresses the unique challenges of programming education by providing assistance with debugging, code generation, and explanations. Despite existing research validating ChatGPT's effectiveness, its application in university-level programming education and a detailed understanding of student interactions and perspectives remain limited. This paper explores ChatGPT's impact on learning in a Python programming course tailored for first-year students over eight weeks. By analyzing responses from surveys, open-ended questions, and student-ChatGPT dialog data, we aim to provide a comprehensive view of ChatGPT's utility and identify both its advantages and limitations as perceived by students. Our study uncovers a generally positive reception toward ChatGPT and offers insights into its role in enhancing the programming education experience. These findings contribute to the broader discourse on AI's potential in education, suggesting paths for future research and application.

Keywords: Generative AI · ChatGPT · Python programming

1 Introduction

The integration of generative artificial intelligence into the educational landscape marks a transformative shift in how teaching and learning processes are conceptualized and delivered. Emerging as a leading model of Large-scale language models (LLMs), ChatGPT is rapidly gaining recognition for its potential to significantly enrich both teaching and learning experiences by enabling users to retrieve explanations for various concepts in just a few minutes via conversation [4,7]. Recently, ChatGPT has seen increasing utilization in education to support teachers and students, including creating educational content, improving student engagement and interaction, as well as personalizing learning experiences [5].

A. M. Olney et al. (Eds.): AIED 2024 Workshops, CCIS 2150, pp. 113–126, 2024.
https://doi.org/10.1007/978-3-031-64315-6_9

As ChatGPT is not limited to natural language but can also handle programming languages, its potential impact on programming education is significant. The use of ChatGPT in program learning has also become widespread lately, as learning programming is a challenging and complex process for most people [11,12,19]. It can aid students through the complexities of programming languages, such as debugging with code, and providing real-time problem-solving assistance [2]. Numerous studies have validated GPT's performance in solving programming problems, including debugging, generating code, and providing explanations [8–10,15,17].

While the existing studies provide initial insights into ChatGPT's potential and challenges, they tend to either focus predominantly on its capabilities or investigate from educators' perspectives rather than students', which often neglects the detailed experiences of learners. Although a few studies have surveyed students' opinions and experiences, their experiments are typically conducted over a relatively short period. In contrast, practical educational settings typically involve longer durations, such as an entire semester. Furthermore, many of these studies fail to capture and analyze the interactions between students and ChatGPT for different learning activities, making it challenging to understand the nuances and dynamics of these interactions comprehensively.

To address this gap, our study aims to examine how first-year university students utilize ChatGPT in an eight-week Python programming course. Through questionnaires and open-ended questions, we investigate students' perceptions of ChatGPT's role in their learning, highlighting its strengths and areas for enhancement. By evaluating conversation interactions for various learning activities, our research offers valuable insights for effectively integrating AI into programming education. Our findings contribute to the ongoing discussion about deploying generative AI tools in educational contexts, providing recommendations for educators and curriculum developers on leveraging ChatGPT in programming courses to maximize learning outcomes while mitigating potential drawbacks.

2   Related Work

In recent years, the application of Artificial Intelligence in Education has gained significant traction, aiming to enhance and innovate learning and teaching methodologies through intelligent technologies. Generative AI tools like ChatGPT have demonstrated immense potential in education, offering personalized learning experiences, student support, and innovative delivery of course content [1,5]. These tools provide real-time interactive feedback based on students' learning progress and needs, greatly improving learning efficiency and engagement [4,6].

In programming education, ChatGPT has shown potential for different coding problems. As an AI language model that can interact using natural language, ChatGPT enables even those without prior programming knowledge to solve coding problems with ease, significantly lowering the barrier to learning programming [16,19]. Recent studies have highlighted ChatGPT's powerful capabilities in various programming tasks. For instance, Tian et al. [17] empirically analyzed ChatGPT's potential as a fully automated programming assistant for code generation, program repair, and code summarization. Their results demonstrated that ChatGPT effectively handles typical programming challenges, such as fixing bugs, providing descriptions, and generating code. Moreover, research has found that ChatGPT's bug-fixing performance is competitive with common deep-learning approaches and notably better than standard program repair techniques [15]. Phung et al. [10] systematically compared GPT models with human tutors using different Python programming problems and real-world buggy programs, revealing that GPT-4 comes close to matching human tutors' performance in several scenarios. Furthermore, ChatGPT assists in providing feedback on programming assignments, supporting students' practical application of theoretical knowledge. Pankiewicz and Baker [8] employed the GPT-3.5 model to address the challenge of generating personalized feedback for programming learning. Their experiments showed that students positively rated the usefulness of GPT-generated hints. Chen et al. [3] proposed GPTutor, a Visual Studio Code extension using the ChatGPT API to provide programming code explanations, and received positive feedback from students and teachers.

While increasing studies have suggested that interactions with ChatGPT could afford personalized learning experiences, enhance motivation, and foster critical thinking and problem-solving skills [13,14,18], these studies have predominantly explored other educational domains rather than specifically investigating the use of ChatGPT in programming education contexts. Existing research in this domain has primarily focused on evaluating ChatGPT's performance in solving programming tasks, with limited emphasis on understanding students' actual experiences and perceptions when using ChatGPT for programming learning. To the best of our knowledge, only a few studies have explored students' programming learning experiences with ChatGPT [2,4,19]. However, these studies either did not investigate such experiences within a real classroom setting [2] or did not analyze specific interaction data between students and ChatGPT [4,19], making it challenging to understand the nuances and dynamics of these interactions comprehensively. To address this gap, our study aims to deeply explore student interactions with ChatGPT within a complete programming course, examining how these interactions affect students' learning experiences and outcomes. By collecting and analyzing dialogue data between students and ChatGPT, as well as student feedback, we comprehensively assess the value of ChatGPT's application in programming education.

3   Method

This study aims to investigate students' experiences with ChatGPT within a university Python programming exercise course. The study employs both quantitative and qualitative methods. The data collection process involved pre-questionnaire and post-questionnaire, including open-ended questions designed

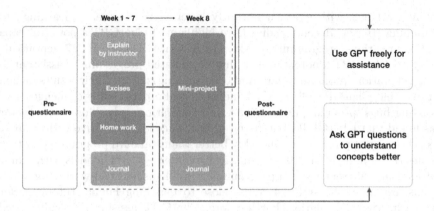

Fig. 1. Workflow of the study.

by the researchers. We set up the questionnaire on Google Forms and shared the online form link with the students, asking them to complete it. Students are also allowed to use ChatGPT freely to complete different activities throughout the class. They can log into a web-based interface that leverages the GPT-4 API, allowing them to engage in conversations directly through this platform. All the conversations between students and ChatGPT were collected for further analysis.

3.1   Participants and Context

The study was conducted among 26 students (69% males) who were enrolled in a Python programming exercises course at our university during Winter 2023, and all participants were undergraduate students from Japan. IRB approval was secured for the study.

The course teaches the foundations of Python programming language over an eight-week period. Each week, students attend two consecutive lessons, each lesson lasting 90 min and structured with the first 45 min dedicated to teaching (presentations and e-books were also utilized), followed by 45 min for students to work on programming assignments related to that week's topic. This setup aimed to enable students to practice the theoretical concepts they learned. Students can submit their tasks through the Learning Management System (LMS), which automatically logs and checks their submissions and keeps a record of all interactions and each code submission.

3.2   Process and Data Collection

Figure 1 shows an overview of this study. The course has three different learning activities focused on programming exercises, concept mastery, and problem-solving. Students are allowed to use ChatGPT freely to complete these activities.

Table 1. Pre-questionnaire items (Originally in Japanese, we translate it into English).

No	Item
Q1	How familiar are you with ChatGPT?
	• I can explain everything about it
	• I can explain it to some extent
	• I somewhat understand it
	• I don't understand it well
	• I don't know it at all
Q2	Do you use ChatGPT?
	• I am using it
	• I used to use it but not anymore
	• I haven't used it but plan to use it in the future
	• I don't use it and have no plans to use it in the future
Q3	What is your level in Python programming?
	• Beginner
	• Intermediate
	• Advanced
Q4	Do you think using ChatGPT to learn Python programming is beneficial or detrimental?
	• I think it is positive
	• I rather think it is positive
	• I am neutral
	• I rather think it is negative
	• I think it is negative
Q5	In your own words, what is ChatGPT? (Open-ended Question)

Pre- and post-questionnaires were also conducted to understand students' experiences with ChatGPT better.

Pre-questionnaire At the beginning of the course, we asked all participants to fill out a pre-questionnaire to collect demographics such as major and gender. As shown in Table 1, we also asked about their programming experiences and their familiarity with ChatGPT. Moreover, we asked the participants to describe ChatGPT in their own words using an open-ended question.

Programming Assignment Students need to complete programming assignments related to that week's topic (e.g., write a specific function based on the requirements) and submit their solutions to the LMS to verify correctness. For this activity, students are allowed to use ChatGPT freely, from obtaining code explanations and code examples to seeking debugging assistance and optimization tips. After this, students were instructed to record their dialogue data with GPT and submit them to LMS.

Table 2. Post-questionnaire items (Originally in Japanese, we translate it into English).

No.	Item
Q1	Is ChatGPT helpful for learning programming? (5-point likert scale)
Q2	Will you keep using ChatGPT for learning programming? (5-point likert scale)
Q3	Will you use ChatGPT often? (5-point likert scale)
Q4	Will you recommend ChatGPT to friends? (5-point likert scale)
Q5	How do you think ChatGPT can help with programming learning? (Multiple answers allowed) • Correct programming code • Answer programming questions • Provide examples of programming code • Offer learning advice and resources • Explain programming concepts
Q6	In your own words, what is ChatGPT after actually using it? (Open-ended Question)
Q7	What are the advantages of using ChatGPT for programming learning? (Open-ended Question)
Q8	What are the limitations or disadvantages of using ChatGPT for programming learning? (Open-ended Question)
Q9	How could ChatGPT be improved to better assist with programming learning? (Open-ended Question)

Homework. At the end of each class, students were assigned homework that asked them to use ChatGPT to investigate topics that were not fully understood during class or explore subjects of personal interest not covered in the lectures. By interacting with ChatGPT, students are expected to gain deeper insights into complex subjects, receive tailored explanations, and expand their knowledge base in areas that intrigue them. In addition, students needed to provide summaries of the topics they explored using their own words.

Individual Mini-projects. In the final week, students were asked to complete an individual project. Students were encouraged to apply the knowledge gained throughout the course to create a unique program, which typically involved writing less than 1,000 lines of code, with the option to explore any theme, ranging from games to chatbots. They were instructed to incorporate various programming concepts covered in the course, including different data objects, conditional statements (if-else), and loops (for, while) to show their comprehensive understanding and creativity. Students can use ChatGPT freely, but the dialog data must be uploaded with their code and report. In activities like these, ChatGPT can be utilized not only for debugging, explaining, and optimization but also as a tool for brainstorming ideas, clarifying project requirements, and deepening understanding of the technical aspects of the project.

Post-questionnaire. As the course progressed, the students' perceptions of using ChatGPT in programming education became more evident. To capture these insights, we conducted a questionnaire in the final week, gathering students' opinions to understand the advantages and challenges of integrating ChatGPT into the programming curriculum from their perspective. A 5-point Likert scale (1-completely disagree, 5-completely agree) is used to measure different aspects of the students' opinions toward using ChatGPT for learning programming. Also, open questions were developed to determine students' viewpoints on the use of ChatGPT for programming learning purposes. All question items are shown in Table 2.

4   Results

4.1   Analysis of Questionnaires

Pre-questionnaire. The pre-questionnaire reveals that all respondents are in their first academic year, with a majority being male (68.4%). All respondents indicated they are beginners in Python programming (Q3). In terms of familiarity with ChatGPT (Q1), a significant portion (68.4%) reported having some knowledge, while 25% reported low or no understanding. However, only 36.8% of respondents are actively using ChatGPT (Q2). Finally, the perceived impact of ChatGPT on learning (Q4) was generally positive, with 78.9% viewing it as beneficial to some extent.

Post-questionnaire. Figure 2 presents an overview of student responses to the post-questionnaire. The results show overwhelming support for ChatGPT as a beneficial tool in learning programming, with all respondents in agreement (Q1, 75% strongly agreeing and 25% agreeing). Similarly, there is a high intention to continue using ChatGPT for learning programming (Q2), with 75% strongly agreeing and 17% agreeing to continue its use. Furthermore, 67% of respondents plan to use ChatGPT frequently in the future (Q3, 42% agree, and 25% strongly agree), and the majority of respondents agree to recommend ChatGPT to their friends (Q4, 42% strongly agreeing and 25% agreeing). This suggests that users find value in ChatGPT's capabilities for their learning needs. Overall, the data reflects a highly positive reception of ChatGPT among learners in programming courses.

Additionally, Fig. 3 illustrates the perceived usage of ChatGPT in programming learning. The highest percentage, 26%, indicates that ChatGPT is helpful for correcting programming code. Equal portions of respondents, at 23% each, believe that ChatGPT aids in answering programming questions and providing code examples, highlighting its role in offering practical coding assistance and clarification on tasks. Analysis of the dialogue data shows that students frequently use ChatGPT as a debugging tool, especially when they are working on programming exercise activities. Explanation of programming concepts is also a noted benefit, with 16% of the respondents finding it useful. Lastly, 12% of respondents value ChatGPT for offering learning advice and resources.

4.2   Qualitative Analysis

Students' Awareness on ChatGPT. In our pre-study questionnaire (Q5 in Table 1), we asked students to describe ChatGPT in their own words. Following the learning period with ChatGPT, we repeated the question (Q6 in Table 2) in the post-study questionnaire. We then created a word cloud to visualize the shift in the students' perceptions, highlighting the most commonly used terms in their descriptions. Notably, all responses were originally in Japanese and have been translated into English for this analysis.

Figure 4 (a) shows the most frequent terms of the responses in the pre-study questionnaire, such as "AI", "conversation", "questions", and "answers". The word cloud effectively captures the essence of ChatGPT as perceived by the respondents: an AI tool designed for interactive, informative, and conversational purposes. In contrast, Fig. 4 (b) from the post-study questionnaire reveals that while "questions" and "answers" remain prominent, terms like "programming", "teacher", "learning", and "understand" now figure more predominantly.

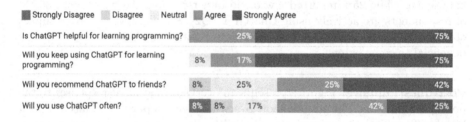

Fig. 2. Students' responses to post-questionnaire (Originally in Japanese, we translate it into English.)

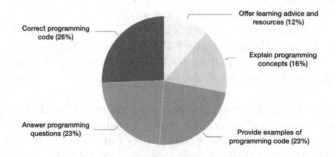

Fig. 3. Students' views on how ChatGPT can help with programming learning (Originally in Japanese, we translate it into English.)

This result reveals a significant shift in students' perceptions of ChatGPT after using it for learning purposes. Students have come to value ChatGPT more for its potential as a learning tool rather than its general conversational abilities. Moreover, students may view ChatGPT as a virtual teaching assistant, reflecting a potential change in learning habits and methodologies.

Benefits of Using ChatGPT in Programming Learning. Students were asked to answer open questions based on their learning experience, and they highlighted several significant advantages of utilizing ChatGPT in programming education (Q7).

(a) Pre-study   (b) Post-study

Fig. 4. Students' perceptions of ChatGPT. (a) Pre-study questionnaire. (b) Post-study questionnaire. (Originally in Japanese, we translate it into English.)

It is noted that the most essential benefit of ChatGPT in this process is that it responds quickly and effectively to questions and reduces time lost in researching solutions to problems. S3 indicated, "It answers my questions about things I don't understand, allowing me to quickly resolve my doubts." S8 mentioned, "My questions are quickly resolved."

Also, many respondents like using ChatGPT to identify errors, guide corrections, and suggest improvements. Six participants explicitly mention the ability of ChatGPT to assist debugging. S1 said, "I can understand where to make corrections in the code", and S6 indicated, "It is said that most of the time spent on programming is for fixing errors, and noticing one's own mistakes is not easy. Asking ChatGPT to correct errors in my program allows me to rewrite it much faster than if I were to do it by myself."

Another aspect often mentioned by the respondents is ChatGPT's ability to provide code examples and explanations, which directly helps learners apply programming knowledge to practical applications. S9's experience, where "actual examples are provided, making it very understandable and useful for application," exemplifies how ChatGPT bridges the gap between learning and doing. S12 indicated, "I thought it would just show me examples of how to do things, but it actually points out where I'm going wrong in my programming and suggests alternative methods when I'm not satisfied with an answer, making it easy to understand."

Limitations of Using ChatGPT in Programming Learning. As for the disadvantages and limitations of using ChatGPT in the programming learning process (Q8), one of the students' points concerns the fact that using ChatGPT makes the students dependent on it, thus reducing self-thinking and learning. S8 indicated, "Because ChatGPT can give plausible answers quickly, getting into the habit of asking it anything can lead to abandoning thinking for yourself and significantly reduce learning effects." S12 mentioned, "You lose the opportunity to think for yourself."

Some students stated that ChatGPT may not always give the correct answers or answers they needed, particularly in the context of programming, where multiple solutions exist. S3 indicated, "You might believe incorrect information, and it's hard to notice mistakes." Also, given that the programming course primarily covers fundamental concepts and involves relatively simple exercises, the solutions provided by ChatGPT often employ advanced and more refined methods, which can fall outside the scope of the course. S10 pointed out, "There isn't just one way to program, so ChatGPT's answers might not always be the best solution."

Suggestions for Improving ChatGPT's Support in Programming Learning. The feedback from users on enhancing ChatGPT for programming education is various (Q9). First, most students have highlighted the significant demand for gradual guidance, such as step-by-step hints, interactive questioning, and student-customized logical paths, to enhance the learning experience with ChatGPT. S3 suggested that "Rather than providing complete answers, giving hints step by step to encourage self-thinking would be more helpful." S8 proposed: "Instead of just answering questions, posing appropriate questions back to the user to ensure understanding would be beneficial. This encourages users not only to receive information but also to engage in their own thought process."

Furthermore, the need for ChatGPT to be seamlessly integrated with development environments and to offer greater accessibility was clearly stated. S7 mentioned, "Integrate with programming applications like JupyterLab to clearly display errors and solutions. Additionally, allow ChatGPT to access files on the computer when composing programs with files."

In addition, respondents expressed the need to ensure code correctness and the provision of diverse solutions. Feedback such as "ensuring the code provided as corrected is indeed correctly amended" (S1) and "offering multiple solutions instead of just one when presenting programming examples" (S9) was suggested to expand learners' understanding and adaptability.

5   Analysis of Interactions

Figure 5 shows the number of requests students have proposed and also the number of tokens generated by GPT models. The red dashed lines mark the days when classes were held, and the blue dashed line marks the final mini-project submission deadline. The data reflects fluctuating interaction patterns

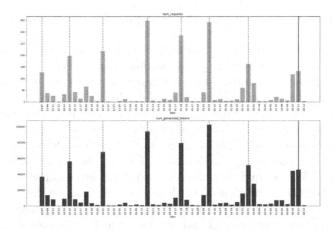

Fig. 5. Student-GPT interaction activity statistics.

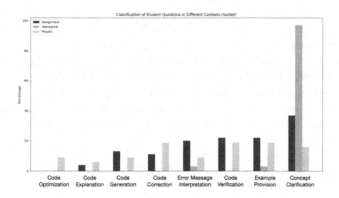

Fig. 6. Comparison of student question categories.

between students and GPT, which are closely aligned with the course curriculum, providing insight into the varying degrees of challenge presented by different topics. For instance, the initial few weeks showed relatively low activity, as only basic concepts were covered. However, the peaks marked periods such as weeks 4–6, where students actively sought GPT's guidance. The spike in activity during week 4 coincides with the introduction of conditional statements-a concept that likely necessitated a higher level of assistance from GPT, as evidenced by the increased number of requests and tokens. This trend was similar in weeks 5 and 6, which focused on 'for loops' and 'while loops' respectively. The complexity of these looping constructs appeared to drive a spike in GPT usage, suggesting that students were seeking more extensive support to master these more complex concepts.

To better understand the utilization of ChatGPT across various activities, we analyzed the dialogue data between students and ChatGPT, and categorized the questions posed by students into different types. These include Code Verification, which involves verifying the correctness of code; Code Explanation, which entails explaining the functionality purpose of the code; Error Message Interpretation, focused on interpreting the meaning and causes of error messages encountered during execution; Code Optimization, aimed at improving existing code; Code Correction, dedicated to debug the code; Concept Clarification, revolving around inquiring about or explaining programming-related conceptual knowledge; Code Generation, involving generating new code snippets or to fulfill specific requirements; and Example Provision, which provides example code to illustrate specific concepts or functionalities. Figure 6 shows the result of student question types. We can find that inquiries about conceptual understanding constitute a significant proportion of all these activities, suggesting that students are utilizing ChatGPT as a virtual teaching assistant, posing questions about concepts they do not understand and receiving timely responses. Furthermore, providing examples and code checking also represent a significant proportion.

Also, the interaction data analysis confirmed that students utilize ChatGPT differently across various learning activities. For exercises, they used ChatGPT as a coding assistant tool for Concept Clarification, Example Provision, Error Message Interpretation, and Code Generation. For homework, ChatGPT primarily served as a conceptual aid tool for concept clarification. For the mini-project, students leveraged ChatGPT as a comprehensive programming companion, utilizing the full range of its capabilities, from concept clarification to code correction. Notably, students frequently requested code optimization to enhance their final project submissions, a feature they rarely used for exercises or survey assignments.

6   Discussion

In this section, we link the questionnaire results to students' interactions and discuss their implications for the design of future LLM-based tools for assisting programming learning.

Questionnaire results suggest that ChatGPT's capabilities could be leveraged to enhance students' learning experiences when incorporated into programming education practices. Responses to the questions indicate that students value ChatGPT as a beneficial tool that helps them correct programming code, answer questions, provide code examples, and explain programming concepts. Most participants said they would use it again and recommend it to their friends. The open-question survey reveals a shift in students' perceptions of ChatGPT. It's no longer seen as a mere conversational tool but has gained recognition for its potential as a learning aid. They highlighted the advantages of utilizing ChatGPT in programming education, such as identifying errors, guiding corrections, suggesting improvements, and providing code examples and explanations. They also mentioned the disadvantages and limitations of using ChatGPT in programming learning, including reducing self-thinking and learning and not always giving the correct answers or answers they needed. In addition, several suggestions are given for improving ChatGPT's support, such as providing step-by-step hints, interactive questioning, and seamless integration with development environments. Interaction data analysis indicated the interaction patterns between students and GPT, which are closely aligned with the course curriculum. This provides insight into the varying degrees of challenge presented by different topics. Also, the results show that students utilize ChatGPT differently across various learning activities. These findings suggest that we may need to provide tailored guidance on effectively utilizing ChatGPT's different roles across various learning scenarios to better support students in leveraging its capabilities appropriately for their needs.

The findings of this work shed light on potential areas for improvement, more effective practices, and considerations for designing future LLM-based learning support systems. Firstly, an effective learning tool should cater to users' diverse information needs across various learning scenarios. Our study revealed that students utilized ChatGPT differently based on the nature of the learning activity. For instance, in programming exercise activities, many students preferred receiving step-by-step hints from ChatGPT rather than direct solutions or examples. Therefore, exploring generating personalized step-by-step hints for programming learning using ChatGPT or other LLM models would be an interesting research direction in the future. Furthermore, a closer examination of the interaction data unveiled that students often struggled to provide high-quality prompts with sufficient contextual information (e.g., accurate problem descriptions), resulting in unexpected or suboptimal responses from ChatGPT. A viable approach could be to provide prompting guidance, templates, or examples to aid students in formulating more effective prompts, thereby enhancing the quality of ChatGPT's responses. Moreover, students expressed a need to integrate ChatGPT seamlessly with other educational systems, such as learning platforms, programming editors, and automatic grading systems. This integrated approach holds the potential to create a cohesive learning environment tailored for different learning activities. By automatically feeding contextual information, such as learning content, exercise problems, student submissions, and debug information, into ChatGPT, the system could offer more personalized and targeted feedback based on students' diverse needs and learning trajectories.

7   Conclusion

This study delves into ChatGPT's role in Python programming learning, focusing on student perceptions and interactions. Through questionnaires and open-ended questions, we explored students' perspectives on the role ChatGPT played in their learning. By evaluating students' interactions with ChatGPT, our research provides insights into effectively integrating LLMs into programming education. Future work will focus on a detailed examination of students' questions and ChatGPT's responses. Evaluating student academic performance and code quality and comparing students' and teachers' views would be interesting.

Acknowledgement. This work was supported by JSPS KAKENHI Grant Number 20H00622 and 24K20903.

References

1. Anagnostopoulos, C.N.: Chatgpt impacts in programming education: a recent literature overview that debates ChatGPT responses. arXiv preprint arXiv:2309.12348 (2023)
2. Biswas, S.: Role of ChatGPT in computer programming.: ChatGPT in computer programming. Mesopotamian J. Comput. Sci. **2023**, 8–16 (2023)
3. Chen, E., Huang, R., Chen, H.S., Tseng, Y.H., Li, L.Y.: Gptutor: a ChatGPT-powered programming tool for code explanation. arXiv preprint arXiv:2305.01863 (2023)
4. Humble, N., et al.: Cheaters or AI-enhanced learners: Consequences of ChatGPT for programming education. Electron. J. e-Learn. (2023)
5. Kasneci, E., et al.: ChatGPT for good? on opportunities and challenges of large language models for education. Learn. Individ. Differ. **103**, 102274 (2023)
6. Luckin, R., Holmes, W.: Intelligence unleashed: an argument for AI in education (2016)
7. Malinka, K., Peresíni, M., Firc, A., Hujnák, O., Janus, F.: On the educational impact of ChatGPT: is artificial intelligence ready to obtain a university degree? In: Proceedings of the 2023 Conference on Innovation and Technology in Computer Science Education, vol. 1. pp. 47–53 (2023)
8. Pankiewicz, M., Baker, R.S.: Large language models (GPT) for automating feedback on programming assignments. arXiv preprint arXiv:2307.00150 (2023)
9. Phung, T., et al.: Generating high-precision feedback for programming syntax errors using large language models. In: Proceedings of the 16th International Conference on Educational Data Mining, pp. 370–377 (2023)
10. Phung, T., et al.: Generative AI for programming education: benchmarking ChatGPT, GPT-4, and human tutors. Int. J. Manag. **21**(2), 100790 (2023)
11. Rahman, M.M., Watanobe, Y.: Chatgpt for education and research: opportunities, threats, and strategies. Appl. Sci. **13**(9), 5783 (2023)
12. Rajala, J., Hukkanen, J., Hartikainen, M., Niemelä, P.: "Call me Kiran"-ChatGPT as a tutoring chatbot in a computer science course. In: Proceedings of the 26th International Academic Mindtrek Conference, pp. 83–94 (2023)
13. Shoufan, A.: Exploring students' perceptions of ChatGPT: thematic analysis and follow-up survey. IEEE Access (2023)
14. Skjuve, M., Følstad, A., Brandtzaeg, P.B.: The user experience of ChatGPT: findings from a questionnaire study of early users. In: Proceedings of the 5th International Conference on Conversational User Interfaces, pp. 1–10 (2023)
15. Sobania, D., Briesch, M., Hanna, C., Petke, J.: An analysis of the automatic bug fixing performance of ChatGPT. arXiv preprint arXiv:2301.08653 (2023)
16. Surameery, N.M.S., Shakor, M.Y.: Use Chat GPT to solve programming bugs. Int. J. Inf. Technol. Comput. Eng. (31), 17–22 (2023). https://doi.org/10.55529/ijitc.31.17.22
17. Tian, H., et al.: Is ChatGPT the ultimate programming assistant-how far is it?. arXiv preprint arXiv:2304.11938 (2023)
18. Tlili, A., et al.: What if the devil is my guardian angel: Chatgpt as a case study of using chatbots in education. Smart Learn. Environ. **10**(1), 15 (2023)
19. Yilmaz, R., Yilmaz, F.G.K.: Augmented intelligence in programming learning: examining student views on the use of ChatGPT for programming learning. Comput. Hum. Behav. Artif. Hum. **1**(2), 100005 (2023)

LANSE: A Cloud-Powered Learning Analytics Platform for the Automated Identification of Students at Risk in Learning Management Systems

Cristian Cechinel[1,3]([✉])(iD), Emanuel Marques Queiroga[4]([✉])(iD),
Tiago Thompsen Primo[2]([✉])(iD), Henrique Lemos dos Santos[1]([✉])(iD),
Vinícius Faria Culmant Ramos[1]([✉])(iD), Roberto Munoz[5]([✉])(iD),
Rafael Ferreira Mello[3]([✉])(iD), and Matheus Francisco B. Machado[1]([✉])(iD)

[1] Universidade Federal de Santa Catarina (UFSC), Florianopolis, Brazil
{cristian.cechinel,v.ramos}@ufsc.br
[2] Universidade Federal de Pelotas (UFPel), Pelotas, Brazil
tiago.primo@inf.ufpel.edu.br
[3] Centro de Estudos e Sistemas Avançados do Recife (CESAR), Recife, Brazil
rafael.mello@ufrpe.br
[4] Instituto Federal de Educação, Ciência e Tec. Sul-rio-grandense (IFSul), Pelotas, Brazil
[5] Universidad de Valparaíso (UV), Valparaiso, Chile
roberto.munoz@uv.cl

Abstract. This article introduces LANSE, an innovative Learning Analytics tool tailored for Learning Management Systems, with the primary goal of identifying student behaviors to predict risks of dropout and failure. The tool uses a cloud-based architecture that supports comprehensive data collection, processing, and visualization. In order to detect students at-risk, the tool offers automated models trained by machine learning algorithms that provide weekly predictions about the risk of the students, together with visualizations about their interactions inside the course. The performances of the models for predicting students at-risk of dropout and failure align with the state-of-the-art in the existing literature. Presently implemented in distance learning courses, initial feedback suggests that the tool effectively optimizes workloads and students behavior tracking. Challenges encountered include ensuring privacy compliance, effective data management, and maintaining real-time processing and security measures.

Keywords: Machine Learning · Educational Data Mining · Learning Analytics in Practice

1 Introduction

The importance of identifying students at high risk of dropping out and failing through automated models has long been recognized as a potential challenge

A. M. Olney et al. (Eds.): AIED 2024 Workshops, CCIS 2150, pp. 127–138, 2024.
https://doi.org/10.1007/978-3-031-64315-6_10

by the International community working in the field of artificial intelligence in education (AIED) [3]. This problem has since become a subject of extensive research nationally, with numerous publications appearing in significant events and journals [10].

While research in AIED (more precisely the subfields of educational data mining and learning analytics) has extensively explored early identification of academic risk, the translation of these findings into practical tools for students and educational professionals has been limited [8]. Published results often stem from experiments with specific datasets, hindering their applicability to diverse contexts and resulting in software maturity still incipient. However, recent studies reporting the longitudinal usage of such solutions indicate a changing landscape [9] This shift is expected to drive the field towards a new cycle of maturity, facilitating the development of practical applications grounded in research discoveries.

This paper aims to present the main features and benefits of a new tool (LANSE) focused on predicting students at-risk in learning management systems and that also offers visualizations about students interactions during their teaching and learning processes.

2   Literature Review

As mentioned earlier, the two subfields of AIED that most deal with the problem of predicting academic risk are Educational Data Mining and Learning Analytics. It is possible to say that both fields currently face a substantial challenge in transforming theoretical applications into practical implementations [1]. Despite having a robust research and theoretical foundation, the fields encounter complexity in effecting large-scale implementation in educational contexts. Factors influencing this challenge include the intricacy of educational data, insufficient infrastructure, and resistance to change within educational institutions.

The variety of research on predicting at-risk academics differs in modeling techniques (neural networks, logistic regression), data sources (VLEs, academic systems), data quantities, and attribute combinations (interactions, demographics). While more extensive literature reviews exist [4], no universal solution has emerged.

Here we mention a few works focused on the problem of identifying academic risk. For instance, [5] proposed an alert system for student performance, utilizing various data types, including demographics, interactions in the VLE, and school aptitude test results. Various prediction models were employed, with logistic regression achieving an overall accuracy of 94.2% and 66.7% accuracy in identifying students at risk of dropping out. Meanwhile, [6] seeks to identify students at risk of performance problems using machine learning models, addressing bias and algorithmic fairness. The proposal ranks students based on machine learning algorithms but also offers recommendations for various actions to reverse potential issues. Moreover, [3] developed a Learning Analytics Dashboards-based tool for predicting the dropout of university students at the time of enrollment, using historical and sociodemographic data. The results highlight an average accuracy ranging from 55% to 73% before the start of the course, emphasizing the usefulness of the approach for preventive interventions at the beginning of the academic journey.

As previously noted, the literature highlights the tangible advantages of integrating predictive tools into dashboards for enhancing teaching and learning processes [9]. The current tool aligns seamlessly with these findings, providing a practical solution that can be easily integrated by Higher Education Institutions (HEIs) using Moodle as their learning management system.

3 The Proposed Solution

The conception of the proposed solution was based on a user-centered design process, prioritizing usability and adaptability. To guide the development of the solution and validate its features, interviews were conducted with 31 stakeholders from January to April 2022, covering a variety of profiles, including higher and technical education teachers, university managers, Distance Education experts, professionals from RNP (National Research and Education Network), and representatives from the Ministry of Education of Brazil (MEC). These interviews highlighted the need to build a solution with a cloud architecture and an important characteristic: that it required minimal effort from the technical teams of client institutions. With the development of a cloud-based solution, clients would require less time for installation and maintenance, facilitating its use and subsequent scalability. Next subsections will explain how the models for automated prediction of students at-risk were developed (Subsect. 3.1), together with the current deployed solution (Subsect. 3.2).

3.1 Models for Detecting Students At-Risk

The LANSE solution offers models tailored to identify students at risk of dropout or failure, depending on available course data. When grades and final status are accessible, models predict failure risk; otherwise, they predict dropout risk. This section outlines the approach for generating these models and provides performance overviews. Deployment occurs at course onset, with periodic retraining as data accumulates.

The methodology for generating the models is based on tracking student interactions within the Learning Management System (LMS). The decision to exclusively utilize data from interaction counts was influenced by prior research that demonstrated satisfactory results through a similar approach [2]. Employing only interaction counts (and derived attributes) enables the development of models that are easily generalizable and applicable to courses regardless of their learning design.

Models for Dropout Prediction. Data from 30 undergraduate courses from the Federal University of Pelotas (first semester of 2021 - during the COVID pandemic) were used to generate the models for dropout prediction. In these courses, a total of 1,901 users had logs inside Moodle for the given period. The number of interactions of unique users for the 30 courses is presented in Table 1.

Table 1. Amount of interactions and unique users per course

Description	Average	Std	Min	25%	50%	75%	Max
Amount of interactions per course	46,359	22,657	7	42,754	52,278	58,994	77,795
Unique users per course	98	55.6	2	60.25	89.5	111.5	255

The data was grouped in such a way that for each user/course pair, a time series was assigned indicating the sum of interactions for that pair each day, from the beginning to the end of the semester. For dropout detection, since we don't have the ground truth for this variable, we opted to establish a cutoff heuristic based on a certain threshold for late dropout. Figure 1 depicts this heuristic. The blue line, whose vertical axis represents the total number of users in our dataset, indicates the proportion of users who stopped interacting on Moodle on or before a given day (horizontal axis). In other words, considering the red lines now, approximately 25.9% of users ceased interacting on Moodle for their respective courses by day 60, which is our cutoff for dropout detection. All users who stopped interacting after this cutoff are considered regular students.

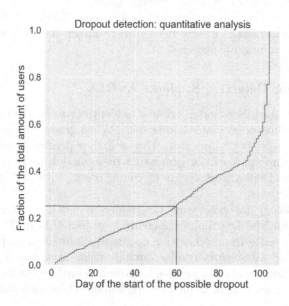

Fig. 1. Dropout detection

We used the scikit-survival library's survival analysis models for predictive modeling, considering the temporal nature and cumulative risk structure of our data. The objective of this type of modeling is to establish a relationship between the independent variables and the time at which an event occurs. In this case, the event is the day on which the student drops out, which is defined as the first day with zero interactions in a sequence of consecutive daily interactions that extends until the end of the observed period.

Four algorithms from the scikit-survival library were used: CoxPHSurvival-Analysis (COXPH), CoxnetSurvivalAnalysis (COXNET), GradientBoostingSur-vivalAnalysis, and RandomSurvivalForest (RSF). Although they have different implementation details, the first three algorithms are based on the Cox proportional hazards regression model. The Cox model assumes that the hazard rates remain proportional over the entire observation period, meaning that the corresponding risk/survival functions for individuals with different covariates are reasonably parallel throughout time.

The general formula for the risk over time in Cox regression-based models is given by:

$$h(t) = h_0(t) \times exp(b_1x_1 + b_2x_2 + ... + b_nx_n)$$

where:

- t represents the time until the event
- $h(t)$ is the hazard function determined by a set of n covariates x_1, x_2, ..., x_n
- the coefficients b_1, b_2, ..., b_n measure the impact of each covariate.
- the term h_0 is called the baseline risk, and it corresponds to the risk value when all covariates are zero, therefore making the result of the exponential operation equal to one

For each algorithm, models were trained and tested using 2, 3, 4, 5, and 6 weeks of accumulated data, always with a stratified cross-validation strategy with five folds. The input data consists of weekly aggregations of average, maximum, minimum, median, sum of interactions, and count of zero days for each user in a given course. Test-level results include three metrics. The first, the model's agreement index, assesses the correctness of ranking churned users. Samples are deemed in agreement if the higher-risk one churned earlier. This metric ranges from 0 to 1, favoring higher values, yet focuses solely on churned users.

As it can be seen in Fig. 2 (upper-left side), as more data is received, all models show an improvement in their agreement performance, so that from the 4th week onward, at least two models consistently maintain a concordance level above 0.82. The second metric is called Integrated Brier Score, representing the average of the squared distances between the occurrence or non-occurrence of dropout (1 or 0) and the prediction/probability of non-dropout at a given time. Integrated because the time variable takes all possible values; as an error metric, a lower value indicates better performance. In contrast to the first metric, the Brier Score considers both dropout and non-dropout users in its calculation.

In this case, as it can be seen in Fig. 2 (upper-right side), It becomes evident that the performance of the model based on the random forest is significantly superior to the others, especially from the 4th week onwards, indicating that this

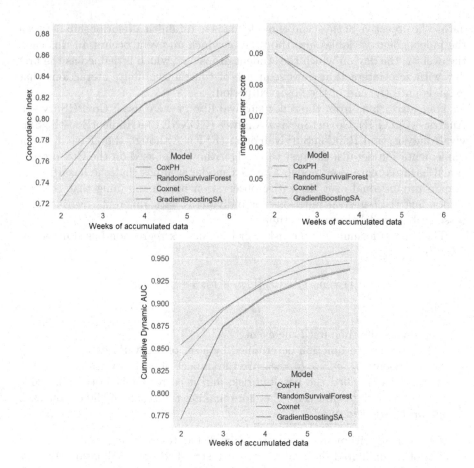

Fig. 2. Dropout detection

model is the most effective in distinguishing between dropout and non-dropout. Finally, we also assessed the models considering the Cumulative Dynamic AUC metric. This metric involves the area under the ROC curve adapted for survival analysis, defining specificity and sensitivity as time-varying measures. Like concordance, this metric evaluates how well the model can distinguish between subjects who dropped out before a certain time $(t_i < t)$ and subjects who dropped out after that time $(t_i > t)$.

With a result quite similar to the concordance (see Fig. 2 lower side), it is noticeable that the Gradient Boosting-based model is more capable of correctly ranking the dropout users up to the 3rd week, and after this period, the random forest-based model is marginally better. This was the implementation uploaded into the solution.

Models for Performance Prediction. For this model, the dataset comprised data from 85 students across 4 classes (2 in algorithms and 2 in programming) - 56 students got approved and 29 failed (either due to attendance or grades). The data included logs for each student on the Moodle platform along with their final status in the course. The generation of the derived attributes followed previous work already validated by the authors [2], where, for each student/course pair, 6 weekly variables were derived: 1) mean, 2) max, 3) min, 4) median, 5) sum of interactions, and 6) count of zero days for each user in that course. With this data, 13 classification algorithms from the scikit-learn library and the XGBoost algorithm were trained and evaluated using an AutoML approach (utilizing the TPOT library). In addition to the classification algorithms themselves, pre-processors and variable selectors were included in the optimization to compose the complete model pipeline. These algorithms and their combinations with pre-processors and variable selectors were trained and tested using a 5-fold stratified cross-validation strategy. The area under the ROC curve (AUC) was used as the target metric for optimization.

Table 2 shows the best results for the datasets from weeks 2 to 6. In the table, results are presented as macro avg./weighted avg. due to class imbalance.

Table 2. Amount of interactions and unique users per course

Week	AUC	f1-score
2	0.64	0.62/0.65
3	0.77	0.72/0.75
4	0.74	0.69/0.72
5	0.78	0.74/0.76
6	0.80	0.77/0.78

As it can be seen from the table, for 2 weeks of accumulated data, the best results were obtained by a pipeline composed of a feature generator based on Radial Basis Function (RBFSampler) coupled with a percentile-based feature selector (SelectPercentile), and finally, a classification model composed of an XGBoost algorithm stacked with a Decision Tree algorithm. Within 3 weeks, the best result came from a pipeline with an L2 normalizer, coupled with a polynomial feature generator (degree 2), and a RandomForest classifier. To achieve the best result with 4 weeks of accumulated data, a GradientBoosting classifier has its predictions stacked on a feature generator for zero counting and a OneHotEncoder. The initial features, the prediction from the first classifier, and the encoded zero counting feature are then inputted into a RandomForest classifier. The best pipelines found for the 5th and 6th weeks have the same characteristics as the best pipeline from the 4th week, except that the first classifier, whose prediction is stacked for the rest of the pipeline, which achieved the best result was a multilayer perceptron classifier for the 5th week and a support vector machine with a linear kernel for the 6th week.

3.2   Final Implemented Solution

The current solution includes a Moodle plugin on the client side that is responsible for: 1) collecting interactions stored in the system logs and 2) sending these logs to be pre-processed by the LA (Learning Analytics) service. For the prediction models to be trained and generated for use in a specific course, the plugin should also relay information related to the course schedule (start and end). The current deployed solution includes only the models for performance prediction (the dropout models will be included in the next versions). In this version the models are retrained after the end of the semester when there is new data available about the performances of the students of the courses.

The Moodle plugin, from a technical perspective, has a mechanism that regularly checks for new information and collects data for processing. This data is then sent to a cloud-based LA service through a push notification. The information is sent in batches, asynchronously, and queued for daily processing. On the cloud-based LA service side, scripts are developed to pre-process data for model training, execute these models, and present the dashboard with visualizations.

In Fig. 3, the "enable" module is responsible for controlling the logic of enabling and disabling the dashboard for a specific course. The security module is positioned at the forefront of the application, with all requests passing through it to validate if they come from authorized sources. Initially, the PyJWT cryptography library is used for this purpose. The Moodle information processing module is responsible for the processing logic for machine learning models and charts. This module primarily utilizes the pandas and scikit-learn Python libraries. The machine learning module is responsible for training the predictive models and saving the results in the database for later presentation on the front-end. Additionally, the visualization module delivers visualization data in the correct format to the front-end, and the logs module is responsible for saving the data received from the Moodle plugin.

The process begins when teachers enable the plugin within their Moodle courses. Once activated, the plugin initiates the systematic collection of data, capturing detailed information about students' access to the VLE. The teacher can access the cloud-based solution by clicking on "Access Dashboard".

Collected data is securely sent to a cloud server for centralized processing, utilizing large-scale data processing techniques to transform raw data into meaningful and measurable information. The architecture enables flexible analysis at different levels of granularity, including daily, weekly, and monthly accesses. Processed data is then presented on an interactive data visualization dashboard, offering detailed metrics on student access patterns, engagement levels with activities, and potential dropout risks. Educators receive alerts to intervene and provide support, such as sending messages to at-risk students. The Moodle course instructor's dashboard provides an overview of access percentages, risk situations, and model confidence. Figure 4 illustrates this initial screen.

The prediction is carried out asynchronously with an architecture containing an event messaging system, ensuring system independence and avoiding the need to recalculate previously made requests. The system can track messages that

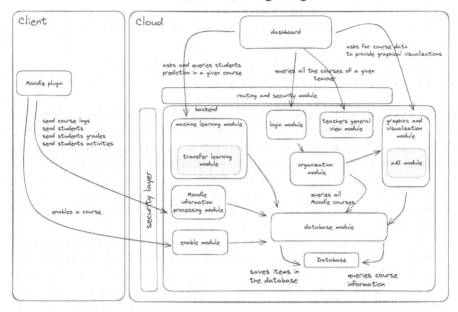

Fig. 3. LANSE architecture

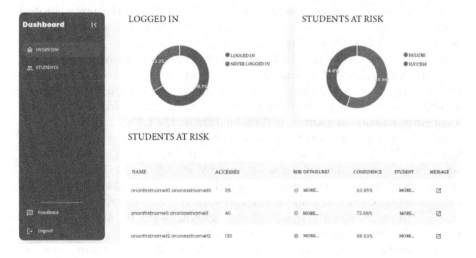

Fig. 4. General view

have already been sent (including failed ones), facilitating the detection of bugs since it is possible to replicate the messages that entered the system. Predictions are calculated and updated weekly, allowing the teacher to monitor changes in the risk associated with their students. In addition to the initial information, other visualizations related to interactions with the course are also provided, such as the total quantities of interactions for each individual student, the cumulative interactions over the weeks, and the heat map of accesses during the days of the week.

4 Discussion

One important point to highlight is the significant evolution in the performances of the models as more data is accumulated over the weeks. This can be attributed to the increased amount of available information, allowing the models to identify more complex patterns in the data. The observation of AUC and F1-score over the weeks demonstrates a positive trend, indicating that the models are improving their ability to distinguish between students at high and low risk. This trend of increasing results over the weeks aligns with findings in the literature, particularly in [5,7,8].

When comparing the quantitative results of this study with the reviewed literature, there is noticeable agreement in terms of values. For example, while we achieved an AUC of 0.78 in the fifth week, previous studies present comparable or slightly different values, often without significant generalization and model utilization, as seen in [5,7]. Furthermore, the use of survival analysis represents a substantial advantage in predicting school dropout. One of the main benefits of employing survival analysis in the context of dropout is the continuous and variable nature of student behavior, allowing for modeling the probability of dropout over specific periods, and furnishing information on the duration until the occurrence of the event of interest.

A significant aspect worth considering is the intricacy involved in the practical implementation of Learning Analytics-based solutions, as highlighted in the literature [1]. This complexity includes some challenges such as model interpretability, computational costs, and seamless integration with actual educational environments. In addressing these challenges, the LANSE solution focuses on optimizing pipelines, shedding light on the intricacies of predicting school dropout in VLE. The approach encompasses the exploration of various techniques for generating derived variables, variable selection, and the use of different classifiers. Notably, the results obtained and their variations across weeks underscore the absence of a one-size-fits-all approach to predicting dropout and failure. This emphasizes the significance of adopting a flexible and personalized strategy when implementing Learning Analytics solutions.

Preliminary interviews were conducted with two distance learning program coordinators who conveyed that their existing process for identifying at-risk students is manual. Here it follows some excerpts regarding the advantages of using the tool:

"The administrative part of Distance Learning is handled from a more distant perspective. The closest involvement comes from the course instructor, who is the one utilizing it the most. Based on this, they can track

down missing students. In distance learning, there is no physical presence, but through the tool, the instructor can monitor accesses and the risk of dropout. They can then get in touch with the pedagogical nucleus, which in turn contacts the students. The instructor accomplishes this through the LANSE visualization.'

"It had an impact. In the administrative aspect, the Distance Learning tools are demonstrated during training sessions at the campuses. And when we do that and show the LANSE, we present one more tool that helps the instructor by providing ease in monitoring the students. I can bring to the instructors a tool that can assist them."

As it is possible to see, the tool has proven instrumental in alleviating the workload of coordinators and teachers. Quantitative feedback and the impacts of this usage will be explored in future studies.

5 Final Remarks

This article presented a Learning Analytics (LA) tool focused on identifying student behaviors in Virtual Learning Environments (VLEs) and formulating models to predict dropout and failure. Its architecture enabled the collection, processing, and display of educational data, providing valuable insights for educators and administrators, enhancing decision-making, and driving academic outcomes.

Creating an analytical and predictive dashboard based on study results is a significant step forward in applying Learning Analytics, especially in Brazil and Latin America. This tool provides an intuitive visual interface for educators and administrators to centrally access vital information on resource utilization, student dropout, and failure risks. Unlike isolated techniques, this integrative dashboard consolidates data from various algorithms and sources, offering a holistic view of the educational environment. It simplifies result interpretation, enhances understanding of student behavior, and supports informed decision-making.

The implementation of these dashboards not only optimizes access to model-derived insights but also facilitates more effective intervention, allowing proactive measures to reduce dropout risks. The interactive nature of these tools encourages educators' active involvement in identifying and supporting at-risk students, strengthening preventive strategies. Concentrating results in a dashboard makes information accessible and action-oriented, serving as a valuable tool to enhance pedagogical practices and improve educational outcomes.

Acknowledgments. This work was funded by the Brazilian National Council for Scientific and Technological Development (CNPq) (grants 409633/2022-4, 305731/2021-1).

References

1. Alwahaby, H., Cukurova, M., Papamitsiou, Z., Giannakos, M.: The evidence of impact and ethical considerations of multimodal learning analytics: a systematic literature review. In: The Multimodal Learning Analytics Handbook, pp. 289–325 (2022)
2. Buschetto Macarini, L.A., Cechinel, C., Batista Machado, M.F., Faria Culmant Ramos, V., Munoz, R.: Predicting students success in blended learning-evaluating different interactions inside learning management systems. Appli. Sci. **9**(24) (2019). https://doi.org/10.3390/app9245523, https://www.mdpi.com/2076-3417/9/24/5523
3. Del Bonifro, F., Gabbrielli, M., Lisanti, G., Zingaro, S.P.: Student dropout prediction. In: Bittencourt, I.I., Cukurova, M., Muldner, K., Luckin, R., Millán, E. (eds.) Artificial Intelligence in Education, pp. 129–140. Springer International Publishing, Cham (2020). https://doi.org/10.1007/978-3-030-52237-7_11
4. Gómez-Pulido, J.A., Park, Y., Soto, R.: Advanced techniques in the analysis and prediction of students' behaviour in technology-enhanced learning contexts (2020)
5. Jayaprakash, S.M., Moody, E.W., Lauria, E.J., Regan, J.R., Baron, J.D.: Early alert of academically at-risk students: an open source analytics initiative. J. Learn. Analytics **1**(1), 6–47 (2014)
6. Khosravi, H., Shabaninejad, S., Bakharia, A., Sadiq, S., Indulska, M., Gasevic, D.: Intelligent learning analytics dashboards: automated drill-down recommendations to support teacher data exploration. J. Learn. Analy. **8**(3), 133–154 (2021)
7. Lykourentzou, I., Giannoukos, I., Nikolopoulos, V., Mpardis, G., Loumos, V.: Dropout prediction in e-learning courses through the combination of machine learning techniques. Comput. Educ. **53**(3), 950–965 (2009)
8. Queiroga, E.M., et al.: A learning analytics approach to identify students at risk of dropout: a case study with a technical distance education course. Appl. Sci. **10**(11), 3998 (2020)
9. Ramaswami, G., Susnjak, T., Mathrani, A.: Effectiveness of a learning analytics dashboard for increasing student engagement levels. J. Learn. Analy. **10**(3), 115–134 (2023). https://doi.org/10.18608/jla.2023.7935, https://learning-analytics.info/index.php/JLA/article/view/7935
10. Sha, L., et al.: The road not taken: preempting dropout in moocs. In: Wang, N., Rebolledo-Mendez, G., Matsuda, N., Santos, O.C., Dimitrova, V. (eds.) Artificial Intelligence in Education, pp. 164–175. Springer Nature Switzerland, Cham (2023). https://doi.org/10.1007/978-3-031-36272-9_14

Parsing Post-deployment Students' Feedback: Towards a Student-Centered Intelligent Monitoring System to Support Self-regulated Learning

Prachee Alpeshkumar Javiya⬚, Andrea Kleinsmith⬚, Lujie Karen Chen⁽✉⁾⬚, and John Fritz⬚

University of Maryland, 1000 Hilltop Cir, Baltimore, MD 21250, USA
{pjaviya1,andreak,lujiec,fritz}@umbc.edu

Abstract. A student-centered intelligent monitoring system collects data from students and provides insights and feedback to students about various aspects of the learning process. It often leverages data collected from educational technology systems, such as Learning Management Systems (LMS), to support students' self-regulated learning. A well-designed system needs to strike a delicate balance between the power of machine intelligence (e.g., automatic characterization and inference of students' behaviors) and the need to promote students' agency (e.g., the desire to be responsible and take control of their own behaviors). This paper presents a comprehensive qualitative analysis of anonymous student survey data collected from over 500 students with experience with a Learning Activity Monitoring System (LAMS), which has been in operation for over a decade in a public minority-serving higher education institute in the US. The study offers valuable insights into the effectiveness and user perceptions of LAMS they use in their learning context. This analysis reveals the sense-making process or lack thereof, and its implication in exploiting the potential utility of the LAMS. The findings also highlight students' varied expectations and requirements, providing critical insights for the ongoing development and refinement of the LAMS system toward an intelligent monitoring system that truly centers students' agency and promotes self-regulated learning. This study contributes to the growing body of work in hearing and understanding students' genuine voices and the sparse literature of large-scale qualitative analysis of students' feedback during the post-deployment phase on student-facing data-driven monitoring systems in an ecologically valid context in higher education.

Keywords: Student-Facing Monitoring System · Self-Regulated Learning · Student Agency

1 Introduction

A student-facing intelligent monitoring system, also known as a Learning Analytics (LA) system, uses learner-produced data and analytical methods to inform

© The Author(s), under exclusive license to Springer Nature Switzerland AG 2024
A. M. Olney et al. (Eds.): AIED 2024 Workshops, CCIS 2150, pp. 139–150, 2024.
https://doi.org/10.1007/978-3-031-64315-6_11

educational practices, enhancing students' learning outcomes. Though students are well-recognized as one of the most critical beneficiaries, they rarely play a central stakeholder role in the system design and development [2,9]. As evidenced in many lines of research [15], the persistent missing of students' active voices and authentic engagement and the lack of transparency in data collection and processing led to students' mistrust. Learning analytics, and intelligent monitoring systems in general, have been viewed as "surveillance analytics" [14], "black box" analytics [12], or as something done to them or about them [25], rather than with them and for them [13]. This lack of student engagement and trust may impede the large-scale adoption and make it challenging to realize the full potential of LA or intelligent monitoring systems in general in higher education [23].

Hearing students' authentic voices is critical to realizing the ultimate vision of a student-centered intelligent monitoring system. There are broadly two main approaches to listening to students' voices. The first is rooted in human-centered design methods, such as participatory design or co-design, typically facilitated through design workshops or focus groups. While this approach can potentially uncover students' in-depth perspectives [16], it often can only reach a limited number of students with specific characteristics. As a result, the resulting design might not be universally effective when deployed across a diverse student population on a large scale, given that design solutions are often not "one-size-fits-all" [21]. A second approach relies on large-scale surveys designed to reach many students. By asking students to rate the perceived utility of various features, these surveys may surface students' general needs [24], thereby providing valuable insights for system design. However, what is missing from the existing literature are perspectives from students who have had concrete experience using the monitoring system, preferably with their own data and over an extended period. Rather than portraying the system as an abstract concept, as is common in survey questionnaires, this approach presents the system as a "tool-in-use." We argue that this perspective is essential for gaining insight into students' experiences, especially considering the complexity of information presented in the monitoring system.

This paper analyzes a dataset of over 500 students' responses to an anonymous optional survey about their experiences, motivations, and perspectives of a norm-based descriptive student-facing learning activity monitoring system (LAMS). This system allows students to compare their grades and activities on a Learning Management System (LMS). It has been operational for over a decade in a medium-sized minority-serving higher education institute in the US. This long-standing deployment offers us a unique opportunity to gather feedback from a diverse group of students with hands-on experience using the system. We will discuss the implications of redesigning this student-facing monitoring system based on the qualitative analysis of the open-ended questions from the survey.

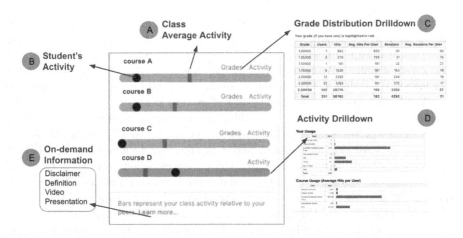

Fig. 1. The Illustration of the Student-facing Learning Activity Monitoring System Dashboard.

2   The History of Deployment and Motivation for This Study

Developed in 2008, this student-facing learning activity monitoring system (Check My Activity or CMA) is a data-driven feedback tool that allows students to compare their LMS Blackboard activity to an anonymous summary of course peers. The design and development of this system is inspired by Thaler and Sunstein's popular work, Nudge (2008) [22], and supported by our empirical study that students earning a D or F have used Blackboard about 40% less than students earning a C or higher [7]. We hope to use the monitoring system to complement our university's robust LMS adoption by nudging students towards self-regulated learning to help scale students' success. Our university is among a few higher education institutions with a long history of deploying student-facing LAMS at a large scale [6].

Figure 1 gives an overview of the key features of the monitoring system. The main display summarizes course activities on the Blackboard, represented by normalized bar meters. The dot labeled (B) indicates an individual student's activity, while the grey bar (A) denotes the average activity of the class. All metrics are norm-referenced. The color coding corresponds to whether students are above, at, or below the average LMS activity of class peers. Students can drill down to view the grade distribution for individual assignments (C). In this view, the activity of students, as measured by the number of hits and sessions, is grouped by grade band so that students may have an idea of the correlation between activity and grade. Additionally, students can drill down in their activity to specific LMS components (D), such as the discussion forum, content, or emails. Students can access further details by clicking the "Learn more" hyperlink. This includes a disclaimer regarding students' rights under the Family Educational Rights and Privacy Act (FERPA) concerning privacy and confidentiality, an explanation of terms like hits and ranks, and a video presentation on the system's background. Over the years, we have conducted quantitative empirical studies to understand the relationship between system usage and academic performance. We note that usage is correlated with a 1.5x improvement in chances of earning a C or higher course grade or 2.0 term grade point average (GPA) [7]. Despite the potential utility of system usage, only about one-third (about 3000–4500 unique users monthly) of all students tend to do so based on usage statistics. We want to understand why they do or don't use the dashboard and how the system could be improved. Over the last decade, we have collected students' feedback via an optional anonymous survey, which includes open-ended questions about students' experiences and perspectives on dashboard usage. The qualitative analysis of those responses constitutes the major focus of this paper.

In the remainder of this paper, we will briefly review existing survey-based research to understand students' perspectives on student-facing learning analytics systems (Sect. 3). We will provide details of the data collection procedure and an overview of the dataset (Sect. 4.1), qualitative analysis method (Sect. 4.2), and results (Sect. 5), followed by a discussion on the implications of our findings for redesigning the system to serve students better (Sect. 6).

3   Literature Review

Student-facing monitoring systems, or Learning Analytics Dashbaord (LAD), typically progress through exploratory research, design, development, and deployment phases. A recent review of the empirical research [4] highlights that literature predominantly covers the exploratory, design, and development phases. In contrast, there is a limited representation of post-deployment empirical studies focusing on students' experiences in authentic educational contexts. This discrepancy might result from the challenges associated with deploying the systems in real-world settings, often due to factors such as funding, infrastructure, and more. In the area of student-facing LADs and LA systems in general, much of the existing literature falls into the exploratory or design phases. For instance, Divjack et al. [3] conducted surveys with 418 students to identify useful LAD features. Similarly, another study [26] collected data using the "Student Expectations of Learning Analytics Questionnaire (SELAQ)" with 417 students from a German university. While such surveys can rapidly gather data from a large pool of students, they often require participants to respond to abstract feature descriptions, creating a disconnect from the actual user experience. Other studies adopt mixed methods, often beginning with qualitative analysis of focus groups or interviews involving a smaller group of students, followed by large-scale quantitative surveys. For example, students in studies by Schumacher et al. [20], Roberts et al. [18], and Jivet et al. [8] were presented with hypothetical scenarios or mock-ups of the system. Although these are more tangible than mere questionnaire items, they still can not replicate the experience of interacting with a live system. Additionally, some design studies, like those described in [17], present the system populated with students' own data. However, these often represent the students' first interactions with the system. It is worth mentioning that both [1,17] employed qualitative methods, utilizing think-aloud or semi-structured interviews with a limited number of participants (N=21 and N=24, respectively). There are indirect methods to gauge student interactions with a system under design. For instance, Jivet et al. [10] detailed a large-scale user study during the LAD design phase, examining the behaviors of 401 students while they interact with the system. While scalable and suitable for answering specific research questions, this passive data collection gives limited insights into students' subjective experiences. In summary, the student-facing monitoring system, specifically the LAD field, has a gap in **large-scale qualitative studies** investigating students' **post-deployment** experiences. Our research fills this gap by soliciting and analyzing student feedback using open-ended survey questions during the system post-deployment phases to understand better their perspectives derived from lived experience with a real system in action. This study complements other approaches to "hear students' voices," such as quantitative analysis of close-ended questionnaire-based surveys or qualitative analysis from interviews and focus groups from human-centered design studies.

4   Methods

4.1   Data Collection and Dataset Overview

The dataset is collected through an optional anonymous survey deployed through a hyperlink on the on-demand information page on the dashboard interface (Fig. 1(E)). Students are informed that this anonymous 10-minute survey helps the university to "better understand how and why" they have used the dashboard and what the university "could do to improve its usefulness." This study was approved by the Institutional Review Board of the university. Over 12 years, from October 2010 to February 2023, we collected approximately 500 survey responses, with an average of 40 responses annually. The survey gathered demographic details of the participants, including gender, race and ethnicity, year of birth, student status (either Undergraduate or Graduate), and their self-reported GPA band. Among the respondents, 62% were female. Regarding ethnicity, 92% identified as "Not Hispanic or Latino," while 6.9% identified as "Hispanic or Latino." In terms of races, 51.5% identified as "White or Caucasian," 23% as "Black or African American," and 21% as "Asian." Of the 442 participants who disclosed their birth year, we estimated their age based on the year they completed the survey. The median age was 23, and the mean was 28.32, with a standard deviation 12.15. Approximately two-thirds of the respondents were undergraduate students, with the remainder being graduate students. A significant majority (around 80%) reported a GPA of 3.0 or higher, while 13% indicated GPAs ranging from 2.0 to 2.99. The survey includes two open-ended questions: (1) Why have you used the CMA tool? (2) Overall, do you think a tool like CMA could help you understand or improve your own engagement in a course? If yes or no, please explain why. We have 362 comments with meaningful content from the collected responses, which was used as the basis for the qualitative analysis described below.

4.2   Qualitative Analysis Method

Qualitative data analysis was performed in two main phases: coding and conceptual modeling. Specifically, in the first phase, armed with a hard copy of the data, the first author performed an initial read-through of all responses to familiarize herself with the data. Line-by-line highlighting and memoing were implemented to identify and keep track of initial categories. An inductive coding process [19] was then carried out to identify the key themes in the data. The themes were reviewed by all authors, and uncertainty or disagreements were resolved through discussions amongst the entire team. In the second phase, we developed a conceptual model to understand the relationships between concepts identified in the coding process. We used affinity diagramming [11] to ensure that the conceptual model accurately reflected students' perceptions.

5   Results

5.1   Students' Motivation for Using the Tool, and Perspectives About How It Could Improve Course Engagement

Comparing with Peers
Without a doubt, the overwhelming majority of students responded that they chose to use the tool to compare their performance in a class with their peers through grade distribution (Fig. 1(C)), a form of anonymous summary. Many participants mentioned using CMA tool to assess their performance on assignments, exams, and overall course engagement compared to their classmates. They wanted to gauge whether they were above or below the class averages to identify areas for personal improvement. For instance, *"To check how my grades compare to those of my classmates; and whether I've [been] as active on Blackboard"* *-P418*

For many respondents, the CMA tool allowed for self-reflection; stating they used it to evaluate their study habits and time management, especially when they observed differences in activity levels compared to others. P307 used it *"To assess how I am doing in comparison with other students in order to determine whether I need to adjust my study habits.* Similarly, P440 remarked, *"Some people are motivated by measurable metrics and I'm definitely one of those people; so for people like me; it is always disappointing to log in and see that you are behind other students in terms of course engagement. It immediately provides additional motivation to study."* P23 astutely observed that, *"If I am struggling on a particular concept and notice I received a lower grade on a particular assignment compared to other students; I know I have to spend extra time studying on that particular concept.*

A few participants noted that in an online learning environment, the tool may be especially beneficial to better understand their performance in the context of the entire class as they may be less likely to receive that information through more social interaction with peers. As noted by P442, *"I am a returning student and this is my first online class. I am unsure of my performance and it is helpful to see how I compare with the other students. ... I learned; for example, that I needed to participate more in the discussion boards. It is feedback; in a way. That is very important in an online class."*

Several participants expressed interest in knowing about how their activity and engagement aligned with their classmates, even if it did not directly impact their academic performance, e.g., *"I just think it is kind of interesting to see how my activity lines up with everyone else in my class! It doesn't really have any effect on how I perform in my classes; but I discovered it by accident and I find it intriguing to check out every once in a while."* -P279

Checking Personal Grades and Progress

Many respondents reported using the CMA tool primarily to check their grades and monitor their progress in the course, indicating that grade transparency and tracking were essential features for students. The platform allowed students to reflect on their study habits, participation levels, and engagement. This self-awareness can lead to improved study strategies and better academic outcomes.

A few participants explained that they viewed it as a challenge to improve their ranking in terms of activity or to ensure they were checking the Blackboard frequently enough. *"...my points on the CMA are always above average which makes me feel like I am working harder at trying to always follow up with class; assignments and grades. If my CMA points ever went low it would definitely help me improve my engagement in a course"* -P451

In some cases, participants specifically mentioned using the CMA tool to track their activity in courses that provided extra credit for participation in discussion forums as it helped them estimate their level of participation. *"I primarily used CMA as [a] means of keeping track of my activity on the Blackboard discussion forums for certain courses. This semester; three of my five courses gave some extra credit for participating in the discussion board; so the tool proved very useful in having a general idea of that number."* -P479

Keeping up with Assignments

Several students reported using CMA tool to stay on top of their assignments and deadlines, finding the tool valuable for checking due dates, and tracking assignments, *"Check assignments; see if there is any action pending from my end; Coursework"* -P64, while other students used it to ensure they were not falling behind, *"I've found that if I start slipping in rank it usually means I am missing some assignment; or underestimated the difficulty of some assignment. Its benefit is in trending."* -P306. P112 stated, *"It helps me see if I am behind on assignments or activity.*

Motivating

The majority of respondents believed that CMA tool as a whole or a specific feature of it positively impacted their engagement in the class. Many respondents mentioned that the CMA tool served as a source of motivation; viewing it as a form of feedback that helped them understand their level of participation and engagement in the course. This feedback was seen as a way to gauge their activity and performance. For instance, according to P114, *"...I believe it could help me understand/improve my engagement in the course. It's quite motivating and it helps me to assess myself and my study habits."* Students were also motivated to be proactive in their learning. For instance, if they found themselves below average, they disclosed they would be more motivated to review materials, participate more, or check the course page more frequently. *"...typically when I've been below average I do poorer on exams and quizzes. It encourages me to review the material more"* -P139.

Some respondents appreciated the visual and "gamified" elements of the tool, suggesting that these features make the learning experience more engaging and interactive, which can further increase motivation. *"... a visual and "gamified" element seems to induce more engagement for me"* -P116. Others explained they were motivated by knowing how they were doing compared to their peers. Similar to the theme of Comparing with Peers, the competitive element of extrinsic motivation drives them to engage more or study harder. *"By knowing what other people's grades are; it will motivate me to try harder."* -P345 and for P347, *"it will motivate me to do better than my classmates and see if I am caught up with all of my assignments."*

5.2   Design Implications

Provide Support for Sense-Making
Many respondents were unclear about CMA tool's purpose or what it actually measured. Terms like "hit" were frequently mentioned, but there was uncertainty around what these terms mean. *"I think it would be better if it count[ed] not only hits. Maybe the time spent etc."* -P153. Similarly, *"I think it would help a lot more if it rated things other than just how many times you visit a class' page."* -P427

There was also confusion around how certain metrics were calculated, such as activity scores or engagement levels. Some respondents found the tool's interface or layout confusing or unhelpful, which affected their ability to interpret the data. P397 argued, *"I don't think it helps me understand OR improve my engagement much at all; primarily because I don't know what constitutes a hit".* I have discussed this with others; who generally agree. P397 also remarked they expected more explanation about the tool's features, *"I think if the tool were to provide some explanation as to what it was actually measuring; as opposed to simply telling its users how many"* hits *"they are generating in certain classes; it would be infinitely more useful."*

Improve Relevance and Utility
A significant portion of the respondents did not find the tool to be beneficial or relevant to their study habits. They believed that engagement with the CMA tool did not necessarily correlate with their academic achievement or understanding of the course material. While some saw the potential value in comparing their activity or engagement with peers, others believed that such comparisons could be misleading or even demotivating. *"...I'm actually looking for something more*

helpful; like grades; but they aren't there... so it's a metric... of some sort. But It's not useful; other than to make me feel neurotic about how often I click the course....." *-P137*. Others believed the metrics presented were incorrect. "*My professor has stated that activity for her class is low. I was curious to find out what mine was. Sadly it is incorrect. It says I've only hit the class 1 time this semester. I can account for hitting the class board at least 10 times.*" *-P424*

Respondents also expressed concerns about the simplicity of the tool interfering and not offering a comprehensive or nuanced understanding of a student's engagement or academic performance. "*It's too simplistic of a model to correlate grades to in any way.*" *-P128*

Consider Alternative Metrics Beyond Hits

Users suggested that the tool could be improved by incorporating other metrics, such as time spent on the platform, instead of just counting "hits". They also felt the tool would be more beneficial if it provided context, explanations, or breakdowns of the data it presented. Some respondents expressed a desire for weekly reports or a clearer understanding of how metrics were computed. A few users highlighted the utility of features like grade distributions and suggested making such features more accessible or prominent. "*...I would like to see some combined score that accounts for usage time and active time as well. If I spam reload; I can fake the score and also potentially affect other students.*" *-P142*.

Consider External Factors that Affect Engagement

Some respondents noted that their use of the system was influenced by external factors, such as instructor preferences or the structure of specific courses. For example, if an instructor primarily communicated via email or did not utilize the Blackboard for grading, a student's activity on the platform would naturally be lower. There were also mentions of technical issues, like the Blackboard not showing courses students were enrolled in or inaccurately reporting activity. Some users felt the tool might create unnecessary pressure or distractions, especially if it placed too much emphasis on metrics not directly tied to academic success. Confused, P460 explained, "*I don't understand what the tool is measuring. Especially since we don't really use Blackboard much to check grades or post on discussion boards. My instructor does everything via email.*" Other users found it unhelpful in smaller classes: "*It's on Blackboard which is my only place to access materials in a virtual environment. As a graduate student; I think the classes are too small for it to be meaningful and find it distracting.*" *-P144*

6   Discussion and Future Work

We will use the insights to guide our tool redesign towards an intelligent monitoring system that centers students' agency and empowerment while promoting self-regulated learning. Based on qualitative feedback from students, it became apparent that many expect the tool to provide feedback on their engagement relative to their peers and value its potential for motivation. While this can initiate the desired self-reflection necessary for the Self-Regulated Learning process, for this reflection to be transformative [5] - meaning, to prompt necessary behavioral changes that yield benefits - students must go further, devising concrete plans to enhance their engagement. This critical step towards action, however, is less frequently noted. Moreover, we have observed that some students misunderstand the tool's intent, dismissing it as irrelevant or attempting to"game" the system by artificially inflating "hits". Such misinterpretations, albeit a minority of the comments we observed, present challenges to realizing the tool's potential to improve students' learning outcomes. To address these concerns, there are several redesign considerations we are contemplating:

1. We need to clarify for students the fundamental difference between engagement, a factor that causally influences academic performance and "hits" (i.e. online activity recorded every time a student views a file, posts to a discussion, or reads an announcement etc.) being one possible proxy measure of engagement. Merely increasing hits without genuinely improving engagement won't lead to improved learning outcomes, even though they are statistically correlated. This distinction between correlation and causation must be explicitly highlighted and explained within this specific context;

2. We need to present students with convincing evidence demonstrating the causal relationship between authentic engagement (as evidenced by proxies) and academic performance. This evidence could be presented in an impactful manner, such as data stories crafted through communicative narrative visualizations, ideally, with students' data from their own communities.

3. We need to design better and more robust proxies for online engagement. Examples might include metrics measuring meaningful engagement, like contributions to discussion forums or session durations, which are not easily manipulated or gamed;

4. We need to recognize that solely measuring online behavior may not capture the full scope of student engagement. Though more challenging to scale, there are numerous offline behavioral indicators worth considering. Given the diverse nature of these indicators and the challenges associated with unobtrusive measurement, students need to be proactive in capturing this additional data, thereby providing a holistic view of their engagement patterns uniquely for themselves by themselves;

5. As some students have pointed out, course design plays a significant role. Specifically, decisions regarding the use of the LMS in a course determine the available engagement proxies. If the necessary information is not available, no students can benefit from it even with a LAD system. It thus is worthwhile to consider a systematic approach to encourage learning design surrounding LMS usage to maximize the capture of engagement data. This step is especially important in large enrollment classes with limited student support, where early alerts on lack of engagement are critical.

7 Conclusion

To support the development of self-regulated learning skills on a large scale, machine intelligence plays a crucial role in "sensing" students' activities in a scalable way. However, realizing a vision of a student-centered intelligent support system requires careful consideration of a human-in-the-loop intelligent system. This approach must balance the power of machine intelligence with human agency. This study offers insights from a deployed "sensing" system, shedding light on what a human-centered intelligent monitoring and support system could look like - one that promotes rather than undermines students' agency, moving toward fostering more self-regulated learners.

Acknowledgement. This work was partially supported by NSF grants #2216633 and #2339674.

Disclosure of Interests.. The authors have no competing interests.

References

1. Bennett, L., Folley, S.: Four design principles for learner dashboards that support student agency and empowerment. J. Appl. Res. High. Educ. **12**(1), 15–26 (2019)
2. Bodily, R., Verbert, K.: Review of research on student-facing learning analytics dashboards and educational recommender systems. IEEE Trans. Learn. Technol. **10**(4), 405–418 (2017)
3. Divjak, B., Svetec, B., Horvat, D.: Learning analytics dashboards: What do students actually ask for? In: LAK23: 13th International Learning Analytics and Knowledge Conference, pp. 44–56 (2023)
4. Ferguson, R., et al.: Aligning the goals of learning analytics with its research scholarship: an open peer commentary approach. J. Learn. Anal. **10**(2), 14–50 (2023)
5. Fleck, R., Fitzpatrick, G.: Reflecting on reflection: framing a design landscape. In: Proceedings of the 22nd Conference of the Computer-Human Interaction Special Interest Group of Australia on Computer-Human Interaction, pp. 216–223 (2010)
6. Fritz, J.: Classroom walls that talk: Using online course activity data of successful students to raise self-awareness of underperforming peers. Internet High. Educ. **14**(2), 89–97 (2011)
7. Fritz, J.: Using analytics to nudge student responsibility for learning. N. Dir. High. Educ. **2017**(179), 65–75 (2017)
8. Jivet, I., Scheffel, M., Schmitz, M., Robbers, S., Specht, M., Drachsler, H.: From students with love: an empirical study on learner goals, self-regulated learning and sense-making of learning analytics in higher education. Internet High. Educ. **47**, 100758 (2020)
9. Jivet, I., Scheffel, M., Specht, M., Drachsler, H.: License to evaluate: preparing learning analytics dashboards for educational practice. In: Proceedings of the 8th international conference on learning analytics and knowledge, pp. 31–40 (2018)
10. Jivet, I., Wong, J., Scheffel, M., Valle Torre, M., Specht, M., Drachsler, H.: Quantum of choice: How learners' feedback monitoring decisions, goals and self-regulated learning skills are related. In: LAK21: 11th International Learning Analytics And Knowledge Conference, pp. 416–427 (2021)

11. Jokela, T., Lucero, A.: Mixednotes: a digital tool to prepare physical notes for affinity diagramming. In: Proceedings of the 18th International Academic MindTrek Conference: Media Business, Management, Content and Services, pp. 3–6 (2014)

12. Kitto, K., Lupton, M., Davis, K., Waters, Z.: Designing for student-facing learning analytics. Australasian J. Educ. Technol. **33**(5) (2017)

13. Ochoa, X., Wise, A.F.: Supporting the shift to digital with student-centered learning analytics. Educ. Tech. Res. Dev. **69**(1), 357–361 (2021)

14. Pardo, A., Siemens, G.: Ethical and privacy principles for learning analytics. Br. J. Edu. Technol. **45**(3), 438–450 (2014)

15. Prinsloo, P., Slade, S.: Student data privacy and institutional accountability in an age of surveillance. In: Using data to improve higher education, pp. 195–214. Brill (2014)

16. de Quincey, E., Briggs, C., Kyriacou, T., Waller, R.: Student centred design of a learning analytics system. In: Proceedings of the 9th international conference on learning analytics and knowledge, pp. 353–362 (2019)

17. Rets, I., Herodotou, C., Bayer, V., Hlosta, M., Rienties, B.: Exploring critical factors of the perceived usefulness of a learning analytics dashboard for distance university students. Int. J. Educ. Technol. High. Educ. **18**, 1–23 (2021)

18. Roberts, L.D., Howell, J.A., Seaman, K.: Give me a customizable dashboard: personalized learning analytics dashboards in higher education. Technol. Knowl. Learn. **22**, 317–333 (2017)

19. Saldaña, J.: The coding manual for qualitative researchers. sage (2021)

20. Schumacher, C., Ifenthaler, D.: Features students really expect from learning analytics. Comput. Hum. Behav. **78**, 397–407 (2018)

21. Teasley, S.D.: Student facing dashboards: One size fits all? Technol. Knowl. Learn. **22**(3), 377–384 (2017)

22. Thaler, R.H., Sunstein, C.R.: Nudge: Improving decisions about health, wealth, and happiness. Penguin (2009)

23. Tsai, Y.S., et al.: The Sheila framework: informing institutional strategies and policy processes of learning analytics. J. Learn. Anal. **5**(3), 5–20 (2018)

24. Viberg, O., Engström, L., Saqr, M., Hrastinski, S.: Exploring students' expectations of learning analytics: A person-centered approach. Education and Information Technologies, pp. 1–21 (2022)

25. West, D., Luzeckyj, A., Toohey, D., Vanderlelie, J., Searle, B.: Do academics and university administrators really know better? the ethics of positioning student perspectives in learning analytics. Australas. J. Educ. Technol. **36**(2), 60–70 (2020)

26. Wollny, S., et al.: Students' expectations of learning analytics across Europe. J. Comput. Assist. Learn. (2023)

How Do Strategies for Using ChatGPT Affect Knowledge Comprehension?

Li Chen[1]([envelope]), Gen Li[1], Boxuan Ma[2], Cheng Tang[1], Fumiya Okubo[1],
and Atsushi Shimada[1]

[1] Faculty of Information Science and Electrical Engineering, Kyushu University, Fukuoka,
Japan
chenli@limu.ait.kyushu-u.ac.jp
[2] Faculty of Arts and Science, Kyushu University, Fukuoka, Japan

Abstract. This study investigates the effects of generative AI on the knowledge comprehension of university students, focusing on the use of ChatGPT strategies. Data from 81 junior students who used the ChatGPT worksheet were collected and analyzed. Path analysis revealed complex interactions between ChatGPT strategy use, e-book reading behaviors, and students' prior perceived understanding of concepts. Students' prior perceived understanding and reading behaviors indirectly affected their final scores, mediated by the ChatGPT strategy use. The mediation effects indicated that reading behaviors significantly influenced final scores through ChatGPT strategies, indicating the importance of the interaction with learning materials. Further regression analysis identified the specific ChatGPT strategy related to verifying and comparing information sources as significantly influenced by reading behaviors and directly affecting students' final scores. The findings provide implications for practical strategic guidance for integrating ChatGPT in education.

Keywords: Generative AI · Knowledge comprehension · Reading behavior · Perceived understanding · ChatGPT strategy

1 Introduction

Advancements in generative AI have transformed the educational landscape, providing unique opportunities for pedagogical approaches and technological support. This transformation significantly affects students' knowledge construction and the outcomes of their comprehension. When faced with challenges in understanding concepts through traditional textbooks and lectures, students use online searches to facilitate comprehension. Generative AI notably enhances the efficiency of this knowledge retrieval process, impacting the effectiveness of both traditional and AI-enhanced learning approaches.

Generative AI, which can create new content based on existing data, is changing education with its advanced prompt processing and human-like responses. [1] systematically reviewed the impact of AI across various educational dimensions, emphasizing its potential in administration, instruction, and learning, such as enhancing the efficiency of administrative tasks like grading or providing personalized learning approaches, adaptive learning systems, and innovative pedagogical tools.

In November 2022, the advent of ChatGPT, a variant of the Generative Pre-trained Transformer (GPT), amplified the demand for education innovation. ChatGPT offers personalized learning experiences by providing services like customized feedback, explanations, translations, and writing in various genres. Studies have shown ChatGPT's potential in various subjects by generating realistic scenarios for analysis and highlighting trends in large datasets. However, ChatGPT's success in educational settings largely depends on its proper use by educators and learners, given concerns about the accuracy and credibility of AI-provided information [2]. Students often struggle to evaluate the credibility of online information [3], which may significantly influence ChatGPT's effectiveness in education. Therefore, it is essential to help learners identify and develop effective strategies for information retrieval and judgment when using ChatGPT, and finally, to improve their understanding of related information.

This study explores how strategies related to using ChatGPT affect knowledge comprehension, by uncovering the interactions between student engagement with ChatGPT and its impact on knowledge comprehension. The goal is to provide practical insights into integrating effective strategies for using generative AI, such as ChatGPT, into education.

In this study, a specialized tool was provided to help students use ChatGPT with strategies, focusing on facilitating information retrieval and judgment. Considering the influence of individual backgrounds and various educational tools, we analyzed the collective impacts of students' digital textbook reading behaviors, the strategies of using ChatGPT, and the final comprehension. The analysis focused on identifying the specific ChatGPT strategies that played critical roles in knowledge comprehension and examined how these strategies interplay with students' prior knowledge background and reading behaviors. The findings provide a foundation for effective incorporation of generative AI into educational settings, suggesting how personalized strategies can enhance the learning experience.

2 Related Work

2.1 Generative AI in Education

Generative AI consists of machine-learning algorithms designed to generate new data samples resembling existing data. Its ability to process complex prompts and produce human-like responses has led to extensive applications in education.

Since the launch of ChatGPT, this transformative tool has been widely integrated into educational practices. For example, [4] explored ChatGPT's functionality, usage effectiveness, critical considerations, and implications, which provided a comprehensive insight into its capabilities and advantages. [5] investigated the application of Chat-GPT in economics and finance. In these fields, ChatGPT was used to create realistic scenarios, which are important in examining various economic and financial strategies. Furthermore, ChatGPT has proven invaluable in interpreting extensive datasets common in these fields. With its ability to analyze data, AI has highlighted nuanced trends and patterns that may have otherwise been overlooked [6].

Despite its advantages, the application of ChatGPT in educational settings is not without challenges. Several studies have highlighted concerns about integrating Chat-GPT into education, such as the reliability and accuracy of the contents generated [2, 4]. Additionally, the impersonal nature of AI also raises concerns about the lack of reflective feedback and the high demand for users' critical thinking skills to evaluate the quality of information provided by generative AI.

2.2 Strategies for Using Generative AI

Although generative AI, including ChatGPT, has been proven useful in educational research, these tools are not universally effective in every scenario. Many researchers have also pointed out the limitations of applying ChatGPT in education. For example, the human-like responses of ChatGPT have the chance to be factually inaccurate or contextually misleading [4, 7]. Therefore, it is necessary for users to actively engage in information judgment to identify the accuracy and reliability of AI-generated content.

Recent studies advocate for comprehensive approaches to critically evaluate AI-generated contents, by conducting information judgment strategies and developing digital literacy [8]. For example, the Currency, Relevance, Authority, Accuracy, and Purpose (CRAAP) test, was designed to critically evaluate web-based information. Some academic institutions adapted this evaluation framework to assess AI-generated content (e.g., [9]). However, considering the complexity of the CRAAP test, learners need to spend more time finding evidence and judging the accuracy of AI-generated content, thereby neglecting the inherent meaning of the contents, which can lead to insufficient investigation of the meaning.

To support students in information retrieval and judgment and promote their knowledge comprehension, we developed a worksheet [10] based on a "fast and frugal" model [3]. This worksheet guides students in improving their information retrieval and judgment strategies through the steps when using ChatGPT. The interface of the worksheet is shown in Fig. 1 (the details can be found in https://limu.ait.kyushu-u.ac.jp/la/ChatGPT_WS_20240418ver.zip). Students need to copy and paste the interactions with ChatGPT and judge the information by the following four steps:

(1) Check for source: Considering whether the answers contained explicit citations or references, checking the credibility of the sources, and comparing the contents from different sources.
(2) Check for expertise: Evaluating the quality of the content by checking the expertise of the sources, such as assessing their track record, credentials and institutional context, and relevant professional experience.
(3) Check for consensus: Considering whether there is a scientific consensus on responses using their previous knowledge, experiences, or accessible domain-specific information to seek scientific consensus regarding ChatGPT's response.
(4) Summarize the concept: Summarizing the concept in one's own words, based on the interactions with ChatGPT.

Fig. 1. The interface and example of ChatGPT Worksheet.

3 Methodology

3.1 Participants and Procedures

This study was conducted in an information technology course at a university, with 105 junior students enrolled.

The course introduced three main tools: a digital learning material reader named BookRoll system, ChatGPT (GPT-3.5), and a specifically designed ChatGPT worksheet. In the BookRoll system, students can use several functional tools while reading the contents, such as highlighting content, adding annotations or bookmarks to materials. The interface of BookRoll system is shown in Fig. 2. All the logs on this system were recorded as cognitive reading behaviors.

ChatGPT was introduced as a support tool in the classroom to facilitate students' investigation of concepts. It helps students search for information not covered in their textbooks. Students can ask ChatGPT various questions to enhance their understanding.

Finally, to ensure the quality and reliability of the information provided by ChatGPT, students were also provided with a ChatGPT worksheet. To understand how students interacted with ChatGPT, all these worksheets were coded to explore how they conducted their information retrieval and judgment strategies.

The duration of the course was eight weeks. Before the course, the students completed a pre-questionnaire to investigate their perceived understanding of key concepts covered in the course. During the course, students were encouraged to use ChatGPT to help them understand difficult concepts and use worksheet to judge the quality of the information. Using ChatGPT and the worksheet was optional, allowing students to choose the optimal support for their learning. As a result, 81 of the 105 students used ChatGPT and the worksheet during this course.

3.2 Data Collection

Data were collected from three sources: questionnaires, e-book reading logs, and the textual content within ChatGPT worksheets.

First, a pre-questionnaire (perceived understanding of 30 keywords) was implemented before the course, using a Likert scale ranging from 1 (strongly disagree) to 5 (strongly agree). The score collected in the pre-questionnaire represents students' prior perceived understanding.

Fig. 2. The interface of BookRoll system.

Second, logs from the BookRoll system were collected to show students reading behaviors. In this study, six types of learning logs were collected and analyzed, including *Page jump* (turning to a specific page of materials), *Get it* (pressing the 'Get it' button when students have understood the content), *Add/Del marker* (adding or deleting highlight markers within a page), *Add bookmark* (adding bookmarks to learning materials), *Bookmark jump*: (jumping to a bookmark). The sum of logs collected from the BookRoll system indicates students' active use of cognitive strategies while reading the learning material.

Third, the textual content of the ChatGPT worksheets was coded to reveal students' information retrieval and judgment strategies while using ChatGPT. A scoring system developed by the Authors [10] was employed to evaluate the specific strategies. The scoring process involved assigning points to each strategy employed by the students, aiming to assess how effectively they engaged with ChatGPT in their information-investigating tasks. In this study, five information judgment strategies were identified through the worksheet.

- *Concept*: Asking for the definition or meaning of a keyword.
- *Source*: Asking for citations or references, inquiring about concepts based on the provided sources, and comparing information from different sources.
- *Example*: Asking for an example related to the keyword.
- *Knowledge building*: Confirming the answer (through yes/no questions), asking for more detailed content and explanations, asking for logic and reasons, and thinking critically about ChatGPT's responses.

- *Questioning method*: Regarding how students ask questions to ChatGPT (i.e., asking for answers in the form of diagrams or formulas; setting roles)

Finally, we calculated the final score of students' summaries to represent their knowledge comprehension. The summaries were collected from the worksheet and scored based on three criteria, which were set through discussions among two researchers and an instructor: *Accuracy* (the summary accurately reflects information, without any incorrect details or misleading expressions), *Clarity* (the summary is written in a clear, readable manner with logical structure), and *Insightfulness* (the summary shows the depth of understanding through specific examples or personal interpretations). The total score of each summary was three points, based on the above three criteria. Due to the varied number of concepts each student investigated, and the total summaries written, an average score per concept was calculated by dividing the total points by the number of concepts, reducing the impact of quantity on evaluation.

3.3 Data Analysis

Among the 105 participants, data from 81 students who used ChatGPT worksheets were analyzed. For this study, although students were instructed to use strategies with the functional tools on the e-book and ChatGPT worksheet, it is difficult to integrate all strategies into one instructional practice. Therefore, it is necessary to identify effective strategies for knowledge comprehension and their optimal use, to further provide actionable suggestions for integrating generative AI into educational practice.

First, a path analysis of Structural Equation Modeling (SEM) was conducted to explore the contributions of students' prior perceived understanding, ChatGPT strategy use, and reading behaviors to the final score. This analysis aimed to assess how these variables collectively affect knowledge comprehension.

Second, to further understand the effective factors of ChatGPT strategy use on knowledge comprehension, mediation analysis, and regression analysis were conducted. These analyses aimed to clarify which specific factors had direct, indirect, or mediating effects on final scores. Clarifying the effects of specific strategic factors is expected to guide educational practice in integrating generative AI tools such as ChatGPT and enhancing educational approaches.

4 Results and Discussion

4.1 Path Analysis of Final Score

To determine how students' prior perceived understanding (PU), ChatGPT strategy use (CS), and reading behavior (RB) affect final score (FS), path analysis was conducted. First, the indicators of model fitting were evaluated. The indicators $\chi^2(2) = 0.03$ (p > 0.1), CFI = 1.000, TLI = 1.265, RMSEA < 0.001, and AIC = 2537.744, indicated that the model fit the data well. All paths met the significance level of $p < 0.05$.

The second step was the evaluation of the structural model. In this phase, the exogenous variables were the score from the pre-questionnaire, the sum score of the ChatGPT worksheet strategy (including five strategies), and the sum of e-book (BookRoll system) logs, while the endogenous variable was the score of students' summaries, representing their knowledge comprehension. As presented in Fig. 3, the results indicated the direct effect of the strategy of using ChatGPT on the final score, indicating that the strategic use of ChatGPT improved the knowledge comprehension of concepts. This result revealed the importance of effectively using ChatGPT in promoting knowledge comprehension.

Additionally, two indirect effects were examined: students' prior perceived understanding and reading behaviors to the final score, meaning these two factors indirectly affect the final score, mediated by the ChatGPT strategy use. Regarding prior perceived understanding, the results indicated that even if students believe they have a deeper understanding of the concept, it does not necessarily lead to higher final scores (and vice versa). Similarly, even if students frequently use cognitive strategies during reading (such as highlighting content they want to focus on), it does not directly increase final scores. However, a higher level of perceived understanding of concepts or engagement in reading behaviors might help students use ChatGPT strategies more actively, thereby improving the final scores.

In this study, considering the impact of two exogenous variables in the path analysis and the presence of five sub-strategies within ChatGPT strategy use, it is difficult to understand the specific impact of each exogenous variable on the mediator and which specific ChatGPT strategies played a crucial role in the mediation effect. Therefore, mediation analysis and regression analysis were conducted in subsequent steps.

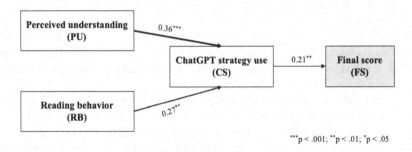

$^{***}p < .001; \, ^{**}p < .01; \, ^{*}p < .05$

Fig. 3. Path analysis results.

4.2 Evaluation of Mediation Effects

First, considering the impact of two exogenous variables, PU and RB, on CS, we verified the mediation effect of CS, under the impact of each exogenous variable. The mediation analysis was conducted using the PROCESS macro Model 4 with 5000 bootstrap samples [11]. The results are presented in two parts to evaluate the indirect effects of PU and RB on FS, respectively, with CS as the mediator. The results are presented in Table 1.

Indirect Effect of Prior Perceived Understanding (PU) on Final Score (FS)
The analysis revealed a significant positive effect of PU on CS ($\beta = .456, p < .001$),

Table 1. Results of mediation effects.

Explanatory Variable	Path	β	t-Value	95% Bootstrap CI (Indirect Effect)
Prior perceived understanding (PU)	PU → CS	.456	3.615***	-
	CS → FS	.008	1.742	-
	PU → FS (Direct Effect)	.001	.155	-
	Indirect Effect	.004	-	−0.0001 to 0.0083
Reading behavior (RB)	RB → CS	.068	2.803**	-
	CS → FS	.008	1.832	-
	RB → FS (Direct Effect)	.0001	0.073	-
	Indirect Effect	.001	-	0.0000 to 0.0013

*** $p < .001$; ** $p < .01$

indicating that students with a higher level of perceived understanding tend to adopt more effective strategies when using ChatGPT. However, no significant effects of PU on FS ($\beta = .001, p = .877$) or CS on FS ($\beta = .008, p = .085$) were found, indicating that neither perceived understanding nor ChatGPT strategy use directly affects knowledge comprehension.

The indirect effect of PU on FS, mediated through CS, was estimated to be .0035, with a 95% bootstrap confidence interval ranging from −0.0001 to 0.0083, including zero. This result showed that the mediation effect is not statistically significant at the .05 level.

From the result, it is indicated that while prior understanding is essential, it is not sufficient for effective knowledge comprehension to use prior understanding alone. However, applying prior understanding through the strategic use of ChatGPT played a crucial role in knowledge comprehension. The non-significant indirect effect CS (with a Bootstrap confidence interval narrowly including zero) indicated that the effectiveness of using ChatGPT as a pathway for transferring prior understanding into knowledge construction may depend on other factors not captured in this model. Educational interventions should focus not only on the content of prior understanding but also on how students can effectively utilize their prior knowledge through strategic learning approaches.

Indirect Effect of Reading Behavior (RB) on Final Score (FS)
As for RB as the explanatory variable, the results indicated a significant positive effect of RB on CS ($\beta = .068, p = .006$), suggesting that active reading behaviors are related to the utilization of effective ChatGPT strategies. Similar to the analysis with PU, the direct effect of RB on FS was not significant ($\beta = .0001, p = .942$), indicating a lack of direct relationship between behaviors and strategies.

However, the indirect effect of RB on FS, mediated by CS, was significant, with an estimated effect of 0.0005 and a 95% bootstrap confidence interval from 0.0000 to 0.0013, indicating a significant mediation effect at the .05 level.

RB, unlike PU, shows a significant indirect effect on FS through CS. This finding reveals the importance of encouraging reading behaviors as a mean to enhance learning strategies when using ChatGPT, and lead to the improvement of their knowledge comprehension. To facilitate this, it is effective to integrate more reading assignments into the curriculum, provide access to diverse reading materials, and create a supportive reading environment.

4.3 Effects of Each ChatGPT Strategic Factor

The above mediation analysis revealed the indirect effect of RB on FS through the mediation of CS. However, considering that ChatGPT strategies comprise five factors, it is crucial to identify which specific factors are related to RB and FS, to offer better strategic guidance regarding integrating generative AI.

A stepwise multiple regression analysis was employed to determine which of the five factors in CS (CS1: *Concept*, CS2: *Source*, CS3: *Example*, CS4: *Knowledge building*, CS5: *Questioning method*) predicted FS, controlling the variable of RB. As a result, the independent variable CS2 accounted for 3.8% of the variance of FS: F = 4.168, p = .045. The results are shown in Table 2. Subsequently, we conducted five separate simple regression analyses, treating each strategy (CS1 to CS5) as the dependent variable and RB as the independent variable. As a result, the independent variable RB shows significant contributions in predicting dependent variables CS2 (β = .286, p = .010), and CS4 (β = .314, p = .004). The results are shown in Table 3.

Table 2. Results of stepwise multiple regression analysis.

Variable	B	SE B	β	t-Value
CS2	.018	.009	.224	2.042[*]

R^2 = .050, Adjusted R^2 = .038, [*]p < .05

Table 3. Results of simple regression analysis with RB as independent variable

Dependent variable	β	SE	t-Value	p-Value	R^2	Adjusted R^2
CS1	.174	.005	1.574	.120	.030	.018
CS2	.286	.012	2.651[**]	.010	.082	.070
CS3	.149	.004	1.341	.184	.022	.010

(*continued*)

Table 3. (*continued*)

Dependent variable	β	SE	t-Value	p-Value	R^2	Adjusted R^2
CS4	.314	.007	2.935**	.004	.098	.087
CS5	.095	.004	.845	.401	.009	−.004

** $p < .01$

The summary of the regression analysis, presented in Fig. 4, showed that RB positively affects CS2 and CS4. This result indicated that when students engage in more reading behaviors, they used the strategies of confirming and searching the sources of information (CS2), and deeply understanding or critically thinking about the information from ChatGPT (CS4) more frequently. Since RB comprises several cognitive strategies during reading, it is essential to further investigate the relationships between specific reading strategies and ChatGPT strategies. This deeper investigation can help instructors effectively design and tailor their instructional approaches.

Not all factors within CS showed a direct effect on FS. Among these five factors, only CS2 demonstrated a positive effect on FS. This indicated that when using ChatGPT, the strategy of verifying the sources and comparing information from different sources, rather than simply accepting the generated information, significantly improves students' knowledge comprehension. This finding emphasizes the importance of critical thinking and source evaluation in the process of interacting with ChatGPT.

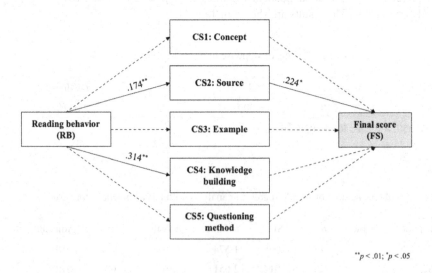

** $p < .01$; * $p < .05$

Fig. 4. Path diagram of regression analysis regarding five factors in CS.

5 Conclusion and Future Work

This study investigates the effects of generative AI on the knowledge comprehension of university students, focusing on the strategies of using ChatGPT.

Data from 81 junior students who used ChatGPT worksheet were analyzed, along with their questionnaire and e-book logs. As the results, first, path analysis indicated a direct effect ChatGPT strategy use on final score, indirect effects of prior perceived understanding and reading behavior on final score, mediated by the ChatGPT strategy use.

As for the mediation effects of ChatGPT strategy use of prior perceived understanding and reading behavior respectively, it was found that only reading behavior had a significant indirect effect on final score through the mediation of ChatGPT strategy use. This result highlighted the importance of engagement with learning materials.

Finally, a regression analysis was conducted to further explore the causal relationships between specific ChatGPT strategies with reading behavior and final score. As a result, among five ChatGPT strategies, the strategy related to the verification and comparison of information sources was not only influenced by reading behavior but also had a direct effect on the final score.

This study provides insights into the effective use of ChatGPT strategies, particularly the strategy of verification and comparison of information sources, which showed a positive effect on knowledge comprehension. These findings offer strategic implications for integrating ChatGPT into educational practices. Notably, the knowledge building strategy of using ChatGPT did not directly affect the final score, even though it was influenced by reading behaviors. However, since reading behaviors include various cognitive strategies regarding reading learning materials, it is essential to conduct a deeper exploration of which specific reading strategies are effective and how they improve learning outcomes when used along with ChatGPT. This exploration could help instructors to determine how to choose effective reading strategies and integrate reading strategies and generative AI-related strategies in their instructional designs.

Furthermore, it is important to provide a more refined guidance on strategies of using generative AI, such as knowledge building strategies, through further qualitative analysis. Additionally, longitudinal studies are essential to assess the long-term effects of generative AI integration, providing a broader understanding of its effective use in higher education.

Acknowledgments. This work was supported by JST CREST Grant Number JPMJCR22D1 and Japan Society for the Promotion of Science (JSPS) [grant number JP22H00551, 24K16759], Japan. We would like to acknowledge the four research assistants who helped us with the coding of the worksheets, Lopez Erwin, Ken Goto, Ryunosuke Hamada, Yuma Miyazaki, as well as all the students who participated in the study.

Disclosure of Interests. Drs. Chen, Ma, and Shimada received research grants from Japan Society for the Promotion of Science (JSPS), and Dr. Shimada also received research grants from Japan Science and Technology Agency (JST).

References

1. Chen, L., Chen, P., Lin, Z.: Artificial intelligence in education: a review. IEEE Access **8**, 75264–75278 (2020)
2. Tlili, A., et al.: What if the devil is my guardian angel: ChatGPT as a case study of using chatbots in education. Smart Learn. Environ. **10**(1), 15 (2023)
3. Osborne, J., Pimentel, D.: Science, misinformation, and the role of education. Science **378**(6617), 246–248 (2022)
4. Qadir, J.: Engineering education in the era of ChatGPT: Promise and pitfalls of generative AI for education. In: IEEE Global Engineering Education Conference (EDUCON), pp. 1–9. Kuwait, Kuwait (2023)
5. Alshater, M.: Exploring the role of artificial intelligence in enhancing academic performance: A case study of ChatGPT. SSRN
6. Wang, D., Shan, D., Zheng, Y., Guo, K., Chen, G., Lu, Y.: Can ChatGPT detect student talk moves in classroom discourse? a preliminary comparison with Bert. In: The 16th International Conference on Educational Data Mining, Bengaluru, India (2023)
7. Cooper, G.: Examining science education in ChatGPT: an exploratory study of generative artificial intelligence. J. Sci. Educ. Technol. **32**(3), 444–452 (2023)
8. Tinmaz, H., Lee, Y.T., Fanea-Ivanovici, M., Baber, H.: A systematic review on digital literacy. Smart Learn. Environ. **9**(1), 21 (2022)
9. Todd, V.: Suggested strategies for referencing generative AI and why your students should take the CRAAP test: Advice from the Library. Tech Europe. https://teche.mq.edu.au/2023/03/suggested-strategies-for-referencing-generative-ai-and-why-your-students-should-take-the-craap-test-advice-from-the-library/. Accessed 8 Mar 2024
10. Chen, L., Shimada, A.: Designing worksheet for using ChatGPT: towards enhancing information retrieval and judgment skills. In: 2023 IEEE International Conference on Teaching, Assessment and Learning for Engineering (TALE), pp. 1–4. Auckland, New Zealand (2023)
11. Hayes, A.F.: PROCESS macro for SPSS, SAS, and R (Version 4.2). [SPSS]. http://processmacro.org/download.html. Copyright © 2012–2022 by Andrew F. Hayes (2022)

Automatic Lesson Plan Generation via Large Language Models with Self-critique Prompting

Ying Zheng[1], Xueyi Li[1], Yaying Huang[1], Qianru Liang[1], Teng Guo[1],
Mingliang Hou[2], Boyu Gao[1], Mi Tian[2], Zitao Liu[1(✉)], and Weiqi Luo[1]

[1] Guangdong Institute of Smart Education, Jinan University, Guangzhou, China
`zhengying@stu2022.jnu.edu.cn, lixueyi@stu2021.jnu.edu.cn,`
`{huangyaying,liangqr,tengguo,bygao,liuzitao,lwq}@jnu.edu.cn`
[2] TAL Education Group, Beijing, China
`{houmingliang,tianmi}@tal.com`

Abstract. In this paper, we utilize the understanding and generative abilities of large language models (LLMs) to automatically produce customized lesson plans. This addresses the common challenge where conventional plans may not sufficiently meet the distinct requirements of various teaching contexts and student populations. We propose a novel three-stage process, that encompasses the gradual generation of each key component of the lesson plan using Retrieval-Augmented Generation (RAG), self-critique by the LLMs, and subsequent refinement. We generate math lesson plans for grades 2 to 5 at the elementary school levels, covering over 80 topics using this method. Three experienced educators were invited to develop comprehensive lesson plan evaluation criteria, which are then used to benchmark our LLM-generated lesson plans against actual lesson plans on the same topics. Three evaluators assess the quality, relevance, and applicability of the plans. The results of the evaluation indicate that our approach can generate high-quality lesson plans. This innovative approach can significantly streamline the process of lesson planning and reduce the burden on educators.

1 Introduction

A lesson plan is the teacher's daily guide for what students need to learn, how it will be taught, and how learning will be measured [20,27,29]. An example of a lesson plan is shown in Table 1. As a fundamental tool for teachers, it serves as a guide that details the course of instruction or "learning trajectory" for a lesson [7,15]. This essential instrument varies in detail, depending on the teacher's preference, the subject being covered, and the specific needs of the students [26]. Lesson planning is an integral part of the teaching process, and demands systematic organization and thoughtful preparation [29,32]. A well-constructed lesson plan not only reflects the interests and needs of students but also incorporates best practices in the educational field.

A. M. Olney et al. (Eds.): AIED 2024 Workshops, CCIS 2150, pp. 163–178, 2024.
https://doi.org/10.1007/978-3-031-64315-6_13

Table 1. An example of a lesson plan.

Parallelogram

Student learning analysis

Considering that our student group is second-year junior high school students, they should have mastered basic geometry knowledge in the first year......

Lesson objectives

1. Knowledge and Skills: Students should be able to understand and master the basic properties of parallelograms, including equal diagonals, ... 2. Process and Methods: Enable students to understand the properties of parallelograms intuitively, for example, through physical models ...... 3. Emotional Attitudes and Values: Stimulate students' interest in learning about parallelograms, and enhance their enthusiasm and participation in learning....... 4. Comprehensive Application: Enable students to apply the knowledge of parallelograms they have learned to......

Key and difficult points

1. Understanding the Properties of Parallelograms: The main content of this lesson is the properties of parallelograms, ... 2. Drawing Parallelograms: Students need to master how to

Materials

1. Teaching Aids: Ensure there are enough rulers and protractors for students to use. Prepare some ...... 2. Textbook Preparation: In addition to being familiar with the content about parallelograms in the textbook, it is necessary to ...... 3. Courseware Preparation: Design a PPT courseware of parallelograms that includes theoretical knowledge, examples, exercises, etc. Ensure that ...... 4. Lesson Plan Preparation: A detailed lesson plan should be developed based on the student's learning situation and teaching objectives. The objectives of ...... 5. Exercise Preparation: Design......

Lesson procedure

1. Reviewing Old Knowledge: Teacher: Let's start by reviewing what we've learned about shapes. Can you tell me what parallel and perpendicular lines are? ...... 2. Introducing New Knowledge: Teacher: Today we are going to learn about parallelograms. Before that, can you try to tell me the definition of a parallelogram and ...... 3. Learning New Knowledge: Teacher: Now, I want you to try to draw a parallelogram with a ruler and protractor, and try to find out its properties. Students: (Students try to draw.) ...... Activity Intent: Through hands-on practice, students can ...... 4. Consolidating New Knowledge: Teacher: Okay, now let's do some exercises about parallelograms, which will help you better understand and master the properties of parallelograms. Students: ...... Activity Intent: Through exercise training, consolidate students' new ...... 5. Expanding New Knowledge: Teacher: In fact, parallelograms can be seen everywhere in our daily life. Now ...... Activity Intent: By guiding students to discover applications in life, enhance students' ......

As a complex yet essential part of the teaching process, making a reliable lesson plan benefits classroom development and management. Various research studies have explored the design of effective lesson plans. For instance, Iqbal et al. primarily focused on enhancing the theory-based lesson plan through key factors [19]. Vdovina et al. integrated critical thinking into the lesson plan, promoting a more interactive and thought-provoking learning environment [31]. Additionally, Ferrell and Barbara explored the use of lesson plans as a tool to measure program effectiveness, demonstrating the multifaceted utility of lesson plans [11].

Despite the importance and benefits, creating high-quality lesson plans remains a challenging task, especially for rural teachers who suffer from limited resources and support [10,16]. While some existing lesson plans are accessible, they often don't align with the specific needs of the teaching environment or student demographics. This mismatch can lead to suboptimal teaching and learning experiences. Moreover, the lack of teacher professional development opportunities for rural teachers hampers their ability to innovate in curriculum design and planning methods.

To address this issue, in this work, we explore the potential of Large Language Models (LLMs) provided by OpenAI for creating customized lesson plans that cater to the unique needs of individual teachers [1,34]. We design a three-stage process to fully utilize the capabilities of LLMs in generating lesson plans. First, we generate each key component of the lesson plan based on Retrieval-Augmented Generation (RAG) to ensure it is comprehensive and detailed. Second, we engage the LLMs to conduct self-critiques to promote continuous improvement and refinement of the generated lesson plans. To ensure the final result is of high quality and meets the intended objectives, the LLMs refine the lesson plan based on these self-critiques. We invite three human evaluators to evaluate the lesson plans generated by LLMs in terms of the quality, relevance, and applicability of these plans.

2   Background and Related Works

2.1   Basics of Lesson Plan

A lesson plan is a guide that outlines the pedagogical goals for a particular course and the strategies students should use to achieve those goals. According to research by Nesari et al. [27], a lesson plan is defined as a written narrative of the educational process that specifies what, when, where, and how learners should learn, and how their learning should be assessed. It is a pivotal tool that aids teachers in maintaining organization and providing a structured framework for instruction, thereby enhancing the effectiveness of teaching and learning. Amininik et al. posited that the preparation of lesson plans by educators is a viable strategy for enhancing the quality of education [2].

The components of a lesson plan can vary widely, depending on the teacher's preferences, the subject matter, and the student's needs. Generally, a basic lesson plan typically includes:

- **Lesson objectives** that clearly state what the students should know or be able to do by the end of the lesson. These objectives provide a focus for the lesson and guide the selection of instructional activities and assessment methods.
- **Lesson procedure** that provides a sequence of activities, from introduction to conclusion, with the main instructional activities forming the lesson's core. These activities, guided by the lesson's objectives, involve direct instruction, guided practice, collaborative learning, or independent work, designed to facilitate students' understanding and mastery of the content.
- **Assessment methods** that include formative assessments, summative assessments, self-assessments, and peer assessments. They provide feedback on student learning and inform instructional decisions and assessment methods.

Aside from the core components above, a lesson plan can include, but is not limited to, the following components:

- **Lesson standards** that are often established by state or national education authorities, to define the knowledge and skills that students should acquire in a particular grade level or subject area.
- **Lesson duration** that specifies the time requirements for each segment of the lesson, ensuring that the pacing of instruction is appropriate and that sufficient time is allocated for each activity.
- **Key and difficult points** that are determined based on the lesson objectives and scientific analysis of the materials. They represent the most fundamental and core content of the lesson, typically embodying the most important principles or laws expounded by the subject.
- **Materials** that list the materials to be used during the lesson. These may include textbooks, handouts, manipulatives, technology tools, and other resources that support instruction and learning.
- **Student learning analysis** that is a critical process to understand the unique learning needs and challenges of each student. This analysis involves assessing students' current knowledge levels, learning styles, and academic progress. It may also include identifying any learning difficulties or barriers that students may be experiencing.

2.2   Lesson Plan Design

The lesson plan serves as a guide for teachers to achieve the desired teaching outcomes. Designing a lesson plan involves considering objectives, subject matter, materials, time allocation, level, methodology, activities, and assessment methods [29]. Previous research on lesson plan design can be categorized three types: First, research-based approach to lesson plan design: teachers rely on the findings of educational research to design theoretically grounded instructional activities and processes [21,31]. Second, technology-supported approach to lesson plan design: teachers heavily rely on educational technologies, such as online teaching platforms and big data analytics, to access efficient and personalized teaching and learning activities and processes [26,32]. Third, collaborative lesson plan design approach: leveraging collaboration among teachers, the collaborative lesson plan design approach, which includes co-designing lesson plans, sharing teaching resources, and evaluating each other's teaching effectiveness. This enables the design of more effective teaching and learning activities and processes [8,27].

2.3   Improving Educational Quality via Large Language Models

LLMs have demonstrated extraordinary capabilities across tasks in numerous fields [4,25,30,33,34]. The use of LLMs in education has recently emerged as a promising direction. For example, Khan Academy, a non-profit educational institution, developed an AI chatbot called Khanmigo as a virtual tutor and classroom assistant using GPT-4 [14]. He et al. proposed combining LLM with an external symbolic solver to solve mathematical applications, which improved performance by 20% compared to previous research [17]. Cao developed an intelligent tutoring system that utilizes GPT3 to assist Chinese students with their CS1 courses [5]. Moreover, there is a growing body of practical experience and specialized tools aimed at leveraging LLMs for lesson planning. Some blogs[1] and online resources provide insights into the practical application of LLMs in generating lesson plans, and specialized tools[2] for lesson planning with LLMs have been developed. In recent research, Lee et al. demonstrated how pre-service elementary teachers from a Korean university successfully integrated ChatGPT into science lesson plans, revealing diverse applications and strategies [22]. Hu et al. conducted a study based on Pedagogical Content Knowledge [6,13] theory that assessed the efficacy of GPT-4 in creating high school mathematics lesson plans [18].

---

[1] https://automatedteach.com/p/using-llms-chatgpt-plan-better-lessons.
[2] https://www.tomdaccord.com/ai-tools-for-lesson-plans.

3 Our Approach

3.1 Overview of the Framework

To enhance the efficiency and quality of lesson plan design, we propose a framework for auto-generating lesson plans utilizing LLMs, drawing inspiration from the iterative improvement method in software development [9,28]. The proposed framework consists of three stages: component generation, self-critique, and refinement (shown in Fig. 1). At the component generation stage, the LLM generates the content of each part of the lesson plan step by step. The model then self-evaluates the generated lesson plan, identifying shortcomings and providing suggestions for improvement. Subsequently, the model refines the generated lesson plan based on these suggestions. The details of each stage are provided in the following sections, respectively.

3.2 Stage 1: Component Generation

Component generation involves the generation of individual components that collectively form a comprehensive lesson plan, as depicted in Fig. 1(a). At this stage, teachers are asked to provide the necessary information that will guide

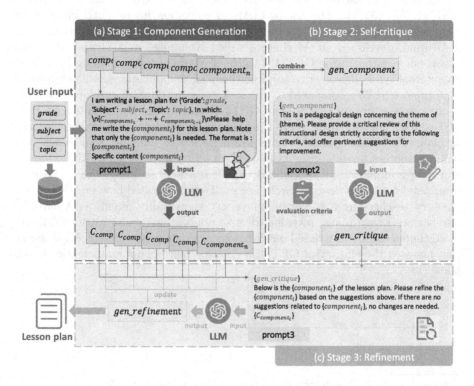

Fig. 1. The framework for lesson plan automatic generation.

the generation of the lesson plan. This includes (1) the lesson topic, which helps to ensure that the content of the lesson plan is relevant and appropriate for the specific course; (2) the grade level, which informs the complexity and depth of the content, ensuring it is suitable for the cognitive abilities and learning needs of students at that grade level; and (3) the subject, which ensures that the lesson plan aligns with the subject matter and covers the necessary knowledge and skills. The teachers can then select the specific components that they wish to include in their lesson plan, i.e., student learning analysis, lesson objectives, and lesson procedure. Additionally, teachers may furnish pertinent information and specifications for each component. For instance, *student learning analysis: Students have gained an initial understanding of three-dimensional shapes in the previous semester. Lesson procedure: The mode of instruction for this lesson is inquiry-based teaching, centered around problem-solving.*

RAG is a technique that combines information retrieval with text generation to enhance the performance of LLMs on specific tasks [12,23]. Utilizing RAG, we first perform a retrieval step based on input from teachers to search within our lesson plan datasets, which have been collected from the real world. This step allows us to identify and extract lesson objectives that align with the teacher's provided information. These objectives then serve as additional information inputs into the model. By using RAG, we enhance the accuracy and practical applicability of the content generated by the model.

Each of these components is independently generated by the model, which incorporates all previously generated components as input to the model before initiating the creation of a new component, to ensure the relevance of each component and the coherence of overall. The step-by-step generation process facilitates the model's meticulous crafting of each component.

3.3   Stage 2: Self-critique

Self-critique [3,24] is an innovative and critical stage that leverages the LLM's ability to evaluate its output. At this stage, the model integrates the components generated in the first stage into a complete lesson plan, to make it easier for the model to evaluate the lesson plan as a whole. Then the model proceeds to evaluate this plan based on evaluation criteria established by senior professional educators. The specific evaluation details are in Sec. 4.1. The prompt, *"Please provide a critical review of this instructional design strictly according to the following criteria, and offer pertinent suggestions for improvement"*, serves as a directive for the model to critically assess the quality of the generated lesson plan. An example of refinement is shown in Table 2. This self-critique process encourages the model to critically analyze its output, identify shortcomings, and consider ways to enhance the quality of the lesson plan. More importantly, it provides the basis and method for the improvement of stage 3.

Table 2. An example of self-critique.

Lesson plan: Two digits plus one digit, two digits

Here are my ratings:

Student learning analysis:

This section is very detailed, covering not only the prerequisite knowledge points that the students have mastered, but also analyzing the characteristics of the knowledge points in this lesson, and considering the students' level of development. However, it should be further specified, for example, by identifying which students have mastered which knowledge points and which students are facing certain difficulties, in order to facilitate more targeted teaching

Lesson objectives:

The lesson objectives are clearly set, including the mastery of knowledge and skills, as well as the application of processes and methods, and the cultivation of emotions, attitudes, and values. However, they could be further refined by breaking down the mastery of knowledge and skills into more specific sub-goals

Key and difficult points:

The key and difficult points are consistent with the lesson objectives, considering the understanding and application of two-digit addition with carrying, as well as solving real-life problems. However, it could be more clearly stated which knowledge points are key and which are difficult, to allow for more targeted teaching

Materials:

The materials are comprehensive, including the preparation of teaching aids, courseware, and exercises, as well as individual tutoring and teaching evaluation preparation. However, the use of teaching aids and courseware, as well as the specific settings of exercises and tutoring content, could be further clarified

Lesson procedure:

The lesson procedure is reasonably designed, including introduction, demonstration and practice, individual tutoring, and solving practical problems, with natural transitions and coherent logic. However, it could consider adding some fun activities, such as games, to further engage students

3.4   Stage 3: Refinement

The refinement stage utilizes the feedback derived from the self-critique stage to augment the quality of the produced lesson plan. This stage has an iterative character, allowing for continual repetition to yield results of superior quality. Figure 1(c) shows that the model refines the lesson plan based on the self-critique ratings and suggestions for improvement. During the refinement process, the model revisits each component of the lesson plan, making modifications as needed to improve its quality and effectiveness. For instance, if the self-critique identifies

Table 3. Information of lesson plan generation for different grades.

Grade	Number	Examples of topic
2nd	23	An Initial Understanding of Multiplication Multiplication Table of 6 Mixed Addition and Subtraction Operations with Parentheses Two-Digit Plus Two-Digit Addition (No Carry) Two-Digit Minus One-Digit Subtraction (No Borrowing)
3rd	25	Three-Digit Minus Three-Digit Subtraction (No Borrowing) Three-Digit Plus Three-Digit Addition (With Consecutive Carrying) Simple Conversion Between Time Units Understanding Fractions 1 Minus a Fraction
4th	24	Parallel and Perpendicular Characteristics of the Parallelogram Understanding Hectares Three-Digit Number Multiplied by a Two-Digit Number Measuring Angles
5th	13	Factors and Multiples Decimal Multiplication - Approximating the Product Decimal Division - Dividing a Number by a Decimal Simple Equations - Representing Numbers with Letters Simple Equations - Solving Equations

a lack of clarity in the lesson objectives, the model would work on refining these objectives to make them more specific, achievable, and relevant. If the critique points out a lack of alignment between the lesson procedure and the key and difficult points of the lesson, the model would adjust the lesson procedure to ensure it adequately addresses these points.

Through the iterative process of self-critique and refinement, we ensure that the generated lesson plan is not a static product but a dynamic one. It undergoes continuous improvement until it meets a high standard of quality. This process also mirrors the reflective practice that educators engage in, constantly evaluating and refining their lesson plans to enhance their teaching effectiveness and improve student learning outcomes.

4   Experiment

We generate lesson plans by calling the GPT-4 API. To ensure the quality and effectiveness of the generated lesson plans, we collect exemplary plans from diverse educational platforms to create our lesson plan dataset, all of which have been meticulously crafted by human educators. We choose mathematics as the subject of focus for the lesson plans in this study, because it is one of the most challenging subjects.

4.1   Evaluation Process

In our study, we generate math lesson plans covering over 80 topics in elementary school stages. The number of lesson plans for each grade and some examples of topics are shown in Table 3. The lesson plans of each topic include three versions:

Table 4. The specific criteria and guidelines for the global and component scores.

Evaluative dimension			Score
Global score (30 points in total)		g1. structural integrity	6
		g2. content accuracy	9
		g3. content consistency	6
		g4. language logicality	9
Component score (66 points in total)	Student learning analysis	c1. pre-knowledge	3
		c2. the knowledge of this lesson	3
		c3. student development level	3
	Lesson objectives	c4. knowledge and skills	3
		c5. process and method	3
		c6. emotion attitude value	3
	Key and difficult points	c7. key points	3
		c8. difficult points	3
	Materials	c9. completeness	3
		c10. relevance	3
	Lesson procedure	c11. completeness of teaching activities	3
		c12. rationality of teaching activities	3
		c13. relevance of teaching activities	3
		c14. fluency of teaching activities	3
		c15. interactivity of teaching activities	3
		c16. achievement of lesson objectives	3
		c17. the solution of the key and difficult points	3
		c18. interest in teaching activities	3
		c19. teaching time allocation	3
	Assessment methods	c20. student assessment	3
		c21. teacher assessment	3
		c22. class assessment	3

- v1: generate once for all, this is a version generated in a single run, demonstrating the model's ability to generate a complete lesson plan at once.
- v2: generate component by component and only one component of the lesson plan is generated at a time, showing the model's capacity for a component-specific generation.
- v3: refine for component-by-component generation, refining and improving the v2 based on feedback and evaluation.

To effectively evaluate the quality and usability of the generated lesson plans, we invite three experienced elementary school educators to develop a comprehensive and detailed set of lesson plan evaluation criteria based on their experience.

These criteria encompass a global score and a component score. The global score, with a maximum of 30 points, pertains to the overall integrity, accuracy, consistency, and logic of the lesson plan. The component score, 66 points in total, evaluates the quality of individual components of the lesson plans, such as the clarity of lesson objectives, the completeness of materials, and the effectiveness of lesson procedures. The evaluative criteria provide a quantifiable measure of the quality and usability of the lesson plans. The specific criteria and guidelines for the global and component scores are detailed in Table 4. Subsequently, based on these evaluation criteria, each lesson plan is scored by the three evaluators. To facilitate a more comprehensive comparison between model-generated lesson plans and their real-world counterparts, in addition to scoring versions v1, v2, and v3 of the model-generated lesson plans, we also score randomly select lesson plans on the same topic from the real world.

4.2   Result Discussion

We evaluate three versions of lesson plans generated by the model, v1, v2, and v3, as well as real-world lesson plans. The evaluation results are presented in Table 5.

Table 5. Average score comparison of four types of lesson plans across 4 grades. The best results of each grade are highlighted in **bold**.

Grade	Version	Global score	Component score	Total score
2nd	human	**28.40**	48.40	76.80
	v1	26.06	41.56	67.63
	v2	24.69	50.19	74.88
	v3	26.90	**53.87**	**80.77**
3rd	human	27.48	44.77	72.35
	v1	**29.17**	42.65	71.74
	v2	26.71	52.11	78.82
	v3	28.80	**55.11**	**83.91**
4th	human	**27.18**	45.22	72.40
	v1	26.16	41.46	67.62
	v2	24.78	48.70	73.48
	v3	27.08	**52.11**	**79.19**
5th	human	26.51	38.48	64.99
	v1	**29.69**	39.18	68.87
	v2	27.21	47.87	75.08
	v3	29.24	**50.52**	**79.76**

As shown in Table 5. (1) In the global score, v2 scores the lowest when compared with v1 and v3, with little difference between v3 and v1. Notably, v3 surpasses v1 in both 2nd and 4th grade lesson plans. This indicates that the component-by-component generation approach does not inherently ensure coherence and relevancy as effectively as the generate once for all, tending to generate redundant content. However, the self-critique and refinement process can mitigate these issues to some extent. In compari-

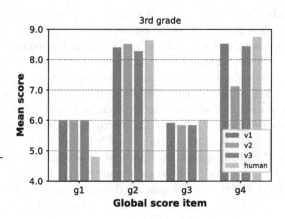

Fig. 2. Average global score of v1, v2, v3 and real-world lesson plans for 3rd grade.

son to human-designed lesson plans, as indicated by "human" in the figures and tables of this paper, the model-generated plans exhibit a more complete structure. In the 3rd and 5th grades, human-designed lesson plans frequently suffer from missing components, primarily within the student learning analysis and materials. For instance, as illustrated in Fig. 2, the human-designed lesson plans for the 3rd grade are only inferior to the model-generated plans in terms of structural integrity (g1). (2) In the component score, v1 is inferior to v2 and v3, with v3 being superior to v2. For example, in the second grade, as illustrated in Fig. 3, the majority of the 22 component score items demonstrate a trend where v1 < v2 < v3. This suggests that the component-by-component generation approach facilitates more targeted contemplation by the model for different components, resulting in more specific content, whereas v1 tends to generate more generalized material. When comparing model-generated lesson plans with human-designed ones, the frequent occurrence of missing components in the latter leads to lower component scores than those of v2 and v3, which are generally above v1. However, when analyzing only the lesson procedure part of the component scores, as depicted in Fig. 4 for 4th grade, the human-designed lesson procedure is predominantly superior to those generated by the model. This may be attributed to the human-designed processes being more aligned with practical teaching, with more detailed and specific activities, yet lacking in teacher-student interaction and time allocation. These findings offer guidance for future optimization of the model's lesson plan generation. (3) In the total score, v3 emerges as the victor across the four types of lesson plans, underscoring the efficacy of our methodological approach. v3 combines the global coherence of v1 with the component specificity of v2, demonstrating the potential of step-by-step generation and refinement in producing high-quality lesson plans.

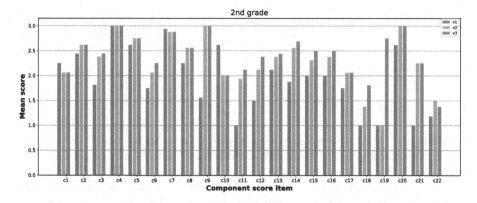

Fig. 3. Average component score of v1, v2, and v3 lesson plans for 2nd grade.

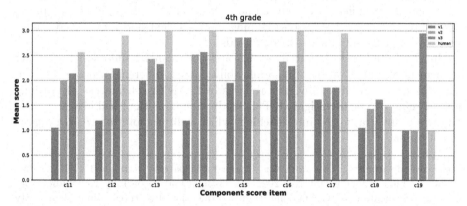

Fig. 4. Average lesson procedure item score of v1, v2, v3, and real-world lesson plans for 4th grade.

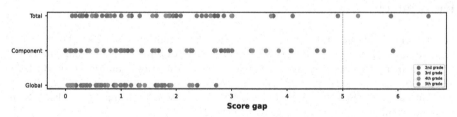

Fig. 5. The difference in scores between each pair of the three evaluators.

To ensure fairness and consistency in scoring, four lesson plans for each topic were evaluated by three evaluators. We calculate the average score for each grade's lesson plan given by the evaluators for pairwise comparisons. A score difference of less than 5 points between any two evaluators was considered consistent; otherwise, the scores were deemed inconsistent. Figure 5 presents the distribution of score differences between each pair of the three evaluators. It is evident that in grading lesson plans for different grades, the three evaluators generally achieve consistency across the global score, component score, and total score, with only a few instances of scoring discrepancies. This indicates that the results of the scoring are reliable.

5   Conclusion

In this paper, we introduce a new method that uses LLMs to create personalized lesson plans for teachers. We design a three-stage process in which LLMs initially create the components of the lesson plan step-by-step through Retrieval-Augmented Generation, then conduct a self-critique based on human-defined evaluation criteria, and finally refine the lesson plan. After rigorous evaluations by three evaluators comparing lesson plans generated by LLMs and real-world ones, the results indicate that our method can generate high-quality lesson plans. This approach can make lesson planning more efficient and improve education by tailoring it to individual teaching needs.

In the future, the potential of this approach can be unlocked by further refinement and exploration of its applicability across different subjects and educational stages. Additionally, the abilities of LLMs in other areas of teaching and learning can be investigated. This study highlights the feasibility and effectiveness of integrating AI into teaching, which will promote future AI research in education.

Acknowledgements. This work was supported in part by National Key R&D Program of China, under Grant No. 2022YFC3303600 and in part by Key Laboratory of Smart Education of Guangdong Higher Education Institutes, Jinan University (2022LSYS003).

References

1. Achiam, J., et al.: GPT-4 technical report. arXiv preprint arXiv:2303.08774 (2023)
2. Amininik, S.: A survey of the implementation of lesson plan in Bushehr university of medical sciences. In: 4th National Medical Education Conference in Bushehr (2000)
3. Antoniou, A., Storkey, A.J.: Learning to learn by self-critique. In: Advances in Neural Information Processing Systems (2019)
4. Brown, T., et al.: Language models are few-shot learners. In: Advances in Neural Information Processing Systems (2020)
5. Cao, C.: Scaffolding CS1 courses with a large language model-powered intelligent tutoring system. In: Companion Proceedings of the 28th International Conference on Intelligent User Interfaces (2023)
6. Chai, C.S., Koh, J.H.L., Tsai, C.C.: A review of technological pedagogical content knowledge. J. Educ. Technol. Soc. **16**, 31–51 (2013)
7. Chen, J., Huang, S., Liu, Z., Luo, W.: DialogID: a dialogic instruction dataset for improving teaching effectiveness in online environments. In: Proceedings of the 31st ACM International Conference on Information & Knowledge Management (2022)

8. Darling-Hammond, L., Richardson, N.: Research review/teacher learning: what matters. Educ. Leadersh. **66**, 46 (2009)
9. Dorn, J.: Iterative improvement methods for knowledge-based scheduling. AI Commun. **8**, 20–34 (1995)
10. Du Plessis, P., Mestry, R.: Teachers for rural schools–a challenge for South Africa. South Afr. J. Educ. **39** (2019)
11. Ferrell, B.G.: Lesson plan analysis as a program evaluation tool. Gifted Child Q. **36** (1992)
12. Gao, Y., et al.: Retrieval-augmented generation for large language models: a survey. arXiv preprint arXiv:2312.10997 (2023)
13. Gess-Newsome, J.: Pedagogical content knowledge: an introduction and orientation. In: Gess-Newsome, J., Lederman, N.G. (eds.) Examining Pedagogical Content Knowledge: The Construct and Its Implications for Science Education, vol. 6. Springer, Dordrecht (1999). https://doi.org/10.1007/0-306-47217-1_1
14. Hadi, M.U., et al.: A survey on large language models: applications, challenges, limitations, and practical usage. Authorea Preprints (2023)
15. Hao, Y., et al.: Multi-task learning based online dialogic instruction detection with pre-trained language models. In: Roll, I., McNamara, D., Sosnovsky, S., Luckin, R., Dimitrova, V. (eds.) AIED 2021. LNCS (LNAI), vol. 12749, pp. 183–189. Springer, Cham (2021). https://doi.org/10.1007/978-3-030-78270-2_33
16. Hatch, L., Clark, S.K.: A study of the instructional decisions and lesson planning strategies of highly effective rural elementary school teachers. Teach. Teach. Educ. **108**, 103505 (2021)
17. He-Yueya, J., Poesia, G., Wang, R., Goodman, N.: Solving math word problems by combining language models with symbolic solvers. In: The 3rd Workshop on Mathematical Reasoning and AI at NeurIPS 2023 (2023)
18. Hu, B., Zheng, L., Zhu, J., Ding, L., Wang, Y., Gu, X.: Teaching plan generation and evaluation with GPT-4: Unleashing the potential of LLM in instructional design. IEEE Trans. Learn. Technol. **17**, 1471–1485 (2024)
19. Iqbal, M.H., Siddiqie, S.A., Mazid, M.A.: Rethinking theories of lesson plan for effective teaching and learning. Soc. Sci. Humanit. Open **4**, 100172 (2021)
20. Jackson, P.W.: Life in Classrooms. Teachers College Press, New York (1968)
21. Kruse, K.: Gagne's nine events of instruction: An introduction. Retrieved the (2009)
22. Lee, G.G., Zhai, X.: Using chatGPT for science learning: a study on pre-service teachers' lesson planning. arXiv preprint arXiv:2402.01674 (2024)
23. Lewis, P., et al.: Retrieval-augmented generation for knowledge-intensive NLP tasks. Advances in Neural Information Processing Systems (2020)
24. Madaan, A., et al.: Self-refine: iterative refinement with self-feedback. Advances in Neural Information Processing Systems (2024)
25. Min, S., Lewis, M., Zettlemoyer, L., Hajishirzi, H.: MetaICL: learning to learn in context. In: Proceedings of the 2022 Conference of the North American Chapter of the Association for Computational Linguistics: Human Language Technologies (2022)
26. Morrison, G.R., Ross, S.J., Morrison, J.R., Kalman, H.K.: Designing Effective Instruction. Wiley, Hoboken (2019)
27. Nesari, A.J., Heidari, M.: The important role of lesson plan on educational achievement of Iranian EFL teachers' attitudes. Int. J. Foreign Lang. Teach. Res. **3**, 25–31 (2014)
28. Salo, O., Abrahamsson, P.: An iterative improvement process for agile software development. Software Process: Improvement and Practice (2007)

29. Sugianto, A.: Applying a lesson plan for a digital classroom: challenges and benefits. Int. J. Engl. Educ. Linguist. (IJoEEL) **2**, 21–33 (2020)
30. Touvron, H., et al.: Llama 2: open foundation and fine-tuned chat models. arXiv e-prints (2023)
31. Vdovina, E., Gaibisso, L.C.: Developing critical thinking in the English language classroom: a lesson plan. ELTA J. **1**, 54–68 (2013)
32. Williamson, B.: Big data in education: The digital future of learning, policy and practice. Big Data in Education (2017)
33. Yang, A., et al.: Baichuan 2: open large-scale language models. arXiv preprint arXiv:2309.10305 (2023)
34. Zhao, J.X., Xie, Y., Kawaguchi, K., He, J., Xie, M.Q.: Automatic model selection with large language models for reasoning. In: The 2023 Conference on Empirical Methods in Natural Language Processing (2023)

Systematic Needs Analysis of Advanced Digital Skills for Postgraduate Computing Education: The **DIGITAL4Business** Case

Carmel Somers[1], Dave Feenan[1], David Fitzgerald[1], Roberto Henriques[2],
Matteo Martignoni[3], Daniela Angela Parletta[3], Eva Cibin[3],
Adriana E. Chis[4], and Horacio González–Vélez[4]([✉])

[1] Digital Technology Skills, Dublin, Ireland
{carmel.somers,dave.feenan,david.fitzgerald}@digitaltechnologyskills.ie
[2] NOVA Information Management School (NOVA IMS), Universidade Nova de
Lisboa, Lisbon, Portugal
roberto@novaims.unl.pt
[3] Akkodis Italy SRL, Rome, Italy
{matteo.martignoni,danielaangela.parletta,eva.cibin}@akkodisgroup.com
[4] Cloud Competency Centre, National College of Ireland, Dublin, Ireland
{adriana.chis,horacio}@ncirl.ie

Abstract. A 15-partner funded project, DIGITAL4Business aims to revolutionise the digital landscape in Europe, fostering strong industry partnerships to empower digital transformation organically and providing postgraduate programmes through a unique multi-academic pan European approach. This paper identifies the needs and gaps in advanced digital skills in countries of the European Union to enable digital transformation. Validated via desk research and surveys, our results indicate that the top two advanced digital skills for companies are Cybersecurity and Cloud Computing, followed by Data Science for small and medium-sized enterprises and Artificial Intelligence (AI) for large companies, respectively. Businesses also indicate notable concerns about the digital divide, digital literacy, the potential impact of automation and job displacement, the role of regulatory frameworks, and training. Our recommendations encompass flexible and accessible education models, *ex-professo* AI & Data Science modules, stakeholder collaboration, and diversity & inclusion initiatives.

Keywords: digital transformation · advanced digital skills · needs analysis · postgraduate education · upskilling · reskilling

1 Introduction

A *needs analysis* identifies the level of skills and training requirements in a given team or organisation. Although human resources departments have long used needs analyses to link training with the results of commercial firms [22], they have only been considered relevant for the holistic development of higher education programmes until recently [16].

A. M. Olney et al. (Eds.): AIED 2024 Workshops, CCIS 2150, pp. 179–191, 2024.
https://doi.org/10.1007/978-3-031-64315-6_14

Advanced digital skills encompass a range of specialised competencies that go beyond basic computer literacy. While it is widely accepted that digital skills (also known as 21st-century digital skills) are structured around technical, information, communication, collaboration, critical thinking, creativity, and problem-solving [13], for them to be deemed advanced, their underlying enablers must arguably entail progressive technologies–e.g., Artificial Intelligence (AI)–to empower individuals with a deeper understanding of digital tools, platforms, and concepts. Thus, advanced digital skills are generally referred to by their enabling technology.

Individuals equipped with advanced digital skills are in high demand, as they possess the ability to harness the power of technology and drive organisational success [2]. DIGITAL4Business is a 20 million euro ground-breaking project with 15 academic, industry, and government partners located in 7 European countries that aims to promote digital transformation and training for Small and Medium Enterprises (SMEs) and businesses in the EU27, that is, the 27 countries of the European Union (EU). As one of the largest initiatives funded by the EU's Digital Europe Programme, it is set to revolutionise the provision of advanced digital skills by furnishing advanced quality education at postgraduate level including an accredited European Master's programme, micro-credentials, and industry certifications, while fostering strong academic, industry, and government partnerships to collaboratively shape the future of advanced digital skills and tackle the growing digital skills gap in Europe.

By upskilling the workforce and fostering digital innovation, DIGITAL4Business supports creating new job opportunities, increased productivity, and overall economic development. DIGITAL4Business must play a crucial role in shaping the digital transformation of Europe's society.

> This work systematically investigates the need for advanced digital skills and identifies skills gaps in the EU27 to ultimately develop and deliver quality post-graduate education to enable digital transformation for SMEs and businesses.

We have conducted extensive desk research and industry surveys, and our analysis indicates that the top two advanced digital skills for companies are Cybersecurity and Cloud, followed by Data Science for SMEs and AI for large companies, respectively.

Recognising the value of capturing a broader perspective, this research has extended its scope to include five non-EU countries (Australia, Canada, Singapore, UK and the USA), enabling the identification of common patterns, differences, and emerging trends in the global digital skills landscape.

2   Related Work

The digital transformation in Europe has undergone significant achievements and strategic changes in the last two decades. The first EU digital agenda, implemented between 2010 and 2020, focused on improving access to digital goods and services, ensuring lower prices for communications, better Internet connectivity, and consumer protection. It aimed to promote digital skills, High-Performance Computing, AI, and modernise public services [19].

The second EU digital agenda, covering 2020 to 2030, is more ambitious, aiming for technological and geopolitical advancements. Key priorities include Quantum Computing, Blockchain, AI, Cybersecurity, gigabit connectivity, 5G and 6G networks, European data spaces, and setting global technology standards. Specific targets for 2030 include promoting digital skills, business adoption of digital technologies, infrastructure development, and the online availability of public services [20].

The demand for advanced digital skills across the EU27 has arguably the potential to drive digital transformation and shape business models and workforce dynamics, as previously reported by the 2020 EU Joint Research Centre [21]. Digitalisation within a country facilitates the digital transformation of its society, leading to increased adoption of basic digital skills among the population, often facilitated by the availability of digital public services [2].

However, proficiency often requires a combination of technical skills, analytical thinking, domain knowledge, and problem solving abilities, highlighting the importance of fostering collaboration and interdisciplinary education to meet skills demands [1], the growing awareness of ethical and legal implications [11], and the relevance of lifelong learning, microcredentials, and adaptability for industry [15].

Companies (and organisations in general) without professionals with advanced digital skills may find it challenging to adopt and implement new technology-intensive projects effectively, potentially leading to decreased competitiveness, and this is especially true for SMEs. The most cited benefits for SMEs reported in the literature are efficiency and effectiveness, cost reduction, productivity growth, customer satisfaction, and competitive advantage when employing technologies such as social networks, websites, cloud computing, and data analytics adequately [17].

In the EU27, the digital skills gap can significantly impact overall economic growth and development. Regions and countries that effectively address their skills needs and reduce the gap will likely experience stronger economic growth, primarily driven by digital industries. In contrast, areas with persistent skills gaps may experience a digital divide, where certain regions or segments of society lag behind in digital capabilities, leading to significant economic disparities [10]. Therefore, addressing the digital skills gap is vital for bridging the digital divide and promoting inclusive economic growth [4]. However, recent research highlights that there is little agreement on a common taxonomy of skills [7], and most policy makers and academics should focus on those related to employability, management, career, and life skills [12,18].

2.1   Research Niche

While multi-institutional accredited Master's programmes in computing disciplines have been previously documented for Europe [3,8,14,23] and the rest of the world [5], the existence of advanced digital skills gaps emphasises the importance of implementing specialised upskilling and reskilling initiatives in the EU27. To address these gaps, European businesses and governments must arguably commit to investing in organically designed programmes that provide flexible accredited educational opportunities based on stakeholder consultations. These programmes must bridge the advanced skills gaps and equip individuals with the necessary competencies, enhancing their employability, and supporting organisations in meeting their digital transformation needs. Consequently, not only does this work furnish extensive FAIR-compliant [9] desk research and surveys to document the skills needed organically, but it also brings together industry, academia, and government to achieve long-term competitiveness and growth through digital transformation and innovation.

> This work therefore furnishes a systematic *Needs Analysis* to underpin the DIGITAL4Business European postgraduate education, which will focus on the practical application of Advanced Digital Skills within European Business.

Our ultimate aim is to seamlessly link four interdependent DIGITAL4Business phases: Needs Analysis, Programme Delivery, Impact Evaluation, and Financial Sustainability.

3   Method

The study employs a mixed methods approach, that is, comprehensive desk research and surveys of industry experts and professionals. Throughout the research process, **292 documents**[1] were analysed, providing comprehensive coverage of advanced digital skills in the EU27. An additional 24 documents were reviewed at the cross-European level, typically referring to advanced digital skill needs and gaps in more than one EU country. To gain insight into the needs and gaps in advanced digital skills in Australia, Canada, Singapore, the UK, and the USA, 69 documents were examined. These documents

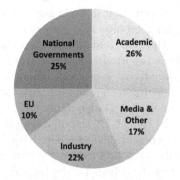

Fig. 1. Breakdown of reviewed publications for the desk research.

included government reports, industry publications, academic research, media articles, and other relevant sources. The general breakdown of the reviewed publications is presented in Fig. 1.

72 national and 29 European reports were identified and reviewed, offering key information and perspectives on the skill requirements and priorities set by government entities throughout the EU27. These reports served as important references for understanding the digital skills needs within each country.

---

[1] Due to page-limit constraints, the entire list of publications has not been included here but it is part of an open-access DIGITAL4Business project deliverable.

In addition, 61 industry publications were examined, providing industry-specific insight into the demand for advanced digital skills. These reports highlighted skill requirements and trends within different sectors, shedding light on skills valued by industries in the digital era. Academic publications played a significant role in providing scholarly research and analysis on advanced digital skills. 74 academic publications were analysed to collect evidence-based insights and deepen understanding of advanced digital skill needs. In addition, 47 media articles were reviewed to capture the broader discourse and public sentiment surrounding advanced digital skills. These articles provided valuable perspectives from journalists, experts, and commentators, enriching the overall analysis. In addition, nine documents were reviewed from international sources, including organisations such as the World Bank and the OECD that provide a global perspective on the needs and gaps in advanced digital skills.

Surveys were administered to European industry representatives from SMEs and Multinational Corporations (MNCs) to address the generalisation bias. They aimed to validate the key findings derived from the desk research, ensuring a more robust and representative analysis. These were i) applied to a targeted industry, academia, and government cohort; and, ii) issued on LinkedIn. The respondents were asked to identify the crucial digital skills for SMEs and businesses in general and rank advanced digital skills in priority order.

4 Findings

Our desk research uncovers a landscape of skills and technologies that have garnered significant attention across the EU27, identifying 18 areas listed here in alphabetic order: 1. AI; 2. Business Intelligence 3. Big Data & Analytics; 4. Cloud; 5. Cognitive Services; 6. Cloud Infrastructure DevOps; 7. Cybersecurity; 8. Data Governance; 9. Data Visualisation; 10. Digital Transformation; 11. Emerging Technologies; 12. Ethics; 13. Information Fusion; 14. Machine Learning; 15. Programming; 16. Project Management; 17. Research Methods; and 18. Statistics Fundamentals.

Our data analysis reveals a clear pattern indicating that **AI is Europe's most highly sought-**

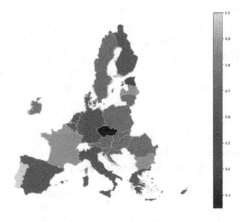

Fig. 2. EU27 heatmap depicting the coverage/need of subjects/clusters of digital skills per country. Higher coverage corresponds to warmer colours (in the yellow spectrum), lower coverage corresponds to colder colours (in the purple spectrum). (Color figure online)

after skill set, closely followed by
Data Science[2].

This observation underscores the significance of these two subjects or groups
and highlights a strong correlation between them. Machine learning also emerges
as a key skill that garners widespread attention in many countries. For Belgium, Cyprus, Denmark, Estonia, France, Germany, Italy, Latvia, Lithuania,
Malta, the Netherlands, Poland, Portugal, Slovakia, Slovenia, Spain, and Sweden, machine learning's significance is indisputable. The focus on Data Analytics and Big Data is strong across multiple countries, with Austria, Belgium,
Cyprus, Czech Republic, Denmark, Estonia, France, Germany, Greece, Ireland,
Italy, Lithuania, Luxembourg, Malta, the Netherlands, Poland, Portugal, Romania, Slovakia, Slovenia, Spain, and Sweden assigning high importance to it. This
finding suggests that organisations across Europe recognise the interdependence
and complementary nature of AI and Data Science, leveraging both skill sets to
drive innovation, enhance productivity, and remain competitive in their respective industries.

> It should be noted that the complete statistical analysis has not been
> included in this work due to space restrictions, but is available upon
> request.

The analysis also highlights other key skill sets, such as Cybersecurity, Cloud,
Blockchain, and Quantum Computing. Cybersecurity emerges as an important
skill set for most European countries, with 24 out of 27 countries citing its importance at least once. The importance of Cybersecurity is evident, attributed to
it by Austria, Belgium, France, Germany, Greece, Ireland, Italy, the Netherlands, Poland, Portugal, Romania, Slovakia, Slovenia, Spain, and Sweden. Cloud
computing, recognised by Austria, Belgium, Cyprus, Estonia, France, Germany,
Greece, Ireland, Italy, Latvia, Lithuania, Luxembourg, Malta, the Netherlands,
Poland, Portugal, Romania, Slovakia, Slovenia, Spain, and Sweden, has emerged
as a critical skill set. The demand for Blockchain and Quantum Computing
skills appears to be more specific, with 19 and 8 EU27 countries, respectively,
acknowledging their importance.

A further categorisation exercise was performed to assign each of the most
prominent advanced digital skills to its respective designated topic area, yielding
the following Subjects/Clusters categorisation listed in order of relevance:

1. *AI:* Ethics, Machine Learning, Natural Language Processing, Chat-
   bots/Robotics, Smart Sensors, and Digital Twins.
2. *Data Science:* Big Data & Analytics, Business Intelligence, Data Visualisation, Data Governance, and Information Fusion.
3. *Cloud:* Cloud Computing and Cloud Infrastructure/DevOps.
4. *Internet of Things (IoT):* Sensors, IoT, and 5G/6G.
5. *Cybersecurity:* Cybersecurity and Network Security with ties to the Cloud
   topic.

---

[2] **N.B.** We consolidated the areas into 8 subjects based on the results of the desk
research, bringing Data Science as one topic–comprising Big Data and Analytics,
Data Visualisation, and Governance–as referenced hereafter.

6. *Quantum Computing*
7. *Blockchain*
8. *Programming:* Python and Automation.

Geo-Spatial Analysis. Based on the Needs Analysis data collected and compiled for this analysis, we have studied the distribution of advanced digital skill needs across the EU27 included in the desk research. Figure 2 shows the coverage of the in-demand advanced digital skills across the EU27 included in the desk research. The coverage depicts the ratio among the number of Subject/Cluster cited at least once and the total number of Subject/Clusters in a given country. In particular, a higher coverage value indicates that most advanced digital skills are needed in those countries. The maximum possible value is 1, indicating that all the Subjects/Clusters are required in that country, although they may have different priorities. We notice that in most countries, there is significant coverage/need for advanced digital skills, with at least 5 Subjects/Clusters of skills cited at least once in each country.

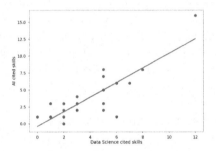

Fig. 3. Correlation between AI and Data Science Skills. It presents the absolute values of AI and Data Science skills per country. Each blue dot on the plot represents the number of times these skills were mentioned across all the documents collected during the desk research for each country. (Color figure online)

Correlation Analysis. As illustrated in Fig. 3, a robust correlation is observed between AI and Data Science skills. This correlation indicates that these subjects are closely related and suggests the possibility of merging them into the same cluster or offering them together. Additional evidence supporting this claim arises from the acknowledgement that modern Data Science heavily relies on Machine Learning, which is an integral component of AI. The desk research collected a total of 292 documents. We manually analysed these documents and counted 136 documents cited at least once, either requirement for AI or Data Science skills. Of these documents, 76 cited the need for both AI and Data Science skills. This denotes a strong statistical correlation between the skill sets; Pearson's correlation coefficient was measured 0.56. The correlation coefficient for the absolute counts is measured at 0.84, indicating a strong positive linear relationship between AI and the data science skills analysed. This implies that there is a tight clustering of data points around the best-fit line, *demonstrating a consistent and predictable increase in skill needs for both AI and Data Science.* As the demand for AI skills rises, the need for Data Science skills also tends to increase relatively consistently and predictably.

4.1 DIGITAL4Business Surveys

The survey sent to the targeted industry, academia, and government cohort yielded 18 responses. The same survey published on LinkedIn yielded an additional 12 curated responses from decision makers. All survey respondents were asked to rank advanced digital skills in priority order. As presented in Table 1, our results show that both the SMEs and the MNC respondents ranked Cyber-

Fig. 4. Transversal skills suggested by survey respondents.

security in the first position, followed by Cloud with Data Science in the third position for the SMEs and AI in third position for MNCs.

In addition, there was a focus on user experience design and the importance of general digital knowledge as a foundation for new technologies. Data Science, Programming, and AI were emphasised as critical skills, and the importance of cloud technologies and strong cybersecurity frameworks was recognised. Other areas mentioned included Blockchain, Quantum Computing, and Machine Learning. In general, respondents acknowledged the importance of understanding and applying cybersecurity to their businesses in conjunction with technological advances.

Table 1. Ranking Advanced Digital Skills for SMEs and MNCs.

Position	SMEs	MNCs
1st	Cybersecurity	Cybersecurity
2nd	Cloud	Cloud
3rd	Data Science	AI
4th	AI	Data Science
5th	IoT (joint position)	Programming
5th	Programming (joint position)	IoT
7th	Blockchain	Blockchain
8th	Quantum Computing	Quantum Computing

As shown in Fig. 4, the surveys also revealed the importance of complementary transversal skills such as critical and innovative thinking, teamwork, communication, self-discipline, and self-motivation. In response to the question, "*Are there any insights or issues related to Advanced Digital Skills you would like to share?*" the respondents touched on several important aspects. Notable concerns include the digital divide, digital literacy, the potential impact of automation and job displacement, the role of regulatory frameworks, and training.

Comparing EU27 with Findings Globally. The comprehensive assessment of the in-demand advanced digital skills in the EU countries provided a solid basis for comparing them with five selected international countries (i.e., Australia, Canada, Singapore, UK, and USA), revealing both commonalities and distinct skill requirements. The results indicated noteworthy similarities between the in-demand advanced digital skills of the EU countries and those sought after in the five global countries, along with a few differences. Although there was a high demand for advanced digital skills in both the EU and globally, specific skill preferences varied across regions. The findings of this research have significant implications for policymakers, educators, and industry leaders, emphasising the importance of adopting a targeted approach to skills development that considers national and global demands.

4.2 Demand, Challenges and Unique Factors in Advanced Digital Skill Needs

The analysis of the demand for advanced digital skills in the EU27 revealed similarities, while also highlighting differences in the level of demand. These variations can be attributed to several factors.

Firstly, disparities in economic development and technological advancement play a significant role. Countries with more advanced digital infrastructure and a higher concentration of technology-driven industries exhibit a greater demand for advanced digital skills such as AI, IoT, Blockchain, Data Science, Cloud Computing, and Cybersecurity. Notable examples include France, Germany, Italy, Sweden, the Netherlands, Finland, Denmark, and Ireland. Differences in national policies, priorities, and investments in digital transformation initiatives also contribute to the divergence in skill needs between countries. These factors collectively shape the unique landscape of advanced digital skills requirements in each of the EU27 countries.

Another factor highlighted in the desk research is evidence that large enterprises are more likely to embrace new technologies compared to SMEs. This discrepancy in the adoption rates of SMEs can be attributed to various factors, including lower financial resources, organisational capacity, and their reduced ability to navigate the complexities of technological implementation [6]. For example, Cloud has emerged as a game changer for enterprises, offering scalable and cost-effective data storage, processing, and software delivery solutions. Large European companies are leading the way in cloud adoption, benefiting from enhanced operational efficiency and agility. However, SMEs have a lower adoption rate, indicating the existence of barriers such as limited awareness, concerns about data security, and a lack of resources to facilitate the transition.

Another example of this is given by Big Data Analytics which has revolutionised the way enterprises extract insights and make data-driven decisions. The current landscape shows that large enterprises have a more significant presence in Big Data processing, while SMEs lag behind. This discrepancy can be attributed to the complexity of implementing Big Data Analytics, the availability of skilled personnel, and the initial investment required. Looking at big data adoption internationally, we observe diverse patterns across countries. In Malta, nearly a third of enterprises analyse big data, indicating a relatively high adoption rate. The Netherlands and Denmark follow closely behind with adoption rates of 27%. However, countries such as Romania, Slovakia, Cyprus and Bulgaria have much lower adoption rates, ranging from 5% to 6%. These disparities reflect differences in digital readiness, infrastructure, and the availability of skilled professionals.

The provision of efficient digital infrastructures plays a crucial role in this digitalisation process and is a key focus for several EU27 countries, as evidenced by the desk research. Countries such as Romania, Luxembourg, Lithuania, Latvia, the Czech Republic, and Bulgaria, among others, have prioritised rolling out digital infrastructures and subsequently expanding their e-government digital services. These initiatives reflect the commitment of these countries to improve their digital landscapes and to promote widespread digitalisation within their societies. Moreover, countries prioritising specific digital domains, such as AI or Cybersecurity, tend to witness a higher demand for the corresponding skills.

5   Recommendations

Flexible and Accessible Education Models. Our research has shown a demand for advanced digital skills in the areas of AI, Data Science, Cloud, IoT, Cybersecurity, Quantum Computing, Blockchain, and Programming in the EU27. Bridging this gap requires promoting flexible and accessible education and training models that cater to individuals of all backgrounds and career stages. This can be achieved through initiatives such as online courses, micro-credentials, and lifelong learning. In addition, fostering international cooperation between academic institutions can provide a platform for flexible and accessible education and training, particularly aimed at SMEs. This cooperation can help promote knowledge exchange on digital skill development, enabling SMEs to learn from best practices implemented in other countries and to collaborate on joint initiatives, research projects, and standardisation efforts.

Ex-professo AI and Data Science Modules. Based on the correlation between AI and Data Science, we consider that both areas of learning should be taught in modules designed *ex professo*–ie in the capacity of an expert–to fulfil commercial needs.

Stakeholder Collaboration. It is necessary to foster collaboration among policymakers, educational institutions, businesses, industry associations, and SME networks and to work together to promote flexible and accessible education and training models that meet the unique needs of all learners, particularly SMEs. Highlighting the importance of lifelong learning and providing resources that enable SMEs to continually up-skill their workforce are also very important. This is in line with the views expressed in the Digital Skills Gap white paper [2, p. 74], "digital competence development necessitates an 'all-government' strategy that encompasses digital competence development in areas such as employment, the labour market, education and training, social services, and economic growth. Governments, industry, education and training providers must collaborate and share information about current and future needs for digital talent."

Diversity and Inclusion Initiatives. It is also necessary to be mindful of diversity and inclusion in digital skill development programmes. It is essential to provide equal opportunities for training and educational resources to individuals of diverse backgrounds, such as women, minorities, and individuals with disabilities. In addition, promoting diversity in hiring practices can lead to innovation and bring different perspectives to the digital workforce.

6   Conclusions

The research findings have significant implications for policymakers, educators, and industry leaders. Adopting a targeted approach to skills development that considers national and European demands is crucial. This includes promoting flexible and accessible education and training models and micro-credentialing to cater to individuals of all backgrounds and career stages.

Collaboration among policymakers, educational institutions, businesses, industry associations, and SME networks is essential to promote accessible and flexible education and training models. Lifelong learning should be emphasised and resources should be provided to enable SMEs to continuously upgrade their workforce. It is also essential to prioritise diversity and inclusion in digital skill development programmes to provide equal opportunities for individuals from all backgrounds.

Although this research has provided valuable information on digital skills needs and gaps in the EU27, areas still warrant further exploration. Future research could focus on specific digital skill requirements and gaps within different industries or sectors. Furthermore, research on the effectiveness of different upskilling and reskilling initiatives in addressing the digital skills gap would benefit policymakers and educators.

In conclusion, advanced digital skills are essential for individuals and organisations to thrive in the digital age. The DIGITAL4Business project is at the forefront of revolutionising digital transformation and upskilling in Europe. By addressing the digital skills gap and fostering collaboration between industry and academia, the project aims to shape the future of advanced digital skills in Europe and promote economic growth and innovation. To remain competitive in the digital age, people must continuously update their skills and embrace lifelong learning. Policymakers, educators, and industry leaders must work together to provide flexible and accessible educational and training opportunities that meet the diverse needs of learners. By doing so, we can ensure that individuals and organisations are equipped with the advanced digital skills necessary to succeed in the digital landscape.

In general, our research findings underscore the transformative potential of AI, Data Science, Cloud, IoT, Cybersecurity, Quantum Computing, Blockchain, and Programming. They highlight the need for individuals, organisations, and society to adapt, collaborate, and embrace lifelong learning to fully leverage the benefits of these emerging technologies while addressing the associated challenges.

Acknowledgements. This work has been developed under the auspice of "DIGITAL4Business: Master's Programme focused on the practical application of Advanced Digital Skills within European Companies" URL: www.digital4business.eu, a project funded from Dec/2022 to Nov/2026 by the European Commission's DIGITAL programme call: DIGITAL-2021-SKILLS-01 grant no.: 101084013.

References

1. Akter, S., Michael, K., Uddin, M.R., McCarthy, G., Rahman, M.: Transforming business using digital innovations: the application of AI, blockchain, cloud and data analytics. Ann. Oper. Res. **308**, 7–39 (2022)
2. All Digital: Strategies to address the digital skills gap in the EU. Whitepaper, Huawei, April 2022. https://www.europeandigitalskills.eu/sites/TDSG/uploads/files/white-paper-eu-digital-skills-gap.pdf. Accessed 25 Feb 2024
3. Barbosa, F., Guincho, H., Leite, F.B., Nunes, J.L., Pereira, C.: European computer science master curriculum development methodology: management and research. In: 2011 NWeSP, pp. 487–492. IEEE, Salamanca, December 2011
4. Bentley-Gockmann, N., Thompson, L.: Disconnected?: exploring the digital skills gap. Worldskills UK report, WorldSkills UK, Learning and Work Institute, and Enginuity, March 2021. https://www.worldskillsuk.org/wp-content/uploads/2021/03/Disconnected-Report-final.pdf. Accessed 25 Feb 2024
5. Bozanic, M., Sinha, S.: A survey of current trends in master's programs in microelectronics. IEEE Trans. Educ. **61**(2), 151–157 (2018)
6. Eller, R., Alford, P., Kallm'nzer, A., Peters, M.: Antecedents, consequences, and challenges of small and medium-sized enterprise digitalization. J. Bus. Res. **112**, 119–127 (2020)
7. Garcia-Esteban, S., Jahnke, S.: Skills in European higher education mobility programmes: outlining a conceptual framework. High. Educ. Skills Work-Based Learn. **10**(3), 519–539 (2020)
8. González-Vélez, H., Dobre, C., Sánchez-Solis, B., Antinucci, G., Feenan, D., Gheorghe, D.: Open science and research data management: a FAIR European postgraduate programme. In: 2022 Big Data, pp. 2522–2531. IEEE, Osaka, December 2022
9. González-Cebrián, A., Bradford, M., Chis, A.E., González-Vélez, H.: Standardised versioning of datasets: a FAIR-compliant proposal. Sci. Data **11**(358), 1–15 (2024)
10. Helsper, E.J., van Deursen, A.J.A.M.: Digital skills in Europe: research and policy. In: Digital Divides: The New Challenges and Opportunities of e-Inclusion, Public Administration and Public Policy, vol. 195, Book Chapter 7, pp. 125–144. CRC Press, Boca Raton (2015)
11. Kendal, E.: Ethical, legal and social implications of emerging technology (ELSIET) symposium. J. Bioeth. Inq. **19**, 363–370 (2022)
12. Khampirat, B.: Relationships between ICT competencies related to work, self-esteem, and self-regulated learning with engineering competencies. PLoS ONE **16**(12), e0260659 (2018)

13. van Laar, E., van Deursen, A.J.A.M., van Dijk, J.A.G.M., de Haan, J.: Determinants of 21st-century skills and 21st-century digital skills for workers: a systematic literature review. SAGE Open **10**(1), 2158244019900176 (2020)
14. Lago, P., Muccini, H., Beus-Dukic, L., Crnkovic, I., Punnekkat, S., Van Vliet, H.: Towards a European master programme on global software engineering. In: CSEET 2007, pp. 184–194. IEEE, Dublin, July 2007
15. Lang, J.: Workforce upskilling: can universities meet the challenges of lifelong learning? Int. J. Inf. Learn. Technol. **40**(5), 388–400 (2023)
16. Pausits, A., Kivisto, J., Pekkola, E., Reisky, F., Mugabi, H.: The impact of human resource management policies on higher education in Europe. In: Research Handbook on Academic Careers and Managing Academics, Handbook Chapter 18, pp. 251–267. Sociology, Social Policy and Education. Edward Elgar Publishing, Cheltenham (2022). https://doi.org/10.4337/9781839102639
17. Pfister, P., Lehmann, C.: Returns on digitisation in SMEs-a systematic literature review. J. Small Bus. Entrep. **35**(4), 574–598 (2023). https://doi.org/10.1080/08276331.2021.1980680
18. Picatoste, J., Pérez-Ortiz, L., Ruesga-Benito, S.M.: A new educational pattern in response to new technologies and sustainable development. Enlightening ICT skills for youth employability in the European Union. Telemat. Inform. **35**(4), 1031–1038 (2018)
19. Publications Office of the European Union: A Digital Agenda for Europe. EUR-Lex J. Small Bus. Entrep. Document 52010DC0245(COM(2010) 245 final), 1–42, May 2010. https://eur-lex.europa.eu/legal-content/en/ALL/?uri=celex:52010DC0245. Accessed 25 Feb 2024
20. Publications Office of the European Union: Shaping Europe's digital future. EUR-Lex Document 52020DC0067(COM(2020) 67 final), 1–16, February 2020. https://eur-lex.europa.eu/legal-content/en/TXT/?uri=CELEX:52020DC0067. Accessed 25 Feb 2024
21. Righi, R., et al.: Academic offer of advanced digital skills in 2019-20. International Comparison. Focus on Artificial Intelligence, High Performance Computing, Cybersecurity and Data Science. JRC Research Reports JRC121680, European Commission Joint Research Centre (Seville site), September 2020. https://ideas.repec.org/p/ipt/iptwpa/jrc121680.html. Accessed 25 Feb 2024
22. Taylor, P.J., O'Driscoll, M.P., Binning, J.F.: A new integrated framework for training needs analysis. Hum. Res. Manag. J. **8**(2), 29–50 (1998)
23. Zwolinski, M., Kunz, W., Svarstad, K., Brown, A.: The European masters in embedded computing systems (EMECS). In: 2016 EWME, pp. 1–6. IEEE, Southampton, May 2016

WideAIED

Predictive Model of School Dropout Based on Undergraduate Course Self-assessment Data

Ronei Oliveira and Francisco Medeiros[(✉)]

Federal Institute of Paraiba, Joao Pessoa, PB, Brazil
ronei.santos@academico.ifpb.edu.br, petronio@ifpb.edu.br

Abstract. School dropout is a daily challenge for educational institutions. In the specific case of higher education, high dropout rates lead to financial losses and a shortage of professionals in the market. This research aimed to develop and evaluate a predictive model to identify students prone to dropout, using data from a semester-based self-assessment model of undergraduate courses at Federal University of Paraiba (UFPB). The study analyzed the relationship between school dropout and institutional self-assessment using educational data mining and the CRISP-DM methodology, followed by exploratory analysis and data preparation for classification. Various modeling techniques, such as Decision Trees, Random Forest, and Support Vector Machines, were applied, with the models evaluated using performance metrics, revealing an accuracy of 87.97%, precision of 91.72%, recall of 91.67%, and F-measure of 91.57% in identifying students with a high probability of dropout. Approximately 59% of active students at UFPB, admitted from 2017 onwards, showed a high probability of abandoning their courses in the proposed predictive model tests. This information can support institutional decisions and the implementation of effective policies and actions against dropouts, aiming to improve academic outcomes. The study contributes to advancements in predicting school dropout, providing valuable insights for decision-making and preventive strategies at UFPB and other higher education institutions.

Keywords: School dropout · Educational data mining · Predictive model · Self-evaluation

1 Introduction

Despite various observations and conceptions regarding school dropout, comprehending its reasons remains challenging for most institutions. To gain a deeper understanding, the Federal University of Paraiba (UFPB) conducts a compulsory and anonymous self-assessment of undergraduate programs each semester. This assessment provides a broad spectrum of data that can be explored to understand the institution's specificities. Self-assessment in Higher Education Institutions (HEIs) is an ongoing process of institutional reflection that extends beyond being accountable to the Ministry of Education. The relationship between evaluation and management varies depending on the mission and characteristics of each HEI. In the case of UFPB, using course self-assessment data can inform the implementation of specific actions to enhance teaching quality and combat school dropout [1].

© The Author(s), under exclusive license to Springer Nature Switzerland AG 2024
A. M. Olney et al. (Eds.): AIED 2024 Workshops, CCIS 2150, pp. 195–207, 2024.
https://doi.org/10.1007/978-3-031-64315-6_15

According to [2], Educational Data Mining (EDM) involves the application of data mining methods and Machine Learning (ML) within the realm of education. Its objective is to unearth knowledge from databases associated with educational contexts [24]. The impetus for this study lies in the potential of employing EDM on undergraduate course self-assessment data as a crucial tool to support senior management in implementing targeted actions that can assist students in completing their programs within the time-frames specified in the pedagogical plan. Subsequently, this aids in students receiving a quality education, facilitating their integration into the labor market or continuing their academic endeavors. This work also aims to make contributions in the realms of education, scientific-technological advancement, and society. In the educational sphere, it seeks to provide insights for institutional management, enabling actions that enhance the quality and effectiveness of the education offered. From a scientific-technological perspective, the research and utilization of computational resources in predictive model development leave a legacy for this field. In terms of societal impact, the reduction of school dropout rates aids students in fulfilling their academic journey, aligning with the expectations of both society and the academic community.

Furthermore, undergraduate course self-assessment can contribute to academic and administrative processes by providing a tool for adjusting goals and objectives. However, it is essential to emphasize that more than merely data collection and dissemination is needed for improving educational quality. It is necessary to process these data using cognitive and computational intelligence to generate valuable insights for the institution. Within this context, this work aims to develop a predictive model that utilizes the self-assessment data of UFPB's undergraduate courses to identify school dropouts early, enabling preventive measures by stakeholders such as senior management, course coordinators, and faculty. To achieve these objectives, the structure of the paper is organized as follows: Sect. 2 presents related works to identify the state-of-the-art and relevant research aligned with the research topic. Section 3 outlines the research methodology used to propose the prediction model. Section 4 presents the obtained results. Finally, in Sect. 5, concluding remarks are made, and directions for future research are provided.

2 Related Work

[11] conducted research that applied EDM techniques and Machine Learning (ML) to identify at-risk students at a South African University of Technology. They used students' final grades, enriching the dataset with 19 attributes derived from the institution's database. To address data imbalance, they performed undersampling, resulting in a dataset of 1.156 records. Subsequently, they applied various supervised classification algorithms, with the Support Vector Machine (SVM) demonstrating the best performance, achieving an F-Measure score of 99.32%. This methodology contributes to current research by providing insights into models with balanced data and highlighting the superior performance of SVM.

[21] research conducted at a Brazilian higher education institution utilized data from seven attributes to construct ten data models corresponding to different semesters. Using the Decision Tree (DT) algorithm, accuracy ranged from 79.31% to 98.25%, with some models achieving 100% accuracy for predicting dropouts. However, the accuracy for

dropout predictions varied across semesters, with the 3rd-semester model exhibiting the lowest accuracy. Precision for the Graduation class varied from 75.68% to 97.22%, with some models achieving 100% recall for graduation predictions. The predictive power of the models improved as semesters progressed due to the inclusion of more attributes. These findings contribute to developing dropout prediction models based on undergraduate course data, shedding light on the relationship between academic performance and dropout or graduation probabilities across different semesters.

[14] study investigates three approaches to school dropout prediction in a Brazilian higher education institution. The first approach uses generic variables applicable to any course, while the second utilizes course-specific data. Both approaches employ Machine Learning algorithms such as Gradient Boosting Tree, Support Vector Machine (SVM), Random Forest (RF), and Naive Bayes. The third approach analyzes the evolution of student data throughout the course, employing the K-Nearest Neighbors algorithm. Results show that the model utilizing global features with the RF algorithm achieves the best accuracy, recall, precision, and F-Measure metrics, highlighting the superior performance of this approach with course-independent generic variables.

This work advances compared to related studies, especially concerning the data context. Predictive research on school dropout based on data from a unique undergraduate course self-assessment instrument can offer substantial scientific advancements by providing a deeper understanding of the factors and patterns leading to dropout. Using data mining techniques and predictive models, this research can identify hidden correlations and risk factors, offering valuable insights for developing more effective intervention strategies. These scientific advancements contribute to the field of education, potentially improving student retention and aiding in developing more personalized and evidence-based approaches to address the school dropout challenge.

3 Methodology

To achieve the established objectives, an Educational Data Mining (EDM) process was conducted, involving the application of data mining and machine learning methods in an educational context to discover knowledge from educational databases [22]. To assist in this process, this research employed the CRISP-EDM methodology (an acronym for CRoss-Industry Standard Process for Educational Data Mining), an adaptation of the CRISP-DM methodology (an acronym for CRoss-Industry Standard Process for Data Mining) for the educational context. Each stage of CRISP-EDM was developed with techniques and approaches suitable for the educational domain under analysis.

3.1 Self-assessment Instrument

The first stage of the process involved understanding the application domain, which means comprehending the context of the educational data to be mined and its correlation with school dropout. The raw data obtained stem from undergraduate course self-assessment, carried out by students at UFPB at the beginning of each semester. The self-assessment is conducted based on the previously completed semester, mandatorily and as a prerequisite for enrollment in the subsequent semester.

The instrument for student self-assessment in higher education comprises four dimensions [5]. In the student dimension, students self-assess their performance in each course, assigning a score from 0 to 10 for their commitment and motivation. In the course dimension, students evaluate each course's perceived importance and difficulty using a 0 to 10 scale. In the faculty dimension, an assessment is made regarding adjustments needed by professors, considering aspects such as course alignment, class interaction, and attendance. Furthermore, students rate their overall satisfaction with each professor's performance, assigning a score from 0 to 10. Lastly, in the course dimension, students can indicate the likelihood of recommending the course to a friend or close relative using a 0 to 10 scale and express their interest in dropping out of the current course using the same scale.

Another relevant aspect of model development is that this instrument was only applied in this format starting from the first semester of 2017 for all undergraduate courses at the institution. In this context, the analyzed data only included students enrolled from that semester onwards. This approach was adopted because students who did not complete assessments throughout their academic cycle generated biased data that hindered the model's efficiency. This fact was verified throughout the CRISP-EDM cycle, leading to refinements in the model.

A total of 1.156.891 records of evaluations for on-site undergraduate courses were analyzed. The primary focus of this stage was on the target variable, STUDENT_STATUS, that presents these possible: ACTIVE, SUSPENDED, CONCLUDED, ACTIVE – GRADUATING and CANCELED. A noteworthy point about the target variable is that its value corresponds to the student's status at the time of data extraction from the system. In other words, the status does not reflect the student's status at the evaluation time. Therefore, there are periods with evaluations for a student who had not dropped out at that time. For example, if a student attended five periods before dropping out, all evaluations for the disciplines in those five periods will have the CANCELED status as the target variable value.

3.2 Data Preparation

Based on previous analyses and refinements made after completing several cycles of CRISP-EDM, some decisions were made to generate a subset of data capable of performing well in the modeling phase.

Filtering
We conducted a filtering process regarding the year of student enrollment. Because self-assessment implementation began in the first semester of 2017, we decided to utilize evaluations from students who enrolled from that period onwards, as they provide a comprehensive representation of their entire academic journey up to the data extraction point. The same cannot be said for enrollments before 2017, as current records do not contain evaluations for several courses taken before the self-assessment instrument was applied. Therefore, the new raw dataset contained 683.634 records, with 473.257 records related to enrollments before 2017 being excluded.

Another filtering step was performed regarding our target variable, STUDENT_STATUS. Since the statuses ACTIVE and SUSPENDED pertain to students still affiliated with the institution, it is impossible to determine whether these students will drop out. Given this situation, only the values CONCLUDED, ACTIVE–GRADUAT-ING, and CANCELED (dropout) were considered for training and testing the developed models because these statuses are conclusive and confirm whether the student completed their academic journey or left the course. With the application of this filter, our raw dataset was reduced to 128.235 records. The other 555.399 records related to non-conclusive statuses are not used in the modeling. Despite the difference between the data used in modeling and those not used, there is still a substantial amount of data to obtain a highly representative model.

All variables were checked for the presence of null values and outliers. After removing all outliers, our subset of data had 120.341 evaluation records. It should be noted that these data still have inconsistencies. A single registration has multiple records in this dataset, meaning that each discipline taken in different periods represents a record. However, as mentioned earlier, all records have the final status of the student at the time of data extraction as the value of their target variable. Given this context, transformations were necessary for this data.

Transformations
The target variable STUDENT_STATUS was transformed into numerical values so that machine learning algorithms could perform better. The statuses CONCLUDED and ACTIVE–GRADUATING assumed the value 0, and CANCELED assumed the value 1. Completing the modifications mentioned above, the need arose to address multiple records for a single registration, corresponding to different self-assessments for each discipline per period. Considering that prediction is executed for each entry in the dataset, the predictive model could deduce both dropout and retention situations for a single registration. To address this challenge, we adopted an approach wherein we calculated an average for each predictive variable, encompassing all self-assessments linked to a particular registration. This enabled the consolidation of a singular record for each student, incorporating the average that summarizes their complete academic progression.

This procedure established an individualized foundation for predicting dropout rates. Furthermore, a critical step entailed anonymizing all registrations. This action was implemented to safeguard the privacy of the students who contributed to building the prediction model. As a result of these transformations, we obtained a final subset containing 6.138 records. These records represent the arithmetic mean per registration derived from the 120.341 records resulting from the initial data processing.

3.3 Modeling

In this phase, data modeling techniques were defined, specifically a set of machine learning algorithms and their parameter adjustments. During the attribute selection process, the Filter, Wrapper, and Embedded methods were employed over several cycles of the CRISP-EDM methodology. The Embedded method using the Random Forest algorithm yielded the best results for all developed models.

The most important attribute corresponds to the student's own expression of their intention to drop out. However, relying solely on this information would be insufficient to predict dropout, as all attributes bear some degree of importance in the classification task.

The definition of weights/importance is crucial for the algorithm to train the machine learning model more effectively. The methodology's results evaluation phase validates the trained model and assesses whether the attribute selection has resulted in a good performance of the classifier model in predicting dropout. This process is cyclical and continuously refined, as demonstrated by the CRISP-EDM methodology.

Data Splitting

To analyze the developed models, the Holdout technique was used on the dataset because it allows for assessing the model's performance based on the evaluation cut. For this analysis, the data was divided into a random test subset with 30% of the records (1.842) and a training subset with 70% of the records (4.296). Out of the 4.296 examples in the training data subset, 3.026 belong to the majority class (CANCELED), and 1.270 represent the minority class (CONCLUDED). The test data is represented by 1.288 (CANCELED) and 554 (CONCLUDED). To validate the generalization power of the developed model, cross-validation of the data was performed. In this way, different configurations of training and test data are ensured, resulting in a total of k repetitions between training and test subsets. A good alternative is to set the value of k to 10 [8].

Data Imbalance

We focused on examining the training data for data imbalance during the classification task. There were more records of dropout students (3.026) than graduates (1.270). Using the Imbalanced-learn library, we implemented models using balanced training data through undersampling and oversampling techniques and compared them to models using unbalanced data. In the case of undersampling, we generated a random undersample from the majority class while preserving the minority class.

In the case of oversampling, a random oversample was generated by replicating records from the minority class (Random Oversampling), and two other oversamples were generated using the SMOTE and ADASYN techniques. With this scenario, it was possible to assess how the algorithms behaved and which approach was more suitable for the research objective: to detect students at higher risk of dropout.

4   Results and Discussions

In the tests conducted in this study using different machine learning algorithms, various result evaluation metrics were employed. These predictions are typically represented in the form of a confusion matrix, where "Yes" corresponds to CANCELED (dropout) and "No" corresponds to CONCLUDED (non-dropout).

Unbalanced Data

The first model developed used unbalanced training data, allowing us to observe how machine learning algorithms performed regarding data imbalances.

As shown in Table 1, the Random Forest algorithm achieved the best results in all metrics when using the Holdout technique for a randomly chosen single training and testing dataset. In the Random Forest algorithm, within a real-world test dataset of 1.288 evaded students (CANCELED), the model had only 68 False Negatives (FN), achieving a recall of 94.72%. The Decision Tree model classified 163 as FN and showed an increase

Table 1. Metrics with unbalanced data using a random dataset.

	Accuracy	Precision	Recall	F- Measure
Decision Tree	0.82193268	0.87209302	0.87344720	0.87276958
Random Forest	**0.88599348**	**0.89574155**	**0.94720496**	**0.92075471**
Support Vector Machines	0.83713355	0.88413685	0.88276397	0.88344988

in False Positives (FP) with 165 cases. The SVM model outperformed the Decision Tree model, with 151 FN and 149 FP.

For a more accurate validation, the k-fold cross-validation technique was applied using 10 folds with the StratifiedKFold algorithm. In this method, the data is divided into k = 10 subsets. Then, the holdout method is repeated k times, where each time, one of the k subsets is used as a validation set, and the other k-1 subsets are combined to form a training set. The average of the results from these evaluations is presented in Table 2, with the Random Forest algorithm achieving the best results in all metrics.

Table 2. Average Results of Metrics with Unbalanced Training Data Using Cross-Validation Technique.

	Accuracy	Precision	Recall	F- Measure
Decision Tree	0.80448055	0.86616020	0.85489494	0.86028856
Random Forest	**0.87827977**	**0.90075786**	**0.93139125**	**0.91542677**
Support Vector Machines	0.83071188	0.88979089	0.86647439	0.87783455

Despite the satisfactory results, where an average Recall of 93.13% was achieved, we must be attentive to the model's specialization regarding the majority class (CANCELED). As we can observe, the model exhibits a significant number of False Positives (FP) relative to the minority class. Therefore, if we had a larger number of test cases for the minority class (CONCLUDED), the tendency is that the Precision metric would decrease proportionally, consequently reducing the F-Measure. In this context, although we have a model with a low number of False Negatives (FN), we could face another issue: having a significant number of students classified as having the potential for school dropout, leading to a significant loss of institution resources by concentrating stakeholders' efforts on mitigating dropout cases with False Positives. The evaluation by cross-validation applied to this model is the same as that applied to the other models and is the metric used for deciding which model to propose.

Balanced Data: Random Undersampling
In this classification task, balanced training data was used through the method of random undersampling, meaning a subset was randomly selected from the majority class

(CANCELED). Thus, the majority class had 1.270 examples, equal to the minority class. Table 3 provides the evaluation of the models developed with training data balanced by the random undersampling method. The model that yielded the best result, using the Holdout technique for a random training data (70%) and test data (30%) set, was the Random Forest.

Table 3. Metrics with data balanced by random undersampling using a random set.

	Accuracy	Precision	Recall	F- Measure
Decision Tree	0.76981541	0.89059674	0.76475155	0.82289055
Random Forest	**0.87133550**	**0.93974895**	**0.87189440**	**0.90455094**
Support Vector Machines	0.82138979	0.93159315	0.80357142	0.86285952

For a more comprehensive evaluation of the models, cross-validation was applied to the models with training data balanced by random undersampling. Table 4 presents the mean of the evaluations.

Table 4. Mean results of metrics with training data balanced by random undersampling using cross-validation.

	Accuracy	Precision	Recall	F- Measure
Decision Tree	0.79339979	0.90131798	0.76475155	0.82289055
Random Forest	**0.86068648**	**0.93974895**	**0.87189440**	**0.90455094**
Support Vector Machines	0.80870923	0.93159315	0.80357142	0.86285952

Analyzing the metrics obtained, a nearly inverse pattern of results between Recall and Precision metrics is observed. This phenomenon can be explained for the same reasons as in the evaluation with unbalanced training data. Due to the loss of data in the majority class, the machine learning algorithm became more inclined to classify overlaps of classes as the minority class. This is in contrast to the previous case, where due to the overfitting of the majority class, overlaps tended to be classified as the majority class. Other undersampling methods, such as Edited Nearest Neighbors (ENN), were tested; however, they showed lower performance than random undersampling and experienced the same classification issues regarding class overlaps. Given this context, the decision was made to explore balanced data through oversampling methods.

Balanced Data: SMOTE Oversampling
Using the Synthetic Minority Oversampling Technique (SMOTE), which generates new instances of the minority class through interpolation between the nearest points, resulted in improved model performance compared to other models. This makes the model with

the Random Forest algorithm the most balanced among the models analyzed. When analyzing the results obtained (Table 5), we observe that this balance is reflected in the Recall and Precision metrics, resulting in an alignment of metrics regarding the classification of class overlaps. This reflects a more accurate model in predicting school dropout.

Table 5. Metrics with data balanced by SMOTE oversampling using a random dataset.

	Accuracy	Precision	Recall	F- Measure
Decision Tree	0.81704668	0.86831913	0.87034161	0.82289055
Random Forest	**0.89033659**	0.91387195	**0.93090062**	**0.90455094**
Support Vector Machines	0.81921824	**0.93212669**	0.80357142	0.86285952

For a more comprehensive evaluation of the models, cross-validation was applied to the models with training data balanced by SMOTE oversampling. Table 6 presents the mean of the evaluations.

Table 6. Average results of metrics with training data balanced by SMOTE oversampling using cross-validation.

	Accuracy	Precision	Recall	F- Measure
Decision Tree	0.80578029	0.88209109	0.84863098	0.86641259
Random Forest	**0.87974371**	0.91724143	**0.91679234**	**0.91574044**
Support Vector Machines	0.80968616	**0.93308219**	0.78580819	0.85346256

The proposed model was applied to the active student database, utilizing the 555.399 records related to non-finalizing statuses that were not used in the modeling process. After applying the same steps as in the modeling phase, 525.694 processed records were obtained. Evasion prediction was then performed on the data containing the averages of these processed records, resulting in 22.560 average self-assessments for each distinct enrollment, using the model with the Random Forest algorithm and training data balanced by SMOTE oversampling.

Out of the 22.560 active enrollments since 2017, the model classified 13.310 as CANCELED and 9.250 as CONCLUDED. Therefore, according to the predictive model, 59% of students have the potential for school dropout. When compared with the higher education enrollment trajectory indicators in the state of Paraiba, "Map of Higher Education in Brazil – 13th Edition (2023)", as shown in Fig. 1, this percentage aligns with the current scenario in the state.

Based on these data, it can be asserted that the proposed model has the potential to be used in the implementation of educational solutions that assist stakeholders in

Fig. 1. Trajectory Indicators in the state of Paraiba.

decision-making, resulting in interventions to improve the educational process and miti-gate dropout, thereby anticipating issues before they become irreversible. By identifying students at risk of dropout based on prediction model indicators, institutions can inter-vene promptly and offer personalized support. This may include additional tutoring, academic counseling, or the implementation of guidance and emotional support pro-grams. However, the effective implementation of these solutions extends beyond the scope of this research, as it needs to be validated and proposed by the institution's top management.

5 Conclusion and Future Work

In this work, various classification approaches for predicting school dropout were pre-sented and implemented, considering the interpretation of data from the self-assessment of undergraduate courses at UFPB. To achieve this goal, the CRISP-EDM methodol-ogy was employed to guide the data mining process. Each of these approaches differed from one another in both the training data balance and the machine learning algorithms they comprised. To validate the different classification approaches, accuracy, precision, recall, and F-measure metrics were used.

Based on the results of these metrics, a prediction method was proposed based on the Random Forest algorithm with data balancing using the SMOTE oversampling technique, achieving 87.97% accuracy, 91.72% precision, 91.67 recall, and 91.57 F-measure. Furthermore, the proposed method was applied to the data from active students' self-assessments, and the results were compatible with current dropout rates in the state when considering the ratio of enrollees and graduates in the same year.

The main conclusions and considerations related to the experiments are that it is possible to predict school dropout at the institution accurately using self-assessment data from undergraduate courses. However, the model did not perform well for students who enrolled before 2017, possibly due to the lack of inclusion of previously taken courses in the self-assessment instrument. It was observed that questions directly related to school dropout were the most important for classification, particularly the question where the student expresses their intention to leave the course, and the question where the student evaluates their personal performance in the discipline. According to the predictive model, of the 22.560 active students who enrolled at the institution from 2017 onwards, 59% are likely to drop out, reflecting the current situation indicated by the Map of Higher Education in Brazil.

The work undertaken raises new challenges and opportunities that drive the need for further investigation. To this end, it is important to explore the possibility of identifying whether specific courses have a significant impact on school dropout, allowing for the creation of more precise and targeted models. Furthermore, it is essential to investigate whether course-specific models can outperform the initially proposed generic model. Another aspect to consider is the incorporation of socioeconomic, academic, cultural, and other data from previous studies to enrich the predictive model and obtain more accurate results.

One way to facilitate data processing and model application is to create an Application Programming Interface (API), enabling the automation of the process and integration with other systems. Additionally, it is crucial to collect and incorporate updated data from subsequent semesters into the model, ensuring its continuous improvement and accuracy over time. To provide relevant real-time insights to stakeholders, integrating the developed API with SIGAA (Academic Management System) of the educational institution is vital. This will offer valuable information for making strategic decisions and implementing specific actions to mitigate school dropout. Lastly, the development of interactive dashboards presenting school dropout predictions for each semester will allow stakeholders to visualize trends and patterns, aiding in policy formulation and process improvement based on this information.

These future works aim to enhance the school dropout prediction model by increasing its accuracy, incorporating relevant data, automating data processing, integrating it with existing systems, and providing actionable insights to combat school dropout effectively.

References

1. Baggi, C.A.D.S., Lopes, D.A.: Evasão e avaliação institucional no ensino superior: uma discussão bibliográfica. Avaliação: Revista da Avaliação da Educação Superior (Campinas) **16**(02), 355–374 (2011)
2. Baker, R., Isotani, S., Carvalho, A.: Mineração de dados educacionais: Oportunidades para o brasil. Revista Brasileira de informática na educação **19**(02), 03 (2011)
3. Batista, G.E., Prati, R.C., Monard, M.: A study of the behavior of several methods for balancing machine learning training data. ACM SIGKDD Expl. Newsl. **6**(1), 20–29 (2004). https://doi.org/10.1145/1007730.1007735
4. Bolón-Canedo, V., Sánchez-Maroño, N., Alonso-Betanzos, A.: A review of feature selection methods on synthetic data. Knowl. Inf. Syst. **34**, 483–519 (2013). https://doi.org/10.1007/s10115-012-0487-8
5. dos Santos Oliveira, R., de Medeiros, F.P.A.: Modelo de Predição de Evasão Escolar com Base em Dados de Autoavaliação de Cursos de Graduação. Revista Brasileira de Informática na Educação **32**, 1–21 (2024)
6. dos Santos, V.H.B., Saraiva, D.V., de Oliveira, C.T.: Uma análise de trabalhos de mineração de dados educacionais no contexto da evasão escolar. In Anais do XXXII Simpósio Brasileiro de Informática na Educação, pp. 1196–1210. SBC (2021). https://doi.org/10.5753/sbie.2021.218167
7. Gamba, E., Righetti, S.: Em crise, universidades federais participam de mais da metade da produção científica. Folha de São Paulo (2022). Recuperado de https://www1.folha.uol.com.br/educacao/2022/12/em-crise-universidades-federais-participam-de-mais-da-metade-da-producao-cientifica.shtml

8. Géron, A.: Mãos à Obra: Aprendizado de Máquina com Scikit-Learn & TensorFlow. Alta Books (2019)
9. Hastie, T., Tibshirani, R., Friedman, J. H., & Friedman, J.H.: The elements of statistical learning: data mining, inference, and prediction, vol. 2, pp. 1–758. Springer, New York (2009). https://doi.org/10.1007/978-0-387-21606-5
10. Joshi, A.V.: Machine Learning and Artificial Intelligence. Springer (2020). https://doi.org/10.1007/978-3-031-12282-8
11. Lottering, R., Hans, R., Lall, M.: A model for the identification of students at risk of dropout at a university of technology. In: 2020 International Conference on Artificial Intelligence, Big Data, Computing and Data Communication Systems (icABCD), pp. 1–8. IEEE (2020). https://doi.org/10.1109/icABCD49160.2020.9183874
12. Louppe, G.: Understanding random forests: From theory to practice. arXiv preprint arXiv: 1407.7502(2014). https://doi.org/10.48550/arXiv.1407.7502
13. Lousrhania, L.: Universidades públicas lideram ranking brasileiro de patentes. Rádio Agência Nacional (2021). Recuperado de https://agenciabrasil.ebc.com.br/radioagencia-nacional/pesquisa-e-inovacao/audio/2021-07/universidades-publicas-lideram-ranking-brasileiro-de-patentes
14. Manrique, R., Nunes, B.P., Marino, O., Casanova, M.A., Nurmikko-Fuller, T.: An analysis of student representation, representative features and classification algorithms to predict degree dropout. In: Proceedings of the 9th International Conference on Learning Analytics & Knowledge, pp. 401–410 (2019). https://doi.org/10.1145/3303772.3303800
15. Mapa do Ensino Superior no Brasil – 13ª Edição. Instituto Semesp, 2023. Recuperado de https://www.semesp.org.br/wp-content/uploads/2023/06/mapa-do-ensino-superior-no-brasil-2023.pdf
16. Pereira, R.T., Zambrano, J.C.: Application of decision trees for detection of student dropout profiles. In: 2017 16th IEEE International Conference on Machine Learning and Applications (ICMLA), pp. 528–531. IEEE (2017). https://doi.org/10.1109/ICMLA.2017.0-107
17. Prestes, E.M.D.T., Fialho, M.G.D.: Evasão na educação superior e gestão institucional: o caso da Universidade Federal da Paraíba. Ensaio: Avaliação e Políticas Públicas em Educação **26**, 869–889 (2018). https://doi.org/10.1590/S0104-40362018002601104
18. Rafiq, M.A., Rabbi, A.M., Ahammad, R.: A data science approach to Predict the University Students at risk of semester dropout: Bangladeshi University Perspective. In 2021 5th International Conference on Trends in Electronics and Informatics (ICOEI), pp. 1350–1354. IEEE, June 2021. https://doi.org/10.1109/ICOEI51242.2021.9453067
19. Ramos, J.L.C., Rodrigues, R.L., Silva, J.C.S., de Oliveira, P.L.S.: CRISP-EDM: uma proposta de adaptação do Modelo CRISP-DM para mineração de dados educacionais. In Anais do XXXI Simpósio Brasileiro de Informática na Educação, pp. 1092–1101. SBC, November 2020. https://doi.org/10.5753/cbie.sbie.2020.1092
20. Saccaro, A., França, M.T.A., Jacinto, P.D.A.: Fatores Associados à Evasão no Ensino Superior Brasileiro: um estudo de análise de sobrevivência para os cursos das áreas de Ciência, Matemática e Computação e de Engenharia, Produção e Construção em instituições públicas e privadas. Estudos Econômicos (São Paulo) **49**, 337–373 (2019). https://doi.org/10.1590/0101-41614925amp
21. Santos, C. H. D., de Lima Martins, S., Plastino, A.: É Possível Prever Evasão com Base Apenas no Desempenho Acadêmico? In: Anais do XXXII Simpósio Brasileiro de Informática na Educação, pp. 792–802. SBC (2021). https://doi.org/10.5753/sbie.2021.218105
22. Saraiva, D., Pereira, S., Gallindo, E., Braga, R., Oliveira, C.: Uma proposta para prediçao de risco de evasao de estudantes em um curso técnico em informática. In: Anais do XXVII Workshop sobre Educaçao em Computaçao, pp. 319–333. SBC, July 2019. https://doi.org/10.5753/wei.2019.6639

23. Sukhbaatar, O., Ogata, K., Usagawa, T.: Mining educational data to predict academic dropouts: a case study in blended learning course. In: TENCON 2018–2018 IEEE region 10 conference (pp. 2205–2208). IEEE (2018). https://doi.org/10.1109/TENCON.2018.8650138

24. Verikas, A., Gelzinis, A., Bacauskiene, M.: Mining data with random forests: a survey and results of new tests. Pattern Recogn. **44**(2), 330–349 (2011). https://doi.org/10.1016/j.patcog.2010.08.011

Expectations of Higher Education Teachers Regarding the Use of AI in Education

Ronald Perez-Alvarez[(✉)] [iD], Cindy Rebeca Chavarría Villalobos[iD],
Melber Dalorso Cruz[iD], and Jorge Miranda Loría[iD]

Universidad de Costa Rica, Puntarenas 60101, Costa Rica
{ronald.perezalvarez,cindyrebeca.chavarria,melber.dalorso,
jorge.mirandaloria}@ucr.ac.cr

Abstract. The integration of Artificial Intelligence in Education (AIED) has emerged as a disruptive innovation promising to transform the educational process. Some research has focused on exploring the potential of Artificial Intelligence (AI) in higher education; nevertheless, the question remains open about the expectations of university educators regarding the use of AI in their classes and their concerns before starting to integrate this kind of tool. This research explores the expectations and challenges of university professors regarding the use of AI in their classes. In order to gather information about their expectations and experiences regarding the use of AIED, a survey was conducted with 27 university professors from several disciplines. The findings show a high interest among teachers in the opportunities offered by AI to improve the quality of teaching and learning in the university classroom. They expect AI to facilitate various educational aspects, from lesson planning to the creation of materials and activities to promote creativity and interactive learning. However, the main concern identified was the risk of plagiarism.

Keywords: Artificial Intelligence · Higher Education · Teachers · innovations

1 Introduction

The use of technology has contributed to enhancing the educational process for many years. For instances, Virtual reality and augmented reality have opened up new possibilities for interacting with real-world elements in a digital environment. Currently, artificial intelligence (AI) is presented as a tool with the potential to transform education even further. The integration of artificial intelligence in education (AIED) brings both challenges and opportunities. AI has the potential to personalize learning experiences, automate routine tasks for teachers, and provide data-driven insights to improve instruction [1]. However, challenges exist such as potential biases in AI algorithms, ethical considerations regarding data privacy, and ensuring equitable access to AI-powered educational resources. For instance, ChatGPT has become a popular tool for teachers and students, facilitating engagement in learning activities [2]. AI has gradually become a valuable resource in education, supporting several academic activities, such as: virtual mentoring, voice assistants, intelligent content generator, translator, automatic assessment of learning, personalizing learning, intelligent tutors, among others [3].

© The Author(s), under exclusive license to Springer Nature Switzerland AG 2024
A. M. Olney et al. (Eds.): AIED 2024 Workshops, CCIS 2150, pp. 208–213, 2024.
https://doi.org/10.1007/978-3-031-64315-6_16

In the AIED learning context, teachers play a determining role so that this integration is successful, and the maximum benefit is obtained [4, 5]. Before starting to integrate AI, it is important to consider the needs of teachers and expectations regarding the use of AI. In the end, they are responsible for using these tools in their educational practices [6]. The success of AIED in higher education goes beyond the potential of the tools themselves. The wide range of experience among teachers can be a critical factor. While some instructors are experienced with using AI to support their classes, a significant portion may have less or no experience. Understanding the needs of teachers in the educational context, as well as their willingness to begin a guided AIED integration process, allows educational institutions to create a personalized training plan and increases the chances of success for the integration of these new tools.

In order to contribute to the understanding of this context, this research explores the expectations of university teachers regarding the use of AI in their classes, as well as identifying the main challenges they face when incorporating this technology into their educational practice. To reach this objective, four research questions were formulated: What are the main challenges that teachers face in their teaching work? What expectations do teachers have regarding the use of AIED? What are the main concerns of teachers to integrate AIED? and What skills do teachers need to enhance before integrating AI into their classes.? This study will provide us with a comprehensive understanding of their perspectives and enable the development of targeted training programs and support structures to facilitate successful AIED implementation.

2   Methodology

This exploratory research investigated university professors' expectations regarding the use of AI in higher education through qualitative analysis of interviews focusing on teachers' expectations and experiences with AI. The participants consisted of 27 university professors from 11 different program such as: English teaching, business informatics, Informatics and multimedia technology, Business management, Industrial Electromechanical Engineering, Cultural Management, Basic Music Stage, Electric engineering, English with Training in Business Management, and Humanities. A total of 13 participants were men and 14 women. Notably, 60% (16) had over 7 years of teaching experience, and 60% (16) reported prior experience using AI tools. Of the teachers who report experience with AI, 11 have used it to support their lesson planning activities, 3 have not used it to support teaching, and 1 has used it to support their research activities.

Data collection was conducted using a semi-structured questionnaire consisting of 14 open-ended questions. These questions explored teachers' needs and challenges in course development, along with their expectations, perceived benefits, concerns, and challenges regarding the use of AI in teaching. The collected data was thematically analyzed based on the similarity of teachers' responses, focusing on qualitative insights rather than quantifying the responses.

3 Results and Discussions

3.1 Main Challenges in Teaching Work

Planning university courses presents a number of challenges that educators must address to ensure an effective educational experience. The most frequent challenge that teachers point out is the lack of learning resources for the development of lessons, such as video, audio, images, activities, technological tools, and materials; that allow the development of more visual, creative, and interactive classes. Another common challenge is the lack of time to carry out their teaching tasks, which makes it difficult to find and select relevant material for students. The lack of time also impacts the ability of teachers to be creative in the design of new academic resources that contribute to the learning of current students, who arrive with broader expectations and greater expertise in the use of technological tools. Therefore, the need to maintain student interest and offer an attractive educational experience becomes another important challenge for teachers.

Likewise, teachers express the need to improve the teaching techniques used in their classes and the effective integration of technological tools. The use of technologies in teaching requires skills for teachers to install the software, acquire licenses, and learn to use the tools appropriately. Teachers need to find and integrate tools to improve the educational experience and adapt to the needs of an increasingly digitized generation. Teachers express the need to train in the use of AI tools. Finally, preserving originality in the work submitted by students within a digitalized environment emerges as a relevant challenge. In this context, teachers face the crucial task of establishing effective strategies to discourage plagiarism, promote academic integrity, and effectively guide students' learning process.

3.2 Teacher Expectations with the Use of AIED

The results show diverse expectations of teachers regarding the integration of AIED, among such as teachers seek to acquire solid knowledge in AI tools, with the aim of enhancing both the effectiveness and efficiency of their teaching work; At the same time, they hope to promote skills among students to effectively take advantage of the AI potential; Furthermore, they expect the same AI tools will collaborate in reducing plagiarism by students, ensuring that these technologies serve as support and not as sources of predefined answers.

Similarly, teachers hope that with the use of AI, they can improve the distribution of learning materials and improve the quality of the teaching process. AI could contribute to the planning of learning activities and resources [7], as well as estimate execution times for each activity. In this way, teachers can better manage the time available in each study session with students and optimize the educational process. The results show positive expectations for integrating this type of tools into their teaching work. The acceptance of AI by teachers is a positive factor when starting a guided AIED integration process. Studies show that teachers with a greater understanding of AI tend to perceive more benefits in the integration of these tools to support their academic activities [4]. The teachers are interested in changing and improving the quality of the education they offer. They seek to integrate creative methodologies, new techniques, and strategies employed to facilitate knowledge acquisition and skill development to motivate student learning. Although some of the benefits expected by teachers were extracted from the responses, 40% of those surveyed do not have experience in the use of AIED, so they are not clear about the expected benefits. According to [1] between 20% and 40% of the tasks performed by teachers can be automated.

Finally, teachers were asked: How do you think AI can help improve your teaching? The results show that teachers recognize the significant impact the use of AIED can have on teaching. Mainly, its contribution stands out in the creation of high-quality teaching materials, the optimization of time in planning, and the preparation of personalized evaluations to the learning needs of students. In addition, the speakers highlight the potential of AI as an assistant to support the professional development of teachers in the organization, planning, and implementation of engaging activities in the classroom.

In the context of programming, it is emphasized that AI acts as an essential tool to solve problems, expand access to information, and facilitate content. The ability to provide detailed information and dose content stands out as a crucial element that thematizes classes and raises the level of projects and research. This specific approach significantly improves the learning experience, making information more accessible and understandable for students.

Additionally, some teachers express the hope that AI will function as a valuable support mechanism to identify students' learning difficulties. If AI is able to anticipate and address student difficulties, teachers can create personalized teaching materials and resources, as well as provide better feedback to students. Together, these results highlight the versatility of AI as an ally in the transformation and continuous improvement of the educational process.

3.3 Concerns or Challenges of Teachers Regarding the Use of AI

Among the concerns expressed, the following stand out: the unethical or improper use of AI, the loss of students' critical sense when using AI, students engaging in plagiarism to present their assignments, the discrediting of learning of the student, that the activities carried out with the support of AI do not generate a real contribution to student learning and that students fall into minimum effort thinking. Ramazan [8] integrated the use of ChatGPT in a systems programming course, at the end of the study, the students perceived that the use of AI leads the student to be lazy, and calm and can cause occupational anxiety, given that the responses generated by AI are not always correct. The results confirm the concern of teachers and call for a planned integration of AIED.

Teachers have positive expectations regarding AIED but are concerned that the potential of these tools is not used appropriately by students. Although it is true that generative AI is a powerful tool for creating text, students do not arrive with good foundations at the level of writing and construction of texts, which is why they may plagiarize the texts. Likewise, the ease with which AI generates texts generates fear that students will lose writing, communication, and critical thinking skills.

At the teaching level, they are also concerned about not having adequate knowledge or the necessary training to use AI appropriately. They worry that students are more familiar with AI and see them as outdated teachers. Additionally, they are concerned about how to create assessments that are not fully resolved by AI and promote learning.

Finally, it is how to integrate AIED without replacing learning. Barrios et al. [9] also find that teachers fear being replaced by AI.

3.4 Skills that Teachers Need to Strengthen Before Integrating AI into Their Classes

The findings show that teachers are not clear about the skills required to integrate AIED. However, they highlight the need to develop specific skills and knowledge about AI-based tools, highlighting the importance of deeply understanding the tools and uses of AI, especially as an effective mediation instrument in the classroom. The importance of training is emphasized, with a particular focus on creating assessments, projects, and resources customized to the type of learning students require. Additionally, teachers considered it essential to receive training to gain knowledge about AI applications.

The survey also highlights the need for training in aspects of the course to be taught, such as programming languages, data quality, English, office tools, and statistical tools, among others. Likewise, specific aspects are mentioned such as strengthening skills for creating digital content, mastering applications for the production of teaching materials, as well as training focused on learning aspects, including the preparation of presentations, tests, and tasks.

4 Conclusions

Teachers expect AI to facilitate various educational aspects, from lesson planning to the creation of materials and activities that promote creativity and interactive learning; They hope to be trained in the proper use of AI tools to obtain greater use of them in the educational context. However, concerns also emerged about the potential negative impact of AI on the educational process, such as the risk of plagiarism and the loss of students' ability to think critically. Finally, there is a fear that the relationship between teachers and students could be negatively affected with the use of AI. The incorporation of AIED prompts us to reflect on our role as educators, the teaching methodologies we have been using, and the ways we evaluate students' learning. Most importantly, it encourages us to stay updated with technological advancements and incorporate them into the educational process.

As future work, we plan to develop a personalized training program that caters to the specific needs of teachers in integrating AI into their classes. Additionally, we aim to conduct research to measure the impact of each AI implementation initiative in classrooms. According to the results obtained, the training program must have a general component of generative AI tools that support the planning and generation of teaching resources, an intermediate component that integrates the tools in the classes for the use of the activities developed by the students, and an Advanced component that allows creating new tools from existing ones.

References

1. Kuleto, V., et al.: Exploring Opportunities and Challenges of Artificial Intelligence and Machine Learning in Higher Education Institutions. Sustainability **13**, 10424. 13, 10424 (2021). https://doi.org/10.3390/SU131810424
2. Costello, E., Donlon, E., Kiryakova, G., Angelova, N.: ChatGPT—A Challenging Tool for the University Professors in Their Teaching Practice. Educ. Sci. **13**, 1056. 13, 1056 (2023). https://doi.org/10.3390/EDUCSCI13101056
3. Chen, L., Chen, P., Lin, Z.: Artificial intelligence in education: a review. IEEE Access **8**, 75264–75278 (2020). https://doi.org/10.1109/ACCESS.2020.2988510
4. Viberg, O., et al.: Teachers' trust and perceptions of AI in education: the role of culture and AI self-efficacy in six countries. ArXiv: Ithaca, NY, USA (2023). https://doi.org/10.48550/ARXIV.2312.01627
5. Viberg, O., et al.: What Explains Teachers' Trust of AI in Education across Six Countries? (2023)
6. Kizilcec, R.F.: To Advance AI use in education, focus on understanding Eeucators. Int. J. Artif. Intell. Educ. **34**, 12–19 (2023). https://doi.org/10.1007/S40593-023-00351-4/METRICS
7. Van den Berg, G., du Plessis, E.: ChatGPT and Generative AI: possibilities for its contribution to lesson planning, critical thinking and openness in teacher education. Educ. Sci. **13**, 998. 13, 998 (2023). https://doi.org/10.3390/EDUCSCI13100998
8. Yilmaz, R., Karaoglan Yilmaz, F.G.: Augmented intelligence in programming learning: examining student views on the use of ChatGPT for programming learning. Comput. Hum. Behav. Artif. Humans. **1**, 100005 (2023). https://doi.org/10.1016/J.CHBAH.2023.100005
9. Barrios, H., Diaz, V., Guerra, Y.: Artificial intelligence and education, challenges and disadvantages for the teacher. 72, 30–50 (2021)

Brilla AI: AI Contestant for the National Science and Maths Quiz

George Boateng[1,2(✉)], Jonathan Abrefah Mensah[1(✉)],
Kevin Takyi Yeboah[1(✉)], William Edor[1(✉)],
Andrew Kojo Mensah-Onumah[1(✉)], Naafi Dasana Ibrahim[1(✉)],
and Nana Sam Yeboah[1(✉)]

[1] Kwame AI Inc., Claymont, U.S.A.
jojo@kwame.ai, kwakuabrefahbusia@gmail.com, quakukevin@gmail.com,
edorwill@gmail.com, kojoakmo@gmail.com, ibrahimnaafi@gmail.com,
nanayeb34@gmail.com
[2] ETH Zurich, Zürich, Switzerland

Abstract. The African continent lacks enough qualified teachers which hampers the provision of adequate learning support. An AI could potentially augment the efforts of the limited number of teachers, leading to better learning outcomes. Towards that end, this work describes and evaluates the first key output for the NSMQ AI Grand Challenge, which proposes a robust, real-world benchmark for such an AI: "Build an AI to compete live in Ghana's National Science and Maths Quiz (NSMQ) competition and win - performing better than the best contestants in all rounds and stages of the competition". The NSMQ is an annual live science and mathematics competition for senior secondary school students in Ghana in which 3 teams of 2 students compete by answering questions across biology, chemistry, physics, and math in 5 rounds over 5 progressive stages until a winning team is crowned for that year. In this work, we built Brilla AI, an AI contestant that we deployed to unofficially compete remotely and live in the Riddles round of the 2023 NSMQ Grand Finale, the first of its kind in the 30-year history of the competition. Brilla AI is currently available as a web app that livestreams the Riddles round of the contest, and runs 4 machine learning systems: (1) speech to text (2) question extraction (3) question answering and (4) text to speech that work together in real-time to quickly and accurately provide an answer, and then say it with a Ghanaian accent. In its debut, our AI answered one of the 4 riddles ahead of the 3 human contesting teams, unofficially placing second (tied). Improvements and extensions of this AI could potentially be deployed to offer science tutoring to students and eventually enable millions across Africa to have one-on-one learning interactions, democratizing science education.

Keywords: Virtual Teaching Assistant · Educational Question Answering · Science Education · NLP · BERT

A. M. Olney et al. (Eds.): AIED 2024 Workshops, CCIS 2150, pp. 214–227, 2024.
https://doi.org/10.1007/978-3-031-64315-6_17

1   Introduction

According to UNESCO, only 65% of primary school teachers in Sub-Saharan Africa possessed the necessary minimum qualifications [19]. Moreover, the average student-teacher ratio at the primary education level in Sub-Saharan Africa stood at 38:1 in 2019, a figure significantly higher than the ratio of 13.5:1 observed in Europe [6]. Consequently, the region requires an additional 15 million teachers by 2030 to meet education objectives, a formidable and costly challenge [20]. The scarcity of adequately qualified educators in Africa undermines the provision of effective learning support for students. Introducing an Artificial Intelligence (AI) teaching assistant for educators holds promise in augmenting the efforts of the limited teaching workforce, facilitating tasks such as one-on-one tutoring and answering student queries. However, the absence of a robust benchmark tailored to real-world scenarios and the African context poses a significant obstacle to evaluating the effectiveness of such AI solutions.

Motivated by this need, Boateng et al. proposed a grand challenge in education - **NSMQ AI Grand Challenge** - which is a robust, real-world challenge in education for such an AI: *"Build an AI to compete in Ghana's National Science and Maths Quiz Ghana (NSMQ) competition and win - performing better than the best contestants in all rounds and stages of the competition"* [2]. In that work, they detailed the motivation for the challenge and key technical challenges that must be addressed for an AI to win the NSMQ. We created an open-source project[1] to conquer this grand challenge and built Brilla AI as the first key output which extends our prior work [3]. Brilla AI is an AI contestant that we deployed to unofficially compete remotely and live in the Riddles round of the 2023 NSMQ Grand Finale, the first of its kind in the 30-year history of the competition.

The NSMQ is an exciting, annual live science and mathematics quiz competition for senior secondary school students in Ghana in which 3 teams of 2 students compete by answering questions across biology, chemistry, physics, and math in 5 rounds over 5 progressive stages until a winning team is crowned for that year [13]. The competition has been run for 30 years and poses interesting technical challenges across speech-to-text, text-to-speech, question-answering, and human-computer interaction. Given the complexity of the challenge, we decided to start with one round, round 5 - Riddles. This round, the final one, is arguably the most exciting as the competition's winner is generally determined by the performance in the round. In the Riddles round, students answer riddles across Biology, Chemistry, Physics, and Mathematics. Three (3) or more clues are read to the teams that compete against each other to be first to provide an answer (usually a word or a phrase) by ringing their bell. The clues start vague and get more specific. To make it more exciting and encourage educated risk-taking, answering on the 1st clue fetches 5 points, on the 2nd clue - 4 points, and on the 3rd or any clue thereafter, 3 points. There are 4 riddles for each contest with each riddle focusing on one of the 4 subjects. Speed and accuracy are key

---

[1] Brilla AI Open-Source Project: https://github.com/brilla-ai/brilla-ai.

to winning the Riddles round. An example riddle with clues and the answer is as follows (see a live example here:[2]. **Question:** (1) I am a property of a periodic propagating disturbance. (2) Therefore, I am a property of a wave. (3) I describe a relationship that can exist between particle displacement and wave propagation direction in a mechanical wave. (4) I am only applicable to waves for which displacement is perpendicular to the direction of wave propagation. (5) I am that property of an electromagnetic wave which is demonstrated using a polaroid film. Who am I? **Answer:** Polarization.

Here are some of the key technical challenges an AI system will need to address. How can it accurately provide real-time transcripts of Ghanaian-accented riddle questions read in English? Speech-to-text systems tend to be trained using data from the West and generally do not work well for African-accented speech [14]. How can it infer the start and end of each riddle and extract only the clues while discarding all statements (e.g., the quiz mistress pausing the reading of the clues after the ring of a bell to listen to an answer from a school, instructions that are read at the start of the round, etc.)? How does the AI know the optimal time to attempt to provide an answer optimizing for speed and accuracy? Providing an answer too early might result in a wrong answer given that earlier clues tend to be vague. Waiting too long could result in an answer being provided by one of the contesting teams. How can the AI say its answer with a Ghanaian accent? Similar to STT systems, several text-to-speech systems tend to be trained using data from the West and generally do not speak with African accents. How do we ensure all these work seamlessly in real time without any noticeable latency?

Similar work on grand challenges that entail question answering include the DeepQA project in which Watson won *Jeopardy!* in 2011 [21], and the International Math Olympiad (IMO) Grand Challenge [7] which has an education focus. The Brilla AI project has some similarities with Jeopardy! given they are both live quiz shows, that entail question answering. It differs, given its science education focus making it arguably more challenging, and provides unique sets of technical challenges discussed previously and also by Boateng et al. [2]. Despite the similarity with the IMO Grand Challenge as both focus on STEM education, it differs by having a scope of both science and math education, and consists of a live competition (IMO is only written). The most important way Brilla AI differs from these 2 challenges though is the African context focus, a rarity in the literature for grand challenges which makes this work the first of its kind.

Our contribution is the first end-to-end, real-time AI system deployed as an unofficial, AI contestant together with 3 human competing teams for the 2023 NSMQ Grand Finale (the first of its kind in the 30-year history of the competition) that (1) transcribes Ghanaian-accented speech of scientific riddles, (2) extracts relevant portions of the riddles (clues) by inferring the start and end of each riddle and segmenting the clues, (3) provides an answer to the riddle, and (4) then says it with a Ghanaian accent.

---

[2] Video Example of Riddle: https://www.youtube.com/watch?v=kdaxoFjiYJg.

2   Background

The task of question answering has two primary paradigms: Extractive QA and Generative QA. Extractive QA involves machine reading, wherein a segment of text within a larger body (referred to as the context) is selected to directly answer a question. The context may be provided directly, as exemplified by the task outlined in the SQuAD dataset. Alternatively, relevant passages can be retrieved from diverse documents through a process known as retrieval. Chen et al. employed this approach in their DrQA system, utilizing TF-IDF for retrieval and LSTM for answer extraction [4]. Presently, BERT-based models are the current state-of-the-art for reader implementation [5]. In contrast, Generative QA employs generative models such as T5 [18] to produce answers given a question as input. These models can generate responses based on various contexts provided alongside the question, a technique termed retrieval augmented generation (RAG) [11]. We use both extractive QA and generative QA in this work and leave RAG for future work.

3   Brilla AI System

Brilla AI is currently available as a web app (built with Streamlit) (Fig. 1) that livestreams the Riddles round of the contest, and runs 4 machine-learning (ML) systems in real-time via FastAPI together running as an AI server on Google Colab: (1) speech-to-text (using Whisper [16]) which transcribes Ghanaian-accented speech of scientific riddles (2) question extraction (using BERT [5]) which extracts relevant portions of the riddles (clues) by inferring the start and end of each riddle and segmenting the clues (3) question answering (using Mistral [8]) which provides an answer to the riddle and (4) text-to-speech (using VITS [10]) which says the answer with a Ghanaian accent (Fig. 2). We used

Fig. 1. Screenshot of Brilla AI web app

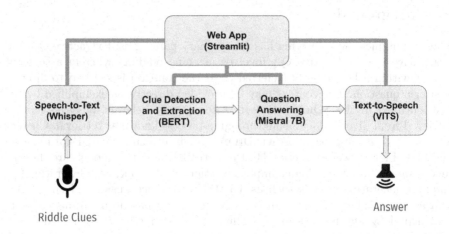

Fig. 2. Brilla AI System

Streamlit due to its applicability to ML and Data science use cases, and quick development cycles.

Here is an overview of the sequential processing and communication between components in the Brilla AI system. The web app extracts audio from the video and chunks it in 5 s using FFmpeg to allow for low-latency processing, and then sends and receives relevant data from the AI server sequentially. First, the audio chunk is sent to the STT API for processing which extracts the transcript, and passes it to the QE service which detects the riddle start and extracts the clues. The clues along with information about whether it is a new riddle are sent to the web app. It displays the transcript and clues and then sends the clues to the QA API which determines whether to provide an answer (details described later). If an answer is generated, it is returned to the web app which then displays it and sends it to the TTS API. It generates an audio of the answer and sends it to the web app which then plays it. This current sequential operation is a bottleneck within the system which caused some latency issues (discussed later). We are exploring parallelizing operations to remove such bottlenecks from the system.

The web app consists of 2 modes: demo and live[3]. The demo mode showcases the AI functionality outside of the quiz season. Functionality has been added for users to experience the AI using curated text, audio, and video content but also allows them to provide real-time content to test the AI. For example, in the demo mode, the user can provide real-time audio input on the Speech-to-Text page. The web app processes this audio and sends an API call to the Speech-to-Text API server. The server returns the real-time transcript displayed in the web app to the user. Also, users can provide a YouTube link of any NSMQ quiz video to the web app and see the AI process the content in real time and compete

---

[3] Brilla AI Demos: https://youtube.com/playlist?list=PLl-Yq3F_E8VJWFLpsnizorhIz OqUqw&si=ddrl7u0DBZcbNt7H.

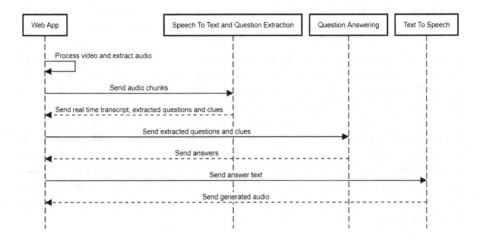

Fig. 3. Sequential Processing in Brilla AI System

for that video. To demonstrate the AI during active quiz season, the web app utilizes a live mode. In this mode, the user can supply a live stream link from YouTube to the web app and see the AI compete in real-time as well (Fig. 3).

4   Data Collection and Preprocessing

We curated a dataset of NSMQ contests from 2012–2022 containing videos of the contest and corresponding metadata, text form of riddles questions, and open-source science textbooks. First, we created a Google sheet with information about each contest such as the contest date, names of the competing schools, and the marks they acquired at the end of each contest. We additionally provided information for the riddles round such as which school answered each riddle and at which clue, which was used to calculate the points obtained by the best human contestants for each riddle. We then used that information to automatically extract video clips of the contests. Next, we manually generated annotations for the riddle round containing information about the start and end timestamps of each riddle and clue. These were then used to automatically extract the video and audio clip segments of riddles.

For the text of riddles, we purchased the digital version of the questions, parsed them, and manually reformatted them into CSVs for usage. Each CSV file had columns "Clue 1" to "Clue 9" for all the clues in per riddle, "Answer" for the ground truth answer, and "Answer 1" to "Answer 4" for alternative ground truth answers, if any, (e.g., hydrogen and h2 as alternative answers). We included additional columns to track information like subject, contest number, and year of the contest. We applied the following preprocessing steps and created riddle-answer mappings: converted all clue texts and answers to lowercase, removed punctuation, fixed whitespace, and removed articles (e.g., "the", "a" "an").

We extracted HTML files from the following open-source Science textbooks on OpenStax - High School Physics, College Physics, College Biology, College Chemistry, and College Algebra - segmenting content into chapters, sections, paragraphs, and passages. We have completed the sheet with information about contests (with some contest information missing for some years as they could not be found online). For information about the performance of contestants in the riddle round, we only have complete information for the years 2019 and 2020. Though we have complete data for the text version of the riddle questions, we only used riddles for these 2 years for evaluation so we could compare the performance of our system with human performances. Consequently, we used a total of 316 riddles (156 in 2019 and 160 in 2020). Work is ongoing to annotate the timestamps of each of the clues for all riddles and also to completely parse the textbooks. Hence, we did not have a complete set of video and audio clips of riddles for this work.

5   Modeling, Experiments, and Evaluation

5.1   Speech-to-Text

The Speech-To-Text (STT) system[4] provides a robust, fast transcription service for Ghanaian-accented English speech containing mathematical and scientific content. We used OpenAI's Whisper model [16] for speech transcription and word error rate (WER) and latency as our evaluation metrics. We previously evaluated all versions of Whisper with a limited sample of 3 audios (approximately 15 s long) from the NSMQ competition which consists of speech with Ghanaian accents, along with their corresponding transcripts [3]. The WERs are quite high, making the case to fine-tune the model with Ghanaian accented speech in the future. None of the models attained the best scores for both metrics, which warranted the need for a trade-off between the model metrics. An ideal model would have the lowest WER and latency values. We selected the Medium (English) model to deploy for use as it had a good trade-off between WER and latency (31.61% and 0.94 s respectively).

5.2   Question Extraction

The Question Extraction (QE) system extracts relevant portions of the riddles (clues) by inferring the start and end of each riddle and segmenting the clues. To infer the start and end of each riddle, we implemented a check for an exhaustive list of specific phrases that tend to be said at the beginning of each riddle such as "first riddle, second riddle, we begin, etc." To segment the clues, we fine-tuned a BERT (Tiny) model to classify automatically transcribed text of audio chunks received from STT as clues or non-clues. We used a dataset of 184 manually annotated clues (n = 81) and non-clues (n = 103) from past riddles. We used 80% as the train and 20% as the test set. We had a separate held-out validation

---

[4] STT System Demo: https://youtu.be/jn4Kh7fNgGs.

set of 173 samples (clues = 119, non-clues = 54) which we used for the final evaluation. We trained it for 10 epochs using an Adam optimizer (with a learning rate of 5e−5) and a batch size of 8. As a baseline, we extracted (1) TF-IDF and (2) SBERT embeddings as features and trained with various ML models such as random forest, support vector machine (SVM), and logistic regression. The BERT approach performed the best with 97% and 91% balanced accuracies for the test set and held-out validation set respectively compared to the best baseline approach of SBERT + SVM of 97% and SBERT + logistic regression of 87% for the test set and heldout validation set respectively. Hence we used the BERT approach for the deployment.

5.3   Question Answering

The Question Answering (QA) system[5] takes as input a riddle that consists of a clue or set of clues, and then attempts to provide answers swiftly and accurately ahead of human contestants. We implemented and evaluated two approaches: Extractive QA and Generative QA.

The Extractive QA approach involves retrieving contexts relevant to the current set of clues from a semantic search engine such as our custom-built semantic search engine consisting of a custom vector database computed with Sentence-BERT (SBERT) [17] over passages from the Simple English Wikipedia dataset or Kwame for Science which used SBERT and a science dataset + Simple English Wikipedia [1]. We passed the concatenated clues into the semantic search engine and retrieved the top three passages with the highest similarity scores. We then passed the concatenated clues into an extractive QA model (DistilBERT) as "questions", while the retrieved passages are provided as the "context" to return a span of text as an answer.

The Generative QA approach uses generative models (Falcon-7b-Instruct and Mistral-7B-Instruct-v0.1) along with a well-developed prompt to generate an answer given a set of riddle clues. We made use of prompt engineering to develop a high-quality prompt that takes the clues as part of the prompt. The prompt (1) asked the model to take on the role of an expert - a science prodigy, (2) used Chain-of-Thought (CoT) prompting by asking it to reason through the clues, (3) stated a penalty for deviating from the instruction to provide a short answer, (4) used few-shots learning by giving an example of a riddle and an answer, and (5) asked for a structured output - JSON.

Furthermore, we implemented a confidence modeling pipeline that produces an estimate of the confidence of the QA system to decide whether or not to attempt to answer the riddle after a clue has been received or wait for more clues. For a riddle, we pass the first clue as an input to the model, generate three answers to the riddle, and keep a running count of the number of times each answer is generated for subsequent input clues and compare against a threshold value which we determine empirically after evaluating different threshold values.

---

[5] QA System Demo: https://youtu.be/VfkxZAdZ2PA.

If any answer's count is equal to or more than the threshold value, we then return that answer as our answering attempt (Fig. 4).

You are a science prodigy currently competing in a National Science competition. You are now in the fifth round, where you must **first reason through the clues of the given riddle and then provide a short answer**. Remember, your answer should consist of just the term the riddle is pointing to, and nothing else. **Adding additional text will result in point deductions.**

Here's **an example to guide you**:

Riddle: You might think i am a rather unstable character because i never stay at one place. However my motion obeys strict rules and i always return to where i started and even if i have to leave that spot again i do it in strict accordance to time. I can be named in electrical and mechanical contexts in all cases i obey the same mathematical rules. In order to fully analyse me you would think about a stiffness or force constant restoring force and angular frequency.

Answer: oscillator

Read the riddle below and provide the three possible correct **answers as a json** with keys: answer1, answer2, answer3

NOTE: You are allowed to include an answer multiple times if your reasoning shows that it is likely the correct answer. Do not provide any explanations.

Riddle: {riddle}

Fig. 4. Prompt for QA Model

We used Exact Match (EM) accuracy as the primary metric and a secondary metric, Fuzzy Match (FM) accuracy. With EM, we search for instances where the generated answer exactly matches the ground truth answer for a given riddle or the alternative answers if there are any. With FM, we perform a more relaxed comparison and check if any of the ground truth answers is a substring of the answer generated by the model. For example, (ground truth: tissue, model answer: tissues, will yield True for fuzzy match). We opt for EM as our primary evaluation metric, given the quiz's nature, which requires that the answers provided by the competing teams precisely match the term the riddle points to. Additionally, we employ FM to assess overall model performance and to gauge generated answers' proximity to the ground truth.

We perform two sets of evaluations. The first evaluation uses all clues concatenated as input to our models and assesses performance (Table 1). This case assumes that the QA system has all the clues and enables us to compare different models. The second evaluation uses the best approach and model from the first evaluation and integrates the confidence modeling approach to simulate the live deployment offline (Table 2). Given that the live mode uses an automatic transcript of 5 s of audio, we estimated how many words are in such a chunk (about 7 words). We break our text of the riddles into those chunks and pass them one after another while appending subsequent chunks as input to the model as it attempts to answer at some point. We compared the performance of our 2 approaches - Extractive QA (using SBERT + Kwame for Science dataset or Simple Wiki for retrieval and DistilBERT for extraction) and Generative QA (using

Mistral-7b-Instruct-v0.1 and Falcon-7b-Instruct) with GPT 3.5 and a human benchmark which we computed using our annotations about the clue number on which the contestanting students answered the riddles. For all experiments and evaluations, we run on either a Tesla T4 GPU with 16 GB VRAM, or a single A100 GPU with 40 GB VRAM on Google Colab.

Our results are shown in Tables 1 and 2. From Table 1, our generative approach using Mistral performed better than the extractive approach but worse than GPT 3.5 and students. Hence, we used Mistral for our second evaluation with confidence modeling and deployed it. GPT 3.5 performed better than our approach but worse than the best human contestants for both evaluations 1 and 2 (Table 2). Overall, these show advancements in generative models for solving science questions. However, the proprietary and closed nature of GPT 3.5, underscores the need to create more accurate open alternatives that will be accessible to students and trainers in Africa and other low-resourced environments. Fine-tuning our open-source models in the future could potentially accomplish that.

Table 1. Evaluation Using All Clues on 2019 (n = 156) and 2020 (n = 160) riddles dataset

Model	EM (%)		FM (%)	
	2019	2020	2019	2020
DistilBERT (Kwame)	0	0	0	0.63
DistilBERT (Simple Wiki)	1.28	0	2.56	1.25
Falcon	22.44	14.37	33.33	36.25
Mistral	38.46	27.5	54.49	45.62
GPT 3.5	40.38	33.75	70.51	68.12
Students	76.3	75	N/A	N/A

Table 2. Mock Live Environment Evaluation for Mistral and ChatGPT on 2019 (n = 156) and 2020 (n = 160) riddles dataset

Model	EM (%)		FM (%)	
	2019	2020	2019	2020
Mistral	12.82	8.75	21.15	24.38
GPT 3.5	28.85	23.75	33.97	35.63
Students	76.3	75.0	N/A	N/A

5.4   Text-to-Speech

The TTS system synthesizes answers generated by QA into speech with a Ghanaian accent[6]. We used VITS [10], an end-to-end model that we fine-tuned using voice samples from three Ghanaian speakers resulting in an output speech similar to those speakers[7]. The training datasets featured audio samples along with automatic transcripts using Whisper (which were manually corrected) from three Ghanaians (with their permission), each with unique recordings extracted from various sources: Speaker 1: TEDX talk (20 min), Speaker 2: podcast interview (22 min), speaker 3: YouTube recording (1 h). We converted our models to the Open Neural Network Exchange (ONNX) format for deployment. ONNX is a file format that allows for easy integration of ML models across various frameworks like Tensorflow and PyTorch [15]. ONNX Runtime optimizes latency, throughput, memory utilization, and binary size and allows users to run ML models efficiently.

We performed two evaluations using the ONNX versions of the models. Evaluation 1 involved synthesizing 30 samples of scientific and mathematical speech from past NSMQ questions and Evaluation 2 involved synthesizing 30 samples of conversational speech. We compared the 3 speaker-specific models. We used Mean Opinion Score (MOS), Word Error Rate (WER), and latency as evaluation metrics. The automatic MOS is an objective evaluation of how 'natural' the synthesized speech sounds with a range from 1 (bad) to 5 (excellent) [12]. The WER of the synthesized speech measures the intelligibility of the synthesized speech, whereas the latency measures the inference speed of the model. The results (Table 3) show that the Speaker 1 model achieves the best WER. The models generally exhibit high MOS scores, indicating a human-like sound. However, they struggle with scientific and mathematical text synthesis as shown by the poor WER. Qualitatively, audios synthesized with these models sound Ghanaian and similar to the original speakers, with a slight robotic undertone and suboptimal intelligibility for single words. We did not use Speaker 1 to avoid confusion in the live deployment since it is the voice of the quiz mistress. Consequently, we deployed Speaker 3 due to its lower latency and WER compared to Speaker 2.

Table 3. Evaluation of TTS Models

Model	Evaluation 1			Evaluation 2		
	Mean Latency (s)	Mean WER (%)	Mean MOS	Mean Latency (s)	Mean WER (%)	Mean MOS
Speaker 1	1.05	35.41	3.00	1.11	6.99	2.81
Speaker 2	1.28	70.45	3.12	1.24	44.17	3.32
Speaker 3	1.08	63.12	2.84	1.05	17.91	2.84

---

[6] TTS System Demo: https://youtu.be/KuOxxAk_Qqk.
[7] TTS Voice Sample: https://youtu.be/dwg7izBMFGA.

6 Real-World Deployment and Evaluation of Brilla AI

In its debut in the NSMQ Grand Finale in October 2023, Brilla AI answered one of the 4 riddles ahead of the 3 human contesting teams, unofficially placing second (tied) and achieved a 25% EM accuracy[8] [9]. For the first riddle, the QE component could not detect the start of the riddle as STT wrongly transcribed "first riddle" as "test riddle". Consequently, no clues were extracted, QA received no data, and thus could not answer the riddle (the answer shown was from a previous test attempt). For the second and third riddles, the start and ends were detected and the clues were extracted but QA attempted too early (perhaps due to the confidence threshold being quite low) and got the answer wrong. For the fourth and final riddle, QE detected the start of the riddle, could not extract clue one, but extracted the second clue and third clue, and then QA generated an answer, resulting in Brilla AI answering correctly before one of the students answered correctly also (which he did after all the clues). Overall, we recorded a success rate of 25% unofficially placing second (tied) as the best-performing team answered 2 riddles correctly (excluding the one the AI answered), one team answered only one and the third answered none. This result is an important milestone, especially given it was our first live deployment and there were a lot of real-world challenges like noise, inaccurate transcripts, undetected clues, and irrelevant data sent along to QA. Our automatic transcripts were not always accurate and some clues were not detected which all compounded and posed challenges for QA. Our real-time transcription began to lag the live stream of the contest because of the sequential processing of our pipeline, which resulted in our AI appearing to have provided an answer after all clues, even though it did so only after receiving the 3rd clue. This issue could have resulted in a late answer.

7 Challenges, Limitations, and Future Work

One key challenge we have is curating the NSQM data. At the end of each contest, the competing schools and their respective scores are shared via images or PDFs on social media and blogs. Sometimes finding these materials is difficult as they may be scattered on the web or may not have been posted by the media channel in charge. Video recording errors due to power surges and Internet connection problems also make getting accurate data difficult.

In the web app, the sequential approach of calling APIs was a bottleneck which caused some latency issues for the live deployment. We plan to explore parallelizing operations to remove such bottlenecks from the system. Also, we plan to deploy the web app and ML models on Google Cloud Platform to ensure that both the front and backends of Brilla AI are running on cloud systems to increase the security and stability of the web app.

We plan to fine-tune our STT model since we did not do so in this work due to a lack of complete NSMQ annotated data, which could improve performance.

---

[8] Brilla AI Debut: https://youtu.be/2AUpiVB6zA4.

To improve the QA performance, we will explore using RAG after completing the curation and processing of the science textbooks. We will also explore using reinforcement learning to improve our confidence modeling approach. For TTS, we will curate a dataset that includes recordings of scientific and mathematical expressions, as well as single-word answers, to address its shortcomings.

8 Conclusion

In this work, we built Brilla AI, an AI contestant that we deployed to unofficially compete remotely and live in the Riddles round of the 2023 NSMQ Grand Finale, the first of its kind in the 30-year history of the competition. Our AI answered one of the 4 riddles ahead of the 3 human contesting teams, unofficially placing second (tied). Our next step is to improve the performance of our system ahead of 2024 NSMQ during which we plan to unofficially compete in both Round 5 (Riddles) and Round 4 (True or False) for all 5 stages of the competition. Aside from being an interesting intellectual challenge, we are building an AI that addresses the unique context of Africans - transcribes speech with a Ghanaian accent, provides answers to scientific questions drawn from a Ghanaian quiz, and says answers with a Ghanaian accent - that could be integrated into an education tool. Imagine a student in a rural part of Ghana calling a toll-free number with a feature phone, asking this AI numerous science questions, and the AI understands her even with her Ghanaian accent and then provides explanations using local context in a Ghanaian accent. That vision is our long-term goal and would radically transform learning support for students across Africa, enabling millions of young people to have one-on-one interactions and support even with limited access to teachers, computers, or even smartphones leading to the democratizing of science education across Africa!

Acknowledgments. We are grateful to all the volunteers that contributed to this project. We are also grateful to Isaac Sesi for donating his voice which we used to train the TTS model for the voice of our AI, and our project advisors for their feedback and support throughout this work: Professor Elsie Effah Kaufmann, the NSMQ Quiz Mistress, and Timothy Kotin, a member of the 2006 NSMQ winning team.

References

1. Boateng, G., John, S., Boateng, S., Badu, P., Agyeman-Budu, P., Kumbol, V.: Real-world deployment and evaluation of Kwame for science, an AI teaching assistant for science education in West Africa. arXiv preprint arXiv:2302.10786 (2023)
2. Boateng, G., Kumbol, V., Kaufmann, E.E.: Can an AI win Ghana's national science and maths quiz? An AI grand challenge for education. arXiv preprint arXiv:2301.13089 (2023)
3. Boateng, G., et al.: Towards an AI to win Ghana's national science and maths quiz. In: Deep Learning Indaba 2023 (2023)
4. Chen, D., Fisch, A., Weston, J., Bordes, A.: Reading wikipedia to answer open-domain questions. arXiv preprint arXiv:1704.00051 (2017)

5. Devlin, J., Chang, M.W., Lee, K., Toutanova, K.: BERT: pre-training of deep bidirectional transformers for language understanding. arXiv preprint arXiv:1810.04805 (2018)
6. Almost 14 pupils per teacher in EU primary schools, September 2021. https://ec.europa.eu/eurostat/web/products-eurostat-news/-/ddn-20210907-1
7. IMO grand challenge. https://imo-grand-challenge.github.io/
8. Jiang, A.Q., et al.: Mistral 7b. arXiv preprint arXiv:2310.06825 (2023)
9. NSMQ 2023: AI answered one riddle correctly ahead of contestants in grand-finale, November 2023. https://www.myjoyonline.com/nsmq-2023-ai-answered-one-riddle-correctly-ahead-of-contestants-in-grand-finale
10. Kim, J., Kong, J., Son, J.: Conditional variational autoencoder with adversarial learning for end-to-end text-to-speech. In: International Conference on Machine Learning, pp. 5530–5540. PMLR (2021)
11. Lewis, P., et al.: Retrieval-augmented generation for knowledge-intensive NLP tasks. Adv. Neural. Inf. Process. Syst. **33**, 9459–9474 (2020)
12. Lo, C.C., et al.: MOSNet: deep learning-based objective assessment for voice conversion. In: INTERSPEECH 2019 (2019)
13. National science and maths quiz. https://nsmq.com.gh/
14. Olatunji, T., et al.: AfriSpeech-200: Pan-African accented speech dataset for clinical and general domain ASR. Trans. Assoc. Comput. Linguist. **11**, 1669–1685 (2023)
15. ONNX runtime. Accelerated GPU machine learning. https://onnxruntime.ai/
16. Radford, A., Kim, J.W., Xu, T., Brockman, G., McLeavey, C., Sutskever, I.: Robust speech recognition via large-scale weak supervision. In: International Conference on Machine Learning, pp. 28492–28518. PMLR (2023)
17. Reimers, N., Gurevych, I.: Sentence-BERT: sentence embeddings using Siamese BERT-networks. arXiv preprint arXiv:1908.10084 (2019)
18. Roberts, A., Raffel, C., Shazeer, N.: How much knowledge can you pack into the parameters of a language model? In: Proceedings of the 2020 Conference on Empirical Methods in Natural Language Processing (EMNLP), pp. 5418–5426 (2020)
19. The international task force on teachers for education. Closing the gap - ensuring there are enough qualified and supported teachers in Sub-Saharan Africa, September 2021
20. The persistent teacher gap in Sub-Saharan Africa is jeopardizing education recovery, July 2021. https://www.unesco.org/en/articles/persistent-teacher-gap-sub-saharan-africa-jeopardizing-education-recovery
21. A computer called Watson. https://www.ibm.com/history/watson-jeopardy

Exploring NLP and Embedding for Automatic Essay Scoring in the Portuguese

Ruan Carvalho[1]([✉]) [iD], Lucas Fernandes Lins[1] [iD], Luiz Rodrigues[2] [iD],
Péricles Miranda[1] [iD], Hilário Oliveira[4] [iD], Thiago Cordeiro[2] [iD],
Ig Ibert Bittencourt[2,5] [iD], Seiji Isotani[5] [iD], and Rafael Ferreira Mello[3] [iD]

[1] Federal Rural University of Pernambuco, Recife, Brazil
ruan.carvalho@ufrpe.br
[2] Center for Excellence in Social Technologies, Federal University of Alagoas,
Penedo, Brazil
[3] Centro de Estudos Avançados de Recife, Recife, Brazil
[4] Federal Institute of Espírito Santo, Vitoria, Brazil
[5] Harvard Graduate School of Education, Cambridge, USA

Abstract. Automated Essay Scoring (AES) presents a promising solution for enhancing the assessment process in education, particularly in standardized tests like Brazil's National High School Exam (ENEM). However, prior research either focuses on the English language or lacks a nuanced consideration of ENEM's particularities, such as competence-based evaluation. This paper investigates AES for ENEM, focusing on a particular competence (C3), which poses challenges due to its subjective nature and high-class imbalance. Leveraging the Essay-BR corpus, which consists of essays aligned with ENEM standards, this research explores traditional Natural Language Processing (NLP) features, contextual embedding representations extracted from BERT models, and a combination of both. Additionally, class weighting techniques are utilized to address class imbalance issues. Results indicate that models based on LGBM and XGBoost, incorporating BERT Embedding alongside NLP features, and augmented by class weighting, yielded the best performance, besides notable enhancements in minority classes accuracy. By presenting a competency-specific analysis, this study contributes towards optimizing AES for ENEM with a contextualized approach to Brazil.

Keywords: Automatic Essay Scoring · ENEM · BERT

1 Introduction

The National High School Examination (ENEM) is a crucial assessment in Brazil, serving as a benchmark for evaluating students' educational competence after completing the foundational education phase. Furthermore, it plays a pivotal role in enabling many students to pursue higher education opportunities[1].

---

[1] https://www.gov.br/inep/pt-br/areas-de-atuacao/avaliacao-e-exames-educacionais/enem.

A. M. Olney et al. (Eds.): AIED 2024 Workshops, CCIS 2150, pp. 228–233, 2024.
https://doi.org/10.1007/978-3-031-64315-6_18

ENEM features a discursive-argumentative essay, wherein students must address a proposed problem. The essay must have up to 30 lines and is assessed based on the five competencies, such as organization (C3) and coherence/cohesion (C5).

Each criteria's scores ranges from 0 (complete lack of mastery) to 200 (excellent mastery). Consequently, ENEM's scores range from 0 to 1000 by summing up the scores for all five competencies. For this, two evaluators review the essays. However, the manual grading process, though indispensable, is recognized for its limitations related to the fatigue evaluators are likely to experience due to its repetitive nature. Furthermore, because it relies on human judgment, the grading process is subject to various inconsistencies and biases, leading to an inherently unreliable assessment.

A potential solution to this challenge is Automated Essay Scoring (AES). AES might partially automate the essay evaluation to improve the efficiency of evaluators while guaranteeing impartial and coherent grades [7]. Most often, AES relies on Natural Language Processing (NLP) and Machine Learning (ML), where regression and classification models are the primary approaches [6]. Particularly within ENEM's context, the emphasis on feature-based approaches is evident, demonstrating promising results on AES tasks given the full essay [4,5]. However, there is a lack of research addressing the assessment of the five competencies individually.

Note that ENEM's evaluators must attribute a score for each competence. Consequently, AES systems that output an overall score, rather than one for each competence have limited practical contribution. Nevertheless, research in this direction is only emerging. To our best knowledge, a single study addressed AES for ENEM's Competence 3 (C3), and the predictive performance in this metric was the worst among the five competencies [4]. To address that gap, this work focuses on C3 of the ENEM, which assesses the student's ability to select, correlate, organize, and interpret information, facts, opinions, and arguments to defend a particular standpoint. Thus, the present study performs an experimental analysis of different strategies for extracting features from ENEM essays.

This paper contributes to AIED research by exploring how to extend AES benefits to an often underserved region: Brazil. Our research contributes to helping Portuguese-speaking people benefit from AES advantages, focusing on a particular aspect of ENEM, C3, presenting a contextualized approach to address a Brazilian challenge. As a continental-sized country featuring millions of students, manually assessing ENEM essays is a prominent challenge. This paper helps address that gap with a targeted intervention for C3.

2   Method

After a manual corpus analysis, a few duplicate essays were identified and removed from the extended Essay-BR [5]. Consequently, the resulting corpus used in this study comprises 6,565 essays, distributed among grades 0 (n = 185; 2.8%), 40 (n = 164; 2.5%), 80 (n = 1601; 24.4%), 120 (n = 3051; 46.5%), 160 (n = 1374; 20.9%), and 200 (n = 190; 2.9%). This demonstrates the imbalance within the dataset, highlighting the need to address it.

This study considered two feature approaches: traditional NLP-based indicators, commonly employed to assess textual cohesion and coherence, and contextual embedding representations extracted from the BERT model.

NLP-Based Features: For each essay, we computed 236 features, organized into seven distinct groups, similar to prior research [7]. The **descriptive** group represents general aspects of the text: number of words; number of words classified as *stop words*; number of sentences; and average of words per sentence.

Coh-Metrix enables the extraction of numerous discourse-level and linguistic attributes that prove useful in assessing cohesion and coherence. These features are categorized into semantic analysis, connectives, lexical diversity, referential cohesion, and syntactic complexity. Eighty-seven metrics were utilized and adapted to the Portuguese language version [1].

Linguistic Inquiry Word Count (LIWC) is a software that, given an essay, compares each word with a set of dictionaries to compute grammatical, psychological, and social aspects within the text. In this study, we adopted the 2007 version [2] to extract 64 features.

Use of connectives is one way to refer to elements previously mentioned in the text. When effectively employed, it ensures a sound logical connection among the ideas presented in the essay. In this group, 33 connectives were computed, 32 following the criteria outlined in [3], and one overall metric.

Lexical Diversity evaluates the student's vocabulary richness by analyzing the frequency of distinct words in the text. For this, we calculate the ratio of various types of words, such as verbs, nouns, adverbs, and adjectives. Besides, three metrics suggested by Palma et al. [8] are also implemented: Hapax Legomena, Yule's K, and Guiraud's Index. In total, we adopted 15 features in this group.

Readability assesses the difficulty of the text from two aspects: sentence length (with longer sentences indicating more incredible difficulty) and word complexity. This study uses five features of this nature [8].

Sentence Overlapping arises when terms mentioned earlier in the text are reintroduced. Six metrics are computed based on the notion that elements shared among adjacent sentences contribute to referential cohesion. Another two metrics are extracted using Term Frequency-Inverse Document Frequency (TF-IDF) across neighboring sentences.

Thematic Coherence is derived from 20 features related to the similarity between the essay and the motivating text. We employed two strategies to represent the essays and their corresponding motivating texts. One strategy utilizes the traditional Term Frequency-Inverse Document Frequency (TF-IDF), while the other leverages embedding representations extracted from BERTimbau-base [9], the Brazilian Portuguese version of the BERT model. Each representation was employed to calculate cosine similarity. The Levenshtein distance was computed using the "fuzzysearch" library[2] and the Jaccard similarity.

BERT Embedding: This approach encodes the essay and its motivating text using the BERTimbau-base [9] to create a combined representation to capture potential relationships. Each element (essay and motivating text) is transformed

---

[2] https://github.com/taleinat/fuzzysearch.

into a 768-dimensional vector (the default hidden layer size of the BERT model). These vectors are then concatenated into a single vector with 1,536 dimensions. This combined representation aims to encapsulate both syntactic and semantic aspects of the essays, potentially revealing underlying patterns. Afterward, this enriched vector is fed as input to the machine learning algorithms considered in this work for estimating the C3 grades.

Next, we selected classic and more recent algorithms that have demonstrated high performance across previous AES works [7]: Logistic Regression, Random Forest, Extra Trees, K-Nearest Neighbors (KNN), Adaboost, Catboost, LGBM, and XGBoost, all in both classifier and regressor versions. These models were trained using the features extracted from the text and the Embedding obtained from the essay text and prompt using BERT. We adopted the default configuration settings defined in the libraries for all algorithms. Following related work, these algorithms were assessed according to accuracy, F1-score, Cohen's Kappa (linear), Quadratic Weighted Kappa (QWK), and Pearson Correlation, and Root Mean Square Error (RMSE) to provide a comprehensive evaluation, encompassing both classification and regression, enabling a qualitative analysis of the results and comparisons to related work.

Lastly, we defined the experimental steps. C3 scores are categorical: 0, 40, 80, 120, 160, and 200. While this fits the classifiers with a straightforward encoding (i.e., 0, 1, 2, 3, 4, 5), regressors expect numeric values (e.g., 0, 0.2, 0.4, 0.6, 0.8, 1.0). Note that a regressor might assess new inputs (test sets) with intermediate scores (e.g., 0.43) during the learning process. As such, intermediate values become feasible, enabling the algorithm to express "uncertainty". However, the regressor's final score must fall into one of the six categories. For this, the output of the regressor is multiplied by 5 (the value of the highest class) and rounded. If the original output of the regressor is less than 0 or greater than 5, the final value is mapped to 0 or 5, respectively. We also employed a weighted approach for each essay during model training based on the *compute_class_weight* from the scikit-learn library[3], assigning higher weights to elements of minority classes to help the algorithms handle class imbalance.

3   Results and Discussion

LGBM and XGBoost consistently yielded superior predictive performance than the other algorithms analyzed. Therefore, this section focuses on LGBM and XGBoost for conciseness. First, we compared the algorithms' performance - with no approach for handling class imbalance - in three training scenarios: features only (F), BERT Embedding only (E), and both (F+E). These results are presented in Table 1. The best results in each evaluation measure are highlighted in bold. The table shows that the optimal values across all metrics consistently stem from configurations involving both features and Embedding, despite Embedding only closely trails behind.

---

[3] https://scikit-learn.org/stable/modules/generated/sklearn.utils.class_weight. compute_class_weight.html.

Table 1. Results achieved by classifiers (c) and regressors (r) considering the features only (F), BERT Embedding only (E) or both (F+E)

Input	Algorithm	Accuracy	F1 macro	Kappa	QWK	RMSE	Pearson
F	LGBM(c)	0.548	0.298	0.341	0.430	0.860	0.470
F	XGB(c)	0.537	0.297	0.327	0.414	0.880	0.448
F	LGBM(r)	0.537	0.295	0.347	0.461	0.841	0.500
F	XGB(r)	0.516	0.289	0.329	0.441	0.882	0.464
E	LGBM(c)	0.582	0.328	0.415	0.502	0.823	0.534
E	XGB(c)	0.589	0.357	0.426	0.511	0.821	0.541
E	LGBM(r)	0.588	0.335	0.449	0.555	0.784	0.585
E	XGB(r)	0.559	0.333	0.420	0.534	0.819	0.554
F+E	LGBM(c)	**0.603**	0.352	0.454	0.540	0.797	0.570
F+E	XGB(c)	**0.603**	**0.388**	0.453	0.537	0.809	0.562
F+E	LGBM(r)	0.596	0.346	**0.473**	**0.586**	**0.767**	**0.611**
F+E	XGB(r)	0.563	0.346	0.433	0.548	0.818	0.563

Second, we investigated the best setup (F+E) with class weighting. These results are shown in Table 2, with values that outperformed those of Table 1 highlighted. The findings demonstrate that, especially for LGBM, class weighting contributed to a slight improvement in predictive performance. For instance, Kappa values increased from ≈0.45 to ≈0.48.

Table 2. Experiments considering (F+E) using class weights

Algorithm	Accuracy	F1 macro	Kappa	QWK	RMSE	Pearson
LGBM(c)	**0.605**	**0.411**	**0.481**	0.574	0.799	0.589
XGB(c)	0.604	**0.427**	**0.475**	0.558	0.829	0.567
LGBM(r)	0.546	0.359	0.457	**0.588**	0.823	0.592
XGB(r)	0.502	0.324	0.398	0.531	0.886	0.533

These results highlight the efficacy of pre-trained language models. Both classifiers and regressors trained with data extracted through BERT outperformed those trained solely on NLP-based features and achieved almost comparable outcomes when these features were combined with Embedding. Moreover, we found that class weighting also contributed to predictive performance, with confusion matrices showing important enhancements within the minority classes. Thus, these findings reveal that combining NLP and embedding features, along with class weighting, holds promising potential to improve AES systems for ENEM's C3.

Mainly, these findings contribute to optimizing ENEM's assessment. The integration of BERT Embedding alongside traditional NLP features, augmented by class weighting techniques, not only enhances predictive performance but also holds the potential to alleviate issues of evaluator fatigue and subjectivity inherent in manual grading processes. Additionally, handling class imbalance provides a particular contribution towards deploying such models. A model that makes predictions only for the majority classes cannot be practically applied, and by incorporating weighting mechanisms, we were able to mitigate this bias.

Moreover, by focusing on competency-specific evaluation, the study contributes a nuanced approach to AES that aligns closely with the multifaceted nature of ENEM's assessment framework. This contextualized approach fills a critical gap in existing research and highlights the importance of tailoring AES systems to address the unique needs of the Brazilian educational system. By bridging regional disparities in adopting AI in education and offering targeted interventions for improving pedagogical practices, this paper contributes towards leveraging technology to empower students and educators in Brazil.

References

1. Camelo, R., Justino, S., de Mello, R.F.L.: Coh-Metrix PT-BR: Uma API web de análise textual para a educação. In: Anais dos Workshops do IX Congresso Brasileiro de Informática na Educação, pp. 179–186. SBC (2020)
2. Carvalho, F., Rodrigues, R.G., Santos, G., Cruz, P., Ferrari, L., Guedes, G.P.: Evaluating the Brazilian Portuguese version of the 2015 LIWC lexicon with sentiment analysis in social networks. In: Anais do VIII Brazilian Workshop on Social Network Analysis and Mining, pp. 24–34. SBC (2019)
3. Grama, D.F.: Elementos coesivos do português brasileiro em córpus de redações nos moldes do Enem: um estudo para a elaboração da CoTex. Phd thesis, Universidade Federal de Uberlândia (2022)
4. Marinho, J., Cordeiro, F., Anchiêta, R., Moura, R.: Automated essay scoring: an approach based on ENEM competencies. In: Anais do XIX Encontro Nacional de Inteligência Artificial e Computacional, pp. 49–60. SBC, Porto Alegre, RS, Brasil (2022). https://doi.org/10.5753/eniac.2022.227202, https://sol.sbc.org.br/index.php/eniac/article/view/22769
5. Marinho, J.C., Anchiêta, R.T., Moura, R.S.: Essay-br: a brazilian corpus to automatic essay scoring task. J. Inf. Data Manag. 13(1) (2022)
6. Marinho, J.C., Cordeiro, F., Anchiêta, R.T., Moura, R.S.: Automated essay scoring: an approach based on enem competencies. In: Anais do XIX Encontro Nacional de Inteligência Artificial e Computacional, pp. 49–60. SBC (2022)
7. Oliveira, H., et al.: Towards explainable prediction of essay cohesion in Portuguese and English. In: LAK23: 13th International Learning Analytics and Knowledge Conference, pp. 509–519 (2023)
8. Palma, D., Atkinson, J.: Coherence-based automatic essay assessment. IEEE Intell. Syst. 33(5), 26–36 (2018)
9. Souza, F., Nogueira, R., Lotufo, R.: BERTimbau: pretrained BERT models for Brazilian Portuguese. In: 9th Brazilian Conference on Intelligent Systems, BRACIS, Rio Grande do Sul, Brazil, October 20–23 (2020, to appear)

Enhancing General Reading Understanding: Measuring Foundational Reading Skills Relevance

Leonardo Marques[1]([✉]), Gilberto Nerino de Souza Jr.[2],
Bruno Almeida Pimentel[1], Álvaro Alvares Sobrinho[3], and Deisy Das Graças de Souza[4]

[1] Federal University of Alagoas, Maceió, Brazil
leonardo.marques@cedu.ufal.br , brunopimentel@ic.ufal.br
[2] Federal Rural University of Amazon, Belem, Brazil
[3] Federal University of the Agreste of Pernambuco, Garanhuns, Brazil
alvaro.alvares@ufape.edu.br
[4] Federal University of São Carlos, São Carlos, Brazil
ddgs@ufscar.br

Abstract. Basic reading skills are crucial for literacy development, and digital tools, like machine learning techniques, offer promising ways to support the acquisition of those skills. This study evaluated the impact of assessing specific foundational reading skills on overall reading acquisition using trial-based direct instruction. We developed and tested three versions of the General Pre-Reading Competence Index (GR-CI) by analyzing data from 1,184 children who completed 228 pre-reading tasks in Portuguese. We found that the GR-CI k++ Filtered performed better, based on the silhouette scores from K-Means++ cluster analysis conversion into weights for each foundational reading skill. Implementing the GR-CI k++ Filtered was highly effective, accurately categorizing 93.8% of participants with distinct reading scores and enhancing the discriminative power of conventional reading assessments (*status quo*) by 52%. This result significantly increased the identification of variations in reading acquisition among students. These findings establish a new pathway for equitable and individualized assessment of reading skills. They also contribute to assisting educators and policymakers in developing more equitable learning strategies. We discuss the proposal of additional trial-based direct instruction parameters to enhance individualized reading assessment.

Keywords: Reading Skills · Matching to Sample · Clusters Analysis · K-means++ · Perceived Competence

1 Introduction

Current studies highlight the importance of intervening early with at-risk students to reduce the lasting impact of delayed reading skills development [5]. Effective reading instruction practices suggest a clear link between knowing words and

A. M. Olney et al. (Eds.): AIED 2024 Workshops, CCIS 2150, pp. 234–241, 2024.
https://doi.org/10.1007/978-3-031-64315-6_19

reading well across different grades [8,12]. At the start, learners connect written words with spoken language. They must turn letter sounds into familiar words and ideas to understand texts. Thus, knowing words in print and speech plays a crucial role in learning to read [3,12]. Moreover, providing direct teaching is essential for building knowledge. In countries with low reading scores, like Brazil, there often needs to be more reading remedial support available [4,6].

Educational technology presents promising solutions for remedial support [7]. For instance, step-based problem-solving systems have proven effective in skill acquisition [13], offering personalized instruction as a crucial aspect of effective teaching. However, the challenge lies in balancing instructional quality with addressing individual needs, which requires continuous formative assessment and skill formalization.

In Brazil, the Learning to Read and Write in Small Steps program (known as ALEPP) [3] has demonstrated effectiveness for children with a history of reading difficulties [9]. The program's core lies in the Matching-To-Sample (MTS) procedure for assessing and teaching reading skills.

Nevertheless, existing algorithms supporting MTS tasks depend on expert opinions rather than empirical data for task weighting [11]. This study enhances the literature by leveraging learners' data and employing Artificial Intelligence (AI) techniques to support these tasks. Thus, we developed an efficient, data-driven approach, representing an initial stride toward an AI-driven reading solution.

2   PRAD Database

Matching-to-sample (MTS) tasks are used extensively in studying human cognitive processes, to teach and evaluate complex behaviors such as symbolic communication and concept formation. It is a choice task where subjects are shown a sample stimulus and then must identify a matching stimulus from several comparison options (differentiating the 'correct' stimuli from the 'distractor' ones).

Our approach assumes that each MTS task is a proxy for different foundational reading skills. To determine the differential weights for each task type, we analyzed a database of 17,760 data points from 238 MTS tasks spread across 11 MTS task types (AC, BB, BD, BC, AB, AE, CC, CE, CB, CF, and AF), organized into blocks of 15 attempts or trials per task type. For example, the AC task can aid in developing phonics skills, in which the student listens to a word, syllable, or letter sound and must choose the correct written form corresponding to what they heard. All data was anonymized and sourced from the Portuguese Reading Assessment Database (PRAD), a commonly used ALEPP pretest ad posttest.

The Phonemic Reading Assessment Ddatabase (PRAD) database datapoints came from 1,184 children's with typical development responses. Each learner completed all 228 trials of a fundamental reading skills assessment (460, 39%, girls and 724, 61%, boys). All children faced difficulties acquiring basic reading skills through standard school instruction, experiencing at least a one-year delay in reading acquisition. The average age was 8.8 years, ranging from 5 to 8 years old, with a standard deviation of 1.52, demonstrating an age distribution resembling a normal curve [2].

3   Proposed Method

It is necessary to group each child trial response by MTS Task Type to assess the specific performance of each reading skill. So, our method adopts the idea that each fundamental reading skill carries a specific weight in determining the difficulty of MTS tasks. Two factors influence this default MTS task weight: (1) the difficulty associated with the characteristics of each trial (MTS type), and (2) the significance of each task in fostering the development of basic reading skills. We derive the General Reading Competence Index (GR-CI) from the data-driven composition of weights assigned to ten primary reading skills available in the PRAD.

As our goal is to devise a unified index for assessing and distinguishing learners' reading skills, a relevant metric is the number of unique values the index generates for performance data. For instance, if a database consisting of 100 students can be ranked using an index that produces three unique values, it implies that approximately 33 students will share the same rank (or score). Consequently, distinguishing possible differences among these 33 students with the same score would not be feasible.

In contrast, if an index categorizes the same database into ten unique values, the uncertainty rate regarding the overall reading level within each group decreases by approximately 70%. Thus, a more significant number of students can be classified more accurately. Suppose the GR-CI score can generate a unique index composition that distinctly separates most of the PRAD students in terms of performance discrimination. In that case, it can validate the specific weight assigned to each MTS task type used on the index composition.

3.1   Reading Performance Unified Index

We propose a baseline version of the GR-CI index (GR-CI BL) where equal weights are assigned to all 10 PRAD reading MTS task types. The GR-CI BL was crucial to ascertain whether MTS tasks significantly impact the overall assessment of pre-reading skills differently. While this assumption may seem evident, particularly considering the cited research data, it represents a necessary empirical test. We can obtain a more accurate aggregated reading score or index by multiplying the weight of each MTS task type by its respective score.

3.2   Pre-reading Competence Index 5 and 10

We also adopted a straightforward approach to propose the initial test index based on the unique values generated by each MTS task type. This approach offers one of our proposal's advantages: after selecting the main MTS tasks, we count the unique values generated by each MTS task in the PRAD database.

Initially, we identified five primary MTS task types (GR-CI 5) that are most relevant to general reading skills based on MTS task-focused research and the *status quo* in the ALEPP Program research: (AC) select among written words from a dictated word cue; (AE) construct syllable-by-syllable word from a dictated word cue; (AF) handwrite a word from a dictated word cue; (BC) match the correct written word from an image as a cue - BC; and (CD) read aloud the written word presented decoding.

Subsequently, we used the task type with the highest number of unique values as a parameter to calculate the absolute weight for the others. Finally, we determined the relative weight of each task type based on the normalization of the absolute weight.

The GR-CI 10 version applied the same procedure as the GR-CI 5 but included just the ten symbolic MTS task types available in PRAD (excluding the BB task). This approach allows us to evaluate how much the standard practice in the research area, which typically focuses on the five main MTS tasks, identifies the learner's pre-reading skills levels.

3.3   Pre-reading Competence Index Based on Cluster Analysis

We consider identifying potential clusters within all symbolic PRAD database tasks. We selected ten features corresponding to performance by type of task: AB, BD, CB, CE, CF, AC, AE, AF, BC, and CD. We normalized these features' values between 0 and 1, representing elements that indicate the student's level of learning in each MTS reading task.

We used the K-means++ algorithm to generate clusters (AI technique), setting ten re-runs and limiting them to 300 steps as the analysis parameters. The cluster centers are selected from the remaining points with a probability proportional to the distance from the nearest center [1]. The Silhouette scores for various numbers of clusters indicate that two clusters have a superior score of 0.611. After identifying two well-defined clusters in the database, the first represents students who performed better (top performers). In contrast, the second cluster comprises students with poorer performance.

We iteratively applied the K-means++ analysis considering task types, which involved subtracting a particular task type for each iteration, resulting in ten re-applications of the algorithm. For example, we removed the *AB* task type in the first re-application, resulting in an analysis with BD, CB, CE, CF, AC, AE, AF, BC, and CD. In a re-application, we removed the BD task type and analyzed with *AB*, CB, CE, CF, AC, AE, AF, BC, and CD, and so on. Subsequently, we normalized the scores, assigning values between 0 and 1, ensuring that the sum of each MTS task type's attributed weight equaled 1 (Fig. 1).

This process computes the average value of the Silhouette Score between the clusters [10], as a higher score indicates better separation and well-defined clusters. Our rationale is that when removing an MTS task type (a feature of the model), if the general Silhouette score remains high, it suggests that

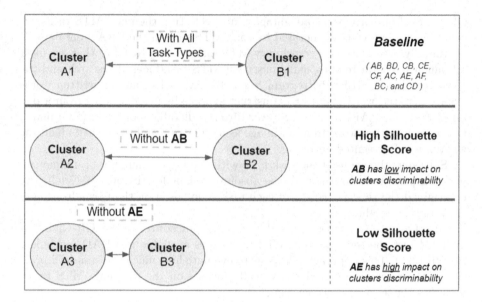

Fig. 1. Cluster difference on repeated K-means++ analysis after removing some task-type

the removed task type has little relevance in cluster separability. Each cluster aggregates students with different performance levels. However, if removing the task type reduces the Silhouette score, we consider this type highly relevant in differentiating student clusters.

For instance, AE is generally considered a more challenging task for students, while CE tends to be easier to understand. If we remove the AE type and execute the procedure, the obtained Silhouette score is 0.423. However, when we remove the CE type (while retaining AE) and repeat the process, the Silhouette increases to 0.621, close to the baseline value of 0.611. This result suggests that the CE type had minimal impact on distinguishing between the two clusters-one for students with high general reading scores and another for those with low scores. By analyzing silhouette values generated by the cluster analysis, we can derive a relative weight for each of the ten MTS reading tasks available in PRAD.

4   Evaluation Results

Our methodology facilitates running all three versions of GR-CI on the PRAD database. Each version generated an index for all learners in PRAD with notable distribution characteristics. Table 1 presents the discriminative power of the GR-CI k++ Filtered method, as evidenced by the number of unique values and its representation of the total student population in PRAD. We demonstrate a significant contrast in score variability between the cluster-based GR-CIs (GR-CI k++ Filtered) and those based solely on the intrinsic discriminative power of each MTS task (GR-CI 5 and GR-CI 10), the *status quo* technique. A key finding of our study indicates that AI analysis better discerns the most crucial MTS-Task-Types for foundational reading skills. The cluster-based version of GR-CI proved more adept at identifying MTS-Task relevances consistent with reading theory. Specifically, while GR-CI 5 and GR-CI 10 underscored the importance of AF (handwritten response to a dictated word) and CB (match the correct image from a written word presented as a cue), the filtered version of GR-CI k++ revealed AE (dictation tasks by construction of letters) and CD (decoding skill, reading at a loud a written word) as the most pertinent task types for discriminating student performance. Moreover, this shift in focus aligns with research on reading and learning and proves highly relevant in distinguishing between students with similar performance levels [2,12]. These findings hold significant implications for further research and development in this field.

Table 1. GR-CI versions weight per MTS reading task and unique values generated score for each GR-CI version (Unique Values = unique general reading score)

Task	Weight per Task Type for each GR-CI version			
Type	GR-CI BL	GR-CI 5	GR-CI 10	GR-CI k++ Filtered
AC	0.1	0	0.0769	0.0401
AE	0.1	0.1789	0.0934	0.2551**
AF	0.1	0.2211**	0.1154**	0.1294
BC	0.1	0.1895	0.0989	0.0774
CD	0.1	0.1895	0.0989	0.204**
AB	0.1	0	0.0824	0.0957
BD	0.1	0	0.1044	0.0846
CB	0.1	0.2211**	0.1154**	0.0678
CE	0.1	0	0.1099	0.046
CF	0.1	0	0.1044	0
Unique values	*163*	*526*	*708*	**1017**
Unique values (%)*	15.04%	48.52%	65.31%	93.82%

*Percentage of the total PRAD sample size.

* **Most relevant task type for the GR-CI version

The GR-CI k++ Filtered was the index version with superior discriminative performance, with nearly 94% of learners receiving a unique general reading score. This version generated 1,017 individual scores, whereas the GR-CI BL produced only 163 individual scores, representing a 16% reduction compared to GR-CI k++. Larger individual score counts enhance discriminative power once a broader range of unique scores enables a more detailed classification of students' general reading skills, allowing for a more individualized assessment.

Furthermore, just 26 learners received a score of 1 from GR-CI k++ Filtered, highlighting the classification sensitivity. In contrast, GR-CI 5 assigned the same score of 1 to 95 learners, three times more students. This significant difference underscores the varying classifications of learners' reading skills across different GR-CI versions. Thus, according to our top-performing GR-CI model, GR-CI k++ Filtered, the most relevant MTS task types for discriminating reading levels are, in descending order: AE, CD, AF, AB, BD, BC, CB, CE, AC, and CF.

The Spoken to Word Composition (AE) and Text to Read Aloud (CD) were the two most significant MTS task types for assessing basic reading skills. Both tasks contributed to approximately 45% of the variability in PRAD, based on the weight assigned to each in the GR-CI k++ Filtered index. The prominence of the Spoken to Handwriting (AF) task as the third most relevant weight aligns with the common perception among teachers regarding the importance of dictation. But, until now, we didn't have an exact number of how much. Conversely, the comparatively low weight values assigned to copy tasks (Text to Word Composition - CE and Text to Handwriting - CF) reinforce the lesser significance of these tasks in assessing basic reading skills, as indicated by the results.

5   Conclusion

The data-driven weighting of reading skill's MTS tasks can impact previous proposals, as evidenced by the effectiveness of adapting to primary reading skills outlined in literature [11]. Previous approaches typically determined task weights through consultation with specialists rather than leveraging a database of learners who underwent the teaching procedure. For instance, in our study, we assumed that reading aloud (CD MTS task type), serving as the reference reading skill, could update the hierarchy of influence of other reading skills.

When considering the relative weight of tasks adopted by GR-CI 5, only one MTS task (CDwor), or possibly two if we include AE, align with the relative weight adopted in the superior index (GR-CI k++). Moreover, the index incorporating the discriminative power of all MTS tasks in its composition, the baseline condition (GR-CI 10), outperforms GR-CI 5 (*status quo*). The former classifies learners' reading skills nearly 17% more accurately than the latter.

However, GR-CI 10 achieved a unique score for 61.35% of the PRAD students, indicating considerable differentiation. Interestingly, its distribution of MTS task weights appeared predominantly flat, with few tasks standing out significantly. This result suggests a need for clearer information regarding the relative impact of each basic reading skill on the learner's general reading competence. Indeed, the weight distribution of GR-CI 10 closely resembles that of GR-CI BL, indicating a similarity in their approach to task weighting.

Our analysis underscores the principle that "the whole is greater than the sum of its parts." Both indexes based on unique values (GR-CI 5 and GR-CI 10) operated on a simplistic and intuitive approach, which can be misleading. As an alternative, we proposed a data-driven analysis provided a weight distribution more nuanced and intricate relationship among basic reading skills. Hence, we introduced a method based on AI that offers higher discriminability regarding the relevance of the initial reading repertoire compared to the typical analysis approach commonly used in MTS reading research.

Future research can analyze parameters such as word complexity (including length, phonetic intricacy, and orthographic complexity) and response complexity (differentiating between single-choice and constructed responses). By considering these additional parameters, educators can further refine teaching methodologies to better cater to the diverse needs of learners.

References

1. Arthur, D., Vassilvitskii, S., et al.: k-means++: the advantages of careful seeding. Soda **7**, 1027–1035 (2007)
2. Castles, A., Rastle, K., Nation, K.: Ending the reading wars: reading acquisition from novice to expert. Psychol. Sci. Public Interest **19**(1), 5–51 (2018)
3. De Souza, D.G., De Rose, J.C., Faleiros, T.C., Bortoloti, R., Hanna, E.S., McIlvane, W.J.: Teaching generative reading via recombination of minimal textual units: a legacy of verbal behavior to children in Brazil. Int. J. Psychol. Psychol. Ther. **9**(1), 19 (2009)
4. Kim, Y.S.G., Lee, H., Zuilkowski, S.S.: Impact of literacy interventions on reading skills in low-and middle-income countries: a meta-analysis. Child Dev. **91**(2), 638–660 (2020)
5. Lovett, M., Frijters, J., Wolf, M., Steinbach, K., Sevcik, R., Morris, R.: Early intervention for children at risk for reading disabilities: the impact of grade at intervention and individual differences on intervention outcomes. J. Educ. Psychol. **109**, 889–914 (2017)
6. Pazeto, T.d.C.B., Dias, N.M., Gomes, C.M.A., Seabra, A.G.: Prediction of reading and writing in elementary education through early childhood education. Psicologia: Ciência e Profissão **40** (2020)
7. Roll, I., Wylie, R.: Evolution and revolution in artificial intelligence in education. Int. J. Artif. Intell. Educ. **26**(2), 582–599 (2016)
8. Roskos, K., Neuman, S.B.: Best practices in reading. Read. Teach. **67**(7), 507–511 (2014)
9. Sella, A.C., Tenório, J.P., Bandini, C.S.M., Bandini, H.H.M.: Games as a measure of reading and writing generalization after computerized teaching of reading skills. Psicologia: Reflexão e Crítica **29** (2016)
10. Shahapure, K.R., Nicholas, C.: Cluster quality analysis using silhouette score. In: Proceedings - 2020 IEEE 7th International Conference on Data Science and Advanced Analytics, DSAA 2020, pp. 747–748 (2020)
11. de Souza, G.N., de Deus, D.F., Tadaiesky, V., de Araújo, I.M., Monteiro, D.C., de Santana, Á.L.: Optimizing tasks generation for children in the early stages of literacy teaching: a study using bio-inspired metaheuristics. Soft. Comput. **22**(20), 6811–6824 (2018)
12. Suggate, S.P.: A meta-analysis of the long-term effects of phonemic awareness, phonics, fluency, and reading comprehension interventions. J. Learn. Disabil. **49**(1), 77–96 (2016)
13. Vanlehn, K.: Educational psychologist the relative effectiveness of human tutoring, intelligent tutoring systems, and other tutoring systems. Educ. Psychol. **46**(4), 197–221 (2011)

Gendered Responses to AI Governance: Insights from a Quantitative National Survey on ChatGPT Usage Among Students and Educators

Lahcen Qasserras(✉) 🆔

University of North Carolina at Charlotte, Charlotte, USA
lqasserr@charlotte.edu

Abstract. This study employs quantitative analysis to investigate how gender differences and educational policies influence students' and educators' willingness to engage with ChatGPT in the face of potential restrictions. Leveraging data from a national survey, which saw participation from 1,500 respondents across the United States, this research offers insights from a varied group encompassing both students and educators, thus ensuring a broad perspective on the subject. The findings reveal significant gender disparities in the inclination to use ChatGPT if it were banned, emphasizing the considerable impact of school policies. Critical Race Structuralism (CRS) was employed as a theoretical lens to delve deeper into these findings.

Keywords: Artificial Intelligence · Education · ChatGPT · Gender · Policy

1 Introduction

The increasing presence of generative AI in education presents both immense opportunities and potential pitfalls, particularly when it comes to gender equity. ChatGPT, a powerful language model, epitomizes this dilemma. While this technology stands at the brink of transforming personal learning and broadening access to knowledge, it concurrently carries a substantial threat of exacerbating and entrenching systemic inequities. In this regard, the rapid incorporation of AI into the educational sphere compels us to scrutinize how these technologies perpetuate or disrupt existing gender dynamics, power structures, and intersectional inequalities among students and educators. In this context, the present study probes into the repercussions of a hypothetical ChatGPT ban within educational settings. Specifically, using the Critical Race Structuralism framework [1], it critically analyzes the intersection of gender with institutional policies and illuminates the ways in which these elements are deeply entrenched in, and shaped by, overarching systems of power and entrenched institutional norms. The study is based on a comprehensive survey conducted by Quizlet, collecting responses from a diverse cohort of 1,500 participants across all American states, including students and educators from various educational levels and backgrounds.

A. M. Olney et al. (Eds.): AIED 2024 Workshops, CCIS 2150, pp. 242–253, 2024.
https://doi.org/10.1007/978-3-031-64315-6_20

2 Related Work

2.1 AI in Education: Enhancing Learning and Inclusivity

AI technology unlocks numerous opportunities as it can enhance the experience and effectiveness for educators and learners alike. AI's ubiquity and profound impact render it an indispensable skill set [2]. As such, it is crucial that AI literacy is integrated into education at the K-12 levels and beyond [3]. Not only do AI tools like ChatGPT facilitate the generation of personalized learning materials and inclusive and interactive lesson plans [4], but they also automate administrative tasks in education, including enrollment and registration, student record management, grading and assessment, and course scheduling [5]. Further, generative AI technologies can improve students' critical thinking, writing, and reading skills [5]. In higher education, ChatGPT facilitates research and writing tasks, promoting problem-solving skills [4]. However, the true capabilities of generative AI technologies can be fully harnessed only "if they are made accessible to all individuals, regardless of their socioeconomic background, geographical boundaries, and gender biasing" [6, p. 1647]. Importantly, AI technologies can aid in developing inclusive learning strategies that cater to marginalized individuals, those with disabilities, and students in remote communities [4].

2.2 Addressing Gender Disparity in AI and STEM Fields

Our society cannot fully reap the advantages of AI if the digital divide persists, with girls and women falling behind boys and men in areas such as STEM and computer science. The underrepresentation of women in AI is a significant concern that must be addressed. The data from a 2019 Nesta report serves as a clear illustration of this inequality. It reveals that a mere 13.8% of authors involved in AI research are women, which is significantly lower than the 15.5% representation of women in the wider STEM fields [7]. Further, the World Economic Forum [8] issued a warning about emerging gender gaps in Artificial Intelligence-related skills. It also highlighted a significant gap between female and male representation among AI professionals as only 22% of AI professionals worldwide are female [8]. In this regard, Young et al. [9] argue that a pronounced gender imbalance is structurally ingrained in the career progressions of professionals within the data science and AI sectors.

2.3 AI and Equity in Education

Gender-based disparities in AI can be ascribed to various factors, one of which includes the discouragement of girls from pursuing AI-related interests [10]. Further, the absence of role models and female representation within the AI field can compound this effect [11]. More concerning, in their study on gender bias in ICT professor recruitment in Finland, Tiainen and Berki [12] found consistent pattern of male dominance. While schools are increasingly implementing policies and guidelines to address ethical concerns, privacy issues, and potential biases inherent in AI tools, the impact of such policies on gender equity remains largely unexplored [13]. One potential concern is that restrictive policies, such as outright bans on ChatGPT, may exacerbate gender disparities as "the strong may become stronger, while the vulnerable risk further marginalization, primarily due to disparities in resource allocation" [14, p. 355].

3 Hypothesis

This study investigates the impact of a school policy regulating AI on the gender usage gap concerning ChatGPT.
Research Hypothesis:

- The presence of a school policy regulating AI increases the gender usage gap in favor of males regarding ChatGPT.

Null Hypotheses:

- There is no significant increase in the gender usage gap in the presence of a school policy regulating AI regarding ChatGPT.
- Gender does not significantly moderate the impact of a school policy regulating AI on the gender usage gap regarding ChatGPT.

4 Methodology

4.1 Participants

The dataset for this study was provided to the author by Quizlet team, who conducted a survey of 1,500 U.S. respondents in June 2023. As seen in Table 1, the dataset comprised 1,000 students between the ages of 14–22 and 500 educators teaching at the high school or college level.

Table 1. Employment Status of the Respondents

Status	Frequency	Percent	Valid Percent	Cumulative Percent
Educator	516	34.0	34.0	34.0
Student	1000	66.0	66.0	100.0
Total	1516	100.0	100.0	

Within the student population, the distribution across educational levels was diverse, including high school freshmen (17.7%), sophomores (10.0%), juniors (11.9%), seniors (13.8%), and college students ranging from freshmen to seniors (49.4%).

As seen in Table 2, among the educators, the majority were high school teachers (67.4%), with the remainder teaching at the collegiate level (32.6%).

Table 2. Level of Schooling Taught

Status	Frequency	Percent	Valid Percent	Cumulative Percent
High school	337	22.2	67.4	67.4
Higher education	163	10.8	32.6	100.0
Total	500	33.0	100.0	

As can be seen in Table 3, the gender composition of the respondents was 40.4% male, 56.0% female, with a small percentage identifying as other (0.3%), preferring not to disclose (0.7%), or non-binary (2.6%).

Table 3. Gender of the Respondents

Status	Frequency	Percent	Valid Percent	Cumulative Percent
Male	612	40.4	40.4	40.4
Female	849	56.0	56.0	96.4
Other	5	.3	.3	96.7
Prefer not to say	10	.7	.7	97.4
Non-binary	40	2.6	2.6	100.0
Total	1516	100.0	100.0	

4.2 Research Instrument and Data Collection

The instrument central to the data collection in this research is the comprehensive survey developed and conducted in June 2023 by Quizlet. The dataset was anonymized and obtained with the proper informed consent of the participants. Quizlet's extensive experience in creating educational tools and platforms should lead credence to the reliability and validity of the survey. For the purposes of this study, Quizlet has granted the author access to their proprietary dataset.

4.3 Data Analysis

For the quantitative analysis of the data, the author utilized IBM® SPSS® Statistics Premium 28 (Windows). This advanced statistical software package enabled the author to conduct a detailed logistic regression analysis, aiming to unpack the determinants influencing students' intentions to use ChatGPT in the event of a school ban, with a particular focus on gender and the presence of a school policy. Logistic regression was the chosen method due to its efficacy in dealing with binary dependent variables, such as the study's primary variable of interest (Q13_Recode), which assesses whether students would continue to use ChatGPT if it were banned.

For the purpose of analysis, the following variables are defined:

Dependent Variable:

Q13: "If ChatGPT or a similar AI technology was banned in your school, would you still use it?" This question was transformed into a binary variable, with non-definitive responses ('definitely not', 'probably not', 'might or might not') coded as '0' (No), and affirmative responses ('probably yes', 'definitely yes') as '1' (Yes), to clearly indicate intent to use ChatGPT despite potential bans.

Independent Variables:

Q14: "Has your school established a code of conduct or an advisory for the use of Chat-GPT and similar AI technologies?" Responses were coded as '1' for 'yes' (indicating an established school policy) and '0' for all other responses ('no', 'discussed but no action', 'does not apply').

Q32: "What is your gender?" This variable was recoded into Q32_Recode for binary classification: '1' for male and '0' for female. Participants identifying as 'Other', 'Prefer not to say', or 'Non-binary' were excluded from this analysis, leading to a final sample size of 1461 participants.

This dichotomous approach in coding Q13 and Q14 accentuates the contrast between definitive and non-definitive attitudes toward ChatGPT usage and the presence versus absence of institutional AI regulations. The binary classification of gender allows for a focused analysis of the determinants of AI technology usage in educational settings, particularly examining how gender interplays with school policies in influencing student intentions.

4.4   Results

Initially, in the absence of other predictors (Block 0), the model's predictive capacity hinged solely on the intercept and exhibited an ability to perfectly predict cases where students would not use ChatGPT if banned. However, it could not predict any of the instances where students indicated they would, leading to an overall prediction accuracy of 62.1%. This starting point, represented by a significant constant (B value of -0.493), established the baseline likelihood of non-use.

Upon introducing gender (Q32_Recode), school policy (Q14_Recode), and their interaction (Q14RecQ32ReInteraction) in Block 1, the model's predictive power was notably enhanced. As seen in Table 4, the Omnibus Tests of Model Coefficients displayed a substantial improvement with a Chi-square of 99.947 (p < .001).

Table 4. Omnibus Tests of Model Coefficients

Chi-square			df	Sig
Step 1	Step	99.947	3	< .001
	Block	99.947	3	< .001
	Model	99.947	3	< .001

As shown in the Table 5, the model's -2 Log Likelihood decreased to 1839.289, with Cox & Snell and Nagelkerke R Square values of .066 and .090, respectively. This indicated that the inclusion of these predictors accounted for between 6.6% to 9% of the variance in the outcome, a significant increment from the baseline model.

Table 5. Model Summary

Step	-2 Log likelihood	Cox & Snell R Square	Nagelkerke R Square
1.0	1839.289[a]	.066	.090

a. Estimation terminated at iteration number 3 because parameter estimates changed by less than .001.

As displayed in Table 6, the predictive accuracy for the 'No' response was high at 93.8%, and the 'Yes' response prediction improved to 22.2%. These figures reflect an augmented ability of the model to anticipate student behavior regarding ChatGPT usage in the context of a ban.

Table 6. Classification Table

Observed			Predicted		
			Q13_Recode		
			.00	1.00	Percentage Correct
Step 1	Q13_Recode	.00	851	56	93.8
		1.00	431	123	22.2
	Overall Percentage				66.7

a. The cut value is .500.

Prior to interpreting the logistic regression outcomes, it was crucial to address the potential issue of multicollinearity among the independent variables. Multicollinearity can significantly affect the stability and interpretation of the regression coefficients. To diagnose this condition, a linear regression analysis was conducted solely for the purpose of assessing multicollinearity through the Variance Inflation Factor (VIF) and tolerance statistics, which are standard measures derived from the linear regression framework. The linear regression diagnostic checks revealed all VIF values to be well below the commonly used threshold of 10 (all VIFs < 2.5), suggesting that the predictors are not overly correlated. Similarly, tolerance levels were all above the threshold of concern (all values > 0.4), providing further confirmation of minimal multicollinearity. This precluded the inflation of the variance of the logistic regression estimates and ensured that each independent variable contributed unique explanatory power to the model.

Having established that multicollinearity was not a concern, the author proceeded with the logistic regression analysis with greater confidence in the reliability of the coefficient estimates. This enabled the exploration of the impact of gender

(Q32_Recode), the presence of a school policy (Q14_Recode), and their interaction (Q14RecQ32ReInteraction) on the probability that students would use ChatGPT if it were banned.

The model's coefficients offered further insight. As shown in Table 7, the interaction term between gender and school policy (Q14RecQ32ReInteraction) bore a B value of .596, which was significant (p = .024). This signified that the effect of a school's code of conduct differed based on the student's gender, with the Exp(B) value of 1.814 indicating an 81.4% increase in the odds of a male student stating they would use ChatGPT if banned when a school policy is present.

Table 7. Variables in the Equation

		B	S.E	Wald	df	Sig	Exp(B)
Step 1[a]	Q14RecQ32RecInteraction	.596	.263	5.112	1	.024	1.814
	Q14_Recoded	.816	.182	20.037	1	< .001	2.261
	Q32_Recoded	.244	.131	3.488	1	.062	1.276
	Constant	−.869	.083	109.477	1	< .001	.420

The school policy variable (Q14_Recode) itself was a strong predictor with a B value of .816 (p < .001), where students from schools with a policy were more than twice as likely to state they would use ChatGPT if banned, as indicated by an Exp(B) of 2.261.

To elucidate these effects further, the author calculated the predicted probabilities for each subgroup using the logistic regression formula. The calculation involved inputting the values of gender and school policy into the equation and considering their interaction:

- Female with no school policy ($X1 = 0, X2 = 0$): The calculated probability of stating they would use ChatGPT if banned is approximately 29.55%.
- Female with school policy ($X1 = 0, X2 = 1$): The probability increases to around 48.68% when there is a school policy.
- Male with no school policy ($X1 = 1, X2 = 0$): The probability for males without a school policy is approximately 34.86%.
- Male with school policy ($X1 = 1, X2 = 1$): The presence of a school policy raises the probability significantly for males, to about 68.72%.

These probabilities demonstrate that males are more inclined than females to state they would use ChatGPT if it were banned, especially when a school policy is in place. The doubling of the probability for males from no policy to a policy scenario underscores the significant impact of institutional policies on student intentions to use banned AI technologies like ChatGPT.

The logistic regression analysis underscores a gender disparity in students' likelihood to use ChatGPT if banned, which is significantly influenced by the presence of school policies related to the use of ChatGPT. The significant interaction between gender and school policy revealed through this model highlights the complex interplay between these factors, affecting males more substantially. These findings are invaluable for educational policymakers, suggesting the need for nuanced, gender-aware policy development to navigate the challenges presented by AI technologies in educational settings. Furthermore, these results confirm the research hypothesis: The presence of a school policy regulating AI increases the gender usage gap in favor of males regarding ChatGPT, thus validating the initial assumptions about the influence of institutional policies on gendered interactions with AI technologies.

5 Discussion

5.1 Critical Race Structuralism

Critical Race Structuralism (CRS) [15] is a theoretical framework introduced to expand our understanding of racial and ethnic relations within social and institutional systems. It examines the complex interplay between race, culture, gender, and social structures, focusing on how these elements interact to shape societal dynamics and institutions. From CRS perspective, "Education often does not provide the safe space needed for deep conversations surrounding race, class, and gender where there may be repercussions or reprisals for raising these issues" [15, p. 2219]. In the analysis of the findings of this study and its broader implications, the author will focus primarily on gender equity and societal structure, using Critical Race Structuralism (CRS) as a guiding lens. The discussion will be informed by the five tenets of CRS [15].

The Five Tenets of Critical Race Structuralism

- Critically analyze societal structure;

- Address dominant cultural indoctrination in education practices and policies;

- Utilize social justice to advocate for equitable representation, access, and resources;

- Synergize institutional change by being a catalyst for deconstructing racism and bias;

- Engage in intercultural collaborative communication and actions of change.

Wiggan et al., 2020

5.2 Critically Analyze Societal Structure

According to Wiggan et al. [1, p. 460], "CRS analyzes racial and ethnic relations in social structures and institutional systems to explore patterns and relationships between race, culture, gender, and social structures". As shown in the findings of this study, policies and guidelines are not mere adjuncts, but they embody the essence of this structure and manifest the spirit of the system [16]. This paper's findings suggest that the existence of a policy or guidelines has catalyzed divergent responses among genders in relation to the use of ChatGPT. It revealed that male students and educators exhibit an intensified pursuit of access, particularly in the face of established regulatory frameworks or guidelines that pose a potential limitation to their interaction with the AI technology. To be sure, systemic structures exert a differential impact on the experiences and actions of different genders. This speaks to the ingrained societal norms and power relations that subtly, yet profoundly, shape gendered interactions with technological systems and policies reflective of the larger societal constructs of power and control [17].

Clearly, the evidence-based disparities in AI technology access due to established policies and regulation becomes a matter of pressing concern. As a case in point, the findings of a Resume Builder survey [18], which underscores that ChatGPT significantly boosts work efficiency, reveals that approximately 25% of employees save up to 10 h weekly. Crucially, 78% of these workers experience an improved work-life balance, and 52% observe a positive impact on their mental health due to using ChatGPT [18]. These findings suggest that women may be disproportionately disadvantaged by inequitably implemented policies governing AI technology usage.

5.3 Address Dominant Cultural Indoctrination in Education Practices and Policies

Given the findings of the current study, we should continue to confront the prevailing narrative that the underrepresentation of females in STEM fields, and by extension in AI, is a byproduct of innate biological differences in personality, academic inclinations, and career ambitions. The divergent responses of males and females to the implementation of ChatGPT policies can be critically examined as a reflection of gendered social conditioning within a patriarchal societal structure. One can venture to assert today's cultural norms that valorize assertiveness and success in males encourage them to eschew established rules or norms [10]. In Tanzania, for example, Matete [19] found that that cultural myths and societal beliefs often discourage women from participating in science-related courses by perpetuating the notion that these fields are more suited for men.

5.4 Utilize Social Justice to Advocate for Equitable Representation, Access, and Resources

Female students and educators, especially those coming from racially marginalized backgrounds, must be acutely conscious of the differential perceptions and applications of laws and regulations. This heightened awareness is crucial, not only as a means of understanding but as a catalyst for active self-advocacy and participation in the regulatory process. The objective here is twofold: to ensure that these regulations are crafted with a keen awareness of potential unequal impacts and to safeguard against their interpretation and enforcement being subject to individual biases. The fundamental aim transcends merely intensifying regulations governing technological tools like ChatGPT. According to Malmström et al. [20], while over sixty percent of respondents in Sweden view the use of chatbots during examinations as cheating, they oppose outright bans on AI in educational contexts. Similarly, Chen [21] reported that while 75% of students acknowledge that it is unethical to use the program for cheating, they continue to do so. As such, it is imperative to approach policy formulation with a heightened sensitivity to the impacts these decisions can have across different genders. It is not enough for policies to appear neutral; they must be critically analyzed for their potential to create unequal access to technology [22].

5.5 Synergize Institutional Change by Being a Catalyst for Deconstructing Racism and Bias

Wiggan et al. [15, p. 2219] assert that "The nation has only progressed as far as the thoughts, language, and actions of its youth, and what they experience in their lives." They suggest that "An engaging educational experience allows for growth and development, while nurturing the intellectual and emotional needs of diverse learners, which is also conducive to improving student learning outcomes in STEM classrooms" [15, p. 2229]. As such, female students and educators should be surrounded by language and narratives that empower them to view themselves as active contributors to the field of AI. As Wiggan et al. [15, p. 2220] posit, "As seen through the lens of CRS, language can be a tool of oppression or liberation, giving voice to social justice issues such as environmental racism".

5.6 Engage in Intercultural Collaborative Communication and Actions of Change

The empowerment of women from every race, ethnicity, and socioeconomic status contributes to the empowerment of all women. White women, as part of the dominant social group, may disproportionately benefit from generative AI policies crafted within a culture and hierarchy dominated by White men. Even in the context of the Diversity, Equity, and Inclusion (DEI) initiatives, such inadvertent biases and exclusions can occur, often subtly undermining the very objectives these programs aim to achieve. In their 2022 study, Nakajima et al. [23] conducted a comprehensive analysis of diversity initiatives in computer science majors at four U.S. higher education institutions. They found that "departments were centering gender diversity work in ways that privileged white women, at the expense of supporting WoC students and staff" [23, p. 5]. It is critical to recognize that injustices against women from minority groups ultimately do not serve the interests of white women or society at large.

6 Implications

This study's quantitative analysis, while providing initial insights, recognizes the limitations of its binary approach. Specifically, the use of only two binary variables- gender and the binary state of policy presence as predictors for AI tool usage may not capture the complex realities of educational settings. Important factors such as the nuanced nature of policy enforcement and the respondents' educational levels were not included in the regression model. Moreover, the variable reflecting usage of the tool could benefit from a more detailed scale than the binary one used. Future studies are encouraged to employ a more refined analysis that incorporates these additional variables and explores the multidimensional aspects of policy impact.

7 Conclusion

This article provides critical insights into the gender dynamics influencing students' intentions to use ChatGPT amidst potential bans. It reveals a complex interaction between gender, school policies, and AI technology usage, emphasizing the need for more gender-sensitive and inclusive policymaking in education. As AI continues to shape educational landscapes, this study highlights the urgent need for policies that are cognizant of and responsive to the nuanced needs of diverse student populations.

Conflict of Interest. The author has no financial or personal relationships with other people or organizations that could inappropriately influence or bias the work presented in this paper.

References

1. Wiggan, G., Teasdell, A., Parsons, T.: Critical race structuralism and Charles Mills' racial contract: pedagogical practices for twenty-first-century educators. Sociol. Race Ethnicity **8**(4), 456–463 (2022)
2. Ng, D.T.K., Leung, J.K.L., Chu, S.K.W., Qiao, M.S.: Conceptualizing AI literacy: an exploratory review. Comput. Educ.: Artif. Intell. **2**, 100041 (2021)
3. Steinbauer, G., Kandlhofer, M., Chklovski, T., Heintz, F., Koenig, S.: A differentiated discussion about AI education K-12. KI-Künstliche Intelligenz. **35**(2), 131–137 (2021)
4. Kasneci, E., et al.: ChatGPT for good? On opportunities and challenges of large language models for education. Learn. Individ. Differ. **103**, 102274 (2023)
5. Pedro, F., Subosa, M., Rivas, A., Valverde, P.: Artificial intelligence in education: challenges and opportunities for sustainable development. UNESCO (2019). https://en.unesco.org/the mes/education-policy-planning/
6. Khan, I.A., Paliwal, N.W.: ChatGPT and digital inequality: a rising concern. Sch. J. App. Med. Sci. **9**, 1646–1647 (2023)
7. Stathoulopoulos, K., Mateos-Garcia, J.: Gender diversity in AI research. Nesta. (2019). https://media.nesta.org.uk/documents/Gender_Diversity_in_AI_Research.pdf
8. World Economic Forum. Assessing gender gaps in artificial intelligence: global gender gap report 2018 (2018). https://www.weforum.org/publications/reader-global-gender-gap-report-2018/in-full/
9. Young, E., Wajcman, J., Sprejer, L.: Mind the gender gap: inequalities in the emergent professions of artificial intelligence (AI) and data science. N. Technol. Work. Employ. **38**(3), 391–414 (2023)
10. Rahmawati, A., Syahriyani, A.: Challenging the patriarchal norms: examining hegemonic masculinity in Dickinson TV Series. Insaniyat J. Islam Hum. **7**, 1–14 (2022)
11. Fraser, S., Mancl, D.: Dimensions of diversity, equity, and inclusion. ACM SIGSOFT Softw. Eng. Notes **48**(2), 18–21 (2023)
12. Tiainen, T., Berki, E.: The reproduction process of gender bias: a case of ICT professors through recruitment in a gender-neutral country. Stud. High. Educ. **44**(1), 170–184 (2019)
13. Villagran, M., Ghosh, S. Understanding sexual and gender minority privacy. In: Proceedings of the ALISE Annual Conference (2022)
14. Li, H.: AI in education: Bridging the divide or widening the gap? Exploring equity, opportunities, and challenges in the digital age. Adv. Educ. Hum. Soc. Sci. Res. **8**(1), 355 (2023)

15. Wiggan, G., Pass, M.B., Gadd, S.R.: Critical race structuralism: the role of science education in teaching social justice issues in urban education and pre-service teacher education programs. Urban Educ. **58**(9), 2209–2238 (2023)
16. Parker, D. C., Conversano, P. Narratives of systemic barriers and accessibility: poverty, equity, diversity, inclusion, and the call for a post-pandemic new normal. Front. Educ. (2021)
17. Sandoval, J.C.B., Figuerêdo de Santana, V., Berger, S. E., Quigley, L.T., Hobson, S.F.: Responsible and inclusive technology framework: a formative framework to promote societal considerations in information technology contexts. ArXiv (2023)
18. ResumeBuilder.com. 1 in 4 workers using ChatGPT started a side hustle with time saved from using the tool (2023). https://www.resumebuilder.com/1-in-4-workers-using-chatgpt-started-a-side-hustle-with-time-saved/
19. Matete, R.E.: Why are women under-represented in STEM in higher education in Tanzania?. FIRE: Forum Int. Res. Educ. **7**(2), 48–63 (2021)
20. Malmström, H., Stöhr, C., Ou, A.W.: Chatbots and other AI for learning: a survey of use and views among university students in Sweden. Chalmers Stud. Commun. Learn. Higher Educ. 1(10.17196) (2023)
21. Chan, C.K.Y.: A comprehensive AI policy education framework for university teaching and learning. Int. J. Educ. Technol. High. Educ. **20**(1), 38 (2023)
22. Ahmad, A.: Men's perception of women regarding the internet usage in the Khyber agency Pakistan: an exploratory study. Global Mass Commun. Rev. **1**, 47–58 (2020)
23. Nakajima, T.M., Karpicz, J.R., Gutzwa, J.A.: Why isn't this space more inclusive?: marginalization of racial equity work in undergraduate computing departments. J. Diversity High. Educ. (2022)

Global Trends in Scientific Debates on Trustworthy and Ethical Artificial Intelligence and Education

Christian M. Stracke[1]([⊠]) [iD], Irene-Angelica Chounta[2] [iD], and Wayne Homes[3] [iD]

[1] University of Bonn, Bonn, Germany
stracke@uni-bonn.de
[2] University of Duisburg-Essen, Duisburg, Germany
irene-angelica.chounta@uni-due.de
[3] University College London, London, UK
wayne.holmes@ucl.ac.uk

Abstract. This paper presents a systematic review of the scientific literature on trustworthy and ethical Artificial Intelligence (AI) and Education (AI&ED), including both AI *applied* in education to support teaching and learning (AIED), as well as education *about* AI (AI literacy). Key interest is the identification of global trends with a special focus on unbalanced disparities. Strictly following the standardised protocol and the underlying PRISMA approach, 324 records were identified and selected according to the pre-defined protocol for the systematic review. Finally, 62 articles were included in the quantitative and qualitative analysis in response to four research questions: Which (i) journals, (ii) disciplines, and (iii) regions are leading scientific debates and sustainable developments in education and trustworthy/ethical AI, and (iv) what are the past trends? The articles revealed an unbalanced distribution across the various dimensions, together with an exponential growth over recent years. Building upon our analysis, we argue for an increase in interdisciplinary research that shifts the focus from the currently dominant technological focus towards a more human-centered (educational and societal) focus. Only through such a development AI can contribute effectively to the UN Sustainable Development Goal no. 4 of a world with equitable and universal access to quality education. The results of our systematic review provide the basis to address and facilitate equality in the future AI&ED progress across regions worldwide.

Keywords: Trustworthy and ethical AI · AI&ED · Web of Science articles · Systematic literature review · Informatics and information technologies · Education and learning sciences · Sustainable digital transformations

1 Introduction

The concept of Artificial Intelligence (AI) has been controversial since the term was first coined [4, 14, 18]. Nonetheless, AI has been introduced in many disciplines, including – for around fifty years – in education [2, 13, 15, 20, 23]. However, it remains the case

A. M. Olney et al. (Eds.): AIED 2024 Workshops, CCIS 2150, pp. 254–262, 2024.
https://doi.org/10.1007/978-3-031-64315-6_21

that AI in education was mostly researched by computer scientists rather than educators in the beginning [33].

2 Background

Educational systems and societies worldwide are increasingly challenged by rapid changes facilitated and caused by globalization, connectivity and new (social) media. In response, Open Education and Open Educational Resources (OER) have been promoted by the United Nations Educational, Scientific and Cultural Organization [29] to help achieve the United Nations' Sustainable Development Goal 4 (equitable and inclusive quality education for all) [25, 29]; and the value of OER was demonstrated during the COVID-19 lockdowns [26, 28, 30]. Meanwhile, it has also been suggested that recent technologies, such as AI, have potential for enhancing education, but they also bring various challenges [5, 6, 9, 13]. Several international agencies have discussed the potential of AI for future sustainable education [8, 19].

"AI in education for sustainable society" is the theme of the AIED 2023 Conference while the AIED 2024 Conference focuses "AI in education for a world in transition". Both themes call for societal considerations and objectives of future AI in education applications. However, to achieve a sustainable society, it is not only necessary to facilitate education *through* AI (the application of AI in education) but also to foster and improve education *about* AI (the teaching of AI in education) [10, 12, 13]. We need students and citizens who have digital competences, including what might be called 'AI Literacy', to understand, support and realize a sustainable society [11, 26, 28, 31]. Thus, we address both directions in our paper: AI applied in education (AIED) and AI taught in education (AI literacy) what we call "AI and Education (AI&ED)". We specifically focus on trustworthy and ethical AI&ED.

Already more than forty years ago, the research on using AI in education (AIED) began mainly focusing school and higher education [3, 13, 16, 27]. There are some systematic literature reviews on the current AIED research providing a first overview [5, 7, 17, 24, 32].

In the global AI&ED research and development community, ethical discussions were launched early but they were not gaining attention and continuation [27]. The identification of the necessity of ethical AI&ED and the development of community-driven proposals and frameworks for ethical AI&ED took twenty years [1, 2, 6, 11, 12].

To the best of our knowledge, the relation between trustworthy and ethical AI and education has not yet been systematically analyzed. This systematic literature review (SLR) aims to fill this gap for a topic that is likely to gain more importance in the near future. In addition, the results might inform future research and be used as a framework to differentiate and classify theoretical concepts and practical approaches. Accordingly, this work aims to explore the latest scientific literature trends concerning the relationship between trustworthy and ethical AI and education. To that end, we set out to answer the following four research questions:

RQ1: Which journals are leading scientific debates and sustainable developments in education and trustworthy/ethical AI?

RQ2: Which disciplines are leading scientific debates and sustainable developments in education and trustworthy/ethical AI?

RQ3: Which geographical regions are leading scientific debates and sustainable developments in education and trustworthy/ethical AI?

RQ4: Which trends are leading scientific debates and sustainable developments in education and trustworthy/ethical AI?

3 Methodology

The systematic review strictly followed the standardized protocol for systematic literature reviews on AI&ED [27] and the underlying PRISMA statement and its procedures [21, 22], which involves four phases for the selection of articles:

(i) *Identification*, (ii) *Screening*, (iii) *Eligibility*, and (iv) *Included*. To ensure reliability, the four phases of the PRISMA process were undertaken by two reviewers (first two co-authors of this paper), each of whom have research experience in AI&ED and educational technology.

For the *Identification* phase, the reviewers reviewed in parallel the titles and abstracts of the records collected from the database Web of Science (Clarivate Analytics) using the search term: "TS = (("artificial intelligence") AND ((trust*) OR (ethic*)) AND (education))". The reviewers agreed on all records but one, and reached consensus on this final record after a discussion.

For the full *Screening* phase, all the titles and abstracts of the records generated by the first phase were reviewed, using the exclusion and inclusion criteria from [27].

For the *Eligibility* phase, the reviewers then reviewed in parallel the full text of the records remaining after the *Screening* phase, following the criteria defined by [27].

In the *Included* phase, all selected records were analysed related to the research questions.

4 Analysis and Discussion

In this section, we present the results of the systematic review and their analysis in terms of the four research questions (RQs).

The *Identification* phase generated a list of 324 records (Table 1). In the *Screening* phase, 43 records were removed based on the analysis of their titles and abstracts and the criteria. In the *Eligibility* phase, 219 records were removed based on an analysis of the full texts and the criteria. This left a total of 62 scientific journal articles all of which were subject to quantitative and qualitative analyses.[1]

RQ1: Which Journals are Leading Scientific Debates and Sustainable Developments in Education and Trustworthy/Ethical AI?

The 62 articles selected and reviewed for this SLR were published in 46 different journals, published by 23 publishing houses. The 'International Journal of the Artificial Intelligence in Education' (IJAIED) published most of the articles (8 in total), followed

---

[1] The selected 62 articles will be published with a DOI under an open and free license.

Table 1. Summary, with numerical results, of the four phases of the systematic review.

Phase	Screened records	Removed records
Identification	Records identified (n = 324)	Duplicates removed (n = 0)
Screening	Records for formal screening (title & abstract) (n = 324)	Records removed by formal reasons (n = 43)
Eligibility	Records for content-related screening (full text) (n = 281)	Records removed by content-related reasons (n = 219)
Included	Papers included for review analysis (n = 62)	

by 'Education and Information Technologies' (EIT) (4 publications) and the 'Journal of Research on Technology in Education' (JRTE) (3 publications). 39 out of the 46 journals published only one article. Based on the Web of Science statistics, the articles published in IJAIED have been cited on average 5 times (SD = 7.5), with 180 days usage count average of 11 (SD = 8). For the articles published in EIT the average 180 days usage count was 14 (SD = 8) but they were cited less (2 citations in total). For the articles published in JRTE the average 180 days usage count was 9 (SD = 10) while no citations were recorded. In terms of citations, the most cited article [32] was published in the 'International Journal of Educational Technology in Higher Education' (197 citations) followed by the article by [12] in the 'International Journal of the Artificial Intelligence in Education' with 23 citations.

RQ2: Which Disciplines are Leading Scientific Debates and...?
To answer RQ2, we analyzed the articles included in this SLR in terms of the discipline of the publishing journal. Mainly, the articles were published in journals focusing on education, medicine and health care, technology and information systems, and business and social sciences. Some journals also had an interdisciplinary focus, such as 'Artificial Intelligence and Society' or 'Technology and Education'. Most articles were published in interdisciplinary journals with an education and technology focus (17 articles), followed by those with a technological focus (15 articles), an education focus (9 articles) and journals with strong medical and healthcare direction (8 articles).

RQ3: Which Geographical Regions are Leading Scientific Debates and...?
The regional distribution of the selected 62 articles is not balanced. The first authors of the articles are affiliated to 27 countries in total, although only 9 countries (USA: 13, UK: 7, China: 6, Australia & South Korea: 4, Finland & Spain: 3, Canada & Germany: 2) were represented by two or more papers. The selected 62 articles are also not equally spread across the globe, as is often seen due to different conditions and opportunities in relation to development and resources. Only five countries with ten articles in total belong to the so-called Global South, six of which are from China. Geographically, only two countries with a total of five articles belong to the southern hemisphere (Australia with four articles and South Africa with one article).

RQ4: Which Trends are Leading Scientific Debates and Sustainable Developments in Education and Trustworthy/Ethical AI?

There are several interesting trends that we can derive from the quantitative and qualitative analysis of the 62 selected articles. Most obviously, there is the large growth of relevant articles during the most recent three years (Fig. 1.). Even though there was no time limit, the first publication only appeared in the year 1999 with a second one not following until 18 years later. Apart from the first article, all the other articles were published during the last five years with an exponential increase during the most recent three years. In short, it is reasonable to suggest that the discussion on trustworthy and ethical AI and education really only started three years ago.

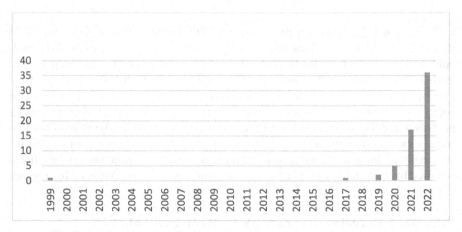

Fig. 1. Publication years of selected 62 articles of AI&ED systematic review.

Furthermore, trends can be identified in the analysis of the different article types. The vast majority of the selected 62 articles are discussions of theories, argumentations and literature (31 articles = 50.0%), followed by ten survey analysis studies (16.1%), seven systematic reviews (11.3%), five mixed methods studies (8.0%), four interview analysis studies (6.5%), three thematic analysis studies (4.8%), one data analysis study and one position paper of an association (1.6% each).

5   Discussion

In this section, we discuss the results and analysis of the selected 62 articles as presented in the sections before. We structure again our discussion along the leading four research questions (RQs).

RQ1: Which Journals are Leading Scientific Debates and Sustainable Developments in Education and Trustworthy/Ethical AI?

Given the basis of 62 analysed articles, the huge number of different journals (46) is indicating a pre-mature research field without any leading publication channels. Only

3 journals (7%) count more than 2 articles while the vast majority (39 of 46 = 85%) has published only one article from the selected 62 items. That underlines that the topic trustworthy and ethical AI&ED is not yet assigned to specific journals which have also not discovered it as potential key focus for distinction from other journals. It is striking that only one article achieved high visibility and citations while even the article on the second position presents a huge distance related citation (197 against 23 citations).

RQ2: Which Disciplines are Leading Scientific Debates and Sustainable Developments in Education and Trustworthy/Ethical AI?

Not surprisingly, most of the selected 62 articles (41 = 66%) are belonging to disciplines summarized in education and technology and all other disciplines are represented only one time except medicine and healthcare (8 = 13%) and (social) sciences (10 = 16%). This huge representation of medicine and healthcare is representing the big efforts to implement and use AI&ED in the health system and education what has to be combined with ethical reflections and statements as mandatory requirement of the discipline. On the other hand, the even higher representation of (social) sciences is remarkable as this discipline is not known for many AI&ED implementations and uses: We can argue that the discussion of trustworthy and ethical concerns is very common for (social) sciences.

RQ3: Which Geographical Regions are Leading Scientific Debates and Sustainable Developments in Education and Trustworthy/Ethical AI?

The distribution in relation to country affiliations of the first authors demonstrates a clear unbalance between developed (52 = 84%) against the rest of the world (10 = 16%) as well as between northern (60 = 97%) and southern (2 = 3%) countries. This suggests again that the development, research and implementation of innovative technologies such as AI are mainly (presumably for socio-economic reasons) driven by countries in the Global North in deep contrast to the Global South. That is independent from the ongoing debate whether China should be assigned to the Global South what it questionable given the economic power and progress of this special country due to its largest population worldwide. Overall, all these conditions are challenging the UN Sustainable Development Goals (SDG) and in particular the achievement of SDG no. 4 which demands equitable and universal access to quality education worldwide.

RQ4: Which Trends are Leading Scientific Debates and Sustainable Developments in Education and Trustworthy/Ethical AI?

The rise of the research on trustworthy and ethical AI&ED started only three years ago as already mentioned in the analysis. However, the almost incredible increase by 2000% within three years is indicating a dramatic disruption and shift in the research focus. One reason could be seen in the similar changes in business investments, funding projects, opened job opportunities (both in enterprises and academia) and established professorships and research positions.

The complete absence of pedagogical, design-based as well as empirical, evidence-based studies on experiments from the practice is notable and underlines the starting point of a scientific debate that is still lacking practical developments and implementations for research.

Overall Discussion

This SLR has revealed that the scientific debate on trustworthy and ethical AI and education is just in its infancy. This is especially noteworthy given the long history of AI&ED research, and the almost ten years of research and publications centered on the ethics of AI in general. Accordingly, we argue that research into trustworthy and ethical AI&ED research requires more effort, in particular in terms of pedagogical, design-based as well as empirical, evidence-based studies of ethical AI in educational practice. In addition, this SLR has shown that the majority of relevant articles have been published in journals focused on technology-oriented education, notable as most of the authors are not from the discipline/field of education. Meanwhile, only one of the 62 reviewed articles addresses AI literacy, which is therefore another research gap that needs to be addressed. Thus, we encourage AI&ED researchers worldwide to undertake studies to test and evaluate practices of ethical AI in education and to explore what a robust AI literacy might mean (addressing both the technological and human dimensions of AI [13]), to help strengthen UN SDG 4.

Limitations

We highlight two limitations of this SLR. First, we collected records only from one database (Web of Science) which revealed only 62 articles. Had we used additional databases, more articles might have been identified. Second, the selection processes were subjective, based on the personal judgements of two researchers (although, to minimize this, as noted earlier we analyzed and compared outputs at two stages of the process, and discussed and resolved any cases that were not clear-cut).

6 Conclusion

This SLR provides the first overview of the published scientific literature on the relationship between trustworthy and ethical AI and education. It addresses four research questions focused on (i) the most common journals, (ii) the most common disciplines, and (iii) the most common regions leading scientific debates and sustainable developments on trustworthy/ethical AI and education, and (iv) past trends. The key finding is that ethical AI&ED analyses are only just starting to appear, which is why we call for more efforts in this increasingly important area.

We argue that research into ethical AI&ED can and should accompany more general AI&ED research and developments, and should involve testing and evaluation of ethical AI&ED practices. In particular, there is a need for pedagogical, design-based as well as empirical, evidence-based studies of ethical AI in educational practice and of AI literacy. In this way, this community can better contribute to sustainable learning practices, to foster and strengthen sustainable equitable and inclusive quality education for all (UN SDG 4) and our future societies and citizens.

References

1. Akgun, S., Greenhow, C.: Artificial intelligence in education: addressing ethical challenges in K-12 settings. AI Ethics **2**, 431–440 (2021). https://doi.org/10.1007/s43681-021-00096-7
2. Borenstein, J., Howard, A.: Emerging challenges in AI and the need for AI ethics education. AI Ethics **1**(1), 61–65 (2021). https://doi.org/10.1007/s43681-020-00002-7
3. Bozkurt, A., et al.: Speculative futures on ChatGPT and generative artificial intelligence (AI): a collective reflection from the educational landscape. Asian J. Distance Educ. **18**(1), 53–130 (2023). https://doi.org/10.5281/zenodo.7636568
4. Chaka, C.: Fourth industrial revolution—a review of applications, prospects, and challenges for Artificial Intelligence, robotics and blockchain in higher education. Res. Pract. Technol. Enhanced Learn. **18**(2), 002 (2023). https://doi.org/10.58459/rptel.2023.18002
5. Chen, L., Chen, P., Lin, Z.: Artificial intelligence in education: a review. IEEE Access **8**, 75264–75278 (2020). https://doi.org/10.1109/ACCESS.2020.2988510
6. Chounta, I.-A., Bardone, E., Raudsep, A., Pedaste, M.: Exploring teachers' perceptions of Artificial Intelligence as a tool to support their practice in Estonian K-12 education. Int. J. Artif. Intell. Educ. **32**(3), 725–755 (2022). https://doi.org/10.1007/s40593-021-00243-5
7. Crompton, H., Jones, M.V., Burke, D.: Affordances and challenges of artificial intelligence in K-12 education: a systematic review. J. Res. Technol. Educ. (2022). https://doi.org/10.1080/15391523.2022.2121344
8. European Commission. Ethical guidelines on the use of artificial intelligence (AI) and data in teaching and learning for educators (2022). https://data.europa.eu/doi/10.2766/153756
9. European Parliament. Report on artificial intelligence in education, culture and the audiovisual sector (2020/2017(INI)) (2021). https://www.europarl.europa.eu/doceo/document/A-9-2021-0127_EN.html
10. Holmes, W.: The unintended consequences of artificial intelligence and education. Educ. Int. Res. (2023). https://www.ei-ie.org/en/item/28115:the-unintended-consequences-of-artificial-intelligence-and-education
11. Holmes, W., Persson, J., Chounta, I.A., Wasson, B., Dimitrova, V.: Artificial intelligence and education. a critical view through the lens of human rights, democracy and the rule of law (2022a). https://rm.coe.int/artificial-intelligence-and-education-a-critical-view-through-the-lens/1680a886bd
12. Holmes, W., et al.: Ethics of AI in education: towards a community-wide framework. Int. J. Artif. Intell. Educ. **32**(3), 504–526 (2022b). https://doi.org/10.1007/s40593-021-00239-1
13. Holmes, W., Tuomi, I.: State of the art and practice in AI in education. Eur. J. Educ. **57**, 542–570 (2022). https://doi.org/10.1111/ejed.12533
14. Huang, R., et al.: Educational futures of intelligent synergies between humans, digital twins, avatars, and robots - the iSTAR framework. J. Appl. Learn. Teach. **6**(2), 1–16 (2023). https://doi.org/10.37074/jalt.2023.6.2.33
15. Ifelebuegu, A.O., Kulume, P., Cherekut, P.: Chatbots and AI in Education (AIEd) tools: the good, the bad, and the ugly. J. Appl. Learn. Teach. **6**(2) (2023). https://doi.org/10.37074/jalt.2023.6.2.29
16. Kent, C., Du Boulay, B.: AI for Learning. CRC Press, Boca Raton, FL (2022).https://doi.org/10.1201/9781003194545
17. Kurdi, G., Leo, J., Parsia, B., Sattler, U., Al-Emari, S.: A systematic review of automatic question generation for educational purposes. Int. J. Artif. Intell. Educ. **30**, 121–204 (2020). https://doi.org/10.1007/s40593-019-00186-y
18. McCarthy, J., Minsky, M., Rochester, N., Shannon, C.: A proposal for Dartmouth summer research project on artificial intelligence (1955). https://www-formal.stanford.edu/jmc/history/dartmouth.pdf

19. Miao, F., Holmes, W.: Guidance for generative AI in education and research. United Nations Educational, Scientific and Cultural Organization (2023). https://unesdoc.unesco.org/ark:/48223/pf0000386693

20. Mills, A., Bali, M., Eaton, L.: How do we respond to generative AI in education? Open educational practices give us a framework for an ongoing process. J. Appl. Learn. Teach. **6**(1) (2023). https://doi.org/10.37074/jalt.2023.6.1.34

21. Moher, D., Liberati, A., Tetzlaff, J., Altman, D.G.: The PRISMA Group: Preferred reporting items for systematic reviews and meta-analyses: the PRISMA statement. PLoS Med. **6**(7), e1000097 (2009). https://doi.org/10.1371/journal.pmed.1000097

22. Page, M.J., et al.: The PRISMA 2020 statement: an updated guideline for reporting systematic reviews. Syst. Rev. **10**, 89 (2021). https://doi.org/10.1186/s13643-021-01626-4

23. Pinkwart, N.: Another 25 years of AIED? Challenges and opportunities for intelligent educational technologies of the future. Int. J. Artif. Intell. Educ. **26**, 771–783 (2016). https://doi.org/10.1007/s40593-016-0099-7

24. Sanusi, I.T., Oyelere, S.S., Vartiainen, H., Suhonen, J., Tukiainen, M.: A systematic review of teaching and learning machine learning in K-12 education. Educ. Inf. Technol. (2022). https://doi.org/10.1007/s10639-022-11416-7

25. Stracke, C.M.: Quality frameworks and learning design for open education. Int. Rev. Res. Open Distrib. Learn. **20**(2), 180–203 (2019). https://doi.org/10.19173/irrodl.v20i2.4213

26. Stracke, C.M., et al.: Responding to the initial challenge of COVID-19 pandemic: analysis of international responses and impact in school and higher education. Sustainability **14**(3), 1876 (2022). https://doi.org/10.3390/su14031876

27. Stracke, C.M., Chounta, I.-A., Holmes, W., Tlili, A., Bozkurt, A.: A standardised PRISMA-based protocol for systematic reviews of the scientific literature on Artificial Intelligence and education (AI&ED). J. Appl. Learn. Teach. **6**(2) (2023). https://doi.org/10.37074/jalt.2023.6.2.38

28. Stracke, C.M., et al.: Impact of COVID-19 on formal education: an international review on practices and potentials of Open Education at a distance. Int. Rev. Res. Open Distrib. Learn. **23**(4), 1–18 (2022b). https://doi.org/10.19173/irrodl.v23i4.6120

29. United Nations Educational, Scientific and Cultural Organization. UNESCO Recommendation on OER (2019). http://portal.unesco.org/en/ev.php-URL_ID=49556&URL_DO=DO_TOPIC&URL_SECTION=201.html

30. UNESCO, UNICEF, The World Bank and OECD.: What's next? Lessons on education recovery: findings from a survey of ministries of education amid the COVID-19 pandemic (2021). http://covid19.uis.unesco.org/wp-content/uploads/sites/11/2021/07/National-Education-Responses-to-COVID-19-Report2_v3.pdf

31. Vuorikari, R., Holmes, W.: DigComp 2.2. Annex 2. citizens interacting with AI systems. In: Vuorikari, R., Kluzer, S., Punie, Y., DigComp 2.2, The Digital Competence Framework for Citizens: With New Examples of Knowledge, Skills and Attitudes, pp. 72–82 (2022). https://data.europa.eu/doi/10.2760/115376

32. Zawacki-Richter, O., Marín, V.I., Bond, M., Gouverneur, F.: Systematic review of research on artificial intelligence applications in higher education – where are the educators? Int. J. Educ. Technol. High. Educ. **16**(1), 1–27 (2019). https://doi.org/10.1186/s41239-019-0171-0

ChatGPT-Based Virtual Standardized Patient that Amends Overly Detailed Responses in Objective Structured Clinical Examinations

Naoki Shindo$^{(\boxtimes)}$ and Masaki Uto[ID]

The University of Electro-Communications, Chofu, Tokyo 182-8585, Japan
{shindo_naoki,uto}@ai.lab.uec.ac.jp

Abstract. Objective structured clinical examinations (OSCEs) are a standardized examination for assessing medical and dental students. OSCEs involve a medical interview task in which examinees are evaluated based on their interactions with standardized patients (SPs), who are trained to respond according to specific clinical scenarios. However, preparing well-trained SPs incurs substantial costs. To overcome this limitation, the use of virtual SPs employing artificial intelligence has attracted considerable attention. In this study, we propose using ChatGPT to create virtual SPs capable of reacting to arbitrary clinical scenarios. However, the direct application of ChatGPT has drawbacks in that it tends to generate overly detailed responses, sometimes mentioning clinical information irrelevant to the examinees' questions. Such behavior is unsuitable for the purpose of OSCEs. To address this limitation, we propose a mechanism that identifies and amends overly detailed responses from ChatGPT and integrates this mechanism into the ChatGPT-based virtual SP.

Keywords: objective structured clinical examinations · educational measurement · natural language processing · large language model

1 Introduction

Objective structured clinical examinations (OSCEs) are a widely used standardized method for evaluating the clinical skills, knowledge, and attitudes of medical and dental students. OSCEs include a medical interview task in which examinees are evaluated based on their interactions with standardized patients (SPs). These SPs are trained to respond according to specific clinical scenarios, as exemplified in Table 1. However, ensuring reliable evaluations can be challenging due to biases from raters and SPs. Several strategies have been explored to address rater biases [7]. Meanwhile, biases from SPs are commonly managed by standardizing the SPs through extensive prior training, although this training can incur substantial costs [6].

One approach to addressing this challenge involves the use of virtual standardized patients (VSPs) powered by artificial intelligence technology, that is,

Table 1. Example of clinical scenario sentences provided to SPs

---

- I have had a persistent cough and phlegm since catching a cold three weeks ago
- For the past week, I've been feeling a slight shortness of breath, especially when climbing stairs at school
- At night, when I lie down, the symptoms become worse, making it difficult to sleep, but sitting up provides some relief

---

Table 2. Example of overly detailed responses from a ChatGPT-based VSP. Italicized sentences indicate overly detailed responses

---

Examinee: And have you woken up at night extremely short of breath?

ChatGPT: Yes, I have been waking up at night feeling extremely short of breath. It's a very distressing feeling, *and I find it difficult to catch my breath even after sitting up and trying to calm myself down. This has been happening consistently for the past week, and it's affecting my ability to get a restful night's sleep.*

---

dialog systems that respond to examinees' questions in place of human SPs. Conventional VSPs rely on rule-based approaches, which utilize a large set of predefined rules created by humans for each clinical scenario. VSPs select responses from a collection of response rules that most closely align with the examinee's posed question. There are two main strategies for this matching process: superficial pattern matching (e.g., [2]) and neural semantic matching (e.g., [4,5]). However, because conventional rule-based approaches require a large set of manually created rules, around 3,000 for an individual scenario, they are prohibitively expensive.

To address this issue, we propose utilizing ChatGPT to create VSPs. In our research, we configure ChatGPT to simulate the responses of human SPs by providing it with specific clinical scenarios. A significant advantage of this approach is its ability to generate flexible responses without the need for manually crafted rules for each scenario. However, the direct application of ChatGPT presents a challenge in that it tends to generate overly detailed responses, often mentioning clinical information irrelevant to the examinees' questions. Table 2 shows an example. Such behavior is unsuitable for OSCEs, which aim to assess examinees' ability to elicit necessary information through appropriate questioning. To address this limitation, we propose a mechanism that identifies and amends overly detailed responses from ChatGPT and integrates this mechanism into the ChatGPT-based virtual SP.

2   Proposed Method

The proposed method detects overly detailed responses from ChatGPT to the examinee's question by combining two independent Bidirectional Encoder Representations from Transformers (BERT) models. One BERT model, named the *Question-Scenario Relevance-Prediction Model*, predicts the relevance of an examinee's question to each individual scenario sentence. Another BERT model, called the *Response-Scenario Relevance-Prediction Model*, predicts whether ChatGPT's responses reflect each scenario sentence. The predictions from these two BERT models are used to detect overly detailed responses from ChatGPT. Subsequently, any identified instances of overly detailed responses are fed back to ChatGPT to refine its responses. The following sections detail each step of the proposed method.

2.1 Question-Scenario Relevance-Prediction Model

The question-scenario relevance-prediction model evaluates the relevance between an examinee's question and scenario sentences, using BERT. The model performs binary classification across all scenario sentences to ascertain their connection to the posed question. This task is akin to next-sentence prediction (NSP), a foundational pre-training task for BERT that determines whether one sentence naturally follows another. Therefore, we presume the BERT model has been pretrained on the NSP task.

The input for the model, which predicts the relevance between the i-th question and the j-th scenario sentence, is formed as "$[CLS]$ *i-th question* $[SEP]$ *j-th scenario sentence*", where $[CLS]$ and $[SEP]$ are the special tokens. As in the NSP task, we consider the output vector corresponding to the $[CLS]$ token as a distributed representation vector. Consequently, letting $\boldsymbol{x}_{ij} \in \mathbb{R}^{768}$ be the distributed representation vector, we input it to the linear layer with sigmoid activation, $v_{ij} = \sigma(\boldsymbol{W}\boldsymbol{x}_{ij} + b)$, where σ is a sigmoid function, and $\boldsymbol{W}$ and b are the trainable parameters. Then, we binarize the continuous value v_{ij}, which ranges between 0 and 1 by setting $z_{ij} = 1$ if $v_{ij} \geq 0.5$ and $z_{ij} = 0$ otherwise. Here, $z_{ij} = 1$ indicates relevance between the i-th question and the j-th scenario sentence, whereas $z_{ij} = 0$ signifies their irrelevance. These processes are repeated for all scenario sentences, thereby obtaining $\boldsymbol{z}_i = \{z_{ij} \mid j \in 1, \ldots, J\}$, where J indicates the total number of scenario sentences. We refer to this vector as the *Q-S Relevance Vector*.

Note that we further fine-tune the question-scenario relevance-prediction model on the NSP task, using the medical interview dataset [3]. This dataset simulates conversations between clinicians and patients, following the OSCE format, and includes dialogues from 272 case studies. For the fine-tuning process, 80% of the dataset serves as the training data, and the remaining 20% is used for early stopping. Within the training data, 50% of the patient responses are randomly shuffled to create negative samples.

2.2 Response-Scenario Relevance-Prediction Model

The response-scenario relevance-prediction model predicts the relevance between ChatGPT's responses and scenario sentences, using another BERT with the same architecture as explained in Sect. 2.1. However, in contrast to the above-introduced BERT model, this one is assumed to be fine-tuned on the recognizing

Table 3. Basic details of the created scenarios

	Scenario 1	Scenario 2	Scenario 3	Scenario 4
Chief complaints	Hernia	Coughing and phlegm	Chest pain	Lower back pain
Number of scenario sentences	38	32	33	39
Number of doctor's questions	24	26	19	30

textual entailment (RTE) task from the general language understanding evaluation (GLUE) benchmark. This fine-tuning process equips the model to ascertain the entailment relationship between two input sentences.

The input for the model is the concatenation of a ChatGPT's response to an examinee's question and each scenario sentence. For example, when a Chat-GPT's response references scenario sentences A, B, and C, the input for the model that predicts the entailment relationship with scenario sentence A is as "[CLS] Response text referring to scenario sentences A, B, and C [SEP] Scenario sentence A". In response to this input, the model should determine the scenario sentence A as an entailment.

In our method, this process is repeated for all scenario sentences, thereby obtaining $r_i = \{r_{ij} \mid j \in 1 \ldots, J\}$, where r_{ij} is a variable that takes 1 when ChatGPT's response to i-th question entails j-th scenario sentence and 0 otherwise. Hereafter, r_i is referred to as *R-S Relevance Vector*.

2.3 Identifying Overly Detailed Responses and Providing Feedback

The proposed method identifies the overly detailed responses of ChatGPT by comparing two vectors, namely, z_i and r_i. Considering that each vector's elements are binary, the case of $(z_{ij}, r_{ij}) = (0, 1)$ signifies an overly detailed response because it means the scenario sentence mentioned by ChatGPT is irrelevant to the given question. Thus, conducting this comparison for every scenario sentence $j \in \{1, \ldots, J\}$ is expected to identify the overly detailed responses of ChatGPT to the i-th question. When an overly detailed response is identified, ChatGPT receives feedback to re-generate an appropriate response. The feedback prompt is as *"Please refrain from mentioning scenarios of {list of identified overly detailed scenario sentences} as a response to this doctor's question."* ChatGPT, upon receiving this prompt, is expected to revise its response, taking the feedback into consideration. The revised response is then re-evaluated using our method. This process is repeated recursively until over-detailing is resolved.

3   Evaluation Experiment

We conducted experiments to evaluate the effectiveness of the proposed method.

For the experiments, we created four scenarios based on the SP scenarios introduced in literature [1], each focusing on different chief complaints. Each

Table 4. Experimental results

	Scenario 1	Scenario 2	Scenario 3	Scenario 4	Avg
Specificity	1.0	0.95	0.98	0.98	0.97
Recall	0.75	0.92	0.67	0.74	0.77
No. of overly detailed responses	20	12	7	16	13.75
No. of predicted overly detailed responses	16	14	4	13	11.75
No. of feedback successes	16	9	1	12	9.50
Feedback success ratio	0.80	0.75	0.14	0.75	0.69

scenario incorporates personal details such as the patient's age and family history, along with scenario sentences that describe specific symptoms the patient is experiencing. Additionally, a list of questions posed by the doctor was prepared according to example dialogs in medical interviews [3]. The number of scenario sentences and the questions for each scenario are shown in Table 3.

In this experiment, we used gpt-3.5-turbo as ChatGPT, with the temperature set to 0.7. Scenario sentences were provided to ChatGPT in the system prompt along with the instruction: *"You'll play the role of the patient. The patient's symptoms and conditions are described in detail."*

3.1 Accuracy in Detecting Overly Detailed Responses

Using the dataset, we evaluated the capability of the proposed method in detecting overly detailed responses. In this experiment, we provided doctor's questions one by one to ChatGPT for each scenario to obtain its responses. Considering the stochastic nature of ChatGPT's responses, we generated responses for each question three times. Consequently, we obtained a total of 297 question–response pairs. Then, the proposed method was applied to classify whether each response generated by ChatGPT was an overly detailed response. Human evaluation was also performed, with the authors evaluating whether or not each response was an overly detailed response. The human evaluation data were used as ground truth to assess the capability of the proposed method.

In this experiment, we calculated two indices: *Recall*, which measures the accuracy of identifying overly detailed responses, and *Specificity*, which measures the accuracy of identifying responses that are not overly detailed. Our research objective is to achieve higher recall while maintaining high specificity.

Table 4 presents the average index values for each scenario. According to the table, the proposed method consistently exhibits high specificity across all scenarios, demonstrating its effectiveness in correctly identifying responses that are not overly detailed. Furthermore, it yields high recall across all scenarios, indicating the accurate detection of responses with overly detailed scenarios.

3.2   Evaluation of Feedback Effectiveness

We also conducted an experiment to evaluate whether the proposed feedback mechanism is effective in improving the responses of ChatGPT. In this experiment, we utilized the responses generated by the previous experiments and collected those identified as overly detailed by the proposed method. Subsequently, for these overly detailed responses, we provided feedback to ChatGPT to refine the responses following the procedures described in Sect. 2.3. We set the maximum number of feedback loops to five. The authors evaluated the presence of over-detailing in the final responses.

The results are shown in Table 4. In the table, the row labeled *No. of overly detailed responses* shows the actual number of overly detailed responses; the row labeled *No. of predicted overly detailed responses* shows the number of times that the proposed model predicted them as overly detailed responses; the row labeled *No. of feedback successes* shows the number of times that the proposed method resolved the overly detailed responses; and the row labeled *Feedback success ratio* indicates the ratio of No. of feedback successes and No. of overly detailed responses. Based on the average column in Table 4, the proposed feedback mechanism successfully addresses 69% of the overly detailed responses. This result confirms the effectiveness of the feedback mechanism in improving the overly detailed responses from ChatGPT.

4   Conclusion

In this study, we proposed a ChatGPT-based VSP for the medical interview task in OSCEs. The developed VSP incorporates a mechanism that identifies and amends overly detailed initial responses from ChatGPT. Through the experiments, we demonstrated that the proposed method accurately detects overly detailed responses and that our feedback mechanism works as designed.

References

1. Ban, N., Suzuki, T., Aomatsu, M., Saiki, T., Abe, K., Kuwabata, A.: An Easy-to-Understand Guide to Medical Interviews and Simulated Patients[In Japanese]. The University of Nagoya Press (2015)
2. Dickerson, R., et al.: Evaluating a script-based approach for simulating patient-doctor interaction. In: International Conference of Human-Computer Interface Advances for Modeling and Simulation, vol. 1, pp. 79–84 (2005)
3. Fareez, F., et al.: A dataset of simulated patient-physician medical interviews with a focus on respiratory cases. Sci. Data 9(1), 313 (2022)
4. Maicher, K.R., et al.: Artificial intelligence in virtual standardized patients: combining natural language understanding and rule based dialogue management to improve conversational fidelity. In: Medical Teacher, pp. 279–285 (2022)
5. Pereira, D.S.M., Falcão, F., Nunes, A., Santos, N., Costa, P., Pêgo, J.M.: Designing and building OSCEBot® for virtual OSCE - performance evaluation. Med. Educ. Online 28(1), 16 (2023)

6. Rau, T., Fegert, J.L.H.: How high are the personnel costs for OSCE? A financial report on management aspects. GMS J. Med. Educ. **28**(1), 16 (2011)
7. Uto, M.: Accuracy of performance-test linking based on a many-facet Rasch model. Behav. Res. Methods **53**(4), 1440–1454 (2021)

Contextual Features for Automatic Essay Scoring in Portuguese

Lucas Galhardi[1], Maria Fernanda Herculano[2], Luiz Rodrigues[2],
Péricles Miranda[3], Hilário Oliveira[4], Thiago Cordeiro[2],
Ig Ibert Bittencourt[2,5], Seiji Isotani[5], and Rafael Ferreira Mello[6]($\boxtimes$)

[1] SENAI PR University Center, Londrina, Brazil
[2] Center for Excellence in Social Technologies, Federal University of Alagoas, Maceio, Brazil
[3] Federal Rural University of Pernambuco, Recife, Brazil
[4] Federal Institute of Espírito Santo, Vitoria, Brazil
[5] Harvard Graduate School of Education, Cambridge, USA
[6] Centro de Estudos Avançados de Recife, Recife, Brazil
rafael.mello@ufrpe.br

Abstract. Automated Essay Scoring (AES) efficacy often varies across linguistic and contextual nuances. This study addresses this gap by proposing and evaluating a contextualized approach tailored for Portuguese. Unlike prior research, which often focused on overall scores or limited to general-purpose features, we explored devising contextualized feature extractors and investigated their impact on predictive performance. Our analysis encompassed the proposed specific features (conjunctions, syntactic quantification, and entity recognition) and two well-established baselines (i.e., TF-IDF and Coh-Metrix). Utilizing the Essay-BR dataset (n = 6,563 essays), we investigated our approach through classification and regression tasks supported by diverse machine learning algorithms and optimization techniques. Mainly, we found that our approach enhanced predictive performance when combined with existing techniques. Our findings reveal the importance of addressing and considering contextual nuances in AES, revealing insights that might help accelerate the evaluation of essays in a large-scale setting.

Keywords: Essay Scoring · Natural Language Processing · Learning Analytics

1 Introduction

The National High School Examination (ENEM) is a key assessment in Brazil. It evaluates students' educational competence after completing the foundational education phase, playing a pivotal role in enabling many students to pursue higher education opportunities, including admission to public universities and access to financial support for private institutions[1]. Among its components,

---

[1] https://www.gov.br/inep/pt-br/areas-de-atuacao/avaliacao-e-exames-educacionais/enem.

© The Author(s), under exclusive license to Springer Nature Switzerland AG 2024
A. M. Olney et al. (Eds.): AIED 2024 Workshops, CCIS 2150, pp. 270–282, 2024.
https://doi.org/10.1007/978-3-031-64315-6_23

ENEM features a discursive-argumentative essay, in which students must address a proposed problem, expressing their viewpoint and concluding by proposing an intervention to mitigate or resolve the stated problem. This essay must have up to 30 lines and its assessment is based on five competencies, such as *proposal intervention*, *organization*, and *coherence and cohesion* [7]. Each of these criteria concerns a competency, and scores are assigned on a scale ranging from 0 (indicating a complete lack of mastery of the mode) to 200 (indicating excellent mastery of the mode). Consequently, ENEM's scores range from 0 to 1,000 by summing up the scores for all five competencies.

Following ENEM's application, two evaluators are responsible for reviewing these essays. Evaluators are tasked to evaluate "*a perspective substantiated by coherent and well-structured arguments with logical consistency.*" In cases where there is a significant divergence in scores, a third evaluator is brought in to reevaluate the essay, to achieve a potential consensus on the final score[2]. However, the manual grading process - although indispensable - is well-recognized for its notable limitations related to the fatigue evaluators are likely experience due to its repetitive nature. Furthermore, because it relies on human judgment, the grading process is subject to various inconsistencies and biases, leading to an inherently unreliable assessment.

A potential solution to this challenge is known as Automated Essay Scoring (AES). In practice, AES might partially automate the essay evaluation to improve the efficiency of evaluators while guaranteeing a consistent, impartial and coherent grading outcome [13]. Most often, AES relies on Natural Language Processing (NLP) and Machine Learning (ML), where regression and classification models are the primary approaches [11]. Particular to ENEM's context, the emphasis on feature-based approaches is evident, demonstrating promising results on AES tasks given the full essay [10]. However, there is a lack of research addressing the assessment of the five competencies individually.

Given that the standard evaluation is split by component, understanding factors affecting each component's score will likely empower evaluators better. Therefore, this work focuses on ENEM's Competence 5 (C5), which is associated with the intervention proposal of the text. Thus, unlike prior research focused on the full essay and overall features, this study investigates the intervention proposal and its distinctive characteristics. For this, we propose three C5-specific features ("approach" hereafter) for the automated assessment of C5. The proposed features explore conjunctions and Named Entities Recognition (NER), informed by the National Institute of Educational Studies and Research (INEP), and analyze the syntactic structure within the essay's conclusion. Thus, we aim to extract informative features from the text to improve the learning model's performance in estimating the C5 score.

For this, we approach estimating the C5 score as both classification and regression tasks based on ML algorithms were applied to the extended Essay-BR dataset [10] (n = 6,563). The algorithms were assessed considering differ-

---

[2] https://www.gov.br/inep/pt-br/assuntos/noticias/enem/conheca-o-processo-de-correcao-das-redacoes.

ent combinations of the proposed and general-purpose features (e.g., TF-IDF e NILC) and the stacking approaches. The results demonstrated that the features extracted by the proposed approach, when combined with general-purpose ones, led to significant enhancements in the AES performance.

Therefore, this paper contributes towards mitigating disparities in AI progress across regions. Whereas research has shown substantial progress in AES for English text, there is a worrying lack of research on other languages [2]. As a continental-sized, Portuguese-speaking country, Brazil and similar countries demand research attention from the AIED community. Hence, given ENEM's relevance for the country, developing AI systems that help optimize essay scoring is paramount. Thus, this paper takes a step towards fulfilling that gap by introducing ENEM-specific features that, based on our experimental results, contribute to AES in the context of ENEM, specifically for C5.

2   Related Work

Research on ML for text evaluation is widely focused on the English language [2], creating a significant disparity in the use of AIED across regions, given that those whose language is not English are left unattended by AIED systems. Consequently, AES for Brazilian Portuguese, as addressed in the recent literature [9], represents a promising field, given the difficulty in efficiently and impartially evaluating a discursive text, as well as the limited research [2].

On the one hand, research related to the assessment of ENEM essays does not analyze scores on a competence basis, considering only the overall score [3]. The issue with this approach is that ENEM's score must be given for each of its five competencies[3]. Therefore, such a general assessment holds the limited potential to contribute to humans evaluators.

Some studies measure the performance per competence but employ the same features for all of them, lacking a specialized proposal [1]. The issue with this approach is that ENEM's evaluation involves contextual nuances, which are detailed for each of its components[4]. Thereby, developing AES systems limited to general-purpose metrics might achieve limited performance by not considering these nuanced insights.

For instance, Santos et al. [14] proposes an approach for Enem's fifth competence (C5). However, it uses the scores from the other four competencies to estimate the C5 grade without addressing its particular characteristics. Moreover, Fonseca et al. [5] personalize the assessment on a competence level but do so experimentally rather than proactively. It employed a feature selection procedure for each competency, retaining only features with a Pearson correlation of at least 0.1. More recently, some studies have directed their efforts towards specific competencies, such as textual cohesion [13], proposing specialized features that include characteristics from established tools like Coh-Metrix [4].

---

[3] https://www.gov.br/inep/pt-br/areas-de-atuacao/avaliacao-e-exames-educacionais/enem.

[4] https://www.gov.br/inep/pt-br/assuntos/noticias/enem/conheca-o-processo-de-correcao-das-redacoes.

As the focus of this study, we concentrated on research that deals with (or at least mentions) C5. One study was found [12] that almost aligns with this aim: a systematic literature review of research identifying and automatically evaluating proposals for intervention in argumentative discursive texts. The review identifies five studies, one of which has been previously presented [1]. While related to the topic, the other four works do not assess C5 according to the scoring levels defined by ENEM; they solely perform binary identification of an intervention proposal. Furthermore, the techniques addressed by these articles focus exclusively on identifying textual markers that indicate the presence of arguments in the text.

With the introduction of the publicly available Essay-BR corpus [10], a new avenue for developing research emerges, mainly focusing on individual competencies. This prospect aligns with ENEM's current assessment framework. The work by [11] presents three assessment approaches: two generic ones based on textual representation (embeddings and recurrent neural networks) and one using feature engineering, with a distinct set for each competency. While achieving favorable results in C5 with the generic approaches, the feature engineering set for C5 exhibited the poorest performance compared to other competencies.

In summary, some studies automatically assess the total score of essays, some focus on other competencies of the ENEM, others relate to the proposal for intervention but fall outside the scope of the ENEM, and finally, studies that propose approaches for C5 but do not adhere to the current official criteria or did not achieve satisfactory performance in the task. Thus, this study aims to investigate C5 based on the official guidelines and evaluation criteria, integrating established and general approaches with aspects motivated by the intrinsic characteristics of intervention proposals. We contribute a contextualized approach towards helping improve AES systems for ENEM, a large-scale assessment with crucial implications for students, besides contributing to research on the Portuguese language, in contrast to prior research often focused on English speakers.

3 Proposal

To enhance the estimation of C5 scores, we conducted an in-depth analysis of its official guideline[5] (guideline hereafter), which offers a comprehensive description of C5's assessment process and serves as the reference for human assessors. Based on that, we propose and investigate multiple features for C5-specific feature extraction to improve the estimation of C5 scores, as detailed below.

The ENEM guideline for C5 describes that evaluating the intervention proposal focuses primarily on identifying the proposal in a general context and five additional elements. Mode/Form, Effect, and Elaboration are described next, whereas Action and Agent are detailed in the subsequent subsections.

In identifying the intervention proposal, emphasis is placed on the use of verbs such as "*dever*" (should) and "*propor*" (propose), with due consideration given to the identified and preserved tense (e.g., "*propõe*" (proposes) signifies

---

[5] https://www.gov.br/inep/pt-br/areas-de-atuacao/avaliacao-e-exames-educacionais/enem/outros-documentos.

a proactive action that has not been developed by the author of the text but rather cites another source proposing something). Note that there is a cautionary note against generic constructions such as "*é preciso*" (it is necessary) and "*é necessário*" (it is necessary), as they indicate that the author suggests the need for a proposal vaguely.

Hence, for the general intervention proposal, we devised two features to model the possible categories of terms: positive and negative. Such strategy considers that the identification of correct terms (*deve propor* - should propose, *proponha* - propose) signifies the presence of the element, whereas the presence of a null counterpart (*propõe* - proposes, *é preciso* - it is necessary) indicates the opposite. The features comprise a list of terms extracted directly from the examples in the assessment guideline.

In identifying the mode/form, there is no division into positive and negative features. Instead, a list of terms may include a mode/form. These terms encompass compound expressions, such as "*por meio de*" (by means of) and "*pelo qual*" (through which), and singular terms, such as "*através*" (through) and "*pelo*" (by). Note that singular terms were tokenized to prevent their inclusion within other words that might contain them.

Concerning "Effect," for which terms such as "*para que*" (so that), "*para isso*" (for this), "*objetivo*" (goal), and similar expressions were selected. Finally, for the "Elaboration" element, a feature was developed that encompasses six possible categories: exemplification, such as "*por exemplo*" (for example); explanation, such as "*na medida que*" (insofar as); justification, such as "*afinal*" (after all) and "*porque*" (because); contextualization, such as "*de forma a*" (in order to); addition, such as "*juntamente*" (together); and uncertainty, such as "*assim*" (thus). These categories were developed to capture various forms of elaboration within the text.

Conjunctions Quantification: Besides features directly based on the elements defined in the assessment guideline, we noted that they used *conjunctions* extensively. These words establish connections between other words and indicate the nature of their relationship. Conjunctions can be categorized into 15 groups, each dealing with a specific aspect (e.g., a cause-effect relationship). Then, extensive lists of Portuguese language terms for each of the 15 conjunction groups were selected and directly used in the extraction process from the essays. The features were extracted through simple counting and normalized by the original text's size (in terms of word count).

Overall Syntactic Quantification: We also investigated the text's syntactic organization based on syntactic quantifications established using the NLTK library[6]. In this analysis, only the essay's last paragraph, which was tokenized into individual words, was considered, thereby identifying the basic units of the text. Subsequently, the words were tagged with their respective part-of-speech (POS) tags to obtain information about the grammatical class of each word. Next, a frequency count of each tag in the paragraphs was conducted. This

---

[6] https://www.nltk.org/.

allowed identifying the most frequent grammatical classes in the text and a better understanding of its linguistic structure.

Named Entities Quantification: Recall that a sound intervention proposal includes action, medium, agent, and effect. Therefore, we developed a model that identifies and quantifies these elements in the text. For this, the spaCy library and the Prodigy framework were used. A new database was created with manual annotations of the location of the elements in the text. Using the Prodigy framework, the last paragraph of 150 essays that received a maximum score in C5 were selected and manually labeled, serving as the basis for creating a Span-Categorizer model, which can automatically predict labels for similar features to those previously marked. Then, the earlier model was applied to the extended Essay-BR text corpus and stored for analysis. Subsequently, these labels were transformed from spans (SpanCategorizer) into entities (EntityRecognizer) to be quantified using methods from the spaCy library. Furthermore, we adopted a second approach that involved replicating the label recognition model with a stratified database containing more than 100 essays with different scores.

4   Proposal Evaluation

The dataset used in this study was the Essay-BR-Estendido ("Essay-BR" for simplicity) [10], which has been widely used by related work [11,13]. Essay-BR comprises 6,579 essays written in the format of the ENEM on 151 socially relevant topics for Brazil, including fake news, deforestation, and healthcare [10]. Its evaluation process employed by human assessors follows the ENEM format, providing a score for each competency. Only the evaluations for C5 were considered in this work. Additionally, 14 duplicate or empty (string) essays were identified and removed, resulting in 6,563 remaining essays.

To assess the proposed feature extractors, we compared our approach to TF-IDF and Coh-Metrix, two well-established, widely used techniques [9]. In this work, the scikit-learn implementation of TF-IDF was used, with the parameters ngram_range set to (1,2) (i.e., unigrams and bigrams) and min_df = 0.01 (only terms that appear in at least 1% of the essays)[7]. Vectorization was trained on only 80% of the data to prevent potential overfitting and enable a fair evaluation. Concerning NILC, we employed a combination and extension of Coh-Metrix [6] and NILC-Metrix[8] (although not solely from NILC-Metrix, for simplicity, this group is referred to in this work simply as "NILC").

Similar to related work [13], we explored classification and regression models. The following algorithms were employed to create models for predicting C5 scores, selected due to their established use in prior research: Logistic Regression, Random Forest, ExtraTrees, and SVM from scikit-learn, eXtreme Gradient Boosting (XGBoost), and Light Gradient Boosting Machine (LGBM). Importantly, despite Marinho et al. [11] presenting a predefined data split, this work opted for traditional cross-validation with five stratified subsets (maintaining the

---

[7] These values were defined empirically.

[8] http://fw.nilc.icmc.usp.br:23380/metrixdoc.

proportion of each class). When the problem is converted to regression, stratification occurs before the conversion to preserve the class distribution.

Similarly, the metrics used to evaluate model performance were selected based on related work [11,13]. We used Root Mean Squared Error (RMSE) and Pearson's Correlation for regression. For classification, we used the F1-score and Cohen's Kappa in two versions: linear and quadratic, as suggested in recent studies on the best way to report Kappa [15]. This is necessary because the linear Kappa has a well-established interpretability [8], while the quadratic Kappa enables complementary interpretation by penalizing disagreements differently between adjacent and distant classes. As Kappa is essentially a classification metric, the outputs of regression models are discretized to the nearest class, following the approach in [11].

In that context, we conducted various experiments to understand how to improve AES for C5. Notably, works focused on automated essay evaluation generally use the entire essay as input for feature extraction or multidimensional modeling. However, according to the official ENEM guidelines, the intervention proposal is often found solely in the conclusion (last paragraph). Therefore, our analysis also investigated how considering the full essay or only its conclusion. Moreover, we also varied individually and in combinations of feature groups. Then, based on the preliminary results, we investigated combining them through feature concatenation, averaging the scoring output, and Stacked Generalization aiming for performance improvements [16].

Furthermore, there are three well-established techniques for model optimization. *Feature selection* was performed on the NILC and TF-IDF groups, which have high dimensionality (325 and +2000, respectively). As some features might be redundant or have no effect, feature selection controls the model's complexity and possibly improve performance. *Class weights* aims to assist with the class imbalance in the dataset's class distribution. To address this, the compute_class_weight function from scikit-learn was employed, which implements a simple heuristic to aid in the classification/regression of minority classes (in the case of this dataset, classes 0, 40, and 200). Finally, the *optimization of 10 hyperparameters* of the selected algorithm was performed using the Optuna library, identifying the best ideal values considering the available dataset. Those techniques are computationally expensive, so we only applied them to the best model (found in previous analyses). Finally, we combined all models, as detailed above.

5   Results

This section presents a series of analyses, as described in Sect. 4, towards improving C5 prediction. Notably, all reported results concern the LGBM algorithm, which consistently outperformed the other algorithms in our experiments.

As a preliminary analysis, we investigated the reliability of our NER feature extraction approach, as shown in Table 1. It demonstrates that the models achieved the best results (F1 between 65% and 71%) in the dataset of conclusions with a maximum score in C5. Furthermore, the entity represented by the *effect* of the intervention proposal obtained the worst results in both models, suggesting potential ambiguity and/or variability of this entity in the text, making its identification more challenging. Overall, given the promising results of the NER approach, we proceeded to investigate along with the other approaches.

Table 1. Results (%) of the entity identification models

Entity	Maximum Score			Stratified Score		
	Precision	Recall	F1-score	Precision	Recall	F1-score
Action	83.72	62.07	71.29	73.58	48.75	58.65
Effect	77.08	56.92	65.49	31.82	16.09	21.37
Form	90.48	51.35	65.52	50.00	26.79	34.88
Agent	88.46	57.50	69.70	59.62	37.35	45.93

First, we compared the proposed group of features specific to C5 (see Sect. 3) to the baseline approaches: NILC and TF-IDF. Regarding classification (clf), Table 2 demonstrates fair-to-moderate results, given the Linear Kappa (just Kappa for simplicity) and Quadratic Weighted Kappa (QWK) values between 0.3 and 0.57. The well-established TF-IDF stood out, whereas NILC also achieved good results. Notably, the concatenation of C5-specific features achieved a performance close to that of NILC, despite its reduced number of features. Furthermore, Table 2 shows that the classification and regression (reg) results were similar. Hence, although the ultimate goal is classification, the subsequent analysis concerns regression, as it has more granular outputs and yields comparable results.

Table 2. Performance of the classifiers and regressors algorithms for each approach

	Kappa	QWK	F1-score	Pearson	RMSE
Proposed (clf)	0.305	0.423	0.306	–	–
Proposed (reg)	0.291	0.423	–	0.478	45.9
NILC (clf)	0.341	0.470	0.331	–	–
NILC (reg)	0.325	0.476	–	0.527	44.3
TF-IDF (clf)	**0.421**	**0.533**	**0.401**	–	–
TF-IDF (reg)	0.388	0.528	–	**0.574**	**42.6**

Second, informed by ENEM's guidelines, we investigated the impact of considering the entire essay (full) compared to only considering the conclusion paragraph (conclusion). As Table 3 demonstrates, the performance in all three groups is better when considering the entire essay. Despite the guideline's suggestion that the conclusion would have the highest importance for C5, this finding suggests that the rest of the essay also strongly influences the evaluation of the competency, even if indirectly.

Table 3. Feature extraction: full essay x only conclusion

	Kappa	QWK	Pearson	RMSE
Proposed (full)	0.291	0.423	0.478	45.9
Proposed (conclusion)	0.249	0.358	0.416	47.8
NILC (full)	0.325	0.476	0.527	44.3
NILC (conclusion)	0.264	0.384	0.438	47.2
TF-IDF (full)	**0.388**	**0.528**	**0.574**	**42.6**
TF-IDF (conclusion)	0.304	0.436	0.486	45.8

Third, we investigated the model optimization approaches. Building upon the previous findings (i.e., advantages of LGBM, regression, and complete essay), the feature selection process was applied to the high-dimensional groups. These results are presented in Table 4, which compares the performance before (white background) and after (gray background) the optimization. Regarding predictive performance, we found no difference for the NILC group, while there was a slight improvement for the TF-IDF group. However, for both groups, the benefit of lower computational cost is evident by reducing the number of features from 325 to 100 for NILC and from 2,287 to 584 for TF-IDF.

Table 4. Models' performance after feature selection in NILC and TF-IDF groups of features

	Kappa	QWK	Pearson	RMSE
NILC	0.325	0.476	0.527	44.3
NILC (FS)	0.327	0.475	0.524	44.4
TF-IDF	0.388	0.528	0.574	42.6
TF-IDF (FS)	**0.407**	**0.557**	**0.601**	**41.5**

Next, hyperparameter optimization and class weights were used as a final step in the performance improvement process. Table 5 presents the results before (white background) and after (gray background) applying these techniques. The performance of all metrics in the three groups improved. The metric with the most significant difference was Quadratic Kappa, explained by being chosen to be optimized by Optuna. Hence, we demonstrate the benefits from our model optimization attempts.

Table 5. Models' performance before and after the hyperparameter optimization

	Kappa	QWK	Pearson	RMSE
NILC	0.327	0.475	0.524	44.40
NILC (optimized)	0.351	0.516	0.540	45.00
TF-IDF	0.407	0.557	0.601	41.50
TF-IDF (optimized)	**0.422**	**0.604**	**0.627**	**41.00**
Proposed	0.291	0.423	0.478	45.90
Proposed (optimized)	0.307	0.454	0.488	46.50

Finally, we combined all features based on the previous findings (i.e., LGBM, regression task, and all feature groups). For this step, we combined the output (prediction) of each of the three base models through i) a simple average of the scores, ii) a stacked generalization with Linear Regression at the 2nd level, or iii) Ridge Regression at the 2nd level. Table 6 presents a comparison between the three proposed combinations (before optimization) and the *final model*, which used the best of our previous findings, including optimization, for prediction. Table 6 shows that Stacking yielded significantly better results than simple averaging and comparable to the 2nd-level regressors. Overall, the performance of the final model was significantly better than all the other results, revealing that combining generic and C5-specific features contributes to estimating C5.

Table 6. Results of models' C5 estimation using the combination of all features

	Kappa	QWK	Pearson	RMSE
Average	0.361	0.507	0.585	42.00
Stacking - LR	0.413	0.570	0.611	41.10
Stacking - Ridge	0.410	0.565	0.609	41.20
Final model	**0.456**	**0.614**	**0.649**	**39.50**

6 Discussion

While AES has been widely explored for English, limited research targets Portuguese. Consequently, countries like Brazil have been unattended by the potential of AES. For instance, Brazil annually applies ENEM, a large-scale assessment that features a discursive-argumentative essay that has to be manually scored by up to three evaluators, a task known to be tiresome and error-prone. To mitigate such a disparity of AIED adoption across regions, this paper presents a contextualized approach for AES in the context of ENEM.

Our study revealed four main findings. First, our approach for identifying intervention proposal elements (i.e., action, medium, agent, and effect) achieved F1 scores around 0.7, given the complete essay as the input. Particularly, the results demonstrated action and effect were the elements in which we achieved the best and worst results, respectively. These findings demonstrate the feasibility of identifying such elements through NER, the importance of exploring the complete essay instead of the conclusion only, and elements in which our approach is more and less likely to succeed.

Second, the findings revealed that our approach enhances C5's prediction performance. Whereas NILC and TF-IDF achieved better results than the approach we compared, we found that concatenating the three groups led to the best predictive performance. Thus, this finding demonstrates that exploring contextualized nuances of ENEM contributes to developing AES for scoring C5.

Third, our findings revealed that approaching AES as classification and regression problems yielded similar results. This finding was consistent when investigating the group feature independently and together. We focused our analyses on the regression task, as it provides more detailed outputs compared to classification. Hence, these findings revealed that classification and regression tasks in AES are likely to yield comparable results, while the latter provides more detailed insights.

Lastly, we found that the model optimization techniques contributed to more robust models for AES in the context of ENEM's C5. Feature selection contributed to model simplification, whereas class weighting, hyperparameter tuning, and Stacking helped enhance the model's predictive performance. These findings demonstrate the relevance of adopting such approaches to improve model robustness, both in terms of complexity and performance.

Thus, these findings hold many main implications. First, *addressing disparities in AI adoption*. We highlighted the potential to bridge the gap in AI adoption across regions like Brazil by showcasing the feasibility of a contextualized approach to AES tailored for examinations such as ENEM. Second, *the importance of contextual nuances*. The research highlights the significance of considering contextual nuances in educational assessments, particularly by dissecting ENEM's competencies and extracting C5-specific features. Third, *enhancing evaluation tools*. The demonstrated improvements in prediction performance through model optimization techniques suggest pathways for enhancing the robustness of targeted AES systems. Fourth, *advancing pedagogical approaches*. By advocating for the integration of AI technologies sensitive to diverse linguistic and educational contexts, the findings help foster more equitable systems.

Despite the promising results of this study, its limitations should be considered. The main issue relates to the extended Essay-BR dataset, which is unbalanced and contains inconsistent data. Through manual inspection of random samples, problems were observed in how the web crawler process was conducted, resulting in repeated words and extra annotations that should not have been present. Additionally, the dataset spans a period that includes changes in the ENEM evaluation criteria, potentially complicating the automated evaluation process when considering different sets of criteria. As future work, we recommend: (i) creating a new database to address some of the mentioned inconsistencies, (ii) investigating a more recent and complex approach like BERT in this context, and (iii) exploring a hybrid approach (text-based and feature engineering-based), with (i) being a general approach and (ii) and (iii) focusing again on C5.

References

1. Amorim, E., Veloso, A.: A multi-aspect analysis of automatic essay scoring for Brazilian Portuguese. In: Proceedings of the Student Research Workshop at the 15th Conference of the European Chapter of the Association for Computational Linguistics, pp. 94–102 (2017)
2. Bai, X., Stede, M.: A survey of current machine learning approaches to student free-text evaluation for intelligent tutoring. Int. J. Artif. Intell. Educ. **33**(4), 992–1030 (2023)
3. Bittencourt Júnior, J.A.S., et al.: Avaliação automática de redação em língua portuguesa empregando redes neurais profundas (2020)
4. Camelo, R., Justino, S., Mello, R.: Coh-metrix pt-br: uma api web de análise textual para a educação. In: Anais dos Workshops do IX Congresso Brasileiro de Informática na Educação, pp. 179–186. SBC, Porto Alegre (2020). https://doi.org/10.5753/cbie.wcbie.2020.179
5. Fonseca, E., Medeiros, I., Kamikawachi, D., Bokan, A.: Automatically grading brazilian student essays. In: Computational Processing of the Portuguese Language: 13th International Conference, PROPOR 2018, Canela, 24–26 September 2018, Proceedings 13, pp. 170–179. Springer (2018)
6. Graesser, A.C., McNamara, D.S., Louwerse, M.M., Cai, Z.: Coh-Metrix: analysis of text on cohesion and language. Behav. Res. Methods Instrum. Comput. **36**(2), 193–202 (2004). https://doi.org/10.3758/BF03195564
7. Instituto Nacional de Estudos e Pesquisas Educacionais Anísio Teixeira (INEP). A redação no Enem 2022: cartilha do participante (2022)
8. Landis, J.R., Koch, G.G.: The measurement of observer agreement for categorical data. Biometrics **33**(1), 159–174 (1977)
9. de Lima, T.B., da Silva, I.L.A., Freitas, E.L.S.X., Mello, R.F.: Avaliação automática de redação: Uma revisão sistemática. Revista Brasileira de Informática na Educação **31**, 205–221 (2023)
10. Marinho, J.C., Anchiêta, R.T., Moura, R.S.: Essay-br: a Brazilian corpus to automatic essay scoring task. J. Inf. Data Manag. **13**(1) (2022)
11. Marinho, J.C., Cordeiro, F., Anchiêta, R.T., Moura, R.S.: Automated essay scoring: an approach based on enem competencies. In: Anais do XIX Encontro Nacional de Inteligência Artificial e Computacional, pp. 49–60. SBC (2022)
12. Nau, J., Haendchen Filho, A., Dazzi, R.L.S.: Identificação e avaliação automática da proposta de intervenção em textos dissertativos-argumentativos: uma revisão sistemática da literatura. In: Anais do Computer on the Beach, pp. 493–501 (2019)
13. Oliveira, H., et al.: Classificaçao ou regressao? avaliando coesao textual em redaçoes no contexto do enem. In: Anais do XXXIV Simpósio Brasileiro de Informática na Educação, pp. 1226–1237. SBC (2023)
14. Santos Júnior, J.J.D., et al.: Modelos e técnicas para melhorar a qualidade da avaliação automática para atividades escritas em língua portuguesa brasileira (2017)

15. Vanbelle, S.: A new interpretation of the weighted kappa coefficients. Psychometrika **81**(2), 399–410 (2016)
16. Wolpert, D.H.: Stacked generalization. Neural Netw. **5**(2), 241–259 (1992)

Breaking the Cycle: AI Boosting Communication Skills of Low-Income Students in Brazil

Renata Miranda de Gama[1], Geiser Chalco[2], Jário Santos[3], Marcelo Reis[3], Álvaro Sobrinho[4(✉)], Seiji Isotani[1,5], and Ig Ibert Bittencourt[3,5]

[1] University of São Paulo, Sao Paulo, Brazil
profrenatamiranda@usp.br, sisotani@icmc.usp.br
[2] Federal Rural University of the Semiarid, Mossoro, Brazil
geiser@alumni.usp.br
[3] Federal University of Alagoas,Maceio, Brazil
ig.ibert@ic.ufal.br
[4] Federal University of the Agreste of Pernambuco, Garanhuns, Brazil
alvaro.alvares@ufape.edu.br
[5] Harvard Graduate School of Education, Cambridge, USA

Abstract. Improving communication skills is challenging for low-income students because they do not have the same opportunities to access quality educational services as other students. To tackle this problem, Artificial Intelligence (AI) solutions have been proposed to improve students' written communication. However, engagement is essential to enhance students' writing skills when using such solutions. We propose a gamified approach based on Csikszentmihalyi's flow theory, evidencing effectiveness in helping students achieve a flow state and enhance their ability to produce texts. We conducted a pre-test/post-test quasi-experiment involving 27 high school students from a low-income Rio de Janeiro, Brazil area. The flow state and learning scores were higher when the solution was gamified. A higher flow state among students may be attributed to the rapid and clear feedback provided. Thus, we provide valuable insights into the gamification of extracurricular activities and the development of educational technologies, such as tutoring systems based on AI. Our approach can assist in planning the use of AI to reduce the risk of promoting inequalities instead of improving engagement and writing skills.

Keywords: flow state · gamification · high education · Portuguese

1 Introduction

The conventional classroom setting often needs to nurture essay-writing abilities adequately. For instance, in public schools in Rio de Janeiro and many other places in Brazil, essays are typically introduced only in the final year of high school, primarily as preparation for the Exame Nacional do Ensino Médio

A. M. Olney et al. (Eds.): AIED 2024 Workshops, CCIS 2150, pp. 283–291, 2024.
https://doi.org/10.1007/978-3-031-64315-6_24

(ENEM) [5]. Moreover, classroom time is predominantly allocated to Portuguese Language classes, focusing on grammar and literature, which leaves limited room for dedicated essay writing and textual production instruction. Essay writing is an example of a valuable extracurricular activity [9].

One significant challenge in these extracurricular activities is providing a clear structure and feedback mechanism [4]. There is often no direct connection to students' interests, and it may not address their individual needs. Additionally, personalized pedagogical feedback should be part of these activities.

Artificial Intelligence (AI) solutions can support students by providing feedback from their text productions. Thus, AI can improve the quality of education, having decades of scientific contributions. However, using AI must be planned because some studies have evidenced that it can enable or inhibit the achievement of sustainable development goals [10].

The introduction of gamification in AI solutions offers a promising way to improve students' performance in essay writing, integrating elements, dynamics, mechanics, and styles commonly found in games into non-gaming contexts [8]. Gamification can transform textual production into a collaborative endeavor, allowing multiple writers to collaborate.

We introduce a gamified design called the Writing Game. This design draws inspiration from Csikszentmihalyi's flow theory [3] and addresses high school students' engagement challenges in extracurricular writing activities. In our research, Information and Communication Technologies (ICT) and social media are resources to support the learning process rather than being the primary goal. To assess whether our gamified design promotes the flow state and improves high school students' essay-writing learning performances, we formulated the following Research Questions (RQ): (RQ1) What is the impact on the state of flow of high-school students? (RQ2) What is the impact on the performance of high-school students?

Thus, we defined the following null hypotheses: (H1) There is no significant difference in flow state between participants in the gamified scenario and participants in the non-gamified scenario; (H2) There is no significant difference in flow state according to participants' player profiles in the gamified and the non-gamified scenarios; (H3) There is no significant difference in learning performance between participants in the gamified and non-gamified scenarios; and (H4) There is no significant difference in learning performance according to participants' player profiles in the gamified and the non-gamified scenarios.

2 Method

2.1 Gamified System

We developed an intelligent system that relies on our gamified design. The activities start with a video presentation of the scenario. Afterward, the students formulate teams, and a presentation on a specific topic is conducted to expose racism, intolerance, and xenophobia against refugees. The participants are involved in a debate initiated by the question, "What is behind all this?".

Thus, the missions are introduced to the students. The first mission incorporates the Dynamics to Maintain Balance (DMB) by instructing students to formulate their theses using four specific textual connectors. The challenge in this mission involves locating these connectors presented clearly before the activity and integrating them into their thesis presentations.

Immediate and direct feedback is provided in the form of color-coded indicators. When the required words are used correctly in the text, they turn green, signifying that the requirements are met. However, if a participant omits one of the specified words, it appears in red, preventing the students from progressing in the game. Thus, students receive a message instructing them to revise the activity. Completing this mission rewards the team with 100 XP.

The second mission incorporates the Dynamics to Avoid Boredom (DAB). This mission involves exploring argumentative strategies on social media platforms. The challenge presented to the students consists of composing written arguments and recording and presenting them in a video format. To complete the mission, participants must write their arguments and record and send the video to the teacher via social media or a WhatsApp classroom group.

Feedback on this mission is derived from the comments received on the video. Upon finishing the written paragraphs, participants earned a reward of 300 XP. An additional 200 XP is added to their scores when they successfully send the video. As a further incentive, participants receive medal badges upon completing this mission. This setup effectively engages students and maintains their interest throughout the activity.

The third mission, encapsulating the Dynamics to Avoid Frustration (DAF), focuses on crafting the essay's conclusion. This mission involves formulating a proposal to address a specific social issue. During this mission, active interaction between two participants is essential as they collaborate to address the following key questions: (1) Who will conduct the action? (2) What will be done? (3) How will this be achieved? (4) Why should this be done?

2.2 Experimental Procedures

The different player profiles were established by assessing preferences for the "achievement," "social," and "immersion" components using the QPJ-Br instrument [1]. This instrument is an adapted and validated Brazilian Portuguese version of Yee's motivation model [11].

We designated the flow state and learning performances (specifically, acquired essay-writing skills) as the dependent variables. The flow state and its corresponding covariate (dispositional flow state) were assessed using the validated Brazilian Portuguese versions of the Flow State Scale-2 (FSS-2) and Dispositional Flow State-2 (DFS-2) [2].

The evaluation of essay-writing skills learning took place in the post-test and pre-test phases (quasi-experiment with a crossover trial design), with the latter as a covariate. Students were tasked with responding to ten questions related to a specific topic. These questions included four multiple-choice questions and six argumentative questions. To minimize the impact of individual variability, the two classrooms were divided into two groups through random assignment: the intervention group and the control group.

To mitigate potential biases further, the activities were implemented in different orders for the two groups. In the first group, participants engaged in the intervention (Scenario 1) on the first day and the control scenario (Scenario 2) on the second. Conversely, the second group experienced the reverse order, participating in the control scenario and the intervention on the subsequent day.

During the pre-test phase, we administered the DFS-2 and the ten-question essay-writing skills test to both groups. Following the experiment's conclusion, we utilized the FSS-2 to evaluate the flow state experienced by participants, and we conducted the essay writing test to measure their learning performance.

We employed a convenience sampling method to select participants. The study included n = 27 students in the final year of high school at the Centro Educacional Integrado (CIEP) Hilda do Carmo, in Duque de Caxias, in Rio de Janeiro. This high school is in a low-income area in the Baixada Fluminense.

Regarding gender distribution among the participants, 51.9% were male, and 48.1% were female. Regarding age, the majority (70%) fell within the range of 15 to 19 years old, while the remaining 30% were between the ages of 20 and 21. For racial demographics, the breakdown was as follows: 80% of participants identified as Black, 15% as Brown, and 5% as White. In terms of residency, most of the students hailed from one of the seven nearby slums, characterized by challenging socioeconomic conditions, elevated rates of violence, involvement in drug trafficking, and a lack of government support.

Many of these students needed more confidence in performing well on the ENEM essay test, a significant aspect of their academic future. These young individuals' daily lives are markedly influenced by the pervasive presence of violence in their communities. Our statistical analysis used parametric Analysis of Covariance (ANCOVA) tests [7]. Before conducting these tests, we applied the Winsorized method [12], which involves reducing the influence of outliers by adjusting extreme values to fall within the 5% to 95% probability range within the collected data. We also assessed the data to determine whether it met the assumptions of symmetry and normality using the Shapiro-Wilk test [6].

3   Results

Tables 1 and 2 provide an overview of the descriptive statistics for the collected data for gamified and non-gamified, respectively. We categorized data based on the type of scenario: gamified (abbreviated as "non.gamified, non") and non-gamified (abbreviated as "gamified, gam") scenarios. We also considered the categorization of the data based on the division of the scenarios according to player profiles, which are defined by high (abbreviated as "U - Upper") and low (abbreviated as "L - Lower") preferences for the achievement (A - Achievement), social (S - Social), and immersion (I - Immersion) components.

The rows of Tables 1 and 2 contain data related to the dispositional flow state scale (dfs) and the flow state scale (fss), as well as pre-test (pre.score) and

Table 1. Descriptive data (gamified).

			gamified		gamified		gamified	
		gam	L(A)	U(A)	L(S)	U(S)	L(I)	U(I)
dfs	N	27	12	13	8	12	11	12
	M	3.30	3.44	3.21	3.53	3.19	3.44	3.12
	SE	0.08	0.10	0.11	0.08	0.14	0.09	0.12
fss	M	3.82	3.94	3.62	3.85	3.66	4.16	3.68
	SE	0.10	0.14	0.13	0.14	0.15	0.10	0.13
adj.fss	M	3.82	3.95	3.60	3.90	3.62	4.17	3.67
	SE	0.09	0.13	0.12	0.16	0.13	0.11	0.11
pre.score	N	25	10	13	7	12	9	12
	M	0.39	0.58	0.22	0.33	0.29	0.42	0.35
	SE	0.07	0.11	0.08	0.14	0.08	0.08	0.12
post.score	M	0.48	0.50	0.44	0.40	0.50	0.57	0.43
	SE	0.03	0.05	0.05	0.05	0.05	0.04	0.05
adj.score	M	0.48	0.46	0.49	0.40	0.51	0.57	0.44
	SE	0.03	0.05	0.04	0.06	0.04	0.05	0.04

L(A): lower achievement, U(A): upper achievement, L(S): lower social, U(S): upper social, L(I): lower immersion, U(I): upper immersion

Table 2. Descriptive data (non-gamified).

			non.gami		non.gamif		non.gam	
		non	L(A)	U(A)	L(S)	U(S)	L(I)	U(I)
dfs	N	27	12	13	8	12	11	12
	M	3.30	3.44	3.21	3.53	3.19	3.44	3.12
	SE	0.08	0.10	0.11	0.08	0.14	0.09	0.12
fss	M	2.98	3.17	2.78	3.03	2.82	3.21	2.91
	SE	0.08	0.11	0.12	0.16	0.12	0.07	0.10
adj.fss	M	2.98	3.18	2.76	3.08	2.78	3.22	2.90
	SE	0.09	0.13	0.12	0.16	0.13	0.11	0.11
pre.score	N	27	12	13	8	12	11	12
	M	0.43	0.64	0.22	0.41	0.29	0.52	0.35
	SE	0.07	0.10	0.08	0.15	0.08	0.09	0.12
post.score	M	0.36	0.37	0.32	0.34	0.35	0.47	0.31
	SE	0.03	0.06	0.04	0.06	0.05	0.05	0.05
adj.score	M	0.36	0.31	0.37	0.31	0.36	0.45	0.32
	SE	0.03	0.05	0.04	0.05	0.04	0.04	0.04

L(A): lower achievement, U(A): upper achievement, L(S): lower social, U(S): upper social, L(I): lower immersion, U(I): upper immersion

post-test (post.score) scores. Additionally, we provide adjusted flow state values (adj.fss) and post-test scores (adj.post.score), which were calculated using the Estimated Marginal Means (EMMs) method.

Additionally, we conducted ANCOVA tests for various scenarios and player profiles. The results of the hypothesis tests are conveyed through statistical significance codes, providing a clear indication of the statistical significance of the observed differences among different scenarios and player profiles.

3.1 Flow State

Based on the results presented in Table 3, it is evident that the null hypotheses H1 and H2 have been rejected. The flow state of students who participated in the gamified scenario was statistically higher than that of participants in the non-gamified scenario. Additionally, these findings indicate that the flow state levels in the gamified setting were significantly higher compared to the non-gamified scenario, regardless of player profile (upper or lower) and preferences.

Table 3. Pairwise comparisons and hypothesis tests for participants' flow state.

	var	(i)	(j)	(i)–(j)	SE	t	p	p.adj	
H1		gamified	non.gamified	0.835	0.13	6.54	<0.001	<0.001	***
H2	lower (A)	gamified	non.gamified	0.769	0.18	4.27	<0.001	<0.001	***
	upper (A)	gamified	non.gamified	0.838	0.17	4.85	<0.001	<0.001	***
	gamified	lower (A)	upper (A)	0.350	0.18	1.93	0.06	0.06	
	non.gamified	lower (A)	upper (A)	0.419	0.18	2.31	0.025	0.025	*
	lower(S)	gamified	non.gamified	0.819	0.23	3.63	0.001	0.001	***
	upper (S)	gamified	non.gamified	0.843	0.18	4.58	<0.001	<0.001	***
	gamified	upper (S)	upper (S)	0.278	0.22	1.29	0.207	0.207	
	non.gamified	upper (S)	upper (S)	0.301	0.22	1.39	0.172	0.172	
	lower (I)	gamified	non.gamified	0.949	0.16	6.08	<0.001	<0.001	***
	upper (I)	gamified	non.gamified	0.769	0.15	5.14	<0.001	<0.001	***
	gamified	lower (I)	upper (I)	0.503	0.16	3.13	0.003	0.003	**
	non.gamified	lower (I)	upper (I)	0.322	0.16	2.01	0.051	0.051	

Significance codes: 0 *** 0.001 ** 0.01 * 0.05 - (A) Achievement, (S) Social, (I) Immersion

3.2 Learning Performance

Based on our results presented in Table 4, it is also evident that the null hypothesis H3 and H4 have been rejected. Regarding H3, the statistical analysis showed a significant difference, suggesting that the post-test scores for the gamified scenario were higher than those of participants in the non-gamified scenario.

Regarding H4, there was a significant difference in post-test scores, considering player profiles and the type of scenario. Among participants with upper social preference, those in the gamified scenario demonstrated higher learning scores than those with the same profile in the non-gamified setting. A similar significant difference in learning scores was observed among participants with two distinct player profiles: upper social preference and lower achievement preference.

Table 4. Pairwise comparisons and hypothesis tests for participants' writing skills.

	var	(i)	(j)	(i)–(j)	SE	t	p	p.adj	
H3		gamified	non.gamified	0.126	0.04	3.05	0.004	0.004	**
H4	lower (A)	gamified	non.gamified	0.149	0.07	2.28	0.028	0.028	*
	upper (A)	gamified	non.gamified	0.117	0.06	1.96	0.057	0.057	
	gamified	lower (A)	upper (A)	-0.026	0.07	-0.37	0.712	0.712	
	non.gamified	lower (A)	upper (A)	-0.058	0.07	-0.85	0.399	0.399	
	lower (S)	gamified	non.gamified	0.088	0.08	1.17	0.249	0.249	
	upper (S)	gamified	non.gamified	0.150	0.06	2.54	0.016	0.016	*
	gamified	upper (S)	upper (S)	-0.107	0.07	-1.55	0.13	0.13	
	non.gamified	upper (S)	upper (S)	-0.045	0.07	-0.67	0.506	0.506	
	lower (I)	gamified	non.gamified	0.119	0.06	1.88	0.068	0.068	
	upper (I)	gamified	non.gamified	0.127	0.06	2.22	0.032	0.032	*
	gamified	lower (I)	upper (I)	0.121	0.06	1.97	0.057	0.057	
	non.gamified	lower (I)	upper (I)	0.130	0.06	2.19	0.035	0.035	*

Significance codes: 0 *** 0.001 ** 0.01 * 0.05 - (A) Achievement, (S) Social, (I) Immersion

4   Discussion

Students who are underserved and belong to minority groups face numerous obstacles, including digital poverty and discrimination. Consequently, the impact of AI solutions on learning opportunities can either exacerbate existing inequalities or contribute to fostering better educational outcomes, depending on the design decisions made.

The observed impact of the gamified scenario on promoting a higher flow state among participants (i.e., RQ1) can likely be attributed to the immediate feedback mechanisms embedded within the dynamics (DMB, DAF, DAB) applied during the experiment. These mechanisms effectively engaged the participants and directed their focus toward the activities, enhancing their overall experience. Language development is intricately tied to social interactions and is acquired by individuals through their relationships with their environment.

Traditional essay-writing classes often emphasize writing within predefined rules, sometimes leading to unproductive outcomes. Students tend to focus on adhering to grammatical structures rather than genuinely engaging with the content of their writing. For textual production to become more meaningful, it necessitates profound reflection, extensive writing, and rigorous revision.

Our gamified design enhanced students' writing skills (i.e., RQ2) and introduced a clear, structured division of steps and missions. This approach involved distinct phases, including planning, textualization, and revision, each with unique challenges and objectives. The interactive elements embedded within our design allowed students to attain a flow state regardless of their player profiles. They were consistently challenged to deepen their understanding of a specific theme and develop both primary and secondary arguments within their essays.

The gamified design facilitated dynamic writing by adopting an approach that emphasized cultivating essential cognitive skills, including planning, negotiation, and discipline. Moreover, this design provided teachers with the means to evaluate students' development and progress by assessing their needs and implementing targeted interventions. The missions embedded within the design sparked discussions that encouraged formulating persuasive arguments, serving as linguistic mechanisms supported and supervised by digital tools.

In the gamified design, collaborative writing is facilitated through ICTs, enabling participants to engage in writing activities from different locations and at their preferred times. Participants with a preference for social interaction particularly benefitted from activities that required them to interact and collaborate in organizing textual elements. On the other hand, participants with a preference for immersion excelled in achieving higher scores due to their skill in effectively managing the challenges presented, leading to more significant progress in the activity. This demonstrates how the gamified approach accommodated diverse preferences and contributed to improved learning outcomes.

5   Conclusion

The study's results demonstrated significant benefits of our gamified intelligent system compared to the non-gamified approach, with higher levels of flow experienced by participants and improved scores on the essay writing test. We designed the system to enhance learning and extend the classroom experience, fostering autonomy, flexibility, and creativity among students. Thus, the significance of our findings revolves around the concept of writing extending beyond the classroom, becoming meaningful, authentic, and practical. Gamification allowed us to consider a real-world theme involving various cultures and languages, influencing the construction of the text's meaning. It also facilitated collaborative activities, fostering the development of strategic thinking, a crucial skill in structuring written arguments.

Disclosure of Interests. The authors have no competing interests to declare.

References

1. Andrade, F., Marques, L., Bittencourt, I.I., Isotani, S.: Qpj-br: questionário para identificação de perfis de jogadores para o português-brasileiro. In: Brazilian Symposium on Computers in Education (Simpósio Brasileiro De Informática Na Educação-SBIE), vol. 27, p. 637 (2016)
2. Bittencourt, I.I., et al.: Validation and psychometric properties of the Brazilian-Portuguese dispositional flow scale 2 (dfs-br). PLoS ONE **16**(7), e0253044 (2021)
3. Czikszentmihalyi, M.: Flow: The Psychology of Optimal Experience (1990)
4. Gan, Z., Liu, F., Nang, H.: The role of self-efficacy, task value, and intrinsic and extrinsic motivations in students' feedback engagement in english learning. Behav. Sci. **13**(5), 428 (2023)
5. Gomes, C.M.A., Amantes, A., Jelihovschi, E.G.: Applying the regression tree method to predict students' science achievement. Trends Psychol. **28**(1), 99–117 (2020)

6. González-Estrada, E., Cosmes, W.: Shapiro-Wilk test for skew normal distributions based on data transformations. J. Stat. Comput. Simul. **89**(17), 3258–3272 (2019)
7. Hedges, L.V., Tipton, E., Zejnullahi, R., Diaz, K.G.: Effect sizes in Ancova and difference-in-differences designs. Br. J. Math. Stat. Psychol. **76**(2), 259–282 (2023)
8. Ismail, U.S., Makhtar, N.I., Chulan, M., Ismail, N.: A model framework for the implementation of gamification in Arabic teaching in Malaysia. Theory Pract. Lang. Stud. **13**(11), 2800–2805 (2023)
9. Onoda, R., Omi, Y.: The value of extracurricular activities to Japanese junior high school students: focusing on the expression of a school's attractiveness in writing. In: Frontiers in Education, vol. 8 (2023)
10. Vinuesa, R., et al.: The role of artificial intelligence in achieving the sustainable development goals. Nat. Commun. **11** (2020)
11. Yee, N.: Motivations for play in online games. Cyber Psychol. Behav. **9**(6), 772–775 (2006)
12. Young, B.D., et al.: Cytokine signaling and matrix remodeling pathways associated with cardiac sarcoidosis disease activity defined using FDG pet imaging. Int. Heart J. **62**(5), 1096–1105 (2021)

Late-Breaking Results

Towards Human-Like Educational Question Generation with Small Language Models

Fares Fawzi, Sarang Balan, Mutlu Cukurova, Emine Yilmaz, and Sahan Bulathwela[✉]

University College London, London, UK
{fares.fawzi.21,sarang.balan.20,m.cukurova,emine.yilmaz,
m.bulathwela}@ucl.ac.uk

Abstract. With the advent of Generative AI models, the automatic generation of educational questions plays a key role in developing online education. This work compares large-language model-based (LLM) systems and their small-language model (sLM) counterparts for educational question generation. Our experiments, quantitatively and qualitatively, demonstrate that sLMs can produce educational questions with comparable quality by further pre-training and fine-tuning.

Keywords: Question Generation · AI in Education · small Language Models

1 Introduction

Large Language Models (LLM) have revolutionised educational applications with Artificial Intelligence (AI). Scalable educational question generation (EdQG) is a direct beneficiary of this trend. While recent studies use Model-as-a-Service (MaaS) products leveraging externally deployed LLMs (eg. ChatGPT) to carry out the educational question/quiz generation [7], such settings pose severe privacy, ethical and control-related issues. Model retraining can heavily affect model behaviour, compromising prompts and all downstream applications dependent on the MaaS LLM [14,15,19]. Limitations also arise during domain adaptation due to substantial training costs. Also, hosting LLMs on-premise is infeasible operationally and financially for the majority of education stakeholders. Small Language Models (sLMs), trained to excel in educational tasks, are a practical alternative that can unlock the quality of service without compromising control and stability. However, objectively comparing sLMs to LLM alternatives is a critical missing piece that this work attempts to address. We define sLMs as models that are easy to store, transfer and deploy (≤250 MB size) [9].

2 Related Work

In EdQG, state-of-the-art (SOTA) systems use pre-trained language models (PLMs) like Google T5 [17]. Recent EdQG research follows i) zero-shot prompt

A. M. Olney et al. (Eds.): AIED 2024 Workshops, CCIS 2150, pp. 295–303, 2024.
https://doi.org/10.1007/978-3-031-64315-6_25

engineering/tuning [3,8] and ii) few-shot fine-tuning [4,21]. Our work focuses on showing that fine-tuned sLMs can match the performance of LLMs on quantitative and human evaluations.

Recent work uses enormous LLMs that require significant computational power and expertise to train and maintain, including MaaS systems (e.g. Chat-GPT [8]). MaaS API services carry the risk of undesirable changes in the behaviour of the host model, API usage limits and pricing changes - all of which pose different risks to the educators with little to no control over the models. Therefore, sLMs can be more desirable and safe in educational applications where the organisation owns and controls the language model (LM) with minimal expert and infrastructural costs. Recent works demonstrate how general-purpose sLM (T5-Small specifically) can be enhanced for EdQG through pre-training [4]. Our work extends their work by comparing the sLM's performance to LLM counterparts while assessing the human-readiness of sLM generations. This critical information affecting the adaptation of sLMs was not covered in [4]. We also measure the effects of post-grammar correction (GC) as sLMs fine-tuned for specific tasks (such as GC) can be used to improve LLM outputs [23].

Leaf (used in [4]), our baseline, is a SOTA LLM system that addresses EdQG by fine-tuning a pre-trained T5 PLM [17] with the SQuAD 1.1 dataset [18]. However, the SciQ dataset [25], a collection of 13,679 crowd-sourced scientific exam questions covering physics, chemistry and other sciences, is better suited for evaluating EdQG systems. [4] uses the S2ORC corpus with English scholarly abstracts [12] to make the model more suited for EdQG. We use EduQG proposed by [4] as the reference sLM in our experiments. Metrics such as BLEU, BERTScore, Human Ratings, Perplexity and Diversity are utilised [8,13,20,24] to measure the quality of EdQG which are also used in this study.

Human evaluation is a reliable way to assess QG models and typical attributes such as fluency, relevance, answerability and usefulness are measured using Likert scales [3,8]. In our study, we measure fluency, answerability and relevance. When collecting measurements, different prior works have used n-point Likert scales to rate the generations, with a 5-point Likert scale being the most common choice [1,11]. We use a 5-point Likert scale from *strongly disagree* to *strongly agree*. We also use preference ratings to measure human preference for AI-generated questions (like [1,3]).

3   Methodology

We aim to answer three main research questions.

- RQ1: Can sLM-based automatic grammar correction further improve EdQG?
- RQ2: How does sLM EdQG quality compare to general-purpose LLMs?
- RQ3: Are sLM generated questions humanly-acceptable?

3.1   Models, Datasets and Evaluation Metrics

We utilised two sLMs in experiments addressing RQ1: i) EduQG [4] which is based on the T5-small model (60.5M parameters) and ii) EduQG + a lightweight

RoBERTa-based GECToR model [16] (127M parameters) for grammar correction (*EduQG + GC*). In RQ2 experiments, we replicate Leaf [21], based on the T5-base (223M parameters), and use GPT3.5-based ChatGPT[1].

To reduce computational costs, a downsampled S2ORC dataset (2.1 million scientific abstracts) was used to pre-train a t5 model to create EduQG. The full SQuAD 1.1 dataset and the test set of the SciQ dataset were used for fine-tuning and evaluation respectively [4]. Furthermore, we randomly selected *9* SciQ contexts with the prompt "*Given text [context], create 5 expert-level questions with multiple choice answers from the text*" and selected the contexts where ChatGPT generates questions with the same answer as the SciQ dataset. Figure 1 illustrates this methodology.

Similar to [4], we use BLEU 1 through 4 (BL-1,..., BL-4) and F1-score (F1) to evaluate the predictive power of the models. We further use BERTScore [26], consisting of BERT Precision (B-Pr.), BERT Recall (B-Re.) and BERT F1 (B-F1.) to assess the semantic similarity of generations to the ground truth as the generative models may not use the same tokens and word order. Perplexity and Diversity are used to measure linguistic quality.

RQ3 was addressed through a questionnaire consisting of 2 parts. Part 1 is a pairwise preference task. Part 2 is a qualitative assessment task. The final part records demography and English fluency. In the Pairwise Preference Task (part 1 of the questionnaire), a pair of questions (A and B) were presented to the participant: i) a Teacher-generated ground truth(human-generated) and ii) an EduQG + GC model-generated (AI-generated) version of the same question. The ordering of the pair is randomised. The participants provide preferences for use in a teaching task based on a 5-level Likert scale (strongly prefer A, prefer A, no preference, prefer B, strongly prefer B). In the Qualitative Assessment (part 2 of the questionnaire), for each AI-generated question, 3 questions are asked about the level of i) Fluency, ii) Answerability and iii) Relevance. Again, a 5-level Likert scale is provided for all 3 questions with detailed descriptions of the definition of each aspect. The candidate questions used in this part were specifically selected to avoid overlap with items in Part 1 to prevent the learning effect and label leakage.

3.2 Experimental Setup

The experimental setup to investigate RQ 1 and 2 is presented in Fig. 1. The EduQG model, and its output through an sLM fine-tuned for English grammar correction (*EduQG + GC*), are analysed to answer RQ1. Leaf, ChatGPT and *EduQG + GC* are compared to answer RQ2. Finally, the outputs from the *EduQG + GC* sLM system are used for the user study (RQ3).

---

[1] https://chat.openai.com.

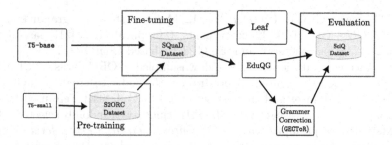

Fig. 1. Methodology for training and evaluating the models to answer RQ 1 and 2.

4   Results and Discussion

Table 1 shows how sLM-based EduQG, EduQG + GC (grammar corrected) and LLM-based systems Leaf and ChatGPT perform on the EdQG task. The perplexity calculation uses `TextDescriptives` [10] with the Spacy `en_core_web_lg` model as the reference PLM. Figure 2 further summarises the key results obtained from the user study.

Table 1. Top section: Comparison of predictive performance between leaf baseline (T5-base-based) and EduQG (T5-small-based sLM) on SciQ testset. Bottom section: Comparison of predictive performance between leaf baseline (T5-base-based), EduQG (T5-small-based sLM), and ChatGPT on 9 randomly selected contexts from SciQ testset. The best and second-best performance is indicated in **bold** and *italic* faces respectively.

Model	Predictive Performance								Language	
	BL-1	BL-2	BL-3	BL-4	F1	B-Pr	B-Re	B-F1	Perp	Div
Leaf LLM (†)	**0.9545**	**0.8176**	**0.6754**	**0.5737**	**0.6528**	**0.9279**	**0.9057**	**0.9165**	1.2942	0.7488
EduQG	0.9468	0.7750	0.6131	0.5016	*0.6044*	0.9145	0.8938	0.9039	**1.2675**	*0.7529*
EduQG + GC	*0.9470*	*0.7796*	*0.6202*	*0.5095*	0.6021	*0.9151*	*0.8944*	*0.9045*	*1.2813*	**0.7555**
Leaf LLM (†)	**0.7522**	**0.5450**	**0.3816**	**0.3080**	*0.4675*	*0.8995*	0.8636	0.8810	*1.3406*	*0.7503*
ChatGPT (†)	0.6071	0.4219	0.3146	0.2630	0.3941	0.8928	**0.8749**	*0.8836*	1.5001	**0.8200**
EduQG + GC	*0.7456*	*0.5018*	*0.3620*	*0.2997*	**0.4838**	**0.9064**	*0.8678*	**0.8865**	1.2819	0.7399

4.1   sLM Vs. LLM-Based EdQG Systems (RQ 1 and 2)

Among the sLMs, Table 1 (top section) shows that the *EduQG + GC* model shows superior performance against EduQG, indicating the value addition of automatic post-grammar correction for this task (RQ1). The same section also highlights that EduQG systems based on sLMs perform similarly to the much larger ($\approx$ 4×) LLM-based Leaf counterpart. sLMs also outperform the Leaf baseline in perplexity and diversity. This is mainly because the sLMs are further pretrained with scientific abstracts. This observation is very insightful as empirical evidence shows comparable performance in STEM-subject-related EdQG can be obtained with significantly lightweight models. While sLMs are intriguing practically and scale-wise, results suggest that sLMs still struggle to capture grammatical structures fully, lending to their limited capacity. However, given that the grammar correction model itself is a sLM, the union of the 2 sLMs (EduQG and GECToR) is still significantly smaller than the larger baseline. Table 1 (bottom section) shows that the *EduQG + GC* model is again comparable to the Leaf baseline while consistently outperforming ChatGPT outputs in the smaller dataset. While the ChatGPT experiment is smaller-scale ($n = 9$), this is promising evidence of the utility of sLMs in place of MaaS-based enormous LLM services like ChatGPT.

4.2   Human Evaluation (RQ3)

The participant set ($n = 9$) which is higher than the median of ($n = 3$) found in the literature [22]. The group consisted of 5 female (55.5%) and 4 male (44.5%) participants. While none of them were native English speakers, they all had post-secondary education to the Master's level at the minimum. 8 participants came from the 25–34 age bracket while the remaining one belonged to the 35–44 age bracket. All but two participants studied Mathematics, Biology, Chemistry, Physics and Earth Science in high school - which are the domains covered by the SciQ dataset.

The summary of results from the pairwise preference assessment study (part 1) is presented in Fig. 2 (i). The majority of participants said that they either prefer AI-generated questions or have no preference between human or AI-generated questions for 12 out of 20 (60%) questions. This shows that the questions generated by the *EduQG + GC* model are perceived to have equal or higher quality compared to human-created questions. The summary results from the qualitative assessment (part 2) are presented in Fig. 2(ii). We observe the median scores for fluency, answerability and relevance factors are above 3, the centre of the 5-level Likert scale (1 to 5). This suggests that the participants express a positive sentiment regarding the quality of the AI-generated questions. Specifically, the fluency score is concentrated around very high values close to 5 suggesting the high linguistic quality of the generated questions. In comparison, answerability scores lie slightly lower with two outliers.

While the qualitative assessment highlights an above-average positive result, the generation quality has significant room for improvement. sLM models that generate education questions hold promise, yet the lack of overwhelming acceptance strongly suggests that the model outputs need to be improved significantly before any kind of deployment of sLMs for EdQG in a mainstream fashion.

4.3   Impact, Limitations and Future Research

Many recent works show zero-shot or prompt-tuned question generation to be operationally feasible using very large language models gated behind private

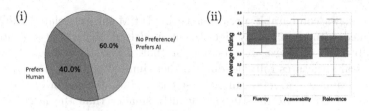

Fig. 2. Summary results from the user study.

APIs of MaaS services [3,7]. Our results contribute to this topic as we demonstrate the utility of openly available sLMs to support EdQG. The proposed models are very lightweight and open-source, giving the stakeholders full control and ownership, a critical feature for quality assurance of the downstream educational systems that rely on these models. Privately owned models carry less cybersecurity and data risk as all network and data interactions would occur within the organisation, as opposed to sending data to an external host. Additionally, the high power consumption needed to use LLMs marks a negative impact on its environmental sustainability. While the proposed models are not yet perfect, our results are positive and indicate that an educator can re-purpose these questions with minimum effort and time. When improved, educational questions can be generated at scale using the proposed model both for existing and new learning resources, adding more testing opportunities for learners/teachers.

Currently, an evident limitation of the EduQG model (even with grammar correction) from the results in Sect. 4 is its inferior performance in comparison to LLMs. The proposed model still needs to be improved significantly to match the performance of LLMs. We hypothesise the sub-par performance attributes to 1) the model size, with 60M parameters and 2) training on 2.1 million out of 81 million available scientific abstracts. Our future work will aim to explore and unblock these bottlenecks. Furthermore, the statistical confidence of the ChatGPT experiment reported here is weak due to the small subset of data points that were used ($n = 9$) at this point. A larger scale comparison with more contexts (using the API) is necessary in the future to derive a better understanding of the sLM behaviour in comparison to ChatGPT. It is also possible to incorporate later versions of ChatGPT in future studies.

At last, we also need to be cautious to avoid the obvious pitfalls of such automatic systems. Intelligent QG models we build tend to exhibit the patterns in the data that we feed them. The pre-trained models we use as a foundation for building these sLMs are already trained with Internet data that is present with many biases. It is sensible to use post-processing tools to detect biases (e.g. [2]) and handle them before questions generated by these models are exposed to learners. Adaptation and assessing the usefulness of sLMs for cross-subject and cross-lingual question generation is another open research question that is under-explored at present.

5   Conclusion

In this work, we compare the performance of LLM-based QG models and sLM-based QG models in the context of EdQG. While the sLM models do not outperform their much larger counterparts, the results show that their generation capabilities are similar, and may be acceptable by humans, while the models being almost four times smaller. Reduced model sizes have significant advantages over larger language models in training and maintaining the models in-house, whilst retaining full ownership to be used in downstream educational services. This improves quality assurance as well as operational and capital costs by enabling complete control and oversight over their behaviours. We see our work being foundational to building a series of tools that can support educators with scalable personalised learning while scaling up question banks and knowledge bases in education [5]. The human-AI collaborative systems emerging initially can also produce valuable data that can be used to further fine-tune models. Ultimately, these models can be improved to the point where an intelligent tutor can create on-demand questions to verify a learner's knowledge state [6].

Acknowledgements. This work is funded by the European Commission-funded projects "Humane AI" (grant 820437) and "X5GON" (grant No 761758). This work was also partially supported by the UCL Changemakers grant.

References

1. Amidei, J., Piwek, P., Willis, A.: The use of rating and Likert scales in natural language generation human evaluation tasks: a review and some recommendations. In: Proceedings of the 12th International Conference on Natural Language Generation. ACL (2019)
2. Bai, Y., Zhao, J., Shi, J., Wei, T., Wu, X., He, L.: FairBench: a four-stage automatic framework for detecting stereotypes and biases in large language models. arXiv preprint arXiv:2308.10397 (2023)
3. Blobstein, A., Izmaylov, D., Yifat, T., Levy, M., Segal, A.: Angel: a new generation tool for learning material based questions and answers. In: Proceedings of the NeurIPS Workshop on Generative AI for Education (GAIED)
4. Bulathwela, S., Muse, H., Yilmaz, E.: Scalable educational question generation with pre-trained language models. In: Wang, N., Rebolledo-Mendez, G., Matsuda, N., Santos, O.C., Dimitrova, V. (eds.) AIED 2023. LNCS, pp. 327–339. Springer, Cham (2023). https://doi.org/10.1007/978-3-031-36272-9_27
5. Bulathwela, S., Pérez-Ortiz, M., Holloway, C., Cukurova, M., Shawe-Taylor, J.: Artificial intelligence alone will not democratise education: on educational inequality, techno-solutionism and inclusive tools. Sustainability **16**(2), 781 (2024)
6. Bulathwela, S., Pérez-Ortiz, M., Yilmaz, E., Shawe-Taylor, J.: Power to the learner: towards human-intuitive and integrative recommendations with open educational resources. Sustainability **14**(18), 11682 (2022)
7. Elkins, S., Kochmar, E., Cheung, J.C.K., Serban, I.: How teachers can use large language models and Bloom's taxonomy to create educational quizzes. Proc. AAAI Conf. Artif. Intell. **38**(21), 23084–23091 (2024). https://doi.org/10.1609/aaai.v38i21.30353

8. Elkins, S., Kochmar, E., Serban, I., Cheung, J.C.K.: How useful are educational questions generated by large language models? In: Wang, N., Rebolledo-Mendez, G., Dimitrova, V., Matsuda, N., Santos, O.C. (eds.) AIED 2023. LNCS, pp. 536–542. Springer, Cham (2023). https://doi.org/10.1007/978-3-031-36336-8_83

9. Fawzi, F., Amini, S., Bulathwela, S.: Small generative language models for educational question generation. In: Proceedings of the NeurIPS Workshop on GAIED

10. Hansen, L., Olsen, L.R., Enevoldsen, K.: Textdescriptives: a python package for calculating a large variety of metrics from text. J. Open Source Softw. 8(84), 5153 (2023)

11. van der Lee, C., Gatt, A., van Miltenburg, E., Wubben, S., Krahmer, E.: Best practices for the human evaluation of automatically generated text. In: Proceedings of the 12th International Conference on Natural Language Generation. ACL (2019)

12. Lo, K., Wang, L.L., Neumann, M., Kinney, R., Weld, D.: S2ORC: the semantic scholar open research corpus. In: Proceedings of the Annual Meeting of the ACL. Online (2020)

13. Lopez, L.E., Cruz, D.K., Cruz, J.C.B., Cheng, C.: Simplifying paragraph-level question generation via transformer language models. In: Pham, D.N., Theeramunkong, T., Governatori, G., Liu, F. (eds.) PRICAI 2021. LNCS (LNAI), vol. 13032, pp. 323–334. Springer, Cham (2021). https://doi.org/10.1007/978-3-030-89363-7_25

14. Loya, M., Sinha, D., Futrell, R.: Exploring the sensitivity of LLMs' decision-making capabilities: insights from prompt variations and hyperparameters. In: Findings of the ACL: EMNLP 2023, pp. 3711–3716. ACL (2023)

15. Lu, Y., Bartolo, M., Moore, A., Riedel, S., Stenetorp, P.: Fantastically ordered prompts and where to find them: overcoming few-shot prompt order sensitivity. In: Proceedings of the ACL (vol. 1: Long Papers). ACL (2022)

16. Omelianchuk, K., Atrasevych, V., Chernodub, A., Skurzhanskyi, O.: GECToR – grammatical error correction: Tag, not rewrite. In: Proceedings of the Fifteenth Workshop on Innovative Use of NLP for Building Educational Applications, pp. 163–170. ACL, Seattle (2020)

17. Raffel, C., et al.: Exploring the limits of transfer learning with a unified text-to-text transformer. J. Mach. Learn. Res. 21(140), 1–67 (2020)

18. Rajpurkar, P., Zhang, J., Lopyrev, K., Liang, P.: SQuAD: 100,000+ questions for machine comprehension of text. In: Proceedings of the 2016 Conference on EMNLP. ACL (2016)

19. Sclar, M., Choi, Y., Tsvetkov, Y., Suhr, A.: Quantifying language models' sensitivity to spurious features in prompt design or: how i learned to start worrying about prompt formatting. arXiv preprint arXiv:2310.11324 (2023)

20. Ushio, A., Alva-Manchego, F., Camacho-Collados, J.: A practical toolkit for multilingual question and answer generation. In: Proceedings of the 61st Annual Meeting of the ACL (vol. 3: System Demonstrations), pp. 86–94. ACL (2023)

21. Vachev, K., Hardalov, M., Karadzhov, G., Georgiev, G., Koychev, I., Nakov, P.: Leaf: multiple-choice question generation. In: Proceedings of the European Conference on Information Retrieval (2022)

22. van der Lee, C., Gatt, A., van Miltenburg, E., Krahmer, E.: Human evaluation of automatically generated text: current trends and best practice guidelines. Comput. Speech Lang. 67, 101–151 (2021)

23. Vernikos, G., Brazinskas, A., Adamek, J., Mallinson, J., Severyn, A., Malmi, E.: Small language models improve giants by rewriting their outputs. In: Proceedings of the 18th Conference of the European Chapter of the ACL (Vol. 1: Long Papers). ACL (2024)

24. Wang, Z., Valdez, J., Basu Mallick, D., Baraniuk, R.G.: Towards human-like educational question generation with large language models. In: Proceedings of International Conference on Artificial Intelligence in Education (2022)
25. Welbl, J., Liu, N.F., Gardner, M.: Crowdsourcing multiple choice science questions. In: Proceedings of the 3rd Workshop on Noisy User-Generated Text. ACL (2017)
26. Zhang, T., Kishore, V., Wu, F., Weinberger, K.Q., Artzi, Y.: Bertscore: evaluating text generation with BERT. In: Proceedings of 8th International Conference on Learning Representations. OpenReview.net (2020). https://openreview.net/forum?id=SkeHuCVFDr

Large Language Models for Career Readiness Prediction

Chenwei Cui[3]($\boxtimes$), Amro Abdalla[2], Derry Wijaya[1], Scott Solberg[1],
and Sarah Adel Bargal[2]

[1] Boston University, Boston, MA, USA
{wijaya,ssolberg}@bu.edu
[2] Georgetown University, Washington, DC, USA
{aaa654,sarah.bargal}@georgetown.edu
[3] Arizona State University, Tempe, AZ, USA
ccui17@asu.edu

Abstract. Large Language Models (LLMs) have recently achieved state-of-the-art performance on many benchmark Natural Language Processing (NLP) tasks. In this work, we are introducing a novel application, career readiness prediction, in the area of NLP for education. We analyze a dataset of student narratives and explore how reliably different LLMs classify them using Marcia's (1980) identity statuses. We explore the capabilities and limitations of LLMs on this new task and find that there is good potential for automated career readiness evaluation, and for improved survey design that enables larger-scale data collection.

Keywords: Large Language Models · Career Readiness Prediction · Natural Language Processing · Classification · GPT

1 Introduction

Large Language Models (LLMs) such as BERT [5], PaLM [3], LLaMA [16], and GPT models [13] recently achieved state-of-the-art performance on many benchmark NLP tasks [1,9,17,19]. This work explores the uses and limitations of LLMs on the novel task of career readiness prediction in school settings. Drawing from recent innovations in career development and vocational psychology, "career readiness" can be defined as one's ability to establish future life and occupational goals that reflect their emerging talent and skills and be proactive in seeking learning opportunities that enable one to pursue those goals [8,14].

This study uses career narratives that were generated in response to 18 open-ended questions derived from Van Esbroeck et al.'s dynamic career exploration model [6]. The responses were classified by experts into one of four identity profiles based on Marcia's (1984) [10] ego identity model [15]. [11] conceptualizes the emergence of identity as the intersection of self-exploration and career exploration which can be described according to four distinct identity statuses: Diffusion, Foreclosure, Moratorium, and Achievement. *Diffusion* identity status refers to one who has not explored one's skills, talent or interests or begun to identify occupational roles. *Foreclosure* identity status refers to one who has

A. M. Olney et al. (Eds.): AIED 2024 Workshops, CCIS 2150, pp. 304–311, 2024.
https://doi.org/10.1007/978-3-031-64315-6_26

selected an occupational role but has not offered evidence that the selection of the occupation emerged through an exploration of skills, talent, and interests. *Moratorium* identity status refers to one who is actively exploring their talent, skills and interests and is continuing to consider a number of potential occupational roles. Finally, an *Achievement* identity status refers to one who has selected an occupational role that is aligned with their awareness of their skills, talent and interests. The raw natural language responses from the students constitutes the data, and the expert-annotated ego-identity classifications is the label. We use deep learning models to re-examine career narratives from ~ 1000 secondary education youth that were previously annotated [15].

Annotating data according to Marcia's ego identity model can be an expensive process that requires the expertise of trained professionals. This was the motivation for exploring LLMs as a potential solution for automating this process. This study will assess the capability of transformer-based NLP models to predict career readiness using a medium-sized dataset. We will explore the ability of LLMs in the classification of career narratives using rubrics that are used to train human experts. Our objective is to explore whether LLMs are able to follow rubrics in inherently ambiguous human text better than previous state-of-the-art due to the way they were pre-trained.

The study will compare the performance of LLMs using several approaches: zero-shot classification, where the model makes a prediction solely based on the data it was pre-trained on; few-shot classification, where the LLM is provided with a limited number of expert-annotated examples; fine-tuning a pre-trained model on the student narrative dataset. Our study seeks to address the question: Can LLM models alleviate the need for collecting and annotating large datasets? In summary, we propose the first application of automated career readiness prediction and investigate the effectiveness of using LLMs for this task.

2   Related Work

Language Models are transforming the education process from different perspectives. Their applications range from automated grading systems and question generation to providing tools for assessing and analyzing educational outcomes.

Bulathwela *et al.* [2] proposed EduQG, a question generation model based on pre-trained language models (PLMs). Fine-tuning was used to generate pedagogically sound questions. Yang *et al.* [18] used language models to develop an Automatic Essay Scoring system (AES) to analyze argumentative essays. Through their research, Yang *et al.*showed that large language models effectively classify argumentative components and highlight the underlying structure of essays, demonstrating their potential for comprehending and analyzing textual pieces. Another application was introduced by Cochran *et al.*[4] to use GPT 3.5 to augment student responses to short essay questions to create a bigger dataset. This work used the dataset to train a BERT model to provide feedback to text-based student answers. This work shows that language models could be used for self-augmentation to improve model performance. This tackles the data scarcity problem in the field of education. Funayama *et al.* [7] propose an approach to reduce the cost of requiring new data and re-training LLMs to the custom prompts in education. Their research demonstrates that a short-answer scoring system can enhance its performance by leveraging cross-prompt training data encompassing a variety of prompts.

In this work, we will explore the ability of Large Language Models (LLMs) in predicting career readiness based on a textual student narrative. We will contrast the use of student narratives, expert defined rubrics, LLM pre-training, and LLM fine-tuning. We are offering insights into how language models can contribute to a holistic understanding of students' abilities and potential professional success.

3   Methodology

This work uses LLMs to create a model that maps students' career narrative responses to the ego-identity status classifications. To approach this, the steps include model selection, text pre-processing, model training, followed by a series of experiments to validate whether the model provides good predictions of the original expert-generated ego-identity classifications.

3.1   Dataset

Data Collection. A dataset consisting of 997 career narratives was used for this study. The original data was collected with 10th and 12th grade high school students attending 14 schools across four states as part of a federally sponsored effort to understand the nature and potential value of personalized career and academic plans [15]. Career narratives were generated by asking each student to provide online natural language responses to 18 open-ended questions that were based on themes identified in the dynamic model of career exploration and planning [6]. In the original study, each narrative was rated by two graduate students using a rubric developed to differentiate the four ego-identity statuses described by Marcia (1981) -These rubrics will later be used as a guide for LLM predictions too. A doctoral student served as auditor to review the results and facilitate consensus when raters differed in their classification ratings for a given narrative. The secondary dataset, therefore, includes the original career narrative responses to each question as well an ego-identity status classification.

Data Splits. We defined five splits for this data. Each split was generated by randomly selecting 20% of the data as test set, 10% as validation set, and 70% as training set. For experiments where we conducted a training/fine-tuning phase, we use 5-fold cross validation to evaluate the models. The split information will be made publicly available upon acceptance. For experiments where there was no training/fine-tuning phase, we set aside 50 examples for initial validation experiments, and used the rest of the 947 examples for testing.

Data Pre-processing. The raw responses from students contain natural mistakes, *e.g.* misspellings. Also, the input text can be too long for mainstream

NLP systems to handle [5]. We implement rule-based text pre-processing procedures. Specifically, we lowercase every letter in the text. Afterwards, we keep only standard characters such as 0 9, a z, commas, and periods. Since each student's response consists of 18 separate answers, we truncate text for each question separately, *i.e.* for each question we truncate the answers longer than the 75th percentile to accommodate for the narrow context window of some models.

3.2 Models

BERT. Bi-directional Encoder Representations from Transformers (BERT) [5] is a deep bidirectional transformer pre-trained using self supervision to understand human language. The main intuition behind such systems can be described by: "you shall know a word by the company it keeps" John Rupert Firth. BERT is trained on a large corpus of web text data in order to model semantics of the human language. BERT is most prominently used for text representation.

Linear Probing. Linear Probing [12] is a pre-trained BERT with an additional layer to interpret the output representation produced by BERT. This is equivalent to freezing BERT's parameters and training only the added layer.

GPT Models. Generative Pre-trained Transformer (GPT) is a causal transformer that uses causal modeling. Causal transformers are trained using autoregressive language modeling [1] *vs.* masked-language modeling for bi-directional transformers. Autoregressive language modeling can be used for generation and masked-language modeling cannot.

3.3 Experimental Setup

BERT. We performed fine-tuning of all parameters of the BERT model using a fixed learning rate of $1 * e^{-5}$. To avoid overfitting, the validation set was used to determine the number of iterations until Early Stopping as a means of regularization. The weights of the BERT model were shared for all the 18 questions. We then performed 5-fold cross validation and reported the model mean accuracy on the classification task.

Linear Probing. We input each student answer into OpenAI's Text-Embedding-3-Large [12] BERT-style model, which returns a $3,072 - D$ vector. We then concatenate the 18 vectors together and feed that into a logistic regression model trained for 100 iterations with the class weights adjusted to be inversely proportional to class frequency. This in turn outputs a vector of dimension 4 representing the class conditional probabilities of the four ego-status classes. This was then contrasted with the ground-truth for the 947 examples.

GPT. We explored the possibility of using the power of pre-training of LLMs, particularly GPT 4.0, together with using one or no examples from our data. We then experimented with providing a rubric, similar to that given to trained human annotators, to help GPT with the classification. We then contrasted both approaches with fine-tuning GPT[1].

---

[1] Please note that fine-tuning has been conducted using GPT 3.5, the only version that currently allows fine-tuning.

The first experiment is a GPT 4.0 zero-shot experiment, where we asked the model to classify each instance in our test data given no fine-tuning or examples from our end. The prompt consists of a system message and a user message. We set the system message to be a brief sentence describing the role of the agent as a helpful assistant that can classify text according to Marcia's ego identity model. The user message for each sample consisted of the 18 questions each followed by their corresponding answers. This prompt is passed to the model to return a prediction in the form of a single word representing one of the four classes. All prompts used will be made publicly available upon acceptance.

The second experiment is a zero-shot with rubrics experiment, where we include rubrics to describe each of the possible classes into the system message. The user message is kept the same in the form of a question followed by its answer. The prompt is passed to the model through the API to get the prediction.

The next experiment is a one-shot experiment, where we not only rely on the training GPT has previously received, but provide *one* example of a student narrative for each of the four classes from our dataset that has received expert annotations. The prompt consists of 3 components: the system message, an example, and the user message. The system message contains the rubric and the survey questions in addition to the brief description used before. The examples contains a user role and an assistant role. In the user role, we employ predefined sample answers from the training set. The assistant role contains the correct class of the provided answers. Finally, the user message only contains the entry that we need to classify (We will provide exact prompt text used to achieve these results in the appendix section upon acceptance). To assess the impact of example selection on model performance, the one-shot experiment was once conducted using an expert-selected example from the dataset, and another time using a randomly-selected example.

Finally, to further explore the limits of LLMs in this task, we fine-tuned a GPT 3.5 turbo on our data using OpenAI's interface. We performed the fine-tuning using the same five splits of the data defined for other experiments. When preparing for this experiment, we used the same prompt we used for the GPT-4.0 zero-shot experiment.

Identifying Questions Contributing Most to Predictions. We designed an experiment to identify the questions, among the 18 in the questionnaire, that contribute most to model predictions. This involved utilizing a single question to predict the class, and subsequently selecting the top-performing questions. Using Linear Probing for this experiment, we identified the top 5 questions with the highest classification accuracy. We then used these questions and answers to perform classification using Linear Probing.

4   Experimental Results

Our experimental design aims to explore the effectiveness and limitations of LLMs in predicting career readiness. Table 1 presents the experimental results.

In the Linear Probing and BERT experiments, we observed varied performance across the four classes. BERT showed excellent accuracy for Foreclosure

Table 1. Per-class and overall accuracy (%) of ego-identity classifications for various LLM setups: BERT, Linear Probing, and GPT. Results presented are on all test samples in the dataset.

	Models	Diffusion	Foreclosure	Moratorium	Achievement	Overall
	BERT	33.3	91.3	3.6	26.0	65.3
	Linear Probing	55.0	81.3	24.5	42.0	65.7
GPT	Zero-shot	53.3	7.6	60.6	75.0	29.6
	Zero-shot +*Rubric*	52.8	7.6	**78.8**	**88.4**	34.1
	One-shot *(Random)*	**61.0**	76.0	27.0	45.0	63.8
	One-shot *(Expert)*	43.0	76.0	25.0	54.0	64.3
	Fine-tuning	43.3	**95.1**	4.5	24.0	**67.9**

(91.3%) but lower accuracy for Diffusion(33.3%) and Achievement(26%) and a very low accuracy for Moratorium(3.6%). Linear Probing, on the other hand, performed well of on Foreclosure (81.3%) and fairly better than BERT on the other classes. We can observe that BERT and Linear Probing with the Text-Embedding-3-Large model have comparable overall accuracies.

Moving to the GPT experiments, we conducted zero-shot evaluations with and without rubrics. As expected, providing rubrics improves the overall classification accuracy (34.1% *vs.* 29.6%). The introduction of a rubric in the zero-shot experiment significantly improved the performance on Moratorium (78.8%) and Achievement (88.35%), while maintaining a similar performance for Foreclosure (7.6%) and Diffusion (52.8%). This led to an increase in the overall accuracy by 3.5%. While the zero-shot experiments do not incur any training of the models, they resulted in an accuracy that is significantly higher than a random guess of one of the four classes, *i.e.* classification accuracy of 25%. The zero-shot accuracies are however significantly lower than accuracies achieved by BERT (training all parameters) and Linear Probing (training parameters of the last layer).

The one-shot experiments, including a rubric and an example, showed an improvement to the zero-shot performance, especially on Foreclosure (76%). We can observe a better distributed per-class accuracy as well as a significant improvement on the Foreclosure accuracy when compared to the zero-shot experiments. Interestingly, an experiment with a random selection of the examples shows slight improvements in the per-class accuracy of Diffusion and Moratorium and a decline in the Achievement accuracy while maintaining the same the Foreclosure accuracy. This shows the model sensitivity to example selection.

We then fine-tuned a GPT 3.5 turbo model on our training data. This achieved the highest overall test accuracy: 67.96%. This is the best performing model surpassing prompting of the GPT 4.0, showing the potential of fine-tuning on model performance. Furthermore, we investigated the impact of individual questions on model predictions. The results of this experiment are presented in Table 2. We used Linear Probing to identify the top 5 (and lowest 5) questions contributing to the highest (and lowest) individual question accuracy and used

Table 2. Per-class and overall accuracy (%) of ego-identity classifications for all Linear Probing (LP) experiments. Results presented are on all test samples in the dataset.

Models	Diffusion	Foreclosure	Moratorium	Achievement	Overall
LP *(all 18 questions)*	55.0	81.3	24.5	42.0	65.7
LP *(top 5 questions)*	51.7	70.4	25.5	56.0	61.5
LP *(low 5 questions)*	51.7	60.9	17.3	45.0	52.4

them combined with Linear Probing. Interestingly, reducing a questionnaire from being 18 questions to only 5 results in a 4.2% decline in accuracy. This indicates that a further study of question combinations has the potential of significantly reducing the questionnaire length, and maintaining comparable accuracy.

5   Conclusions and Limitations

In summary, our experiments revealed that while LLMs showed promise in career readiness prediction, additional annotated data are still required to improve their performance to be ready for deployment. This solution has major implications as an assessment tool to monitor individual youth progress through secondary education as well as a tool that post-secondary institutions can use to assess applicants. This solution mechanizes evaluation of career readiness in schools by solely using student narratives (answers to an 18-question questionnaire), *i.e.* requiring minimal time and resource investments from schools and teachers. Such systems could provide means of assessing the need of allocating resources that promote career readiness to schools, leading to equitable education.

However, some classes remain challenging to predict, either due to imbalanced data, or the classes being more subtle than others. Model sensitivity to example selection and the importance of specific questions were evident. Fine-tuning, as explored with GPT 3.5 turbo, demonstrated potential for further improvement. Our findings demonstrate the complexity of predicting career readiness and the need for refined approaches to address specific challenges within the task.

References

1. Brown, T., et al.: Language models are few-shot learners. Adv. Neural. Inf. Process. Syst. **33**, 1877–1901 (2020)
2. Bulathwela, S., Muse, H., Yilmaz, E.: Scalable educational question generation with pre-trained language models. In: Wang, N., Rebolledo-Mendez, G., Matsuda, N., Santos, O.C., Dimitrova, V. (eds.) Artificial Intelligence in Education, pp. 327–339. Springer, Cham (2023). https://doi.org/10.1007/978-3-031-36272-9_27
3. Chowdhery, A., et al.: Palm: scaling language modeling with pathways. J. Mach. Learn. Res. **24**(240), 1–113 (2023)

4. Cochran, K., Cohn, C., Rouet, J.F., Hastings, P.: Improving automated evaluation of student text responses using GPT-3.5 for text data augmentation. In: Wang, N., Rebolledo-Mendez, G., Matsuda, N., Santos, O.C., Dimitrova, V. (eds.) Artificial Intelligence in Education, pp. 217–228. Springer, Cham (2023)
5. Devlin, J., Chang, M.W., Lee, K., Toutanova, K.: Bert: pre-training of deep bidirectional transformers for language understanding. arXiv preprint arXiv:1810.04805 (2018)
6. Esbroeck, R.V., Tibos, K., Zaman, M.: A dynamic model of career choice development. Int. J. Educ. Vocat. Guid. **5**, 5–18 (2005)
7. Funayama, H., Asazuma, Y., Matsubayashi, Y., Mizumoto, T., Inui, K.: Reducing the cost: cross-prompt pre-finetuning for short answer scoring. In: Wang, N., Rebolledo-Mendez, G., Matsuda, N., Santos, O.C., Dimitrova, V. (eds.) Artificial Intelligence in Education, pp. 78–89. Springer, Cham (2023). https://doi.org/10.1007/978-3-031-36272-9_7
8. Guichard, J.: Reflexivity in life design interventions: comments on life and career design dialogues. J. Vocat. Behav. **97**, 78–83 (2016)
9. Lin, X.V., et al.: Few-shot learning with multilingual generative language models. In: Proceedings of the 2022 Conference on Empirical Methods in Natural Language Processing, pp. 9019–9052 (2022)
10. Marcia, J.E.: Citation classic-development and validation of ego identity status (1984)
11. Marcia, J.E., et al.: Identity in adolescence. Handb. Adolesc. Psychol. **9**(11), 159–187 (1980)
12. Neelakantan, A., et al.: Text and code embeddings by contrastive pre-training. arXiv preprint arXiv:2201.10005 (2022)
13. Radford, A., Narasimhan, K., Salimans, T., Sutskever, I., et al.: Improving language understanding by generative pre-training (2018)
14. Savickas, M.L.: Reflection and reflexivity during life-design interventions: Comments on career construction counseling. J. Vocat. Behav. **97**, 84–89 (2016)
15. Solberg, V.S., Wills, J., Redmon, K., Skaff, L.: Use of individualized learning plans: a promising practice for driving college and career efforts. Findings and recommendations from a multi-method, multi-study effort. In: National Collaborative on Workforce and Disability for Youth (2014)
16. Touvron, H., et al.: Llama: open and efficient foundation language models. arXiv preprint arXiv:2302.13971 (2023)
17. Workshop, B., et al.: Bloom: a 176b-parameter open-access multilingual language model. arXiv preprint arXiv:2211.05100 (2022)
18. Yang, B., Nam, S., Huang, Y.: "Why my essay received a 4?": A natural language processing based argumentative essay structure analysis. In: Wang, N., Rebolledo-Mendez, G., Matsuda, N., Santos, O.C., Dimitrova, V. (eds.) Artificial Intelligence in Education, pp. 279–290. Springer, Cham (2023). https://doi.org/10.1007/978-3-031-36272-9_23
19. Zhang, S., et al.: Opt: open pre-trained transformer language models. arXiv preprint arxiv:2205.01068 (2022)

Difficulty-Controllable Multiple-Choice Question Generation for Reading Comprehension Using Item Response Theory

Yuto Tomikawa$^{(\boxtimes)}$ ⓘ and Masaki Uto ⓘ

The University of Electro-Communications, Chofu182-8585, Japan
{tomikawa,uto}@ai.lab.uec.ac.jp

Abstract. In recent years, there has been increasing interest in automatically generating reading comprehension questions with controllable
difficulty levels for educational purposes. A recent study proposed a
method for generating reading comprehension questions with levels of
difficulty suitable to a learner's ability level involving the use of large
language models and item response theory. However, the conventional
method targets only the extractive question format, where the answer
can be found directly in the reading passage, and does not support the
multiple-choice question format that is widely used in educational settings. To address this issue, this study develops a method for automatically generating multiple-choice questions with controllable difficulty
levels by extending the conventional method. We evaluate the performance of difficulty controllability of the generated questions based on
the correct answer rate of responses and the nominal response model,
a polytomous item response theory model. The results confirm that the
proposed method can generate multiple-choice questions that accurately
reflect the intended difficulty level.

Keywords: Automated Question Generation · Item Response
Theory · Deep Neural Networks · Adaptive Learning · Natural
Language Processing

1 Introduction

In recent years, the automatic generation of reading comprehension questions has
attracted significant attention in the field of education, and methods using deep
neural networks have achieved high performance [2,5]. For effective learning support, it is desirable that the question generator be able to offer questions at any
difficulty level [16]. Previous studies thus have focused on difficulty-controllable
question generation [6,18]. For example, a recent study [18] proposed a method
that uses pretrained transformer-based large language models and item response
theory (IRT) [9] to generate pairs of questions and answers with the desired difficulty levels. However, this method targets only the extractive question format,

A. M. Olney et al. (Eds.): AIED 2024 Workshops, CCIS 2150, pp. 312–320, 2024.
https://doi.org/10.1007/978-3-031-64315-6_27

where the answer can be found directly in the reading passage, and cannot be directly applied to the multiple-choice question format, which is widely used in educational settings. To address this issue, this study develops an automatic question-generation method for multiple-choice questions with controllable difficulty levels. The proposed method is realized by fine-tuning LLaMA 2 [15], a pretrained large language model, to generate multiple-choice questions and sets of options from a reading passage with the desired difficulty levels.

This study evaluates the difficulty controllability of the proposed method based on the correct answer rate and the nominal response model, a type of IRT model [1]. Consequently, it was confirmed that the proposed method is capable of generate multiple-choice questions reflecting the specified difficulty level.

Additionally, to evaluate the effectiveness of fine-tuning for difficulty control, we investigated the performance of a few-shot approach using GPT-4 [11]. It was revealed that controlling difficulty is a challenge, suggesting that the proposed method utilizing fine-tuning is effective.

2   Item Response Theory

IRT is a testing theory that formalizes characteristics such as a learner's ability level and the difficulty of questions. It is utilized in various high-stakes exams [17, 19]. Here, we introduce two IRT models: the Rasch model [14] and the nominal response model [1].

The Rasch model, a binary IRT model, defines the probability P_{ij} of learner j correctly answering item i with the following equation:

$$P_{ij} = \frac{1}{1 + \exp(-(\theta_j - b_i))}. \tag{1}$$

In this equation, θ_j represents the ability level of learner j, and b_i represents the difficulty of item i. Both parameters are estimated from the set of correct and incorrect answers made by learners in response to each question. To estimate the difficulty of questions, we use the Rasch model.

The nominal response model, a polytomous IRT model, is frequently used to analyze the characteristics of options in multiple-choice questions. This model defines the probability P_{ijk} that learner j selects option k out of K options for item i as follows:

$$P_{ijk} = \frac{\exp(\alpha_{ik}\theta_j + \zeta_{ik})}{\sum_{k'}^{K} \exp(\alpha_{ik'}\theta_j + \zeta_{ik'})}. \tag{2}$$

In this equation, α_{ik} represents the discrimination parameter for option k of item i, and ζ_{ik} represents the location parameter for option k of item i. Both parameters are estimated from the set of options the learners can select in response to each question. To investigate the characteristics of the options generated, we use the nominal response model.

3   Difficulty-Controllable Extractive Question Generation

This study builds upon the conventional IRT-based difficulty-controllable question-generation method [18].

The conventional method utilizes the SQuAD dataset [13], a benchmark dataset for extractive question answering and question generation. SQuAD comprises reading passages c_i and corresponding questions q_i and answers a_i, which are found within the passages (i.e., $a_i \subseteq c_i$). Thus, the dataset can be represented as $\{c_i, q_i, a_i | i \in 1 \ldots I\}$, where I denotes the number of data points. The purpose of the conventional method is to generate sets of questions and answers from a reading passage with an arbitrary difficulty value.

However, the SQuAD dataset does not include the difficulty level of questions, making it challenging to construct a difficulty-controllable generator. Thus, the conventional method estimated the difficulty level of each question within SQuAD by using IRT and QA (Question Answering) systems. Specifically, this method involves administering each question q_i in SQuAD to a variety of QA systems and collecting data on the correct and incorrect responses. From these data, the difficulty value b_i for each question is estimated using the Rasch model. By incorporating b_i into SQuAD, a dataset that includes difficulty levels is constructed.

Utilizing the SQuAD dataset with difficulty values, the conventional method can be applied to fine-tune the following two models:

1. **Difficulty-Controllable Answer Extraction Model:** A model that extracts answers given a reading passage and a specified difficulty level. Specifically, it is designed as a BERT model [4] that takes a concatenated string of the difficulty b_i and the reading passage c_i as input and then outputs the start and end positions of the answer a_i within the reading passage.
2. **Difficulty-Controllable Question Generation Model:** A model that generates questions given a reading passage, an answer, and a specified difficulty level. Specifically, it is designed as a T5 model [12] that takes a concatenated string of the difficulty b_i, a reading passage c_i, and an answer a_i as input and then outputs the question q_i.

This methodology is specifically designed for extractive question formats and does not directly apply to multiple-choice questions, which are prevalent in educational settings.

4   Proposed Method

In this study, we propose an automatic method for generating multiple-choice questions with controllable difficulty levels. We use the RACE dataset [7], which has been commonly utilized in research on the automatic generation of multiple-choice questions for training models and evaluations of their performance [10,20]. RACE includes reading passages c_i, corresponding questions q_i,

Table 1. Input and Output for Question Generation

Input

Create a question and four options having a difficulty level of {b} based on the Context. Option 1 is the correct answer, and Options 2, 3, and 4 are the distractor options. Difficulty level -3.0 is the easiest, and 3.0 is the most difficult.

Context: ⟨tag⟩ {context} ⟨/tag⟩

Output

Option 1 (Correct Option): ⟨tag⟩ {correct_option} ⟨/tag⟩

Question: ⟨tag⟩ {question} ⟨/tag⟩

Option 2 (Distractor Option): ⟨tag⟩ {distractor_options_1} ⟨/tag⟩

Option 3 (Distractor Option): ⟨tag⟩ {distractor_options_2} ⟨/tag⟩

Option 4 (Distractor Option): ⟨tag⟩ {distractor_options_3} ⟨/tag⟩

---

the correct answer option a_i, and three distractor options d_{i1}, d_{i2}, d_{i3}, represented as $\{c_i, q_i, a_i, d_{i1}, d_{i2}, d_{i3} | i \in 1 \ldots I\}$. Like the SQuAD dataset, RACE does not include question difficulty levels. Thus, we construct a dataset having multiple difficulty levels, using a method akin to that of the conventional method. The proposed method utilizes LLaMA 2, a popular open-source pre-trained large language model that enables fine-tuning, to generate the question, correct answer option, and three distractor options, given a reading passage and a difficulty value. The specific steps for constructing the proposed method are as follows:

1. We construct 400 QA systems with varying capability. Specifically, we first train four deep learning models (bert-base-uncased, roberta-base, deberta-v3-large, albert-base-v1), which are publicly available on huggingface[1]. Each model uses the architecture proposed in [8], which is designed to select the correct option from the offered options, considering the given reading passage and question. The model training is performed using RACE's validation data. After this training, we prepare 400 QA systems by applying a dropout layer with different dropout rates for each model. Specifically, we create 100 QA systems for each model by adjusting the dropout rate from 0.00 to 0.99 in increments of 0.01 during testing, resulting in a total of 400 QA systems.

2. We estimate the question difficulty level b_i for each question in the RACE training dataset, using the responses from the 400 QA systems. By incorporating these difficulty values into the RACE training dataset, we create a training dataset having multiple difficulty levels.

3. We fine-tune the LLaMA-2-7b[2] with the dataset created in Step 2 in order to generate the question, correct answer option, and three distractor options from the given reading passage and difficulty level. The input/output format for the model is outlined in Table 1, where {b} is the

---

[1] https://huggingface.co/.

[2] https://huggingface.co/meta-llama/Llama-2-7b.

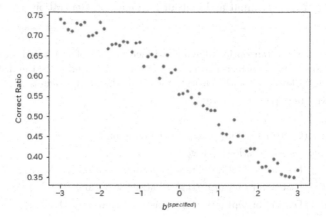

Fig. 1. Average correct answer rate for each set of questions by difficulty

estimated question difficulty from Step 2, {context} is the reading passage, {correct_option} is the correct answer option, {question} is the question, and {distractor_options_1}, {distractor_options_2}, {distractor_options_3} are the three distractor options, each replaced with respective data from the dataset. This prompt format is designed based on the prompt engineering guidelines[3]. Additionally, in the prompt, the correct answer is placed before the question text because our preliminary experiments showed that this order tended to improve generation quality.

5 Experiments

This section describes the experiments conducted to evaluate the performance of the proposed method.

5.1 Experimental Procedure

In our experiments, we first assigned difficulty levels ranging from -3.0 to 3.0 in increments of 0.1 to each of the 300 reading passages in the RACE test dataset. Then, using the dataset, we generated a total of 18,300 questions with the proposed model. Subsequently, the generated questions were answered by the 400 QA systems developed in Sect. 4 and the response data were collected. The analysis of this response data was performed based on correct answer rates and nominal response models.

---

[3] https://www.promptingguide.ai/.

5.2 Evaluation of Difficulty Controllability by Correct Answer Rate

To verify whether the generated questions reflected the specified difficulty levels, the collected response data were binarized into correct and incorrect answers, and evaluated based on the correct answer rates. Figure 1 shows the average correct answer rate of the 400 QA systems for each group of questions by difficulty level. The horizontal axis represents the specified difficulty levels, which can take 61 value points ranging from -3.0 to 3.0, with a step size of 0.1, and the vertical axis represents the average correct answer rate for the group of questions corresponding to each difficulty level. The results indicate that as the specified difficulty level increases, the correct answer rate decreases. Moreover, with a correlation coefficient of -0.97 between the specified difficulty levels and the average correct answer rates, we confirmed that the questions' difficulty was accurately reflected.

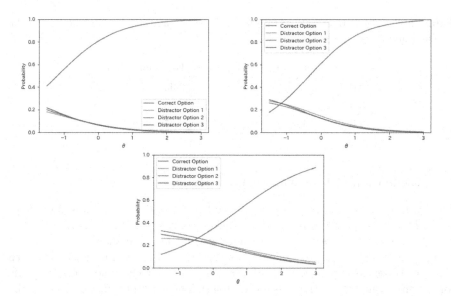

Fig. 2. Item characteristic curve of the nominal response model for questions generated with the specified difficulty $b^{(\text{specified})} = -3.0$ (upper left), $b^{(\text{specified})} = 0.0$ (upper right), $b^{(\text{specified})} = 3.0$ (bottom).

5.3 Evaluation of Generated Options by Nominal Response Models

We also investigated the quality of the generated options, using the nominal response model. Specifically, we applied the nominal response model to the collected response data for estimating the parameters α_{ik} and ζ_{ik} for each option across all questions. We then calculated the average of the parameters α_{ik} and ζ_{ik} for each option across the questions generated at each specified difficulty level of $b^{(\text{specified})}$.

As examples, the item characteristic curves derived from these difficulty values, $b^{(\text{specified})} = -3.0, 0.0$, and 3.0, are shown in Fig. 2. The horizontal axis represents the learner's ability θ, and the vertical axis represents the probability of choosing option k. We adjusted the horizontal axis range from -1.5 to 3.0 because the estimated ability values for the QA systems ranged from -1.2 to 2.6.

From these figures, it can be seen that as the ability value increases, the probability of choosing the correct answer option increases while the probability of choosing the distractor options decreases. Additionally, as the difficulty level increases, the probability of choosing the correct answer option decreases while the probability of choosing the distractor options increases. This indicates that both the correct answer and distractor options are generated in a manner that aligns with the intended difficulty levels. From these results, it was confirmed that the specified difficulty level is reflected in each option.

Table 2. Average correct answer rate for each set of questions by difficulty, using the few-shot approach

Difficulty	Fine-tuning (LLaMA 2)	Few-shot (GPT-4)
−2.0	0.73	0.43
0.0	0.55	0.43
2.0	0.39	0.45

5.4 Evaluation of Fine-Tuning Effectiveness

Finally, to evaluate the effectiveness of the fine-tuning in the proposed method, we evaluated the performance of the few-shot approach using GPT-4, which has recently achieved top performance among various large language models across many natural language processing tasks [3]. Few-shot question generation with GPT-4 utilizes prompts augmented with three output examples, following the format shown in Table 1 (i.e., 3-shot learning). The three output examples in the prompt correspond to the difficulty levels -2.0, 0.0, and 2.0, which are randomly selected from the dataset constructed in Sect. 4. Using this prompt format, we generated questions for 212 reading passages from the test data of the RACE dataset, while changing the specified difficulty levels to -2.0, 0.0, and 2.0. The generated questions were answered by the 400 QA systems created in Sect. 4, and the correct answer rates for each specified difficulty level were calculated.

Table 2 shows the results. The column "LLaMA 2" shows the values obtained from the experiments in Sect. 5.1. The results shown in the table suggest that the few-shot approach failed to control the difficulty level in the generated questions. This shows that few-shot learning might be insufficient for controlling difficulty and that fine-tuning with a substantial amount of data may be necessary.

6   Conclusion

In this study, we developed an automatic method for generating multiple-choice questions that enables control of difficulty levels by utilizing IRT and LLaMA 2. Furthermore, we analyzed the choice response data for the questions generated by the proposed method, using correct answer rates and the nominal response model. The results confirmed that the questions generated accurately reflect the specified difficulty levels. Additionally, it was confirmed that question generation using fine-tuning can control the difficulty more precisely compared with question generation using few-shot learning.

References

1. Bock, R.D.: Estimating item parameters and latent ability when responses are scored in two or more nominal categories. Psychometrika **37**, 29–51 (1972)
2. Chan, Y.H., Fan, Y.C.: A recurrent BERT-based model for question generation. In: Proceedings of the 2nd Workshop on Machine Reading for Question Answering, pp. 154–162 (2019)
3. Chiang, W.L., et al.: Chatbot arena: an open platform for evaluating LLMs by human preference. arXiv (2024)
4. Devlin, J., Chang, M.W., Lee, K., Toutanova, K.: BERT: pre-training of deep bidirectional transformers for language understanding. In: Proceedings of the 2019 Conference of the North American Chapter of the Association for Computational Linguistics, pp. 4171–4186 (2019)
5. Du, X., Shao, J., Cardie, C.: Learning to ask: neural question generation for reading comprehension. In: Proceedings of the 55th Annual Meeting of the Association for Computational Linguistics, pp. 1342–1352 (2017)
6. Gao, Y., Bing, L., Chen, W., Lyu, M., King, I.: Difficulty controllable generation of reading comprehension questions. In: Proceedings of the Twenty-Eighth International Joint Conference on Artificial Intelligence, pp. 4968–4974 (2019)
7. Lai, G., Xie, Q., Liu, H., Yang, Y., Hovy, E.: RACE: large-scale reading comprehension dataset from examinations. In: Proceedings of the 2017 Conference Empirical Methods in Natural Language Processing, pp. 785–794 (2017)
8. Liu, Y., et al.: RoBERTa: a robustly optimized BERT pretraining approach. arxiv (2019)
9. Lord, F.M.: Applications of Item Response Theory to Practical Testing Problems. Routledge (1980)
10. Offerijns, J., Verberne, S., Verhoef, T.: Better distractions: transformer-based distractor generation and multiple choice question filtering. arxiv (2020)
11. OpenAI: Gpt-4 technical report. arXiv (2023)
12. Raffel, C., et al.: Exploring the limits of transfer learning with a unified text-to-text transformer. J. Mach. Learn. Res. **21**(140), 1–67 (2020)
13. Rajpurkar, P., Zhang, J., Lopyrev, K., Liang, P.: SQuAD: 100,000+ questions for machine comprehension of text. In: Proceedings of the 2016 Conference Empirical Methods in Natural Language Processing, pp. 2383–2392 (2016)
14. Rasch, G.: Probabilistic Models for Some Intelligence and Attainment Tests. The University of Chicago Press (1981)
15. Touvron, H., et al.: Llama 2: open foundation and fine-tuned chat models. arXiv (2023)

16. Ueno, M., Miyazawa, Y.: IRT-based adaptive hints to scaffold learning in programming. IEEE Trans. Learn. Technol. **11**(4), 415–428 (2018)
17. Uto, M.: A Bayesian many-facet Rasch model with Markov modeling for rater severity drift. Behav. Res. Methods **55**, 3910–3928 (2022)
18. Uto, M., Tomikawa, Y., Suzuki, A.: Difficulty-controllable neural question generation for reading comprehension using item response theory. In: Proceedings of the 18th Workshop on Innovative Use of NLP for Building Educational Applications, pp. 119–129 (2023)
19. Uto, M., Ueno, M.: Empirical comparison of item response theory models with rater's parameters. Heliyon **4**(5), e0062 (2018)
20. Zhou, X., Luo, S., Wu, Y.: Co-attention hierarchical network: generating coherent long distractors for reading comprehension. Proc. AAAI Conf. Artificial Intell. **34**, 9725–9732 (2019)

Can VLM Understand Children's Handwriting? An Analysis on Handwritten Mathematical Equation Recognition

Cleon Pereira Júnior[1,2](✉) , Luiz Rodrigues[3] , Newarney Costa[2] ,
Valmir Macario Filho[4] , and Rafael Mello[1,4]

[1] Centro de Estudos Avançados de Recife, Recife, Brazil
[2] Instituto Federal Goiano, Iporá, Brazil
`cleon.junior@ifgoiano.edu.br`
[3] Universidade Federal de Alagoas, Maceio, Brazil
[4] Universidade Federal Rural do Pernambuco, Recife, Brazil

Abstract. Handwriting Mathematical Expression Recognition has several applications, including the potential to make Intelligent Tutoring Systems (ITS) more accessible to underserved regions. However, young children's handwriting pose several challenges, even for instructors, calling for advanced approaches to understand their writings. This paper explores the potential of pre-trained Vision-Language Models (VLM) for recognizing numbers or mathematical expressions handwritten by children by comparing GPT-4V, LLaVA 1.5, and CogVLM on a dataset of 251 images. Results indicate that while pre-trained models offer promise, their performance without fine-tuning or zero-shot learning remains inadequate. Results reveal the challenges of utilizing pre-trained VLMs for recognizing children's handwriting, particularly in educational settings. Issues such as poor handwriting and partial erasures pose difficulties for existing models, with GPT-4V's safety system limitations hindering its efficacy. In general, CogVLM presented the best performance. GPT-4V exhibited superior performance in recognizing equations, but still struggles with handwritten content, highlighting the need for model refinement and data policies. This study contributes insights into the potential of existing pre-trained models in educational contexts and demonstrates the importance of, for example, fine-tuning to domain-specific datasets. Continued research is necessary to enhance VLM capabilities for educational support, particularly in children's handwriting recognition.

Keywords: Handwriting · VLM · Mathematical Education

1 Introduction

Intelligent Tutoring Systems (ITS) play an important role in teaching and learning, providing educational support across various contexts. Among other factors, these systems consider issues of performance, engagement, and personalization

A. M. Olney et al. (Eds.): AIED 2024 Workshops, CCIS 2150, pp. 321–328, 2024.
https://doi.org/10.1007/978-3-031-64315-6_28

[14]. The mathematics education is an example of how ITS are widely used at different levels of learning. However, it is noteworthy that many of these tools still rely predominantly on keyboard and mouse input [13]. This might be a problem, once in k-12 education, much of educational practice involves handwriting.

One way to improve ITS focused on teaching math to children and teenagers is to allow handwritten input. Currently, some researches focus on handwritten input from touchscreens and digital pens [13]. Nonetheless, when considering the context of developing countries, technologies such as tablets and digital pens are often not integrated into educational practices [7]. In such scenarios, an alternative worth investigating could be scanned images of handwritten as an input [17]. Here, in the mathematics education context, one of the challenges lies in math equation recognition [1].

Research has pointed out the potentials of Large Language Models (LLM) to operate in the field of Natural Language Processing (NLP). An example of this is the use of GPT, which, through the ChatGPT interface, can receive text inputs in different languages, process, and return consistent outputs [2]. In addition to their potential with texts, language models have also advanced in the fields of image generation [8], data extraction from documents, and interpretation of image content [12]. Models capable of extracting content from images (e.g., generating captions) are known as Vision-Language Models (VLM) or, in the most cases, Large Multimodal Models (LMM). LMM extend LLM to multimodal scenarios involving multiple inputs. With the advancement of these models, research in Artificial Intelligence in Education (AIED) is also exploring their potential and seeking solutions to support the educational process [10].

When proposing an ITS in k-12 education, especially in elementary school, as mentioned, it is important to anticipate that input may consist of images with handwritten content. In this sense, analyzing VLM considering their ability to extract handwritten content from digitized images may yield results that will contribute to future research aiming at educational support. Therefore, we propose an analysis of GPT-4V (GPT-4 with Vision) [15] based on a dataset of images with mathematical content written by children. Additionally, we also compare the results with two other open-source models (LLaVA 1.5 [11] and CogVLM [21]) to verify differences in behavior from similar prompts. The challenge of interpreting the content of the image in this work is significant because they were produced by children in the developmental stage of handwriting skills. The results of this research should contribute to the possibility of using VLM with zero-shot to collaborate with educational design components.

2   Literature Review

Since the Covid-19 pandemic, interest in research involving the digitization of mathematical activities and the recognition of handwritten equations has taken a new look. There was a greater need for systems that could take image files as input and perform processing to facilitate the work of people involved in the educational process [1]. With the end of the COVID-19 pandemic and the return of face-to-face activities, a new problem that has arisen is the insertion of tools to minimize the impact on learning caused by the period of social isolation. In this regard, developing countries must be given due attention and, consequently, the insertion of technology must take into account different realities [7]. Consequently, ITS or another educational system has to consider, among other things, scanned images of handwritten mathematical equations.

There are several research initiatives that train specific machine learning models to recognize handwriting mathematical equations [1,4]. In these studies, the validation and testing phase usually finds a high accuracy (in some cases, above 90%). However, there are training costs and they are used to good effect in the specific scenario in which the model was trained. Furthermore, the dataset is not always built from handwriting images of children in the early stages of learning. A promising solution to the problem of handwritten mathematical equation recognition in different contexts could be VLM.

When dealing with VLM, at least three types of models are expected: i) multi-modal to text generation, which takes as input an image and a text and returns as output a response to the input set; ii) image-text matching, which seeks to associate a text with an image; and iii) text-to-image generation, which creates an image from a text specification [6]. Additionally, there are categories of prompts. Prompts are extra information submitted to the model that helps guide towards a better outcome. Different research categorizes prompts in specific ways. In general terms, prompts can be manually created or automated using fine-tuned techniques employing gradient-based methods [6,20].

One example of handwritten mathematical equation recognition without VLM is the research of Chevtchenko et al. (2023) [4]. Authors emphasize the importance of handwritten recognition to the Intelligent Tutoring System (ITS), but the dataset used in the work was constructed by the authors and consisted of adult's handwritten, despite the operations being basic and solved step by step as in early childhood. On the other hand, the work of Feichter et al. (2024) [5] conducts a study on the automated transcription of mathematical formulas and the use of language models. In this case, the study presents results from the combination of existing models for mathematical expression recognition with LLM and compares with the use of VLM alone. The best overall result is achieved by combining existing models with LLM. In this scenario, the LLM plays the role of converting text into LaTeX. Regarding the use of VLM, the best model was trOCR, but it only yielded interesting results after fine-tuning.

From the literature review, it is evident that the problem of handwriting recognition remains open, with a transition occurring from traditional machine learning models to language models, such as LLM or VLM. The research focused specifically on digit recognition due to the nature of the study. It is noted that the input of text through images in ITS is crucial. However, literature on testing Language Models with real handwritten images from children in the domain of mathematics was not found. Therefore, this study poses the following research question: **Is it feasible to employ VLM to extract handwritten mathematical content from images?**

3   Method

We aim to explore the potential of pre-trained Vision-Language Models for recognizing numbers or mathematical expressions written by children. In this particular instance, we expect a multi-modal to text generation model. Figure 1 shows an overview of this approach. The first stage of the research consisted of selecting the dataset. In this case, we were interested in investigating the potential of VLM in recognizing simple mathematical expressions performed by children. Thus, a set of 251 images extracted from a real scenario of children between six and eight years old was selected. Figure 2 shows examples present in the dataset. It is noted that the images can appear in different formats. In some cases, it is not well erased and one number overlaps another. The challenge of interpreting the content exists even in a manual process. The next step concern defining the text input to the model once we have the dataset.

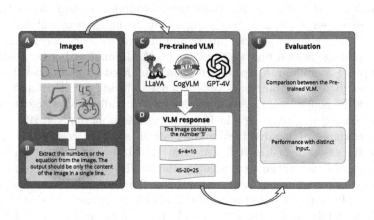

Fig. 1. Approach Overview

Fig. 2. Dataset Example

Prompt engineering is an important task when working with language models and has the potential to maximize the usefulness of these pre-trained models. In relation to LLM, certain parameters are expected for the construction of a good prompt [18]. In the same way, LMM have distinct research to discuss prompt engineering [20]. In our case, after testing and verifying the results of several prompt combinations on smaller sets and considering distinct type of image input, we chose to construct a text prompt that had the objective of the task (content extraction from the image), the type of content to be extracted (number or expression), and the output format (a single line containing the content). In general, for mathematical problems, the output is expected in LATEX [5]. In this study, since it is about simple equations, this step was not added. Figure 1 shows an example with image (bullet A) and text (bullet B) input.

The next stage consisted of selecting pre-trained models to compare the results with GPT-4V. The reason for comparing with GPT-4V is that it is a pre-trained model considered more comprehensive so far when analyzing different domains [15]. In this case, two open-source models were selected: LLaVA [11] and CogVLM [21]. Currently, there are more than 20 VLMs, and for each type of analysis, results may vary [9]. We focused on those three models because they are considered well established options in the field [3]. It is important to emphasize that GPT-4v has been cited as the best model in different contexts [16]. In this case, our research focused on this model, while the other two were used to observe the situation of some open-source models.

Finally, tests were performed with GPT-4V and compared with the other two pre-trained models. It is important to emphasize that, at the time of writing, GPT-4v does not allow fine-tuning. So, we compare the other models in the same condition to prevent favoring one alternative over another. The temperature was adjusted to 0.1 for all models. The results allow us to discuss the limitations of pre-trained models for specific domains and the next research paths involving children's handwriting, number recognition, and VLM. The following section presents the results of this work.

4   Results

Table 1 provides a summary of the results obtained. The first row presents the overall performance, considering the entire available dataset. In this case, COGVLM was the only model to achieve more than 50%, and consequently, it had the best result among the models. Since the concern of the study is not necessarily to analyze the best model but rather to verify the possibility of using VLM in ITS to support the educational process, some more detailed analyses should be observed.

From Table 1, it is noticeable that GPT-4V had the worst overall performance compared to the other two VLM in the entire dataset. One reason for the significantly lower performance is a problem detected by the model in 104 out of the 251 images of the dataset. When attempting to process the request, those images were flagged as violating rights, resulting in the following error message: "Your input image may contain content that is not allowed by our safety system." Figure 3 shows some examples of images that received this classification.

When detailing the results, we observed superior performance from GPT-4V even with the issue detected in several images. For equations, both horizontally and vertically (Fig. 1 at mark A presents an example of each), the performance is

Table 1. VLM comparison

	Dataset	GPT-4V	CogVLM	LLaVA 1.5
Total	251	74 29.88%	127 **50.60%**	83 33.07%
Horizontal Equations	33	14 **42.42%**	9 27.27%	1 3.03%
Vertical Equations	25	10 **40.00%**	3 12.00%	0 0.00%
Only numbers	193	51 26.42%	115 **59.59%**	82 42.49%

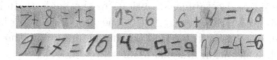

Fig. 3. Example of images that is not allowed by GPT-4V.

much better than the other models. On the other hand, COGVLM shows interesting performance in detecting only numbers that do not involve expressions. LLaVA 1.5, in this context, did not yield satisfactory results in any situation. Given that the analyzed context has a relatively small dataset, aiming for a qualitative discussion, the images that were not correctly classified by the best model of each category (horizontal, vertical, or number-only) were then examined and categorized as presented in Table 2.

Table 2. Summarization of errors presented by image type

Type/VLM	Error	Amount	Type/VLM	Error	Amount	Type/VLM	Error	Amount
Horizontal GPT	overwritten	15.79%	Vertical GPT	carry	11.76%	Number CogVLM	circled value	1.12%
	partial erasing	15.79%		operand disorder	5.88%		clear writing	4.49%
	poor handwriting	5.26%		operator disorder	11.76%		image blur	3.37%
	unread image	63.16%		overwritten	5.88%		image cut out	4.49%
				partial erasing	17.65%		overwritten	10.11%
				poor handwriting	29.41%		partial erasing	25.84%
				unidentified	11.76%		poor handwriting	43.82%
				unread image	5.88%		unidentified	6.74%
	—	—		—	—		—	—
	TOTAL	19		TOTAL	17		TOTAL	89

When it comes to horizontal equations, the primary issue detected was unreadable images (due to the GPT safety system), as discussed previously. Additionally, across all types, another recurring problem involved partially erased and overwritten digits. In the former scenario, children often fail to erase content properly before rewriting. The second instance occurs when children attempt to write new content but opt to write over existing text instead of erasing it. While early childhood education specialists are adept at identifying and addressing these challenges, VLM still struggle with such nuances.

Another issue detected during the investigation of errors is that the children are still developing their writing skills. In these cases, handwriting still presents common issues typical of early age but uncommon in adults. Some examples include numbers written in reverse order or extra strokes in certain digits. These problems were classified as poor handwriting. The classification difficulty likely stems from being a specific domain that still lacks extensive studies and datasets.

5   Discussion

The task of recognizing children's handwriting is sometimes complex, even for humans. The dataset shows that children often write numbers backwards (marked as poor handwriting), for example, during the learning process. There is also difficulty in correcting a potential error, which leads to overlapping of writing. Another observed factor is writing in spaces already occupied by the explanation of the exercise, exceeding the limit intended for the solution. There are other experiments that compare the behavior of GPT4-V in different contexts and observe human superiority through diversification [19].

An important point to discuss is the number of images not accepted by GPT-4V. It is essential for pre-trained models to have a policy and algorithms for automatic classification of images that may contain sensitive content. However, in the studied context, such limitation made the use of GPT-4V for transcribing handwritten numbers and mathematical expressions by children inefficient. Out of 104 unread images, 91 contained only numbers. When analyzing the 147 images read by the GPT, we would have an overall performance of 51.02% for the GPT and 52.23% for the COGVLM. In this case, the difference between the GPT and the COGVLM would be slightly small.

This study discusses a new possibility for support in educational environments, mainly ITS. Since handwriting is fundamental and integrated into the educational process of children, investigating the potential of Vision-Language Models to recognize numbers and mathematical expressions can assist in designing new educational tools. However, the results suggest that pre-trained models without fine-tuning and with zero-shot, even with text prompts specifying the activity, are still insufficient to achieve results comparable to those from traditional approaches for handwriting math equation recognition.

References

1. Arya, M., Yadav, P., Gupta, N.: Handwritten equation solver using convolutional neural network. In: Intelligent Systems and Applications in Computer Vision, pp. 72–85. CRC Press (2023)
2. Baidoo-Anu, D., Ansah, L.O.: Education in the era of generative artificial intelligence (AI): Understanding the potential benefits of ChatGPT in promoting teaching and learning. J. AI **7**(1), 52–62 (2023)
3. Chen, D., et al.: Mllm-as-a-judge: assessing multimodal llm-as-a-judge with vision-language benchmark. arXiv preprint arXiv:2402.04788 (2024)
4. Chevtchenko, S., et al.: Algoritmos de reconhecimento de dígitos para integração de equações manuscritas em sistemas tutores inteligentes. In: Anais do XXXIV Simpósio Brasileiro de Informática na Educação, pp. 1442–1453. SBC (2023)
5. Feichter, C., Schlippe, T.: Investigating models for the transcription of mathematical formulas in images. Appl. Sci. **14**(3), 1140 (2024)
6. Gu, J., et al.: A systematic survey of prompt engineering on vision-language foundation models. arXiv preprint arXiv:2307.12980 (2023)
7. Kibirige, I.: Primary teachers' challenges in implementing ICT in science, technology, engineering, and mathematics (STEM) in the post-pandemic era in uganda. Educ. Sci. **13**(4), 382 (2023)
8. Koh, J.Y., Fried, D., Salakhutdinov, R.R.: Generating images with multimodal language models. Adv. Neural Inf. Process. Syst. **36** (2024)
9. Kostumov, V., Nutfullin, B., Pilipenko, O., Ilyushin, E.: Uncertainty-aware evaluation for vision-language models. arXiv preprint arXiv:2402.14418 (2024)
10. Lee, G.G., Latif, E., Shi, L., Zhai, X.: Gemini pro defeated by GPT-4v: evidence from education. arXiv preprint arXiv:2401.08660 (2023)
11. Liu, H., Li, C., Li, Y., Lee, Y.J.: Improved baselines with visual instruction tuning (2023)
12. Liu, H., Li, C., Wu, Q., Lee, Y.J.: Visual instruction tuning. Adv. Neural Inf. Process. Syst. **36** (2024)
13. de Morais, F., Jaquies, P.A.: Does handwriting impact learning on math tutoring systems? Inf. Educ. (2021). https://doi.org/10.15388/infedu.2022.03
14. Mousavinasab, E., Zarifsanaiey, N., R. Niakan Kalhori, S., Rakhshan, M., Keikha, L., Ghazi Saeedi, M.: Intelligent tutoring systems: a systematic review of characteristics, applications, and evaluation methods. Interact. Learn. Environ. **29**(1), 142–163 (2021)
15. Achiam, J., et al.: Gpt-4 Technical Report. OpenAI (2024)
16. Qian, Y., Zhang, H., Yang, Y., Gan, Z.: How easy is it to fool your multimodal llms? an empirical analysis on deceptive prompts. arXiv preprint arXiv:2402.13220 (2024)
17. Rodrigues, L., et al.: Mathematics intelligent tutoring systems with handwritten input: a scoping review. Educ. Inf. Technol. (2023)
18. Santu, S.K.K., Feng, D.: Teler: a general taxonomy of llm prompts for benchmarking complex tasks. arXiv preprint arXiv:2305.11430 (2023)
19. Wadhawan, R., Bansal, H., Chang, K.W., Peng, N.: Contextual: evaluating context-sensitive text-rich visual reasoning in large multimodal models. arXiv preprint arXiv:2401.13311 (2024)
20. Wang, J., et al.: Review of large vision models and visual prompt engineering. Meta-Radiology **1**(3), 100047 (2023)
21. Wang, W., et al.: Cogvlm: visual expert for pretrained language models (2024)

To Kill a Student's Disengagement: Personalized Engagement Detection in Facial Video

Egor Churaev[1]([✉])[ID] and Andrey V. Savchenko[1,2][ID]

[1] HSE University, Laboratory of Algorithms and Technologies for Network Analysis, Nizhny Novgorod, Russia
echuraev@hse.ru
[2] Sber AI Lab, Moscow, Russia

Abstract. This work deals with engagement detection for e-learning systems based on a student's facial video. We propose a novel, accurate technique that adapts the neural network-based binary classifier by using a short set of videos of a concrete user. The facial features are extracted with pre-trained lightweight deep networks that can be launched on a person's device without a privacy leak. Moreover, we introduce the publicly available dataset that can be used to test the model's efficiency. Experimentally, it is demonstrated that the approach with model adaptation can significantly (by 10–15%) improve the accuracy of predicting user engagement in the video.

Keywords: E-learning · engagement detection · facial analysis in video

1 Introduction

Due to COVID-19, a fundamental reconfiguration of academic communication unfolded, with lectures, conferences, and meetings migrating to online formats in response to a paramount commitment to health and safety [1]. As virtual classrooms became the new norm, educators faced an unprecedented challenge: gauging and enhancing student engagement in a digital environment [2]. It is known that the quality of material understanding and knowledge retention is intricately linked to the level of engagement students exhibit during online classes [3]. However, in contrast to offline education, where students begin to communicate and provide some noise, it is not easy for a professor to detect the disengagement of a particular student in online education.

Nowadays, various approaches exist for automatically predicting a person's engagement [4,5], including sensor-based [6], behavior-based [7], and computer-vision-based methods [8]. The latter have become increasingly popular due to their capacity to comprehensively analyze non-verbal cues like facial expressions [9], eye movements, and body language, providing a more nuanced understanding of engagement compared to sensor-based or behavior-based methods [10]. Additionally, these methods eliminate the need for individuals to wear

A. M. Olney et al. (Eds.): AIED 2024 Workshops, CCIS 2150, pp. 329–337, 2024.
https://doi.org/10.1007/978-3-031-64315-6_29

devices, enhance comfort and versatility, and enable deployment in diverse environments. Most modern computer vision applications in engagement detection are based on deep neural networks (DNNs) [11,12]. Recent studies show that modern state-of-the-art video-based engagement prediction approaches achieve an accuracy range of approximately 60% to 80% [8,13,14]. However, such quality may be too low for real-world practical applications.

Thus, in this paper, we propose to apply a personalized approach [15] to accurately improve student engagement recognition by adapting a classifier to a concrete student using a small number of their facial videos. To evaluate the efficiency of the proposed approach, we collected a special testing dataset with short videos where the person is engaged or distracted. It is experimentally demonstrated that our personalized models can significantly increase the overall accuracy (up to 90+%).

2   Methodology

2.1   Proposed Approach

The task of this paper is the video binary classification: a sequence of $N > 1$ video frames with the student's face $\{X_i\}, i = 1, 2, ..., N$ is assigned to one of $C > 1$ engagement categories. In practice, $C = 2$, i.e., only engaged and distracted classes are essential. However, public datasets can contain broader categories like distracted, barely engaged, engaged, and highly engaged [16]. We assume that the engagement level remains the same throughout the video. Hence, small parts (1–5 s) of a video stream should be analyzed.

This work proposes a novel personalized approach for engagement estimation (Fig. 1). The algorithm is divided into three parts: 1) training the universal engagement detection model; 2) fine-tuning the student-independent model to the concrete student and creating the student-adapted models; and 3) video-based engagement estimation in online video. Let us consider it in details.

In the first stage (top left rectangle in Fig. 1), the universal (or student-independent) neural network classifier for predicting a person's engagement on a video is trained on an available engagement dataset, e.g., EngageWild [16] and DAiSEE (Dataset for Affective States in E-Environments) [17]. An appropriate facial detector, such as MTCNN or RetinaFace, is used to find and crop faces from video frames. Then, facial features are extracted using lightweight emotional DNNs [18], such as EmotiMobileNet [19] or EmotiEffNet [12,20]. Using lightweight models allows us to run this pipeline in real time on various devices (smartphones, laptops, HPC servers, etc.). Finally, the video binary classifier is trained based on a frame-attention network, such as single attention or self-attention [21]. The final layer contains a sigmoid activation function, and the binary cross-entropy loss function is used. In addition, we borrow the idea from [15] and concatenate the frame embeddings obtained by a pre-trained DNN with the aggregated descriptor of the whole video. The latter is computed using the component-wise statistical (STAT) functions. This paper used the standard deviation (std) that shows the best quality for engagement prediction [12].

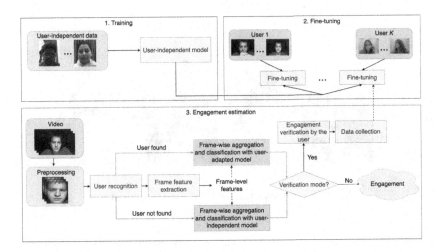

Fig. 1. Proposed approach for personalized student engagement estimation.

In the second stage (top right rectangle in Fig. 1), we gather a small set of labeled videos for a student. Hence, it is possible to adapt the universal attention-based video classifier by fine-tuning it to the video data of the selected user. It is important to emphasize that the feature extraction DNNs are not fine-tuned, so the number of required video samples for each user can be relatively small.

In the third stage (bottom rectangle in Fig. 1), the student's engagement is predicted by an online facial video. The input video is split into frames; a facial detector is used to select the facial area on the frame, crop it, and prepare for the input into the DNN-based feature extractor. The face identification model can be applied if the student's identity is unknown. If the face is found in the database of registered students and there is a fine-tuned (student-adapted) model for them, this model will be used for engagement prediction. Otherwise, a pre-trained classifier (student-independent model) is used. An added verification mode allows the user to improve the accuracy of engagement determination by correcting model prediction. If the validation mode is activated, the user can confirm or correct the result of the model's work after predicting the user's engagement on the video. These data are accumulated, transferred to the "fine-tuning" stage, and used to improve and re-train the personalized model. In a real-world application, validation mode may be enabled by a teacher or her assistant, who can use recorded videos after an online lesson and correct the model responses for the concrete student.

2.2 Testing Dataset

To evaluate the personalized engagement prediction approach, we collected a small engagement dataset of MS and PhD students from online lectures (Fig. 2). The videos in the dataset include five persons, and videos were recorded in

Fig. 2. Example of frames from our testing dataset.

different orientations. In total, 160 videos were recorded, 80 videos per class (distracted and engaged), and 16 videos for each person. The length of each video is 4–6 s. This dataset is publicly available[1]. It includes only two classes due to the following reasons. First, the labeling with more categories is very complicated as there is not strict definition of the differences between, e.g., engaged and highly engaged classes for a short video. Secondly, it is usually much more easier for the teachers to interpret the binary decisions if the student was distracted or not.

In the experiments, five videos for each class (engaged/distracted) were put into the training part, and the remaining videos were used to validate the final accuracy of the model. We examined the dependence of accuracy on the size of the training set, so from 1 to 5 videos from the training part were randomly selected to form the training set for our adaptation procedure. Thus, the formula to determine the total number of video files in the training set is $M \times C$, where M represents the number of videos per engagement class and C is the total classes. In our scenario, the latter is equal to 2 (distracted and engaged). The personalized models were fine-tuned on this subset for 20 epochs using the Adam optimizer, with a learning rate of 0.0001.

3 Experimental Results

The algorithm described in Sect. 2.1 was implemented using TensorFlow 2.5. The universal frame attention models (single attention and self-attention) were trained on two datasets: EngageWild [16] and DAiSEE [17]. The standard split of the datasets contains four classes (distracted, barely engaged, engaged, and highly engaged); the attention models were trained to classify these four classes and compared them with other works. In most works related to the EngageWild dataset with four engagement categories, the Mean-Square Error (MSE) is used. In addition, we computed the Unweighted Average Recall (UAR), which works well for imbalanced data. Finally, we examined a reduced case with two classes, namely, engaged ("engaged" and "highly engaged") and distracted ("distracted"

---

[1] https://drive.google.com/drive/folders/1e5_H8FSYhpioTb_u_AgEaMKDiXhcqR5t.

Table 1. Experiments results of universal models

Dataset	Model	4 classes		2 classes		
		MSE	UAR	UAR	Recall	Precision
DAiSEE	3D-CNN (baseline) [17]	–	0.3066	–	–	–
	3D-CNN + TCN [22]	–	0.3146	–	–	–
	ResNet+TCN+weighted loss [23]	–	0.3711	–	–	–
	ResNet+TCN [24]	–	0.3355	–	–	–
	EmotiMobileNet, single attention	0.1152	0.3507	0.6509	0.4471	0.1377
	EmotiMobileNet, self-attention	0.0992	0.2911	0.6275	**0.4588**	0.1046
	EmotiEffNet, single attention	0.0823	0.3795	**0.6376**	0.3412	**0.2117**
	EmotiEffNet, self-attention	0.0748	**0.4137**	0.5843	0.2235	0.1743
EngageWild	OpenFace+LSTM (baseline) [25]	0.1	–	–	–	–
	LGCP [26]	0.0884	–	–	–	–
	VGG [27]	0.0653	–	–	–	–
	ED-MTT [28]	**0.0427**	–	–	–	–
	EmotiMobileNet, single attention	0.1327	0.3322	0.6702	0.4286	0.6667
	EmotiMobileNet, self-attention	0.1331	0.2079	0.6492	0.3571	0.7143
	EmotiEffNet, single attention	**0.0957**	0.3065	0.6912	0.5	0.6364
	EmotiEffNet, self-attention	0.0906	**0.3737**	**0.7563**	**0.5714**	**0.8**

and "barely engaged"). Our models are compared with existing approaches in Table 1. Here, we referenced to four best-known results for each dataset and compare them with our results. As one can notice, our attention models show near state-of-the-art results.

Next, the universal attention-based models trained for binary classification were fine-tuned to predict the engagement of a concrete user. As our testing set is balanced, the accuracy metric is used. Its dependence on the number M of training examples per class for our models is presented in Fig. 3. If $M = 0$, the universal model from Table 1 is used. The value $M = 1$ means the training dataset contained only one video file for each engagement class. As one can notice, the personal models demonstrate 10–15% higher accuracy in recognizing user engagement than the general model, requiring little data to achieve good results. The EmotiEffNet facial features with self-attention classifier pre-trained on the DAiSEE dataset showed the best accuracy of 91.82%. An accuracy degradation can be noticed with an increase of M, which can be explained by smaller size of the testing set.

The confusion matrices for this experiment are presented in Fig. 4. It is important to note that EmotiEffNet and MobileNet require 5–30 ms to process one face using the CPU of our laptops, so it is possible to run our pipeline in real-time without sending facial videos to a remote server with multiple GPUs. Hence, our pipeline appears preferable from the data security perspective.

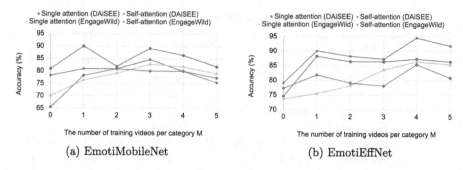

Fig. 3. Accuracy of the adapted model concerning the number of videos M per category in the training set.

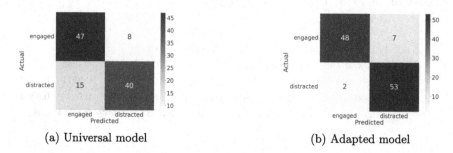

Fig. 4. Confusion matrices, self-attention classifier (DAiSEE), EmotiEffNet.

4 Conclusion

In this work, we proposed a high-precision approach to automatically determine the student's engagement based on facial video (Fig. 1). Its usage can significantly improve the quality of online education, as it will be possible to provide the teacher with relevant information about students' engagement in a material. The pipeline can be implemented on laptops or mobile devices in real-time without sacrificing privacy. Indeed, the collected data are not shared with any third party companies or applications. The source code of our pipeline and all experiments are publicly available[2]. Our current implementation expects that engagement recognition and fine-tuning will be executed on the teacher laptop by using a screen capture application. As a result, it will be possible to provide handy personalized analytics to a teacher after the lecture or students listen to the pre-recorded material to improve course materials and the academic performance of online learners. Moreover, it is even possible to integrate our method with online conference tools to estimate the average level of distraction of all students, so the teacher can adjust the explanations in real time.

We demonstrated that even two examples of short videos (one for engagement and one for distraction) are enough to improve the quality of universal

---

[2] https://github.com/echuraev/EngagementEstimator.

engagement detectors significantly. Finally, we introduced a novel video dataset (Fig. 2) available for testing engagement prediction techniques. Experimental results (Fig. 3) show that using proposed adaptive models for determining user engagement allows achieving an accuracy improvement of more than 20% compared to the universal model. Thus, the proposed adaptation of the universal model can deal with diversity, equity, and inclusion principles as it tries to fit the facial characteristics of each person better. In the future, it is necessary to significantly extend our current dataset to make it more diverse, include persons of various ages, races, etc.

Acknowledgements. The article was prepared within the framework of the Basic Research Program at the National Research University Higher School of Economics (HSE).

References

1. Adedoyin, O.B., Soykan, E.: Covid-19 pandemic and online learning: the challenges and opportunities. Interact. Learn. Environ. **31**(2), 863–875 (2023)
2. Goldberg, P., et al.: Attentive or not? toward a machine learning approach to assessing students' visible engagement in classroom instruction. Educ. Psychol. Rev. **33**, 27–49 (2021)
3. Coates, H.: The value of student engagement for higher education quality assurance. Qual. High. Educ. **11**(1), 25–36 (2005)
4. Peng, Y., Kikuchi, M., Ozono, T.: Development and experiment of classroom Engagement evaluation mechanism during real-time online courses. In: Wang, N., Rebolledo-Mendez, G., Matsuda, N., Santos, O.C., Dimitrova, V. (eds.) AIED 2023. LNCS, pp. 590–601. Springer, Cham (2023). https://doi.org/10.1007/978-3-031-36272-9_48
5. Yadav, S., Siddiqui, M.N., Shukla, J.: EngageMe: assessing student engagement in online learning environment using neuropsychological tests. In: Wang, N., Rebolledo-Mendez, G., Dimitrova, V., Matsuda, N., Santos, O.C. (eds.) AIED 2023. LNCS, pp. 148–154. Springer, Cham (2023). https://doi.org/10.1007/978-3-031-36336-8_23
6. Henry, J., Tang, S., Mukhopadhyay, S., Yap, M.H.: A randomised control trial for measuring student engagement through the internet of things and serious games. Internet of Things **13**, 100332 (2021)
7. Balti, R., Hedhili, A., Chaari, W.L., Abed, M.: Hybrid analysis of the learner's online behavior based on learning style. Educ. Inf. Technol. **28**(10), 12465–12504 (2023)
8. Liao, J., Liang, Y., Pan, J.: Deep facial spatiotemporal network for engagement prediction in online learning. Appl. Intell. **51**, 6609–6621 (2021)
9. Ruan, X., Palansuriya, C., Constantin, A.: Affective dynamic based technique for facial emotion recognition (FER) to support intelligent tutors in education. In: Wang, N., Rebolledo-Mendez, G., Matsuda, N., Santos, O.C., Dimitrova, V. (eds.) AIED 2023, pp. 774–779. Springer, Cham (2023). https://doi.org/10.1007/978-3-031-36272-9_70
10. Dewan, M., Murshed, M., Lin, F.: Engagement detection in online learning: a review. Smart Learn. Environ. **6**(1), 1–20 (2019)

11. Pabba, C., Kumar, P.: An intelligent system for monitoring students' engagement in large classroom teaching through facial expression recognition. Expert. Syst. **39**(1), e12839 (2022)

12. Savchenko, A.V., Savchenko, L.V., Makarov, I.: Classifying emotions and engagement in online learning based on a single facial expression recognition neural network. IEEE Trans. Affect. Comput. **13**(4), 2132–2143 (2022)

13. Bulathwela, S., Pérez-Ortiz, M., Lipani, A., Yilmaz, E., Shawe-Taylor, J.: Predicting engagement in video lectures. arXiv preprint arXiv:2006.00592 (2020)

14. Abedi, A., Khan, S.S.: Affect-driven ordinal engagement measurement from video. Multim. Tools Appl. **83**(8), 24899–24918 (2023)

15. Churaev, E., Savchenko, A.V.: Multi-user facial emotion recognition in video based on user-dependent neural network adaptation. In: Proceedings of the VIII International Conference on Information Technology and Nanotechnology (ITNT), pp. 1–5. IEEE (2022)

16. Kaur, A., Mustafa, A., Mehta, L., Dhall, A.: Prediction and localization of student engagement in the wild. In: Proceedings of the Digital Image Computing: Techniques and Applications (DICTA), pp. 1–8. IEEE (2018)

17. Gupta, A., D'Cunha, A., Awasthi, K., Balasubramanian, V.: DAiSEE: towards user engagement recognition in the wild. arXiv preprint arXiv:1609.01885 (2016)

18. Savchenko, A.: Facial expression recognition with adaptive frame rate based on multiple testing correction. In: Proceedings of the 40th International Conference on Machine Learning (ICML), vol. 202, pp. 30119–30129. PMLR (2023)

19. Demochkina, P., Savchenko, A.V.: MobileEmotiFace: efficient facial image representations in video-based emotion recognition on mobile devices. In: Del Bimbo, A., et al. (eds.) ICPR 2021. LNCS, vol. 12665, pp. 266–274. Springer, Cham (2021). https://doi.org/10.1007/978-3-030-68821-9_25

20. Savchenko, A.V.: MT-EmotiEffNet for multi-task human affective behavior analysis and learning from synthetic data. In: Karlinsky, L., Michaeli, T., Nishino, K. (eds.) ECCV 2022, Part VI, pp. 45–59. Springer, Cham (2023). https://doi.org/10.1007/978-3-031-25075-0_4

21. Meng, D., Peng, X., Wang, K., Qiao, Y.: Frame attention networks for facial expression recognition in videos. In: Proceedings of the IEEE International Conference on Image Processing (ICIP), pp. 3866–3870. IEEE (2019)

22. Geng, L., Xu, M., Wei, Z., Zhou, X.: Learning deep spatiotemporal feature for engagement recognition of online courses. In: Proceedings of the Symposium Series on Computational Intelligence (SSCI), pp. 442–447. IEEE (2019)

23. Zhang, H., Xiao, X., Huang, T., Liu, S., Xia, Y., Li, J.: An novel end-to-end network for automatic student engagement recognition. In: Proceedings of the 9th International Conference on Electronics Information and Emergency Communication (ICEIEC), pp. 342–345. IEEE (2019)

24. Abedi, A., Khan, S.S.: Improving state-of-the-art in detecting student engagement with ResNet and TCN hybrid network. In: Proceedings of the 18th Conference on Robots and Vision (CRV), pp. 151–157. IEEE (2021)

25. Dhall, A.: EmotiW 2019: automatic emotion, engagement and cohesion prediction tasks. In: Proceedings of the International Conference on Multimodal Interaction, pp. 546–550 (2019)

26. Zhang, Z., Li, Z., Liu, H., Cao, T., Liu, S.: Data-driven online learning engagement detection via facial expression and mouse behavior recognition technology. J. Educ. Comput. Res. **58**(1), 63–86 (2020)

27. Zhu, B., Lan, X., Guo, X., Barner, K.E., Boncelet, C.: Multi-rate attention based GRU model for engagement prediction. In: Proceedings of the International Conference on Multimodal Interaction (ICMI), pp. 841–848 (2020)
28. Copur, O., Nakıp, M., Scardapane, S., Slowack, J.: Engagement detection with multi-task training in E-learning environments. In: Sclaroff, S., Distante, C., Leo, M., Farinella, G.M., Tombari, F. (eds.) ICIAP 2022, Part III, pp. 411–422. Springer, Cham (2022). https://doi.org/10.1007/978-3-031-06433-3_35

Automated Detection and Analysis of Gaming the System in Novice Programmers

Hemilis Joyse Barbosa Rocha[1,2]([✉]) [iD], Evandro de Barros Costa[3] [iD],
and Patricia Cabral de Azevedo Restelli Tedesco[2] [iD]

[1] Federal Institute of Alagoas, Viçosa, Brazil
[2] Computer Center-Cin, Federal University of Pernambuco, Recife, Brazil
hemilis.rocha@ifal.edu.br, pacrt@cin.ufpe.br
[3] Institute of Computing-IC, Federal University of Alagoas, Maceió, Brazil
evandro@ic.ufal.br

Abstract. There is growing interest in "gaming the system" behavior, where students in online learning environments seek to progress without engaging in authentic learning processes. This study addresses this issue, aiming to automatically recognize this behavior in beginning programmers from a public and rural school in Northeast Brazil, as well as analyzing the demographic context of these students identified with this behavior. With the participation of 67 students, we collected data through student interactions with a programming learning environment, developing an automatic detection model. As a result, our detector based on the decision tree algorithm provided the best performance. Our findings highlight a significant difference between the group of students who exhibit "gaming the system" behavior and those who do not. Furthermore, younger students are more likely to exhibit such behavior.

Keywords: gaming the system · novice programmer · context

1 Introduction

Within computer-based environments, the exploration of unproductive help-seeking behavior [4] and "gaming the system" behavior [2] reveals numerous instances that have been scrutinized and studied [2,11]. These behaviors are intrinsically related, and one manifestation of gaming the system involves the inappropriate seeking of assistance by students. The comprehension of these attitudes is frequently associated with low prior knowledge and demographic characteristics [13,14]. However, identifying the various nuances associated with gaming the system remains a substantial challenge across diverse contexts.

Especially in the domain of introductory programming, where many students have reported high rates of gaming the system behavior [14], understanding the manifestation of gaming attitudes represents a particularly relevant challenge for a deeper understanding of this behavior, and there is still few studies reported in the literature. Thus, one of the previous studies in CS1 programming courses

A. M. Olney et al. (Eds.): AIED 2024 Workshops, CCIS 2150, pp. 338–346, 2024.
https://doi.org/10.1007/978-3-031-64315-6_30

investigated the interaction between affective states, observable behaviors, and student performance. Within this context, gaming the system emerged as an observable behavior [17]. In another study, the authors analyze one of the gaming the system attitudes, unproductive help-seeking, in novice programmers by investigating log data from students working in a programming environment that offers automated hints and propose a taxonomy of unproductive help-seeking behaviors in programming [12]. This taxonomy reports attitudes such as requesting hints before starting to program a task and a high frequency of requesting hints within a short period. On the other hand, the authors studied students in a flipped CS1 course, examining the survey data to identify factors that contribute to novice programmers' involvement in three inappropriate behaviors, two of which are related to attitudes of the gaming the system [14] behavior. This study found no correlation between gaming attitudes and students' prior knowledge, contradicting previous reports in the literature [6].

The behavior of gaming the system is not fully explored in the literature, especially in introductory programming, where some underlying characteristics are to be investigated. Considering that previous research indicates that demographic factors are often related to differences in educational outcomes more generally [9] and to constructs associated with motivation more specifically [10], exploring these relationships in the context of gaming the system becomes relevant. For example, in studies in the field of mathematics, it was found that increased use of hints was associated with better math performance in urban schools but correlated with lower performance in suburban/rural schools. Other demographic categories, such as the number of economically disadvantaged students or those with limited English proficiency, also showed significant differences [13]. Thus, this study investigated the gaming behavior of novice programmers at a public and rural school in northeast Brazil, a region historically marked by high social inequality [15]. The research aims to understand the attitudes of novice programmers characterized by gaming the system, detect such behavior, and identify this audience's demographic context. To this end, in this study, we investigate the following research questions, considering novice programmers in a rural school in northeastern Brazil:

RQ 01: Is the automatic detector developed effectively in identifying instances of gaming the system behavior among novice programmers?

RQ 02: What is the demographic context of novice programmers, particularly students from public and rural schools, who exhibit "gaming the system" behaviors?

To investigate these research questions, we conducted an experimental study in a public and rural school with novice programmers. Hence, overall, the main contributions of this work can be summarized as follows: (i) we developed an effective automatic detector model to identify this behavior, considering the exploration of three relevant machine learning algorithms: Decision Tree, K-Nearest Neighbors, and Multilayer Percepton Neural Network, where the constructed decision tree model shows best performance. In addition, (ii) we deepened the data analysis to characterize the demographic context of the students.

This integrated approach allows for a more comprehensive understanding of the contexts in which "gaming the system" behavior occurs.

2 Research Methods

2.1 Materials

The ADA[1] environment was developed as a learning environment in introductory programming courses. ADA was designed and developed following the conceptual structure of elements: class, sessions, problems, alternatives, attempts, tips, code, and subjects. A class is associated with some students and problem-solving sessions. Each problem has four tips and alternatives related to it. Each student can make several attempts to solve the problems. The environment has several essential features to help students learn programming. One of these resources is the problem solver, which allows the student to present partial or complete solutions for each proposed problem. When submitting partial solutions, students answer specific questions about solving the problem, while when submitting complete solutions, they have the opportunity to write programming code to solve the problems. Another important feature is providing tips to students. The tips are organized into different levels, with each level offering progressively specific suggestions on how to proceed at a particular stage of the problem. The structure of the tips starts with more abstract levels and progresses to more specific tips until the answer direct. Students can request hints at any point during the problem.

2.2 Experiment

The experience carried out in this research took place in the first semester of 2023 and involved students from two second-year technical high school classes located in the interior of Northeast Brazil. These students were new to programming and had recently completed an introductory course on the subject. They were currently enrolled in a second course, focused on Java programming (CS2). CS2 students were chosen because they were new to programming and had some familiarity with basic concepts. The study was carried out at this time, as the students were about to start learning object-oriented programming in the CS2 course. The experiment was carried out in four sessions, with two classes each, totaling 67 participating students, 29 from the afternoon class and 38 from the morning class, who participated in all problem-solving sessions. Of these participants, 35 were female and 32 were male, aged between 16 and 21 years. Of these, 61% were white and 39% were of African descent. The experiment was carried out for each class in three stages: pre-test, intervention, and post-test. In the pre-test, the first session was subdivided into 30 min for presentation and 70 min for problem-solving. During this period, the objectives of the study were explained to the students, forms of consent for participation in the experiment

---

[1] https://ada-projeto.vercel.app/.

were distributed, following the guidelines of the ethics committee, and a comprehensive explanation of the environment's functionalities was provided. The following sessions were held for 100 min each.

2.3   Data

Through the experiments detailed in the previous section, we acquired a dataset with 4,492 records of student interactions with the learning environment. Each instance in our dataset was labeled as belonging to one of two possible classes: Class0, if there was no manifestation of "gaming the system" behavior (2,814); Class1, if there is a manifestation of this behavior (1,678). To label the dataset, we adopted a model composed of thirteen previously validated "gaming the system" behavior patterns [7]. However, to use this model it was necessary to understand the concept of "text replays", which are groupings of up to five sequences of student actions when interacting with the environment. Thus, if a pattern is identified in the analysis of the data set, all actions involved are labeled as "gaming the system". These text repetitions facilitate efficient classification and have acceptable reliability [7], and have also been previously used to train detectors of the same behavior. In the first labeling step, coders had a relatively low inter-rater agreement, reflected by a Cohen's Kappa coefficient of 0.42. To improve consistency, a second round of labeling was carried out, where divergences were analyzed and discussed, leading to a more aligned interpretation of behaviors. This collaborative approach resulted in a significant increase in Cohen's Kappa coefficient to 0.71. However, it was still considered necessary to improve the agreement. Therefore, a third round was carried out to resolve the remaining conflicts, culminating in a very satisfactory final coefficient of 0.93, indicating a high level of agreement between coders.

In the data preprocessing phase, we divide the dataset into training (75%) and testing (25%) subsets. For classification, we employ three algorithms: decision tree, k-nearest neighbors (KNN), and Multilayer Perceptron (MLP) neural networks, based on previous studies such as those by Baker et al. (2008). The implementation was carried out using the scikit-learn library. The decision tree was configured with the CART algorithm, KNN with 3 neighbors, and MLP neural network. For evaluation, we analyzed the performance of each algorithm on the training and testing subsets using six metrics: area under the ROC curve (AUC), accuracy, precision, recall, F1 score, and Kappa. A significance level of $p < 0.05$ was applied. AUC evaluates the model's classification accuracy, while precision, recall, and F1 provide a comprehensive performance assessment. Kappa ensures that the model's correct predictions are not random.

3   Results and Discussion

As mentioned in Sect. 2.3, we used the patterns described in the model of a previous work to annotate the data [7]. However, when we analyzed the application of this model to the data and functionalities of the ADA environment, we found that it was not compatible with some standards. This incompatibility stems from the difference between the types of tasks offered by the environment in which the model was developed and the problem-solving tasks available in the ADA environment. Despite this, as the patterns described in the previous work are actually data annotation rules, we realized that patterns 1, 5, 6 and 11 were suitable for our environment. We therefore used them to annotate the data, analyze performance in our data context and performance of our classifiers.

3.1 RQ 01: Automatic Detector

To investigate our first research question, we first compared the overall performance of the algorithms about "gaming the system" behavior, considering the four standards. In Table 1, we present the results of comparing algorithms in two phases: training and testing, using various evaluation metrics. The metrics include AUC (Area under the ROC Curve), precision (PRE), recall (REC), F-measure (FMe), Kappa coefficient (KAP) and the ROC curve itself. In the training phase, the KNN algorithm obtained a high accuracy (ACU) of 0.93, indicating a good generalization capacity. The decision tree (DT) algorithm achieved an even higher accuracy of 0.96, with excellent precision (PRE) and F-measure (FMe). On the other hand, the neural network (MLP) performed less well than the other two algorithms, with an accuracy of 0.72 and lower values in other metrics. In the test phase, the results remained consistent, with DT remaining the leader with an accuracy of 0.96 and robust scores in other metrics. KNN also maintained a good performance, albeit with a slight drop compared to the training phase. However, MLP, despite improving slightly in some metrics, still underperformed compared to the other algorithms.

Table 1. Results of the algorithm comparison

	Training						Test					
	ACU	PRE	REC	FMe	KAP	ROC	ACU	PRE	REC	FMe	KAP	ROC
KNN	0.93	0.93	0.94	0.94	0.86	0.93	0.86	0.85	0.88	0.87	0.72	0.86
DT	0.96	0.97	0.92	0.96	0.96	0.92	0.96	0.95	0.95	0.95	0.92	0.96
MLP	0.72	0.85	0.55	0.67	0.44	0.72	0.72	0.86	0.55	0.67	0.45	0.73

Using the Wilcoxon test, we compared the performance of the algorithms at a significance level of $\alpha = 0.05$. In the training phase, we observed a significant difference between the performance of CART and KNN ($p = 0.022$), while there was no significant difference between CART and MLP ($p = 0.061$). In the test phase, there are significant differences between CART versus MLP ($p = 0.021$) and CART versus KNN ($p = 0.021$). Therefore, the CART decision tree algorithm was the best performer, which is why we will analyze the performance of the decision tree on all patterns.

Table 2. Results obtained from applying the decision tree algorithm to pattern "gaming the system"

	Training					Test					
Pattern	ACU	PRE	REC	FMe	KAP	ROC	ACU	PRE	REC	FMe	KAP	ROC
PA01	0.80	0.75	0.61	0.67	0.53	0.75	0.64	0.45	0.34	0.39	0.14	0.57
PA05	0.80	0.73	0.62	0.67	0.54	0.75	0.61	0.41	0.40	0.41	0.12	0.56
PA06	0.79	0.72	0.64	0.68	0.53	0.76	0.59	0.39	0.38	0.39	0.08	0.54
PA11	0.85	0.81	0.67	0.69	0.57	0.83	0.72	0.37	0.42	0.59	0.35	0.67

Table 2 shows the results of applying the decision tree algorithm to identify "gaming the system" patterns. Divided into training and testing, the results reveal a general consistency in pattern performance. During training, patterns PA01, PA05, and PA06 exhibited similar metrics, with an accuracy of around 0.80 and a good combination of precision and recovery. The PA11 pattern stood out with a slightly higher accuracy (0.85) and robust precision (0.81). However, in the test phase, there was a downward trend in performance metrics, especially for the PA11 standard, where precision dropped considerably to 0.37.

Comparing the results of this research with previous work, in previous work the authors state that the complete model was applied to the training set, where it accurately detected 340 (64.03%) game clips, misdiagnosed 551 (7.07%) non-game clips, and obtained a Kappa coefficient of 0.430 [7]. In the test set, the model achieved an accuracy of 93 (52.54%) gaming clips, while misdiagnosing 210 (8.67%) non-gaming clips, resulting in a Kappa coefficient of 0.330. This performance was comparable to Baker's [1] model. In this research, the best algorithm evaluated was the decision tree CART, achieving a hit rate between 98% and 100% in all the metrics evaluated. This algorithm identified 30.92% true positives, 68.57% true negatives, 0.35% false positives, and 0.92% false negatives.

3.2   RQ 02: Demographic context

In this section, we present the results of the analysis of three demographic variables: gender, age, and race/ethnicity. To analyze the age variable, three age groups were created: group 1, which includes participants aged 16 and 17; group 2, aged 18 and 19; and group 3, aged 20 and 21. Firstly, we found that there was no difference between the frequency of "gaming the system" in the gender and ethnic-racial groups ($p < 0.05$). However, students in group 1 exhibited the behavior more than the other groups ($p < 0.05$). In addition, we compared the learning gains between the same gender, age group, and ethnicity, subdivided into two groups: No-game and Game. Table 3 displays the results of the Mann-Whitney test, comparing learning gains in different groups based on ethnicity (Afro-descendant and white/yellow) and gaming the system behavior (without gaming and with gaming). Significant differences in learning gains were observed between the groups analyzed, with notably higher gains among individuals who did not engage in gaming the system behavior compared to those who did, regardless of ethnic origin. Considering gender (female and male) and gaming the system behavior (Without gaming and with gaming). Statistically significant learning gains were found among women who did not engage in gaming the system behavior, as well as among all other groups who refrained from this behavior compared to those who did. Analyzing learning gains based on ethnicity and behavior. Notably, for Track 1, students without abusive behavior made significantly greater learning gains compared to all tracks in the abusive group. Similarly, for Track 3, greater learning gains were observed among students without abusive behavior compared to the abusive group in the same track and Track 1. However, no significant gains were observed when comparing the groups in Track 2. These results suggest that, in the analyzed context, younger novice programmers are more likely to engage in gaming the system behavior

Table 3. Results of Mann-Whitney test for comparison of learning gain by gender, age and race/ethnicity and gaming the system behavior.$\alpha = 0.05$

No game/game	Female	Male	
Female	121.5; $p < 0.05$	104; $p < 0.05$	
Male	201; $p < 0.05$	306; $p < 0.05$	
No game/Game	**Afrodescendant**	**White/Yellow**	
Afrodescendant	81.2; $p < 0.05$	156; $p < 0.05$	
White/Yellow	331; $p < 0.05$	106; $p < 0.05$	
No game/game	**Track3**	**Track2**	**Track1**
Track1	201; $p = 0.05$	104; $p < 0.05$	143; $p < 0.05$
Track2	111; $p = 0,31$	362; $p = 0,35$	-
Track3	61; $p < 0,05$	–	–

4   Conclusion

This study investigated gaming the system behavior in experimental research in a rural school in Northeast Brazil with beginner programmers. We developed an effective automatic detector to identify this behavior and analyzed three variables from a demographic context. As a future perspective, we intend to carry out a comparative analysis of data from students from different areas or cities. This study offers an analysis of the influence of the rural context; however, to expand the generalization of these results, we consider it crucial to replicate the research in other schools, especially in urban environments with different conditions. Repeating the study in varied contexts will allow a more comprehensive understanding of the observed dynamics.

References

1. Baker, R.S.J.D., de Carvalho, A.M.J.A.: Labeling student behavior faster and more precisely with text replays. In: Proceedings of EDM, vol. 2008, pp. 38–47 (2008)
2. Baker, R.S.J.D., Corbett, A.T., Roll, I., Koedinger, K.R.: Developing a generalizable detector of when students game the system. User Model. User Adapt. Inter. **18**, 287–314 (2008)
3. Baker, R.S.J.D., Mitrović, A., Mathews, M.: Detecting Gaming the System in Constraint-Based Tutors. In: De Bra, P., Kobsa, A., Chin, D. (eds.) UMAP 2010. LNCS, vol. 6075, pp. 267–278. Springer, Heidelberg (2010). https://doi.org/10. 1007/978-3-642-13470-8_25
4. Aleven, V., McLaren, B.M., Roll, I., Koedinger, K.R.: Toward meta-cognitive tutoring: a model of help seeking with a cognitive tutor. Int. J. Artif. Intell. Educac. **16**, 101–130 (2006)
5. Baker, R.S., Corbett, A.T., Koedinger, K.R.: Detecting student misuse of intelligent tutoring systems. In: Lester, J.C., Vicari, R.M., Paraguaçu, F. (eds.) ITS 2004. LNCS, vol. 3220, pp. 531–540. Springer, Heidelberg (2004). https://doi.org/ 10.1007/978-3-540-30139-4_50
6. Baker, R.S., et al.: Do performance goals lead students to game the system?. AIED (2005)
7. Paquette, L., de Carvalho, A.M.J.B., Baker, R.S.: Towards understanding expert coding of student disengagement in online learning. CogSci. (2014)
8. Paquette, L., Baker, R.S.: Comparing machine learning to knowledge engineering for student behavior modeling: a case study in gaming the system. Interact. Learn. Environ. **27**(5–6), 585–597 (2019)
9. Childs, D.S.: Effects of math identity and learning opportunities on racial differences in math engagement, advanced course-taking, and STEM aspiration. Temple University (2017)
10. Zeldin, A.L., Britner, S.L., Pajares, F.: A comparative study of the self-efficacy beliefs of successful men and women in mathematics, science, and technology careers. J. Res. Sci. Teach.: Official J. Natl. Assoc. Res. Sci. Teach. **45**(9), 1036–1058 (2008)
11. Richey, J.E., et al.: Gaming and confrustion explain learning advantages for a math digital learning game. In: Roll, I., McNamara, D., Sosnovsky, S., Luckin, R., Dimitrova, V. (eds.) AIED 2021. LNCS (LNAI), vol. 12748, pp. 342–355. Springer, Cham (2021). https://doi.org/10.1007/978-3-030-78292-4_28
12. Marwan, S., Dombe, A., Price, T.W.: Unproductive help-seeking in programming: what it is and how to address it. In: Proceedings of the 2020 ACM Conference on Innovation and Technology in Computer Science Education (2020)
13. Karumbaiah, S., Ocumpaugh, J., Baker, R.S.: Context Matters: differing implications of motivation and help-seeking in educational technology. Int. J. Artif. Intell. Educ. **32**(3), 685–724 (2022)
14. Solyst, J., et al.: Procrastination and gaming in an online homework system of an inverted CS1. In: Proceedings of the 52nd ACM Technical Symposium on Computer Science Education (2021)
15. Oliveira, D.A., Clementino de Jesus, A.M.: As políticas de avaliação e responsabilização no Brasil: uma análise da Educação Básica nos estados da região Nordeste. Revista Ibero-americana de Educação (2020)

16. PNUD. Atlas do desenvolvimento Humano no Brasil (2018). https://bit.ly/3dnx1QL
17. Rodrigo, M.M.T., et al.: Affective and behavioral predictors of novice programmer achievement. In: Proceedings of the 14th Annual ACM SIGCSE Conference on Innovation and Technology in Computer Science Education (2009)

Implementation and Evaluation of Impact on Student Learning of an Automated Platform to Score and Provide Feedback on Constructed-Response Problems in Chemistry

Cesar Delgado$^{(\boxtimes)}$ ⓘ, Marion Martin ⓘ, and Thomas Miller ⓘ

North Carolina State University, Raleigh, NC, USA
cdelgad@ncsu.edu

Abstract. Introductory undergraduate chemistry courses are gatekeepers for STEM careers, particularly for underrepresented students. Findings from implementation of a novel platform for STEM problem-solving that automatically scores and gives students feedback for enhanced learning are presented. The platform uses classic AI with recursive algorithms and computer algebra processing. The platform tracks and scores the students' solution pathway and displays the scores and descriptions for each correct step on the rubric as feedback for students to use on their next attempt. Teachers track student progress on a dashboard. Descriptions of the platform, student learning and perceptions, and differences from status quo solving and scoring are provided. 286 students used the platform. Scores increased with a large effect size. The platform distinguishes correctness of solutions beyond capabilities of human graders of students' handwritten work. The types of errors students can make differ from manually solved problems.

Keywords: chemistry · feedback · automated scoring · constructed response

1 Introduction

STEM education is critical for economic advancement. However, the US STEM workforce lacks diversity [1]. First-year general chemistry courses (GC1) required for most STEM majors act as gatekeeper [2], with a 30% D/F/withdrawal/incomplete rate [3]. Research shows that GC1 leads to underrepresented students abandoning STEM majors [4]. Improving learning in GC1 courses could preferentially help underrepresented groups in Chemistry, diversifying the STEM workforce. GC1 topics such as stoichiometry, equilibrium, and gas laws require students to solve quantitative constructed-response (CR) problems. It is essential to improve student quantitative problem solving.

Learning processes and outcomes are improved by specific and timely feedback [5, p. 154]. Multiple-choice (MC) tasks or platforms for automated assessment of CR problems that consider only the final answer do not provide detailed feedback. Students may then need help, but first-generation and underrepresented students seek help at lower rates than peers [6, 7]. Methods to provide feedback on CR problems would be particularly relevant in large classrooms where individualized attention is not available.

Students solve quantitative STEM problems using paper, pencil, and calculator to construct their solution. CR tasks can be graded for partial credit against a rubric based on the student's pathway. CR problems (also known as "word problems"), are pervasive in the curriculum across all levels [8–10]. Compared to MC items, word problems measure more complex skills [11], are more valid and reliable means of assessing understanding [12], and have lower differential performance by gender (less bias) [13].

Manually scoring and providing feedback on students' solution pathways on CR problems is very time-consuming. Thus, students do not receive the feedback in a timely manner. Therefore, there is a pressing need for a system that can provide timely, accurate, and detailed feedback on chemistry CR problems.

Grade-It is a computer application to produce, administer, and score quantitative CR problems, where students receive instant, detailed feedback to support learning. Students log in using an access code, then solve problems on the system using a web browser. They drag-and-drop or click parameters (displayed as numerical values) in the problem statement along with mathematical operators to construct their solution (see Fig. 1). Every component is of known origin so their work can be tracked and analyzed.

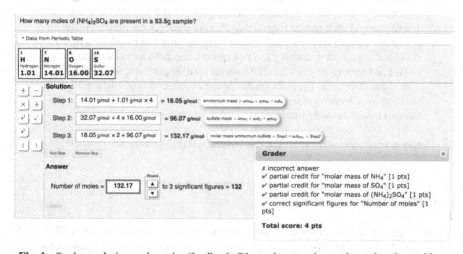

Fig. 1. Student solution and scoring/feedback. Blue values can be used to solve the problem.

The system utilizes reactive machine artificial intelligence (RM-AI), using algorithms to reproducibly follow rules to deliver consistent results in a transparent manner. The grading algorithm analyzes and assigns a grade by algebraically parsing the submission (using the Python-based SymPy computer algebra system) and recursively searching the submitted solution's expression tree for mathematical constructs defined in the rubric. E.g., the rubric for the problem shown in Fig. 1 assigns 1 point for correctly determining the molar mass of NH_4: $amu_N + amu_H \times sub_H$. If that operation is present (Step 1) a point is assigned. When a student submits a solution, the grading algorithm compares the parameters, operators, and their relationships for every step on all strategies in the rubric. If all steps are found for any given strategy and the final answer's combination of parameters and operators is correct, the student is awarded full credit. If not fully correct, the system assigns the appropriate points based on the highest-scoring strategy in the rubric. If the final answer's combination of parameters and operators is correct but the steps are *not* all present, the solution is flagged as an "unanticipated strategy". The educator can add that strategy to the rubric for future use.

The deep search pattern matching employed by the grading algorithm uses mathematical identities. E.g., if a rubric awards points for the expression $A + C$ and a node in the solution contains the expression $C - B + A$, the pattern matching logic in our grading algorithm will find $A + C$ by applying both the commutative and associative properties of addition. The platform displays to the student a report generated by the algorithm that identifies each correct operation on their solution pathway (See Fig. 1). The *descriptions* of each step, built into the rubric by the problem author, are also displayed, e.g., "molar mass of NH_4". This naming and identification of correct steps is timely, detailed feedback to support student learning. Assignments can be set up in "practice mode" with a teacher-determined number of attempts with feedback after every submission; or in "graded" mode, with feedback available only after the assignment closes. Users create classes and assignments, and view student work and scores, on the system.

This study is guided by the following research questions:

RQ 1: What is the impact on student learning of the Grade-It platform?

RQ 2: How did students perceive the Grade-It platform?

RQ 3: How are solving and scoring CR problems by hand different than on the Grade-It platform?

2 Methods

2.1 Context and Assignment

This study took place in a US public research university, in two sections of GC1 each with over 240 students. Demographic information for the participants is not available but the university (including graduate students) is composed of 8.11% Asian, 7.01% Hispanic, 6.17% Black or African American, 3.96% two or more races/ethnicities, and around 74% White non-Hispanic. Undergraduates are 49% female and 51% male [14].

A 12-problem extra credit final exam review assignment was built on Grade-It, drawing from previous semester's review materials (PDFs with problems and answer key). Students also had access to the PDFs, assigned exams and answer keys from earlier in the semester, and a multimedia presentation with review problems.

2.2 Design and Deployment of Chemistry Problems on the Platform

The course review materials were provided to the first author, who identified the problems suitable for the platform. He then changed some values and chemical species so they would differ from the original problems. This allowed for students to practice both by using the platform and by solving the PDF problems with paper and pencil. E.g., the

MC problem "Which sample has the fewest atoms" with answer choices of the number of grams of an element, the number of moles of an element, the number of moles of a compound, and the number of molecules of a diatomic element, was revised to "How many atoms are there in 100.0 g of Fe_2O_3? Avogadro's number $= 6.02 \times 10^{23}$." This involves the same skills (gram-to-mole and mole-to-atom conversions, and use of subscripts in formulas). Two strategies were built into the rubric: calculating the number of atoms per formula unit of Fe_2O_3 first, and calculating the number of atoms of each element. The platform scores the student's work using both strategies and assigns the maximum number of points earned.

The professor of the course (second author) created the assignment on Grade-It and posted the link to the assignment on his course management system, along with links to a 4-min instructional video and a 4-page how-to PDF. In addition, he played part of the video in class. The assignment was open for two weeks prior to the exam.

2.3 Data Sources

Google Analytics provided data on usage. The platform's dashboard stored all student work (solution pathway and answer submitted) and scores for each of their attempts on every problem. A Qualtrics survey asked students about ease of use, perceptions of learning, and desire to use in future classes; the link was placed on the course website.

2.4 Analysis

For RQ 1, student scores on the first attempt at each problem were obtained from the dashboard, as were their highest scores. Averages were calculated for each problem across all the students' first and highest-scoring attempts. The pattern of scores on each student's scores on a given problem was examined and coded as one of four types (Table 1): F for full credit on first attempt (no further learning necessary or possible); L for increased scores relative to first attempt, indicating improved performance (interpreted as learning); N for no increase in scores (no learning); and 1 for a single submission resulting in less than full credit (interpreted as student making no attempt to learn). The within-subjects effect size was calculated using the paired-samples t-test (preliminary analysis established normality of distribution, $D = 0.061$, $p = 0.22$). For RQ 2 we employed descriptive statistics. For RQ 3, we qualitatively examined how scoring might differ between manual scoring based on a rubric and platform scoring.

Table 1. Coding of patterns on students' scores on a given problem

Code	F	L	N	1
Description	Full credit on first attempt	Learned: score increased	No learning: no increase	No attempt to learn
Example	View detail 10 / 10 points Solution	Attempt 1 4 / 7 points Attempt 2 6 / 7 points Attempt 3 7 / 7 points Solution	Attempt 1 1 / 4 points Attempt 2 1 / 4 points Attempt 3 1 / 4 points Solution	View detail 1 / 5 points Solution
Problems	1129	1346	264	235

3 Results

Students were on the platform for an average of around 40 min over two sessions. Platform data showed that a total of 286 students solved a total of 2975 problems (10.4 of 12 problems on average) and generated a total of 5474 submissions.

3.1 RQ 1: Student Learning

The average student score on the first attempt was 57.7 (SD = 23.7) and on the highest attempt 72.5 (SD = 28.5), with a difference of 14.8 points. 1129 out of 2975 (38%) problems were solved fully correctly on the first attempt and coded F (see Table 1). These problems did not allow for additional learning. The remaining 1846 problems had room for learning. 1346 (73% of 1846) had scores that increased from first attempt to subsequent attempt, coded as L, and interpreted as learning taking place. 264 problems (14% of 1846) had no increase in scores over multiple attempts and coded N for "no learning". Finally, 235 problems (13% of 1846) were coded 1: a single attempt, earning partial credit. These students did not use the feedback to inform a subsequent trial. There is a statistically significant difference between first attempt (M = 57.7, SD = 23.7) and highest attempt (M = 72.5, SD = 28.5), t(285) = 26, p < .001), with a very large within-subjects effect size, Cohen's d = 1.54.

3.2 RQ 2 How Did Students Perceive the Grade-It Platform?

32% of users responded to the three-item, five-level Likert survey. We grouped "strongly agree" and "agree" responses ("A" in Table 2) as well as "disagree" and "strongly disagree" ("D"). Neither agree nor disagree was labeled "N". Between 2.7 and 6.4 times more students agreed than disagreed with the items, reflecting ease of use, perceived value for learning, and desire to use again in future classes.

3.3 RQ 3 Differences Between Paper-And-Pencil and Platform Problem Solving

Solving Problems. Solving problems on the platform requires making explicit some steps that students and instructors may carry out implicitly or skip, such as simple SI

Table 2. Student perceptions of the Grade-It platform

Item	A	N	D
Easy to use once I learned how it works	78%	6%	16%
Learned more than with MC questions or platforms that only evaluate the final answer	70%	19%	11%
Would like to be able to use in my future STEM classes	59%	19%	22%

conversions (e.g., gram to milligram). Another difference is that the platform automatically calculates the results of steps (Fig. 1). This precludes entire classes of errors, such as entering numbers incorrectly on the calculator, or transferring results incorrectly to paper. These errors are construct-irrelevant – not strictly related to the chemistry content being assessed. Instructors who want to assess whether their students do make these mistakes can use some paper or MC problems to complement Grade-It. Another difference is having to show explicit values of "1" in balanced equations or formulas for students to use in stoichiometric ratios. The rest of the process is the same as when manually solving problems: constructing the solution in a series of steps, submitting the answer, and determining the correct number of significant figures. The problem author can include units with each quantity that the users can click or drag and drop; if they do so, they will appear next to the numerical value throughout the problem. Results of steps do *not* include units; rather, the student can optionally label the step and include units. The platform does not process or cancel units for the student.

Scoring Problems. One important difference from hand solving and scoring problems is that quantities that are conceptually distinct but numerically identical can be tracked in the platform but not with paper and pencil problem solving. For example, there are two subscripts with the value of 4 in Fig. 1: the number of oxygen atoms in the sulfate ion (sub_O) and the number of hydrogen atoms in the ammonium ion (sub_H). When a student writes a 4 on paper, the grader cannot know which of the two subscripts the student selected, but Grade-It can, and treats them differently. Thus, the platform can achieve greater resolution in scoring than skilled human graders in this respect. Another difference concerns rubrics. Rubrics developed for manual scoring tend to be simple, with only a small number of steps awarded points and steps weighted the same. With the platform, it is possible to have fine-grained scoring with points awarded for each granular step, and they can be weighted to reflect difficulty or conceptual importance.

4 Discussion

The data show that students accepted the platform, with close to 60% of students using it and engaging in extended fashion (solving 10.4 of 12 problems over 40 min on average). Feasibility and usability were established, with student surveys showing ease of use, perceived support for learning, and desire to use in future classes. Student learning was supported, presumably due to the feedback: score reports show what steps on their solution are correct, supporting procedural knowledge, and name each step, supporting students' conceptual understanding. In some cases, students' increase on subsequent attempts might be due to guessing or making random changes, but the overall large score increase supports the inference that students learned. The platform's approach of providing information on students' work rather than providing hints on how to solve the problem or the correct answer may foster meta-cognition and self-assessment among students, encouraging students to become agents of their own learning. Students who did not learn across their three attempts (coded N in Table 1) or who did not attempt to learn (coded 1) could still view the correctly solved problem and grading rubric after their third attempt, or if they made fewer than three submissions, after the assignment closed. Viewing "worked examples" has been found to be effective in supporting learning as it imposes a low cognitive load [15].

The platform's ability to track conceptually distinct quantities that are visually identical (e.g., the two subscripts of 4) might help mitigate biases in grading, offering fairer assessments for students from diverse backgrounds.

In addition to providing the timely, specific feedback that research has established supports learning processes and outcomes [2], innovative pedagogical approaches could be developed that go beyond one-to-one models. Matching students by patterns of complementary strengths and weaknesses, based on data stored by the platform, could lead to highly effective peer-to-peer teaching and learning, and position all or most students as experts or tutors on the problems they excelled in.

The platform can be used beyond chemistry, for all STEM subjects and classes that have quantitative CR problems; and beyond undergraduate, with a successful pilot test in high school chemistry and ongoing tests in middle school math.

Setting up problems on Grade-It takes more time than typing problems and preparing a PDF, but there will be large time savings relative to hand scoring. The ability to provide feedback to students in large classes that currently use binary scoring platforms or MC questions is valuable, especially for underrepresented and first-generation college students who are less likely to ask for help, fomenting greater equity in STEM.

4.1 Limitations

The comparison of first and highest attempts is in effect a single-group, pre- and post-test quasi-experimental design. It can detect and characterize student change in performance (learning) but does not have a control and thus cannot inform about the effectiveness of the platform compared to traditional paper-and-pencil or MC practice problems. The first and third authors are involved in the company that developed the platform, and so might be biased. As an early-stage, pre-revenue company, there was not an option to have external researchers conduct this study. To mitigate reader and reviewer concerns, data is available upon request.

5 Conclusion and Future Directions

The Grade-It platform demonstrated in its early implementation in a GC1 classroom that it is usable and supports substantial student learning. Partial-credit scoring of CR problems had to be done manually prior to this platform, and the automation of this process has different implications for different practices. For classes where partial-credit scoring is done by the instructor or teaching assistants, the platform can relieve them of hours of work. For large classes where manual grading is impractical and the instructor resorts to MC or final answer right/wrong scoring, the platform presents the possibility of scoring and feedback that is individualized and supports student learning.

Future work will involve examining interrater agreement across human graders and between human graders and the platform, as well as more robust research studies on student learning. We are working on a quasi-experimental control-intervention analysis comparing two self-selected groups – students who did and did not use the platform. We intend to conduct focus groups or interviews to garner students' more detailed impressions. We will seek to involve more highly diverse populations across a wider variety of educational settings and study the effects on underrepresented groups.

Disclosure of Interests. Authors A and C own stock in Grade-It.

References

1. Pew Research Center: STEM jobs see uneven progress in increasing gender, racial and ethnic diversity (2021). https://www.pewresearch.org/social-trends/2021/04/01/stem-jobs-see-uneven-progress-in-increasing-gender-racial-and-ethnic-diversity. Accessed 04 Feb 2024
2. Sevian, H., et al.: Addressing equity asymmetries in general chemistry outcomes through an asset-based supplemental course. JACS Au. **3**(10), 2715–2735 (2023)
3. Koch A., Drake B.: Digging into the disciplines: the impact of gateway courses in accounting, calculus, and chemistry on student success. https://umaine.edu/provost/wp-content/uploads/sites/14/2018/11/Gardner-Institute-Digging-into-the-Disciplines-The-Impact-of-Gateway-Courses-in-Accounting-Calculus-and-Chemistry-on-Student-Success-.pdf. Accessed 04 Feb 2024
4. Arnaud, C.H.: Freshman chemistry is an exit point for many underrepresented STEM students, study shows. Chem. Eng. News **98**, 23 (2020)
5. Shute, V.J.: Focus on formative feedback. Rev. Educ. Res. **78**(1), 153–189 (2008)
6. Williams-Dobosz, D., et al.: Ask for help: Online help-seeking and help-giving as indicators of cognitive and social presence for students underrepresented in chemistry. J. Chem. Educ. **98**(12), 3693–3703 (2021)
7. White, M., Canning, E.A.: Examining active help-seeking behavior in first-generation college students. Soc. Psychol. Educ. **26**, 1369–1390 (2023)
8. Daroczy, G., et al.: Word problems: a review of linguistic and numerical factors contributing to their difficulty. Front. Psychol. **6**, 1–13 (2015)
9. College Board: AP exams overview/Exam timing and structure. https://apstudents.collegeboard.org/ap-exams-overview/exam-timing-structure. Accessed 04 Feb 2024
10. National Governors Association Center for Best Practices, and Council of Chief State School Officers: Common Core State Standards for Mathematics. Wash., DC (2010)
11. Livingston, S. A.: Constructed-response test questions: why we use them; how we score them. https://www.ets.org/Media/Research/pdf/RD_Connections11.pdf. Accessed 04 Feb 2024
12. McKenna, P.: Multiple choice questions: answering correctly and knowing the answer. Interact. Technol. Smart Educ. **16**(1), 59–67 (2019)

13. Weaver, A.J., Raptis, H.: Gender differences in introductory atmospheric and oceanic science exams: multiple choice versus constructed response questions. J. Sci. Educ. Technol. **10**, 115–126 (2001)
14. North Carolina State University: About Us: Students. https://catalog.ncsu.edu/about/. Accessed 04 Feb 2024
15. Atkinson, R.K., et al.: Learning from examples: Instructional principles from the worked examples research. Rev. Educ. Res. **70**(2), 181–214 (2002)

An Interpretable Approach to Identify Performance Indicators Within Unstructured Learning Environments

Ethan Prihar[✉][iD], Bahar Radmehr[iD], and Tanja Käser[iD]

École Polytechnique Fédérale de Lausanne, Rte Cantonale, 1015 Lausanne, Vaud, Switzerland
{ethan.prihar,bahar.radmehr,tanja.kaeser}@epfl.ch

Abstract. As educational technology becomes increasingly integrated into curricula, more students are engaging with online learning platforms, interactive simulations, and MOOCs. These unstructured environments record students' behaviors, providing a rich source of data on their learning processes. Researchers can model this data to gain insights into which behaviors and interactions most significantly affect students' performance. However, most current methods for interpreting these models rely on manually designed features, which may not generalize across different scenarios or research questions. Conversely, post-hoc explainability methods for more complex models exist, but they lack consensus. To overcome these challenges, this study introduces the Transformer-based Identification of Performance Indicators (TIPI), a novel approach for identifying student behavior patterns that influence performance, emphasizing interpretability by design. TIPI employs a transformer-based deep learning model to convert students' click-stream data into action tokens, periodically predicts students' overall learning outcomes, and identifies the actions most indicative of performance. We conducted a comprehensive analysis using data from two versions of an interactive simulation for trade-school students to demonstrate TIPI's effectiveness. The results reveal that the actions TIPI identifies as performance indicators align with subject matter experts' insights. Furthermore, actions deemed similar by TIPI correspond to their context and location within the interactive simulation. Looking forward, TIPI has the potential to enhance personalized learning experiences in online environments by guiding students towards behaviors that most positively impact their predicted performance.

Keywords: Interpretability by Design · Early Performance Prediction · Unstructured Learning Environments · Transformers

1 Introduction

Interactive educational environments play a crucial role in modern curricula, offering diverse learning experiences. However, understanding and interpreting

A. M. Olney et al. (Eds.): AIED 2024 Workshops, CCIS 2150, pp. 356–363, 2024.
https://doi.org/10.1007/978-3-031-64315-6_32

student behaviors in these environments, especially in unstructured settings where traditional performance metrics are unclear, present significant challenges. The lack of consensus on key student behavior features indicative of learning outcomes further complicates the issue, evident in both computational post-hoc interpretability methods [10] and educators' perspectives [9].

In structured environments like intelligent tutoring systems, researchers have measured latent knowledge [6] or emotional states [2] by extracting informative student features from sequential data on task performance and content information. These methods are effective in environments with clear objectives and available content metadata. Conversely, unstructured learning environments like open-ended simulations present unique challenges. In these settings, students have freedom in interacting with learning content, making it difficult to determine performance indicators. Prior research has developed feature sets representing student behavior in these environments [5], but interpretability is limited to these specific features, hindering the exploration of new behavioral patterns without retraining. The absence of a post-hoc ground truth for performance indicators in unstructured environments complicates interpretability method validation. This highlights the need for interpretability by design, focusing on inherently interpretable models that provide insights into predictions.

To address these challenges, we propose the Transformer-based Identification of Performance Indicators (TIPI), a model designed to interpret student behavior in unstructured environments from click-stream data. By leveraging self-attention mechanisms [3,11], TIPI offers early performance predictions and interpretable insights without relying on specific features. TIPI aims to identify performance indicators through raw behavioral data analysis, eliminating the need for extensive feature engineering or inconsistent interpretability methods. This approach enables the investigation of behavioral patterns on performance, aligning with educators' views on meaningful student engagement.

This work explores TIPI's effectiveness through the following research questions using data from online interactive simulations:

1. How can an attention-based model (TIPI) predict students' performance interpretably?
2. Does TIPI's estimation of specific behaviors' impact align with ground-truth in pre-defined relationships?
3. Does TIPI's estimation of behavior similarity align with their presence in the learning environment?

2   Methodology

In this study, we address the challenge of identifying behavioral patterns that indicate student performance by framing it as a time-series regression task. Each time-series corresponds to a click-stream sequence of actions taken by a student within a learning environment. For each time-series, there is a corresponding measure of performance, represented by a continuous value in the range [0, 1]. To prepare the data for modeling, TIPI first partitions each click-stream sequence of actions into uniform time periods. These sequences of time periods are then used to train a self-attention-based model, which predicts the final measure of learning after each time period. In essence, for each sequence, the model makes a series of predictions equal to the sequence's length. Each prediction attempts to forecast the same performance value, but with each successive prediction incorporating information from an additional time period compared to the previous prediction. After training the model, we measure the impact and similarity of different actions using the model's weights and the predictions corresponding to each action.

2.1 Data Collection

This study employs data from PharmaSim (PS), an interactive simulator designed for training aspiring pharmacy technicians, where they assist virtual customers. PS captures detailed click-stream data, documenting each user's actions and their timing, alongside their final learning outcomes. These click-streams, representing user interactions such as medication lookups or customer inquiries, are analogous to sequences of tokens in large language models [3]. However, due to the sparse nature of data in unstructured learning environments, this study aggregates actions into discrete time periods to simplify the data structure and reduce model complexity, thus preventing overfitting.

The initial phase of the Temporal Interaction Pattern Identification (TIPI) process involves transforming these click-stream sequences into a uniform array of time intervals, counting the occurrences of each action within these intervals. This format is crucial for transformer-based models, which require equidistant sequence elements [11].

PharmaSim features static images with interactive components, starting with a customer presenting a problem for the student to diagnose using various tools. The simulation offers 66 potential questions, medication information, medical literature, and pharmacist assistance. It records each action-totaling 119 unique actions-as students navigate through two distinct customer scenarios, each with its own diagnostic process. These scenarios, one involving a baby with diarrhea (PS-A) and another concerning breastfeeding issues (PS-B), require students to ask essential questions for accurate diagnosis. The learning measurement in PS is based on the proportion of essential questions asked within the first 10 inquiries, reflecting the realistic constraint of limited customer patience. This approach was chosen over analyzing post-simulation diagnoses due to the observation that students often reach a diagnosis without exploring all essential questions, risking incomplete diagnostic conclusions. Data for this study was collected from vocational school students enrolled in pharmacy technician programs, with 138 students completing PS-A and 214 completing PS-B within a 24-minute time frame. The data was segmented into one-minute intervals for subsequent analysis.

2.2   Model Design

The TIPI model is based on the transformer model by [11] with notable adaptations. Instead of processing a sequence of actions directly, the TIPI model preprocesses students' click-streams into normalized action count vectors within specific time periods. The initial layer in the TIPI model is a linear layer, not a conventional embedding layer, with nodes equal to the desired embedding dimension. The weights of this linear layer form an array (# of actions, embedding dimension), where each row represents the embedding for a specific action. Subsequently, a sinusoidal positional encoding is added to the token embeddings, following [11]. The resulting embeddings are then input to a series of decoder blocks without cross-attention. Unlike the encoder-decoder structure in [11], this model employs causally masked decoders without cross-attention, similar to [3]. While typically used for next-token prediction, in this study, the model predicts the same value after each step in the sequence, enabling performance estimation after each time period and impact assessment of individual actions or action sequences.

To prevent data leakage and ensure robust model evaluation, 10-fold cross-validation was conducted. Each fold involved hyperparameter tuning on the training data and model training for predicting the held-out fold, ensuring no overlap between training and testing data from the same student. Bayesian optimization was utilized for hyperparameter tuning, with the best set determined based on the lowest mean squared error across 100 models from 10 trials of 10-fold cross-validation. Once hyperparameter optimization converged, the best set was trained on all training data and used to predict the held-out fold 100 times. This process generated 100 predictions for each time period in the dataset for each model architecture, providing a distribution of AUC scores for each data subset and transfer learning approach. Separate hyperparameter searches and model fittings were conducted for each Pharmasim Scenario, detailed in Sect. 2.3.

2.3   Transfer Learning

Due to limited data availability and the success of fine-tuning attention-based models in prior studies [7], this research investigates pre-training (PT) and fine-tuning (FT) the TIPI model. Three models are developed for each PharmaSim scenario to evaluate the impact of pre-training and fine-tuning. The first model (No PT or FT) is solely trained on the specific subset's training data. The second model (PT w/out FT) is trained on the training data for the specific subset and all data from the other subset within the same environment. The third model (PT and FT) uses the same data as the second model, with fine-tuning of the last layer solely on the specific subset's training data. Following the methodology in Sect. 2.2 results in 6,000 models trained (10 folds × 100 trials × 3 training approaches × 2 PharmaSim scenarios). The models' predictions are analyzed to determine the AUC score distribution for different transfer learning approaches at various points before students complete their activities in these environments. For PharmaSim, the AUC score is computed based on predictions made after each minute of student data, comparing predicted success to a binary indicator of whether all essential questions were asked within the first 10 questions.

2.4   Identification of Performance Indicators

The main objective of TIPI is to identify actions that strongly correlate with students' performance. While an accurate model is essential, interpreting the model is equally critical. This study introduces a simple method to interpret the TIPI model and extract insights into impactful actions. Single-action click-stream sequences are generated and processed through the trained TIPI model. The resulting score predictions are aggregated and normalized, providing a clear understanding of each action's influence on the final score prediction relative to other actions.

To evaluate the reliability of this impact measure, models trained in Sect. 2.3 are used to generate impact distributions for each action in each data subset. The PharmaSim (PS) data serves as a test case to assess how well this impact measure aligns with real-world insights. In PS, essential actions crucial for determining student performance are predefined. These essential actions are expected to have the highest impact. A Mann-Whitney U test [8], a non-parametric test comparing two random samples' similarity, is conducted on all models trained with PS-A or PS-B data. This test compares the impact distribution of essential actions to non-essential actions. A significantly higher impact distribution for essential actions across all model trials compared to non-essential actions indicates the model's successful recognition of these actions' importance.

2.5   Identification of Similar Behaviors

In addition to assessing individual actions' impact, this study investigates using linear embeddings from the TIPI model to identify similar actions. To evaluate this approach's efficacy, the Calinski-Harabasz score [4] is computed for the action embeddings of each model trained in Sect. 2.3. This score measures the clarity of cluster distinction based on provided labels for each embedding's cluster membership. Cluster labels are assigned according to the simulation page where the action occurred and whether the action was essential (compendium, diet, drawer, home, pharmacist, shelf, discuss non-essential, or discuss essential). To determine if the Calinski-Harabasz score significantly exceeds what would be expected by chance, indicating that clusters of similar actions are more closely grouped than random assignment, a Fisherian approach is adopted. For each model, action clusters are randomized 100 times, and the Calinski-Harabasz score is computed for each randomization. Subsequently, a Wilcoxon signed-rank test [12], a non-parametric test comparing paired random sample similarities, is used to compare the actual Calinski-Harabasz score distribution with the corresponding randomly-clustered Calinski-Harabasz scores.

3   Results

The results of this study reveal the Transformer-based Identification of Performance Indicators (TIPI) model's effectiveness in identifying behavioral patterns that significantly impact student performance. TIPI effectively distinguished between essential and non-essential actions, aligning its estimations with expert insights and predefined relationships between specific behaviors and student performance. Additionally, TIPI identified which actions are associated with each other through cluster analysis of the action embeddings.

3.1   Model Fitting and Transfer Learning

To address the first research question, this study conducted a comprehensive analysis of the TIPI model, exploring three transfer learning approaches for each PharmaSim scenario. Table 1 displays the mean and standard deviation of the AUC score for each model in early performance prediction, specifically five minutes after each student began. In Table 1, PT indicates pre-training, and FT indicates fine-tuning. The model with the highest AUC score for each data subset is highlighted in bold.

Table 1. Model Fitting Results

Data Source	No PT or FT	PT w/out FT	PT and FT
PS-A	0.499 ± 0.005	$\mathbf{0.525 \pm 0.033}$	0.503 ± 0.012
PS-B	$\mathbf{0.882 \pm 0.009}$	0.728 ± 0.045	0.793 ± 0.002

TIPI demonstrated strong early prediction performance for PS-B but only marginally outperformed random for PS-A, likely due to the limited sample size of 138 students completing PS-A, posing challenges in learning behavioral patterns. Pre-training on combined PS-A and B data enhanced the AUC score for PS-A, but this improvement diminished post fine-tuning on PS-A data. Even with pre-training, fine-tuning the model for estimating PS-A student performance struggled to overcome data limitations. These findings suggest training one large model with all available data for an environment (PT w/out FT) and smaller models for each data subset (No PT or FT) may be the most effective strategy, as fine-tuning was, at best, comparable to training separate models for each subset.

3.2   Identification of Performance Indicators

To address the second research question, the impact of each action was assessed following the methodology in Sect. 2.4. After applying the Benjamini-Hochberg procedure [1] for multiple hypothesis testing correction, all statistical tests comparing essential and non-essential impact distributions were highly statistically significant ($p \approx 0$). Across all model permutations, a significant difference in impact between essential and non-essential actions was observed. Furthermore, the top three actions consistently identified were essential actions. Instances where non-essential actions had higher impact than essential ones often involved additional questions that contributed to a more comprehensive diagnosis, though not technically required. For example, in PS-A, inquiring about symptom duration or intensity, while not essential, had a greater impact on performance than the essential action of asking about the mother's medications. This suggests that students who delve deeper with their questioning tend to perform better by the simulation's end compared to those skipping these in-depth questions. These results affirm that TIPI consistently identifies actions most critical to student performance, proving valuable in data-limited scenarios, regardless of its early performance prediction accuracy.

3.3 Identification of Similar Behaviors

To address the third research question, the methodology described in Sect. 2.5 was implemented to evaluate if the models' embeddings could effectively identify inherent data clusters. Following multiple hypothesis testing corrections, all cluster validity tests were statistically significant ($p < 0.05$), indicating TIPI's ability to discern similar actions through embeddings. Furthermore, the Spearman correlation between action embedding magnitudes and their impact was positive and statistically significant ($p < 1 \times 10^{-4}$) for all models after adjusting for multiple hypotheses, highlighting TIPI's capacity to embed pertinent information effectively.

4 Conclusion

Transformer-based Identification of Performance Indicators (TIPI) demonstrates a promising method for analyzing student behavior in unstructured learning environments by using transformer-based deep learning to interpret click-stream data. This approach has proven effective in predicting student performance and identifying key behavioral patterns across various educational settings, including interactive simulations.

TIPI's ability to predict student performance early, particularly in short-term exercises, opens up possibilities for personalized interventions, such as identifying students who may need additional support based on their online behavior. This could lead to more targeted and effective teaching strategies. However, its application in long-term settings and handling extensive sequences of actions requires improvement.

The potential of TIPI to enhance personalized learning through real-time analysis and intervention suggests a significant step towards more adaptive and responsive educational environments. As TIPI evolves, its contribution to understanding and improving student learning processes is expected to grow, highlighting its importance in the future of education.

Acknowledgements. This project was substantially co-financed by the Swiss State Secretariat for Education, Research and Innovation (SERI). The authors would like to thank Peter Bühlmann and Hugues Saltini who designed and implemented PharmaSim and collected the data necessary for this work, as well Paola Mejia for the idea to bin actions into time-periods.

References

1. Benjamini, Y., Hochberg, Y.: Controlling the false discovery rate: a practical and powerful approach to multiple testing. J. Roy. Stat. Soc.: Ser. B (Methodol.) **57**(1), 289–300 (1995)
2. Botelho, A.F., Baker, R.S., Heffernan, N.T.: Improving sensor-free affect detection using deep learning. In: André, E., Baker, R., Hu, X., Rodrigo, M.M.T., du Boulay, B. (eds.) AIED 2017. LNCS (LNAI), vol. 10331, pp. 40–51. Springer, Cham (2017). https://doi.org/10.1007/978-3-319-61425-0_4
3. Brown, T., et al.: Language models are few-shot learners. Adv. Neural. Inf. Process. Syst. **33**, 1877–1901 (2020)
4. Caliński, T., Harabasz, J.: A dendrite method for cluster analysis. Commun. Stat. Theor. Methods **3**(1), 1–27 (1974)
5. Cock, J.M., Marras, M., Giang, C., Kaser, T.: Generalisable methods for early prediction in interactive simulations for education. In: Mitrovic, A., Bosch, N. (eds.) Proceedings of the 15th International Conference on Educational Data Mining, pp. 183–194. International Educational Data Mining Society, Durham, United Kingdom (2022). https://doi.org/10.5281/zenodo.6852968
6. Corbett, A.T., Anderson, J.R.: Knowledge tracing: modeling the acquisition of procedural knowledge. User Model. User-Adap. Inter. **4**, 253–278 (1994)
7. Devlin, J., Chang, M.W., Lee, K., Toutanova, K.: BERT: pre-training of deep bidirectional transformers for language understanding. arXiv preprint arXiv:1810.04805 (2018)
8. Mann, H.B., Whitney, D.R.: On a test of whether one of two random variables is stochastically larger than the other. Ann. Math. Stat., 50–60 (1947)
9. Swamy, V., Du, S., Marras, M., Kaser, T.: Trusting the explainers: teacher validation of explainable artificial intelligence for course design. In: LAK23: 13th International Learning Analytics and Knowledge Conference, pp. 345–356 (2023)
10. Swamy, V., Radmehr, B., Krco, N., Marras, M., Käser, T.: Evaluating the explainers: black-box explainable machine learning for student success prediction in MOOCs. arXiv preprint arXiv:2207.00551 (2022)
11. Vaswani, A., et al.: Attention is all you need. In: Advances in Neural Information Processing Systems, vol. 30 (2017)
12. Wilcoxon, F., Katti, S., Wilcox, R.A., et al.: Critical values and probability levels for the Wilcoxon rank sum test and the Wilcoxon signed rank test. Sel. Tables Math. Stat. **1**, 171–259 (1970)

Masked Autoencoder Transformer for Missing Data Imputation of PISA

Guilherme Mendonça Freire$^{(\boxtimes)}$ ⓘ and Mariana Curi$^{(\boxtimes)}$ ⓘ

Instituto de Ciências Matemáticas e de Computação, Universidade de São Paulo,
Av. Trab. São Carlense, 400 - Centro, 13566-590 São Carlos SP, Brazil
`guilhermemfreire@usp.br, mcuri@icmc.usp.br`

Abstract. This study introduces a scale-down transformer model to address the challenge of missing responses in educational assessments for psychometric evaluation. Traditional estimation methods for Item Response Theory (IRT) models are frequently computationally inefficient in generating estimates for a higher number of dimensions. The challenge becomes more pronounced when dealing with missing responses. We propose a Masked Autoencoder Transformer model for discrete input (DiT-MAE) to impute missing answers for OECD-PISA response data. The model learns context information from the unmasked parts and reconstructs it using a decoder. For evaluation purposes, we estimate item and person parameters from two different approaches, (i) an adapted Variational Autoencoder that incorporates the Item Response Theory (IRT) method; (ii) the traditional statistical tool, Joint Maximum Likelihood (JML), that can produce estimates in occurrence of missing values.

Keywords: Item Response Theory · Missing Data · Neural Networks · Transformer Model · Variational Autoencoder

1 Introduction

Exams are commonly used to assess students' performance. Item Response Theory (IRT) is a set of mathematical models widely employed in psychometrics for educational and psychological testing. The primary objective of measurement in IRT is to position an individual's ability on a continuous rating scale based on their responses to the test items [15, 19]. However, in real-world situations, test items often require a combination of multiple skills in areas of knowledge for a correct response. Therefore, it's crucial to consider a modeling approach that involves the multidimensionality of abilities (latent variables). The mathematical model for such an approach is an extension of the unidimensional Item Response Theory (IRT) models and is referred to as Multidimensional Item Response Theory (MIRT) [19].

Parameter estimation has traditionally relied on Expectation-Maximization (EM) algorithms when applying unidimensional IRT methods. Methods such as

© The Author(s), under exclusive license to Springer Nature Switzerland AG 2024
A. M. Olney et al. (Eds.): AIED 2024 Workshops, CCIS 2150, pp. 364–372, 2024.
https://doi.org/10.1007/978-3-031-64315-6_33

Monte Carlo (MC), within Metropolis-Hastings Robbins-Monro (MHRM) algorithm, and Joint Maximum Likelihood have been applied to obtain numerical approximations for MIRT models [2,3]. However, these methods have severe limitations in a multidimensional latent trait space due to the computational complexity of evaluating high-dimensional integrals in likelihood functions [2,19]. Recently, researchers have found consistent results incorporating MIRT models into Deep Neural Network (DNN) architectures to overcome computer complexity while dealing with high-dimensional datasets. Results are efficient in parameter estimation and achieve results competitive to traditional techniques with much faster speed [5,7,20].

Incomplete or missing data can make difficult parameter estimations within MIRT models, fostering significant computational challenges. DNNs have also been investigated to address this issue. One approach is to treat missing data as noise and use denoising techniques within Autoencoder architectures to extract high-quality features [4,13,14,18]. [9] have proposed Convolutional Neural Networks (CNNs) models to extract neighboring features for better imputation results. Meanwhile, [11] proposed a scaled-down transformer Masked Autoencoder model to denoise masked parts and learn the context from unmasked parts of the image for classification and recognition tasks.

The paper is organized in Sect. 2 which introduces the concepts of IRT, Missing Data, and Variational Autoencoders. Section 3 explains a two-step process for (i) imputing missing values in the response dataset and (ii) estimating MIRT parameters from the imputed values. Section 4 compares VAEQ architecture and the Joint Maximum Likelihood (JML) method results.

2   Background

Missing data is a major issue and can compromise statistical analysis. Traditional methods often lack empirical evidence and are chosen for their ease of implementation. However, it is essential to recognize that they can significantly compromise the effectiveness of statistical analysis, especially in parameter estimation [1,16].

Multiple Imputation (MI) and Maximum Likelihood (ML) are considered to be "state-of-the-art" techniques for addressing missing data and exhibited the absence of biased interference even when dealing with Missing Completely at Random (MCAR) or Missing at Random (MAR) patterns and outperform traditional techniques by eliminating the need to discard any data [1,8].

The selection between MI and ML depends on various factors when estimating parameters with Multidimensional Item Response Theory (MIRT). It includes the number of items, latent traits, imputation count, and the specific methods chosen for both imputation and estimation. Besides, computational complexity may be implicated as the combination of these factors increases [15,19].

This work centers on the class of dichotomous responses, focusing on the Multidimensional Two Parameters Logistic (M2PL) model. Mathematically, the model is represented by respondents i and items j, where $i = 1, ..., N$ and $j = 1, ..., M$. Each respondent possesses a vector of K abilities, characterized as latent traits, denoted as $\theta_i = (\theta_{i1}, \theta_{i2}, \theta_{i3}, ..., \theta_{iK})$. Additionally, each item is characterized by two parameters: a vector $a_j = (a_{j1}, ..., a_{jK})$ discriminating the probability of success based on the examinee's ability to answer the item correctly, and b_j that represents the difficulty of the item [15,19]. Equation 1 provides the mathematical representation:

$$P(u_i = 1|\theta_i) = \frac{e^{a'_j \theta_i + b_j}}{1 + e^{a'_j \theta_i + b_j}}, \tag{1}$$

where $P(u_i = 1|\theta_i)$ is the probability of respondent i provide a correct response to item j.

The Joint Maximum Likelihood (JML) method is a well-known and common approach employed to estimate probability distribution parameters by maximizing the likelihood function of observed data in a limited scenario [2,3]. When missing values are present, it is treated as unobserved random parameters and estimates of item and person parameters are done simultaneously by solving an optimization problem.

Deep Learning (DL) has shown a major potential to solve several problems and has been investigated to bypass computer complexity while processing large and high-dimensional datasets. Researchers have delved into DL integration with traditional statistical methods, such as Item Response Theory (IRT) methods, aiming to enhance the extraction, estimation, and interpretability of IRT parameters, particularly in high-dimensional scenarios [6,7,17]. In the subarea of DL, generative models such as Variational Autoencoders [12] have been adapted to incorporate IRT methods due to the possibility of compressing information to latent space and extracting interpretable information based on students test response [6,7,10].

The imputation model for missing values draws inspiration from the work proposed by [11]. Their approach is based on a scaled-down transformer model [21] with self-supervised learning for computer vision using masked autoencoders (MAE). The approach consists of randomly masking patches of the input image and reconstructing the missing pixels using an asymmetric encoder-decoder architecture. The encoder operates only on the visible subset of patches (without masked tokens), while the decoder reconstructs the original image from the latent representation and masked tokens. The authors have discovered that masking 75% of the input image is a crucial aspect in creating a significant self-supervisory task. The approach enables efficient and effective training of large models, resulting in improved accuracy and transfer performance in downstream tasks.

3   Methods

This section will cover the methods and process for imputing missing data in OECD-PISA datasets and estimating item and person parameters with a MIRT model adapted to a neural network architecture. The PISA cognitive science

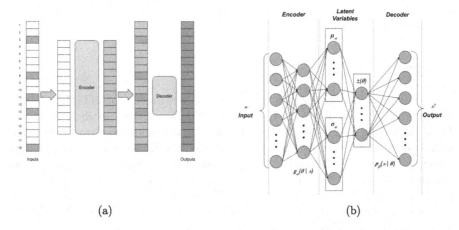

(a)   (b)

Fig. 1. (a) Discrete Transformer Masked Autoencoder. Adapted from [11]. (b) VAEQ architecture proposed by [7].

dataset evaluates student cognitive knowledge from countries members of the OECD[1].

Initially, the imputing process is done by the Dit-MAE (Discrete Transformer Masked Autoencoder) model. Firstly, the input set of eighteen values, which are item responses, will be randomly masked, simulating missing values. the masking portion of discrete data was tested for 10%, 25%, and 50%. The unmasked subset is fed to the encoder model, a scaled-down version of the transformer model, containing self-attention mechanism to learn context information from unmasked data [21]. Secondly, the encoder's learned representations and masked values are combined forming again eighteen values to feed the decoder for reconstruction. Figure 1a illustrates the DiT-MAE model. The model consists of 18 items of input data, represented by rectangles. The red rectangles indicate missing responses.

A main feature of transformer models is the ability to learn data semantics through positional encoding. Every learning step is updated into the unmasked positional embedding. This allows the model to capture the spatial relationships between different elements in the data. In our model, the learned values do not update at a positional layer. This design choice is influenced by the nature of binary values, which inherently carry limited information. In a binary sequence input, where missing answers may vary in index position, the model adapts its learning based on the context provided by unmasked answers from each student. Our code is available at Github repository[2].

The next step is replacing the original dataset's missing values with imputed data from the DiT-MAE model. This will ensure that we have a complete and reliable dataset to work with. The imputed dataset can then be confidently fed

---

[1] https://webfs.oecd.org/pisa/PUF_SPSS_COMBINED_CMB_STU_COG.zip.
[2] https://github.com/guilhermemfreire/DiT-MAE/tree/main.

into the VAEQ (Fig. 1b) model and JML method to evaluate item and person parameters estimation.

4   Results

As part of our evaluation process, we tested the performance of DiT-MAE in predicting missing responses across five large, high-dimensional artificial datasets. These datasets have 21 dimensions, consist of 10,000 examinees, and vary in the number of items of 90, 180, 270, and 360. We fed the datasets containing missing values to both DiT-MAE and JML, since JML can produce IRT estimates from incomplete datasets. We then used the imputation results from DiT-MAE to feed into the VAEQ model from [7] for IRT parameter estimation. The last step was to correlate VAEQ and JML estimates with the true values of each IRT parameter for evaluation purposes.

Table 1. Correlation between artificial true values and VAEQ estimated parameters of discrimination (left), difficulty (center), and ability (right).

	90	180	270	360		90	180	270	360		90	180	270	360
10%	0.69	0.87	0.90	0.93		0.97	0.97	0.97	0.96		0.70	0.79	0.81	0.83
25%	0.64	0.81	0.87	0.92		0.97	0.96	0.96	0.96		0.67	0.75	0.78	0.81
50%	0.47	0.70	0.72	0.76		0.96	0.95	0.96	0.94		0.59	0.71	0.74	0.74

Table 2. Correlation between artificial true values and JML estimated parameters of discrimination (left), difficulty (center), and ability (right).

	90	180	270	360		90	180	270	360		90	180	270	360
10%	0.43	0.59	0.51	0.42		0.78	0.77	0.87	0.94		0.60	0.71	0.78	0.83
25%	0.44	0.58	0.53	0.42		0.79	0.81	0.83	0.87		0.57	0.67	0.75	0.74
50%	0.41	0.50	0.59	0.55		0.84	0.86	0.85	0.83		0.50	0.59	0.66	0.73

Tables 1 and 2 present three ranges of missing values in the dataset and the correlation between artificial true and estimated values of discrimination, difficulty, and ability from VAEQ model and JML method, respectively. Table 2 shows a notable decrease in correlation between true values and estimates as the number of items and the range of missing values increases.

To evaluate the real-world scenario, we used the educational assessment PISA dataset for training, validating, and testing the DiT-MAE model comprises 2,154 students with four dimensions (abilities). The imputation process during training and validation has yielded satisfactory results. Notably, the model's performance was evaluated under three scenarios of missing values. For a scenario with 10%

missing values (2 missing values in 18 items for each individual), the validation process returned a 4% loss and a 98% binary accuracy. In the case of 25% missing values (4 missing values in 18 items), the validation yielded a 10% loss and a 95% binary accuracy. Under the condition of 50% missing values (9 missing values in 18 items), the validation exhibited a 17% loss and a 93% binary accuracy. Figure 2 illustrates a cross-table chart presenting the number of successes of the imputation process using DiT-MAE model.

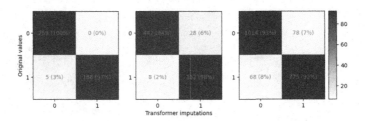

Fig. 2. 10%, 25%, and 50% missing values, respectively, imputation success from DiT-MAE model.

We also impute real missing values from PISA dataset, selecting examinees with missing answers ranging from 1 to 6. The dataset size has 1,427 persons and is aimed at correlating the estimated parameters from the VAEQ and JML method and comparing the results of these parameters between the dataset with simulated (masked) and real missing values.

The correlation between the JML and VAEQ estimates of the test dataset (which includes masked data to simulate missing values) is illustrated in Fig. 3a. When the range of missing values is small, there is a visible correlation between the two approaches. However, as the range of missing values increases to 50%, the correlation becomes scattered, with most of the effect being horizontal and influenced by the JML estimates. It is worth noticing that the test dataset has the same number of missing values for all examinees. In opposition, the number of missing values ranges for each person in the real incomplete dataset. This issue can induce more scattered values in the correlation chart, such as shown in Fig. 3b. Nevertheless, from both figures, it is observable that VAEQ maintains a vertical grouping of values, concluding a more solid result of abilities' estimation probabilities.

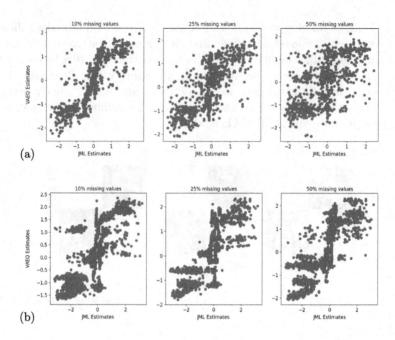

Fig. 3. (a) Scatter plot for all theta estimations in different ranges of test missing values. (b) Scatter plot for all theta estimations in different ranges of real missing values.

5   Conclusion

In this study we presented a two-model process for the psychometric analysis of an educational assessment, the OECD-PISA response dataset. The first model focused on the imputation process, utilizing a test dataset with known missing responses to validate the viability of the DiT-MAE model. The second model involved estimating item and person parameters, comparing the outputs of a traditional statistical method, Joint Maximum Likelihood (JML), with a neural network adaptation of Item Response Theory (IRT) methods using the VAEQ model.

The imputation process demonstrated satisfactory results, as reported. Further investigation may be warranted to comprehensively understand the factors contributing to the observed variations in the scatter plots across different missing value rates. Future work could explore techniques adapting the self-attention mechanism to a relative self-attention, considering that encoding position may not be crucial for binary input of missing values. Additionally, identifying better hyperparameters for the model could enhance its performance. Another avenue for future research involves evaluating the outputs of the VAEQ model with real values of item parameters for the PISA test, providing a more realistic and context-specific validation.

This study contributes to the evolving field of psychometric analysis in educational assessment by integrating traditional statistical methods with advanced neural network architectures, offering insights into the imputation of missing values and the estimation of latent abilities.

Acknowledgments. This study was financed in part by the Coordenação de Aperfeiçoamento de Pessoal de Nível Superior - Brasil (CAPES) - Finance Code 001.

References

1. Allison, P.D.: Missing data. The SAGE Handbook of Quantitative Methods in Psychology, pp. 72–89 (2009)
2. Cai, L.: High-dimensional exploratory item factor analysis by a metropolis-hastings robbins-monro algorithm. Psychometrika **75**, 33–57 (2010)
3. Chen, Y., Li, X., Zhang, S.: Joint maximum likelihood estimation for high-dimensional exploratory item factor analysis. Psychometrika **84**, 124–146 (2019)
4. Chen, Z., Liu, S., Jiang, K., Xu, H., Cheng, X.: A data imputation method based on deep belief network. In: 2015 IEEE International Conference on Computer and Information Technology; Ubiquitous Computing and Communications; Dependable, Autonomic and Secure Computing; Pervasive Intelligence and Computing, pp. 1238–1243. IEEE (2015)
5. Converse, G., Curi, M., Oliveira, S.: Autoencoders for educational assessment. In: Isotani, S., Millán, E., Ogan, A., Hastings, P., McLaren, B., Luckin, R. (eds.) AIED 2019. LNCS (LNAI), vol. 11626, pp. 41–45. Springer, Cham (2019). https://doi.org/10.1007/978-3-030-23207-8_8
6. Converse, G., Curi, M., Oliveira, S., Templin, J.: Estimation of multidimensional item response theory models with correlated latent variables using variational autoencoders. Mach. Learn. **110**, 1463–1480 (2021)
7. Curi, M., Converse, G.A., Hajewski, J., Oliveira, S.: Interpretable variational autoencoders for cognitive models. In: 2019 International Joint Conference on Neural Networks (IJCNN), pp. 1–8. IEEE (2019)
8. Enders, C.K.: Applied missing data analysis. Guilford Publications (2022)
9. Gad, I., Hosahalli, D., Manjunatha, B., Ghoneim, O.A.: A robust deep learning model for missing value imputation in big NCDC dataset. Iran J. Comput. Sci. **4**, 67–84 (2021)
10. Hasan, M., Deng, L.Y., Sabatini, J., Bowman, D., Yang, C.c., Hollander, J.: Effect of q-matrix misspecification on variational autoencoders (VAE) for multidimensional item response theory (MIRT) models estimation. In: Proceedings of the 15th International Conference on Educational Data Mining, p. 811 (2022)
11. He, K., Chen, X., Xie, S., Li, Y., Dollár, P., Girshick, R.: Masked autoencoders are scalable vision learners. In: Proceedings of the IEEE/CVF Conference on Computer Vision and Pattern Recognition, pp. 16000–16009 (2022)
12. Kingma, D.P., Welling, M., et al.: An introduction to variational autoencoders. Found. Trends® Mach. Learn. **12**(4), 307–392 (2019)
13. Lin, J., Li, N., Alam, M.A., Ma, Y.: Data-driven missing data imputation in cluster monitoring system based on deep neural network. Appl. Intell. **50**(3), 860–877 (2020)
14. Lin, W.C., Tsai, C.F., Zhong, J.R.: Deep learning for missing value imputation of continuous data and the effect of data discretization. Knowl.-Based Syst. **239**, 108079 (2022)

15. Linden, W.J., Hambleton, R.K.: Handbook of modern item response theory. Springer Science & Business Media (2013). https://doi.org/10.1007/978-1-4757-2691-6

16. Little, R.J., Rubin, D.B.: Statistical Analysis with Missing Data, vol. 793. John Wiley & Sons (2019)

17. Liu, T., Wang, C., Xu, G.: Estimating three- and four-parameter MIRT models with importance-weighted sampling enhanced variational autoencoder. Front. Psychol. **13** (2022). https://doi.org/10.3389/fpsyg.2022.935419, https://www.frontiersin.org/journals/psychology/articles/10.3389/fpsyg.2022.935419

18. Pereira, R.C., Santos, M.S., Rodrigues, P.P., Abreu, P.H.: Reviewing autoencoders for missing data imputation: technical trends, applications and outcomes. J. Artif. Intell. Res. **69**, 1255–1285 (2020)

19. Reckase, M.D.: 18 multidimensional item response theory. Handbook Statist. **26**, 607–642 (2006)

20. Urban, C.J., Bauer, D.J.: A deep learning algorithm for high-dimensional exploratory item factor analysis. Psychometrika **86**(1), 1–29 (2021)

21. Vaswani, A., et al.: Attention is all you need. In: Advances in Neural Information Processing Systems, vol. 30 (2017)

Effective and Scalable Math Support: Experimental Evidence on the Impact of an AI-Math Tutor in Ghana

Owen Henkel[1]([✉]), Hannah Horne-Robinson[2], Nessie Kozhakhmetova[2], and Amanda Lee[3]

[1] University of Oxford, Oxford, UK
owen.henkel@education.ox.ac.uk
[2] Rising Academies, Accra, Ghana
hannah.horne-robinson@risingacademies.org
[3] J-PAL North America, Cambridge, USA
alee@povertyactionlab.org

Abstract. This study is a preliminary evaluation of the impact of receiving extra math instruction provided by Rori, an AI-powered math tutor accessible via WhatsApp, on the math performance of approximately 500 students in Ghana. Students assigned to both the control and treatment groups continued their normal classes with identical curricula and classroom hours. Students in the treatment group were given access to a low-cost smartphone for one hour a week – during a monitored study hall period – and used Rori to independently study math. All other aspects of the groups' in-school experience were the same. We find that the math growth scores were substantially higher for the treatment group and statistically significant ($p < 0.001$). The effect size of 0.36 is considered large in the context of educational interventions: approximately equivalent to an extra year of learning. Importantly, Rori works on basic mobile devices connected to low-bandwidth data networks, and the marginal cost of providing Rori is approximately $5 per student, making it a potentially scalable intervention in the context of LMICs' education systems. While the results should be interpreted judiciously, as they only report on year 1 of the intervention, they do suggest that chat-based tutoring solutions leveraging AI could offer a cost-effective and scalable approach to enhancing learning outcomes for millions of students globally.

Keywords: LLMs · conversational agents · mobile-learning

1 Introduction

Fewer than 15% of students in Africa achieve minimum proficiency in math by the end of middle school [1]. Most students in the region are taught content that surpasses their ability level, have limited opportunities to practice new skills, and often do not receive the necessary feedback due to large class sizes and under-resourced teachers [1]. Research has long suggested that high-quality one-on-one instruction can significantly improve educational outcomes [2–4]. However, in many Low and Middle-Income

Countries (LMICs), including West Africa, the low supply and high cost of quality tutors mean that many learners cannot access this type of support [5]. Thus, research is needed to identify affordable and feasible interventions appropriate for this space.

In recent years, interest has been increasing in technology-enabled approaches such as adaptive learning environments and virtual tutors to offer additional support to learners where personal tutors are not available. These approaches have been used to varying degrees of success in supplementing traditional instruction, but, when they work, they are extremely cost-effective, a consideration that is particularly important in LMIC contexts. However, a major hurdle to extending the use of these types of tools to LMICs, has been the lack of widespread access to personal computers and reliable, high-speed internet that these learning environments typically require. In West Africa, for instance, less than 20% of the population has access to home computers and robust internet connections [6]. However, mobile phone usage is remarkably high in the region, with data indicating that around 90% of West Africans own a mobile phone and reside in areas with 3G coverage [6, 7]. This trend suggests the possibility of implementing interactive educational experiences on mobile platforms, circumventing the need for expensive personal computers and high-speed internet.

Rising Academies, an educational network based in Ghana, has created Rori, a free AI-powered math tutor available on WhatsApp. Inspired by successful elements of Teaching at the Right Level and Intelligent Tutoring System, Rori has students work their way through a curriculum of over 500 micro-lessons, based on the Global Proficiency Framework, an internationally recognized set of math learning standards. Each micro-lesson includes a brief explanation and a series of scaffolded practice questions tied to a specific learning standard. If a student struggles with a question, they receive a hint first and then a worked solution. A key component of Rori is the use of various natural language processing (NLP) methods, including specialized language models (LLMs). This enables students to chat with Rori using natural language, rather than simple navigating a pre-defined button-based dialog. The natural language interface of Rori positions it to both better interpret students' answer attempts and questions, and to engage students in more open-ended conversations about math goals and metacognitive skills. For more context you can watch this 2-min video.

To evaluate if receiving extra math instruction provided by Rori impacts math learning, students in grades 3–9 from Rising Academies' schools in Ghana participated in this study. The schools were divided into treatment and control groups. The treatment group, as a supplement to regular math instruction, engaged in two 30-min weekly sessions with Rori during their study hall period. The control group continued regular math instruction and did not receive access to Rori during study hall. Students in both groups were given a math assessment twice over 8 months which made it possible to assess the impact of Rori on math performance while controlling for confounding variables through the comparison of similar schools (i.e., a difference-in-difference measure between the two groups).

2 Prior Work

2.1 ITSs and Adaptive Learning Environments

While high-dosage tutoring is a highly effective tool for enhancing student learning [2, 4], its scalability is hindered by operational and financial constraints. Advancements in technology have led to attempts to replicate this effect by delivering personalized education through Intelligent Tutoring Systems (ITSs) which aim to adapt the learning experience to individual student needs, customizing the content and pace, thus emulating one-on-one tutoring [8]. These systems have been implemented in various educational contexts, such as K-12 classrooms and universities, and there is evidence that they can be effective, particularly for the teaching of mathematics and science [9]. In addition to ITSs, other related approaches have emerged in the field of educational technology. Adaptive learning systems, for example, use algorithms to tailor content and pace to students' needs and typically allow students to move through the curriculum in a flexible and personalized manner [10]. While opinions vary on their effectiveness, these approaches hold potential in addressing several educational challenges, particularly in the context of resource-constrained education systems where increasing instructional time, or providing high-dosage tutoring would not be possible [11].

2.2 Teaching at the Right Level

Despite a successful push to increase school enrolment in LMICs, this has not always translated into improvements in learning for all students [1]. Practically, many national curricula primarily cater to high-achieving students, neglecting most children who lag behind. Various factors at both school and home contribute to this predicament [12]. One effective approach to this challenge has been Teaching at the Right Level (TaRL) [13]. First, children's learning levels are evaluated, then they are grouped based on those levels rather than age or grade, and taught at their current ability level, using a curriculum that prioritizes foundational skills [12]. Randomized evaluations conducted in India over 2001–2020, Western Kenya in 2005, and Botswana in 2020 consistently show that programs employing these strategies improve learning outcomes. For example, research by Nobel laureates Abhijit Banerjee and Esther Duflo demonstrated a 0.28 SD improvement in learning outcomes, equivalent to an additional 1.5 years of schooling [13]. While extremely effective, inventions such as TaRL typically require a non-trivial amount of teacher training as well as requiring operational changes at the school, which present challenges to wide-spread adoption. As result, there has been interest in seeing if technology-supported inventions can partially emulate the successful components.

2.3 Technology Supported Learning in LMICs

The personalization inherent in Intelligent Tutoring Systems has substantial similarities with the TaRL approach, and some interventions have tried to leverage technology to implement individualized teaching strategies [13, 14]. A study of a computer-assisted intervention using similar principles resulted in a 0.47 SD improvement, approximately

2.5 years of schooling, and a recent randomized control trial on a computer-assisted program estimated a 0.6 SD improvement in math scores over a year, or approximately an additional three years of schooling [14]. However, students in resource-constrained settings often don't have access to home computers and the internet, prompting exploration for more cost-effective and accessible educational solutions. There have been attempts to explore whether phone-based interventions can offer a viable alternative for delivering educational content and support, particularly for cases of disruption in traditional educational settings [15]. For instance, in Kenya, M-Shule offers SMS-based micro-courses and personalized tuition tools to support primary school students in English, Kiswahili, and Math [16]. However, other studies found that another mobile-based educational intervention did not produce learning gains [17]. Considering the diverse findings, our research seeks to contribute to the growing body of literature on mobile learning in LMICs by examining the impact of Rori on math learning gains. We aim to shed light on the potential of such interventions to provide a cost-effective and accessible means of delivering educational content and support to students in these low-resource settings.

3   Current Study

3.1   Overview

Students from 11 of Rising Academies' schools in Ghana were involved in the study. These schools were selected based on their similarities in geography, demographics, curricula, and teaching methodologies. Schools were randomly assigned to two groups: a control group, which included students from six schools; and a treatment group, which included students from five schools. The assignment was done as the school level, as assignment at the student level was not feasible for a variety of reasons. The treatment group participated in two 30-min sessions with Rori every week, during their study hall period. While teachers were present during these sessions to help resolve potential technical issues, and students spent the period working independently on Rori; watch a clip of a Rori session. Meanwhile, the students in the control group did not have access to Rori during their study hall period and continued to receive their regular math instruction. Both groups completed math assessment in early February (baseline) and late August (endline) of 2023. The same math assessment was used at both timepoints and all grade levels. The assessment consisted of 35 questions, each worth one point, with a mixture of multiple-choice and open-response questions covering numeracy and algebra skills from grades 3 to 5 on the Global Proficiency Framework. Rori is an intervention that is characterized by its low cost and minimal disruption to school operations. Other potentially efficacious interventions such as extended instructional time, smaller classroom sizes, or high-dosage tutoring would simply not be feasible due to the associated costs. Resultingly, we compare the incremental impact of Rori relative to normal schooling, which allows us to understand the specific impact of potentially scalable intervention in enhancing educational outcomes.

3.2   Participation

Initially, 637 students in grades 3–8 participated in the baseline assessment and out of that group, 477 students also completed the endline. Due to having access to the on-platform data, we were able to confirm that students in treatment group used Rori on a regular basis, though have not yet calculated dosage. The attrition of 160 students was primarily due to inconsistent school attendance; it is common for students to miss school periodically due to personal reasons. Of the 477 students who participated in both tests, 241 were in the control group and 236 were in the treatment group. Statistical examinations were conducted to assess any systematic differences between the control and treatment group at baseline and to determine if there were distinctions between students who completed the study and those who did not. We found no systematic difference in control and treatment group, and only small differences in the students who took the baseline but not the endline, which is discussed in detail in the subsequent section.

4   Results

4.1   Estimated Learning Gains

To assess the impact of using Rori on learning, growth scores were computed by subtracting baseline raw scores, the number of questions answered correctly, from endline raw scores for each student who completed both tests. An independent samples t-test between the control (M = 2.12, SD = 6.30) and the treatment group (M = 5.13, SD = 7.03) revealed that the 3.01 difference in growth scores was highly statistically significant (p < 0.001) (Table 1).

Table 1. Descriptive statistics for each condition.

Condition	Control (N = 241)		Treatment (N = 237)	
Test	Baseline	Endline	Baseline	Endline
Mean	20.20	22.32	20.29	25.42
St Dev	8.81	8.06	8.72	7.25
Growth	2.12		5.13	

To understand the magnitude of the difference in learning gains between the treatment and control groups, we calculated the effect size using the pooled standard deviation from both conditions and time points, as suggested by Morris [18]. The calculated effect size between the baseline and endline, expressed as Cohen's d, was found to be 0.36. This effect size would be considered a moderate to large effect size in educational research. Hattie et al. would categorize 0.29 as moderate, and similar to the magnitude of the effect of a good teacher [19]. Whereas Kraft proposes argued that educational interventions with an 0.20 SD of over should be considered as large [20]. Another way to get a sense of the magnitude of the learning gains is to compare the learning gains of the control group (2.12) – which is indicative of the amount of improvement due to normal classes and potential retest effects – to that of the treatment group (5.13), which suggests the differential impact of Rori might be equal or greater than a year of schooling.

4.2 Comparing Characteristics of Treatment and Control Group

To establish baseline equivalence and ensure comparability of the two groups, baseline scores, gender and age were compared. No statistically significant difference was found in the baseline scores of the two groups (t(475) = -0.17, p = 0.87), suggesting that the two groups had equivalent levels of math knowledge at the start of the evaluation. Gender distribution was compared across the two groups. The chi-square test for the association between gender and condition yielded a non-significant result (χ^2 = 0.93, p = 0.33), suggesting no statistically significant difference in gender distribution between the control and treatment groups. Age variation was also examined, and an independent samples t-test was conducted, and the results indicated no statistically significant difference in mean age between the control and treatment groups (t (475) = 1.23, p = 0.22).

Overall, these analyses indicate that the control and treatment groups had statistically similar baseline scores and were balanced across age and gender. The similarity of the control and treatment group combined with the fact that the results report a difference in difference score (i.e., a "growth" score), suggest that the improvement in learning observed is attributable to the intervention (i.e., using Rori) rather than other measured factors (e.g., prior aptitude, gender, age).

4.3 Comparing Students Who Completed Both Tests vs Those Who Did not

The baseline was taken by 637 students in grades 3–8 (336 in Treatment, 301 in Control), of which 477 took the endline and 160 did not. The mean age for the students who took both tests was 12.10 (SD = 1.95) and those who dropped out was 11.93 (SD = 2.15) and there was no statistically significant difference between the two groups (t(635) = 0.88, p = 0.38). The gender distribution was compared between the students who completed both assessments and those who dropped out. The chi-square test for the association between gender and test completion yielded a non-significant result (χ^2 = 0.766, p = 0.38). The mean raw baseline score of those that completed both tests was 22.26 (SD = 7.57) and the mean score of those that dropped out was 18.53 (SD = 7.70). While there was a difference in baseline ability between these two groups of students, the reported effect size of the intervention is calculated based on growth scores of students who completed both tests, so it would not be impacted by this difference. None-the-less, this difference needs to be investigated as there are many possible explanations that are not within the scope of this paper.

5 Discussion

5.1 Implications

The initial evidence reveals that receiving extra math practice with Rori led to substantial learning gains across multiple grade levels. The intervention's effectiveness in improving math skills is not confined to a narrow age band but is instead spread across a broad spectrum, indicating that it can adapt to different learning stages without the need for considerable financial investment. While the results should be interpreted with caution, they do suggest that chat-based tutoring solutions leveraging AI could offer a cost-effective and operationally efficient approach to enhancing learning outcomes for students globally.

In a policy context, the effect size of an intervention like Rori is favorable in relation to its cost to implement. The cost structure (i.e. LLM API, mobile device, 3G data costs) presents a scalable model, with initial setup expenses offset by the possibility of achieving an annual cost as low as $5 per student. Therefore, it could be a financially viable solution for resource-constrained educational settings. In addition to cost, interventions can struggle to scale because of the difficulties of replicating technical and support infrastructure in new geographies. Relative to changing curricula or teaching practices, Rori is relatively straightforward to implement once schools have access to mobile phones, as it does not require additional teaching staff or extensive training. Relying on widely used and accessible devices, such as budget phones and chat platforms like WhatsApp, minimizes dependence on existing infrastructure, avoiding the limitations posed by the lack of personal computers and reliable high-speed internet.

5.2 Limitations

While the experimental design of this study was robust and the learning gains were both large and statistically significant, there are certain limitations. With educational interventions there is always a potential for unobserved differences between the treatment and control group. While schools were randomly assigned to treatment and control groups and both had the same curriculum, lesson plans, and timetable, it is possible that the school administrators and teachers at the treatment schools changed their behavior because of increased observation (i.e. Hawthorne effect). However, schools in the Rising Academies Network are accustomed to routine testing and observation, so this is less likely to be a potential confound. Relatedly, while multiple analyses were conducted to investigate baseline equivalence of the two groups, there may have been unobserved differences in students' prior ability, or subtle differences in their socio-economic background that could have influenced the results. Another potential limitation was the design of the assessment. To make tracking student progress comparable across grades, the same assessment was used for all students. This led to some students, particularly those in higher grade levels, receiving perfect scores at baseline, so it was not possible to observe their improvement over time. While this would suggest that growth scores might have been *understated*, it is also possible that Rori is only effective in helping students with relatively easier math topics, and that the learning gains might decrease if older students were tested on more difficult topics.

5.3 Further Research

There are several opportunities for future research, as mentioned an examination of the observed attrition in this study is warranted. Another area of investigation that was outside the scope of this study was the specific relationship between time spent using Rori and improved math ability. While we were able to confirm that students in the treatment group used Rori weekly, at the time of writing, we have not yet been able to analyze exactly how much time each student spent on Rori, and how this impacted overall learning gains. It is plausible that student learning was an approximately linear function of the amount of time students spent using Rori, but it is equally plausible that the relationship is nonlinear and that the beneficial effect of Rori plateaus after a certain amount of usage. Subsequent research investigating dosage over time might also reveal interactions with learning gains, such as identifying an optimal amount of interaction time before reaching diminishing returns in performance or potential interference with learning in other subjects. Future studies could also employ more diverse and comprehensive assessment tools, reducing the ceiling effects observed in this study and enhancing the validity and reliability of student evaluation. Finally, future studies could investigate the impact of Rori in other schools outside of Rising Academies Network or in other countries. Rori could also be used by children at home. This would contribute to the generalizability of this study's findings that using Rori improves math learning.

6 Conclusion

The present study explored the impact of the artificial intelligence conversational math tutor accessible through WhatsApp on the math performance students in Ghana. We found that the treatment group's math growth scores were markedly higher, with an effect size of 0.36, and data exhibiting statistical significance ($p < 0.001$). Given its ability to operate on basic mobile devices on low-bandwidth data networks, the intervention exhibits substantial potential in supporting personalized learning in other LMICs. In such regions, the possession of laptops and access to high-speed internet connectivity, requisites for many video-centered learning platforms, remain significantly limited. However, it is prudent to interpret these results with caution. As this study provides insight from just the first year of intervention in a school-based context, it underlines the necessity for additional, future research to ascertain the conditions essential for ensuring its successful implementation, but also identifies the importance of a personalized, adaptive approach to enrich students' learning experience. Nevertheless, the results presented here highlight the potential of chat-based tutoring solutions leveraging artificial intelligence to offer a cost-effective strategy to boost learning outcomes for disadvantaged students globally.

References

1. World Bank, UNESCO, UNICEF, USAID, FCDO, Bill & Melinda Gates Foundation, The State of Global Learning Poverty: 2022 Update (2022)
2. Bloom, B.S.: The 2 sigma problem: the search for methods of group instruction as effective as one-to-one tutoring. Educ. Res. 13(6), 4 (1984). https://doi.org/10.2307/1175554
3. Vadasy, P.F., Jenkins, J.R., Antil, L.R., Wayne, S.K., O'Connor, R.E.: The effectiveness of one-to-one tutoring by community tutors for at-risk beginning readers. Learn. Disabil. Q. 20(2), 126–139 (1997). https://doi.org/10.2307/1511219
4. Chi, M.T.H., Siler, S.A., Jeong, H., Yamauchi, T., Hausmann, R.G.: Learning from human tutoring. Cogn. Sci. 25(4), 471–533 (2001)
5. Bray, M.: Shadow education in Sub-Saharan Africa: scale, nature and policy implications. (2021)
6. Overview of state of digital development around the world based on ITU data. International Telecommunication Union (ITU) (2022)

7. Silver, L., Johnson, C.: Internet Connectivity Seen as Having Positive Impact on Life in Sub-Saharan Africa. Pew Research Center

8. Vanlehn, K.: The relative effectiveness of human tutoring, intelligent tutoring systems, and other tutoring systems. Educ. Psychol. **46**(4), Art. no. 4 (2011)

9. Steenbergen-Hu, S., Cooper, H.: A meta-analysis of the effectiveness of intelligent tutoring systems on college students' academic learning. J. Educ. Psychol. **106**(2), 331–347 (2014). https://doi.org/10.1037/a0034752

10. Phobun, P., Vicheanpanya, J.: Adaptive intelligent tutoring systems for e-learning systems. Procedia - Soc. Behav. Sci. **2**(2), 4064–4069 (2010)

11. Nye, B.D.: Intelligent tutoring systems by and for the developing world: a review of trends and approaches for educational technology in a global context. Int. J. Artif. Intell. Educ. **25**(2), 177–203 (2015). https://doi.org/10.1007/s40593-014-0028-6

12. Abdul Latif Jameel Poverty Action Lab (J-PAL), Teaching at the Right Level to improve learning.' J-PAL Evidence to Policy Case Study (2018)

13. Banerjee, A.V., Banerji, R., Berry, J., Duflo, E.: Evidence from Randomized Evaluations of 'TARL' in India. MIT Dep. Econ. Work. Pap. No 16–08 (Nov. 2016)

14. Muralidharan, K., Singh, A., Ganimian, A.: Disrupting education? experimental evidence on technology-aided instruction in India. Am. Econ. Rev. **109**(4), 1426–1460 (2019)

15. Kizilcec, R.F., Chen, M., Jasińska, K.K., Madaio, M., Ogan, A.: Mobile Learning During School Disruptions in Sub-Saharan Africa. AERA Open **7**, 233285842110148 (202)

16. K. JordanLancaster University, Lancaster, UK and Learners and caregivers barriers and attitudes to SMS-based mobile learning in Kenya," Afr. Educ. Res. J. **11**(4), 665–679 (2023), https://doi.org/10.30918/AERJ.114.23.088

17. Kizilcec, R.F., Chen, M.: Student Engagement in Mobile Learning via Text Message In: Proceedings of the Seventh ACM Conference on Learning @ Scale, Virtual Event USA: ACM, Aug., pp. 157–166 (2020). https://doi.org/10.1145/3386527.3405921

18. Morris, S.B.: Estimating effect sizes from pretest-posttest-control group designs. Organ. Res. Methods **11**(2), 364–386 (2008)

19. Hattie, J.: Visible learning: a synthesis of over 800 meta-analyses relating to achievement, Reprinted. Routledge, London (2010)

20. Kraft, M.A.: Interpreting effect sizes of education interventions. Educ. Res. **49**(4), 241–253 (2020). https://doi.org/10.3102/0013189X20912798

Interpret3C: Interpretable Student Clustering Through Individualized Feature Selection

Isadora Salles[✉][iD], Paola Mejia-Domenzain[✉][iD], Vinitra Swamy[✉][iD],
Julian Blackwell[iD], and Tanja Käser[✉][iD]

EPFL, Lausanne, Switzerland
isadorasalles@dcc.ufmg.br,
{paola.mejia,vinitra.swamy,julian.blackwell,tanja.kaeser}@epfl.ch

Abstract. Clustering in education, particularly in large-scale online environments like MOOCs, is essential for understanding and adapting to diverse student needs. However, the effectiveness of clustering depends on its interpretability, which becomes challenging with high-dimensional data. Existing clustering approaches often neglect individual differences in feature importance and rely on a homogenized feature set. Addressing this gap, we introduce Interpret3C (Interpretable Conditional Computation Clustering), a novel clustering pipeline that incorporates interpretable neural networks (NNs) in an unsupervised learning context. This method leverages adaptive gating in NNs to select features for each student. Then, clustering is performed using the most relevant features per student, enhancing clusters' relevance and interpretability. We use Interpret3C to analyze the behavioral clusters considering individual feature importances in a MOOC with over 5,000 students. This research contributes to the field by offering a scalable, robust clustering methodology and an educational case study that respects individual student differences and improves interpretability for high-dimensional data.

Keywords: XAI · Clustering · MOOCs · Feature Selection

1 Introduction

Clustering a student population into educationally relevant groups facilitates a deeper understanding of student behaviors and learning patterns, enabling educators to optimize curriculum design and implement group interventions [5, 11,12,15]. The effectiveness of these strategies hinges on the interpretability of the clusters as it directly influences the quality of insights educators can extract from them [5,17]. In large-scale educational settings such as Massive Open Online Courses (MOOCs), where individualized attention is challenging due to the sheer number of participants, clustering provides a feasible approach to understand and address the needs of different student groups.

P. Mejia-Domenzain and V. Swamy—Equal contribution.

A. M. Olney et al. (Eds.): AIED 2024 Workshops, CCIS 2150, pp. 382–390, 2024.
https://doi.org/10.1007/978-3-031-64315-6_35

In these online contexts, abundant and varied features have been used to study student behavior [1–3,7–10,13,21]. A broad spectrum of features is necessary for a comprehensive study of complex student behavior. However, it simultaneously poses substantial challenges in terms of robustness and interpretability due to the *curse of dimensionality*. In high-dimensional spaces, the sparsity of data makes most distance measures less effective, leading to less robust results [6]. In addition, a large number of features can lead to intricate yet less intuitive clustering outputs, thereby complicating the translation of these results into practical educational applications, such as supporting educators to identify and address areas where groups of students may require assistance.

To target the issues of interpretability and robustness, previous works have used a subset of expert-selected features [4,11,12], and data-driven feature selection methods [5,14] before clustering. The learning science-driven approach has been to select a limited number of setting-specific features [2,4,9,11,12]. For example, [4] followed the domain modeling step of the evidence-centered design framework to engineer the features used for clustering. However, the process of hand-picking features is heavily reliant on the expertise and perspectives of the researchers or educators involved. This reliance can inadvertently introduce subjective biases, as the selected features may reflect the specific hypotheses or expectations of the individuals involved, rather than the full spectrum of student behaviors and learning patterns. On the other hand, there are multiple data-driven approaches for feature selection in unsupervised settings, including filter, wrapper, hybrid, and embedded approaches [6]. A significant limitation within these approaches is that the features are chosen averaging over all students, thereby neglecting individual differences in feature importance. This oversight results in an aggregation process that identifies features beneficial on a global scale but fails to account for unique student characteristics.

Addressing this challenge, neural networks (NNs) could be adapted to act as *embedded* methods to perform individual feature selection as part of their training process. Compared to traditional ML methods, NNs are exceptionally good at capturing complex, non-linear relationships and interactions between features. This characteristic could enable NNs to identify important features in datasets where the significance of a feature is not obvious or is dependent on interactions with other features. Moreover, NNs are well suited for high-dimensional data, as they are robust to irrelevant or redundant features. However, NNs face interpretability issues due to their complex, multi-layered "black box" operations [15–17]. To address this issue, [16] proposed an interpretable NN architecture (InterpretCC) that uses a dynamic feature mask to enforce sparsity regularization on the number of input features. This method enables feature importances vectors per student without compromising classification or regression accuracy.

While interpretable-by-design NNs offer a promising solution for dealing with high-dimensional data and individualized feature selection, their application has predominantly been in supervised tasks [13,16,17]. The extension of these methods to unsupervised settings, particularly for interpretable clustering, remains unexplored. This could address the existing gap in the consideration of individual differences in clustering. Traditional feature selection techniques, typically employed before clustering, opt for a global feature selection strategy [6]. Such a strategy creates a homogenized feature set that fails to recognize and incorporate the distinctive attributes of individual students.

This work bridges the aforementioned gaps by developing Interpret3C (Interpretable Conditional Computation Clustering), a clustering pipeline that leverages the strengths of interpretable NNs in an unsupervised learning context. This approach is tailored to adaptively select features per student, thereby facilitating easier and more meaningful interpretation of clusters without compromising the quality and robustness of the clustering process. This is achieved through a deep feature selection method where the individual feature importance masks are extracted from interpretable NNs trained to predict academic performance. Clustering is then performed on the important features. We evaluate our pipeline on a large MOOC with over 5,000 enrolled students and hundreds of thousands of interactions to address the following research question: What kind of student behavioral clusters are identified through inherently interpretable clustering?

2   Methodology

We contribute an inherently interpretable clustering pipeline[1]. Our approach involves the adaptive selection of the most pertinent features for individual students, and uses only these identified features for the clustering process.

2.1   Student Interactions and Feature Extraction

We begin by collecting clickstream data from MOOCs on student interactions from videos and problems, and their associated actions, which include video-related actions. We extract 45 behavioral features from the student interactions[2] that have been found to be predictive in MOOCs [2,3,7,9,18]. The feature sets encompass a wide array of online learning behaviors, including clickstream patterns, study regularity, quiz performance, and video interaction metrics. [2] evaluates study pattern regularity and [3] explores clickstream activities. Video interactions in MOOCs are analyzed in detail by [7], while [9] focuses on features related to quiz performance.

Each feature is computed on a weekly basis, resulting in a time series vector for every student. This transforms each student's behavior into a multivariate time series, where one dimension represents the progression of weeks and the other encapsulates the diverse set of features. We then use unit-norm scaling to normalize the features, preventing vanishing or exploding gradients in the deep feature selection step.

2.2   Deep Feature Selection

We extract the most important features for each student using an interpretable NN architecture. We define feature importance as the degree to which individual

---

[1] https://github.com/epfl-ml4ed/interpretable-clustering/.

[2] The full list of features is available at https://github.com/epfl-ml4ed/interpretable-clustering/blob/main/docs/features-description.pdf.

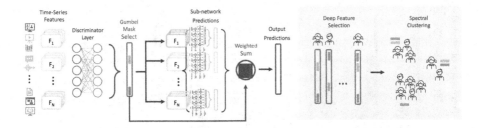

Fig. 1. Interpret3C pipeline with deep feature selection and clustering.

features contribute to the predictive accuracy of the model. Thus, to extract the individual feature importance masks, we train NNs on the supervised task of predicting students' academic performance (pass or fail).

The architecture, an extension of the predictive model presented by [16], leverages a feature gating mechanism that dynamically selects relevant input features for making predictions. As shown in Fig. 1, the features initially pass through discriminator layers, followed by a sigmoid function that produces a feature mask. Features with a sigmoid output of 0.5 or higher are considered activated. Each activated feature is then processed by a dedicated BiLSTM sub-network to predict the student's likelihood of passing or failing. The overall prediction is the average of these individual predictions, weighted by the feature activations.

To encourage interpretability through sparsity, we integrate an annealed mean-squared error regularization on the feature mask. Moreover, the incompatibility of discrete feature gating with backpropagation is overcome by using the Gumbel-SoftMax technique. This technique approximates discrete choices through differentiable functions by adding Gumbel noise and applying softmax as done in [16].

2.3 Clustering

To obtain the clusters, we use as input the masked feature matrix per student containing only the important feature vectors. For instance, if a student s has a feature mask $m_s = (0, 1, 1)$ and their corresponding feature matrix is $F_s = (f_s^1, f_s^2, f_s^3)$, with f_s^i representing the time series vector for each feature i, the resulting masked feature matrix would be $F_m = (-, f_s^2, f_s^3)$. Next, for each feature f, we calculate D_f the Euclidean pairwise distance between students' time series. To calculate distances between vectors in the presence of incomplete data, we impute a vector of zeros. This allows non-missing features to have a stronger contribution to the distance calculation. Following, we compute D as the average of all the features' distance matrix (D_f). Next, we apply a Gaussian Kernel to compute S_f, the similarity matrix of D_f. We use Spectral Clustering to cluster S_f with the number of clusters n as a hyper-parameter. We considered from 3 to 10 clusters and choose n according to the Eigengap heuristic as

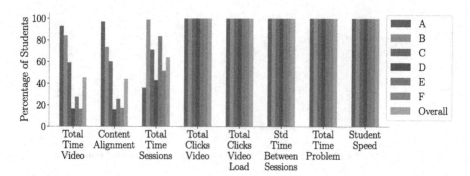

Fig. 2. For each of 8 important features (*x-axis*), the percentage of students (*y-axis*) from each cluster (*color*) that selected the feature as important.

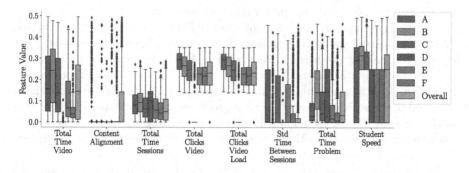

Fig. 3. For each of 8 important features (*x-axis*), the feature value distribution (*y-axis*) from each cluster (*color*).

suggested in [20]. We deliberately choose n to be greater than two to prevent replicating the outcomes of the binary classification task (pass or fail) and obtain additional insights into student behavior.

3 Experimental Evaluation

In our analysis, we trained an `Intepret3C` pipeline to examine the clusters obtained over thousands of students in a MOOC on Digital Signal Processing. In particular, the course is a Computer Science Master's course taught in English with 5,611 active students for 10 weeks. The choice of this course was based on its prior use in other studies [9,18,19].

We use the first four weeks in an early prediction setting to identify six behavioral clusters labeled A through F and ordered according to academic outcomes. Cluster A is the group with the highest percentage of students passing the course, while students in F have the highest failure rate. The distribution of students across these clusters is as follows: Cluster A contains 786 students, equating to 14% of the course's population, with a passing rate of 60%. Clusters B through F comprise 15%, 14%, 14%, 18%, 25% of the student body, respectively, with passing rates of 42%, 31%, 17%, 15%, 11%.

Figure 2 illustrates the distribution of important features within each cluster. For each cluster, it shows the percentage of students for whom each feature was deemed relevant. One initial observation is that out of the 45 original features, only eight were selected as important for at least one student. Out of these eight features, five emerged as significant across all clusters. The features are related to video activities (*Total Clicks Video Load, Total Clicks Video*), quiz interactions (*Time in Problem Sum*, and *Student Speed*) and regularity patterns (*Time Between Sessions Std*). The other three features (*Total Time Video, Content Alignment* and *Total Time Session*) vary considerably between clusters. Only in the best-performing clusters (A, B, and C), *Total Time Video* and *Content Alignment* were selected as important for the majority of the students.

To gain a deeper understanding of the differences between the clusters, Fig. 3 presents a comparison of the average feature values for each important feature across clusters. It also includes the "Overall" category, which shows the average feature values for all students. Different from Fig. 2 where there were slight variations in the percentages of feature importances, in Fig. 3 the distributions vary considerably between clusters. This variation suggests that while a feature may be considered important across clusters, the degree to which it is manifested in student behavior could differ. One specific feature of interest is "Content Alignment", which measures schedule adherence. In Fig. 3, this feature consistently exhibits one of the highest levels of variability, with both the 25th and 75th percentiles at zero for all clusters. This indicates that the vast majority of students did not complete the video content within the designated timeframe. This contrasts with the *Overall* category distribution of the feature where the definition of outlier is above 0.35 (instead of zero). Thus, it seems that *Content Alignment* is mostly considered as an important feature for the students who are behind schedule. For example, in Clusters A and B, it was an important feature for more than 70% of the students. Interestingly, in Cluster B, all students had a value of zero for this feature, while for Cluster A, there were multiple outliers with values greater than zero. This suggests that while this feature was important for both clusters, students in Cluster B were universally off-track, whereas in Cluster A, a subset of the students managed to adhere to the schedule.

An additional observation is the difference in engagement between the best-performing clusters (A and B). As seen in Fig. 3, students in Cluster B dedicate more time to videos and problem-solving than those in Cluster A. However, students in Cluster A exhibit more active engagement, as evidenced by lower *Student Speed* values related to quiz attempt frequency as well as higher values in video-related features such as *Total Video Clicks*.

Furthermore, Fig. 3 reveals a general downward trend in feature values, with high-performing clusters (A, B, and C) having higher values compared to low-performing clusters (E and F). Cluster D, however, deviates from this trend by exhibiting the lowest median values (zero) for all video-related features. Despite this, students in Cluster D are not disengaged from the course as they exhibit the longest time online, attributed primarily to their extensive interaction with quizzes (*Total Time Problems*). The passing rate for Cluster D is 16.8%, which is marginally higher than the 15.2% passing rate of Cluster E. Nevertheless, their interaction with the course material was very different. The students in Cluster D spent a lot of time online interacting with the quizzes, while the students in Cluster E showed lower levels of engagement across both videos and quizzes.

4 Discussion and Conclusion

In this work, we presented a novel methodology and initial evaluation for an interpretable-by-design clustering pipeline, `Interpret3C`.

We performed an in-depth cluster analysis based on student behavior from the first four weeks of a large MOOC course. Our pipeline revealed six diverse behavioral clusters, each characterized by varying levels of interaction with course materials and platform features. The contrast between Cluster D, highly engaged in quizzes, and Cluster E, generally disengaged, illustrates that while the outcomes may appear similar, the underlying behaviors can differ significantly; thus, effective interventions should address the specific needs of each group. Moreover, we found five features important for all students and three features with a varying percentage of importance across clusters. Global feature selection methods [5,6,14] could have also identified the five universal relevant features but would likely have overlooked the nuanced variations in the importance of *Total Time Video*, *Content Alignment*, and *Total Time Sessions* across different clusters.

The quality and relevance of the selected features depends strongly on 1) the signal of the input features and 2) the performance of the discriminator network and subsequent time series networks, which require a relatively complex architecture. Although the utilization of these resources for feature selection may initially appear excessive, the number of parameters is much smaller than a LLM or other models used for educational deep learning [13,18]. In these settings, the training can serve a double purpose: to predict and also to learn about the group behaviors in the class. Furthermore, the deep feature selection process has a bias towards the initial predictive task of the discriminator NN. In our case, the selection mechanism identifies important features related to academic outcomes. This bias towards the initial task is not inherently negative, as the prediction task can be adapted to fit other educational goals (e.g., retention or self-regulated learning skills). Including the outcome measure in unsupervised settings mirrors that of other methodologies, such as supervised PCA and sparse supervised CCA [22], helping to identify features associated with the outcomes and correlations between datasets. Extending the Interpret3C approach to other tasks and interpreting the clusters over a more generalizable set of courses is left as future work.

In conclusion, `Interpret3C` introduces an innovative clustering pipeline by emphasizing the interpretability of high-dimensional data and respecting individual student differences. Through adaptive feature selection, our approach enhances the relevance and clarity of clustering outcomes. Our findings illustrate the potential of `Interpret3C` to identify insightful clusters in MOOC environments and paves the way for more effective and personalized group interventions.

Acknowledgements.. We kindly thank the Swiss State Secretariat for Education, Research and Innovation (SERI) for supporting this project.

References

1. Akpinar, N.J., Ramdas, A., Acar, U.: Analyzing student strategies in blended courses using clickstream data. EDM (2020)
2. Boroujeni, M.S., Sharma, K., Kidziński, Ł., Lucignano, L., Dillenbourg, P.: How to quantify student's regularity? EC-TEL (2016)
3. Chen, F., Cui, Y.: Utilizing student time series behaviour in learning management systems for early prediction of course performance. JLA (2020)
4. Choi, H., Winne, P.H., Brooks, C., Li, W., Shedden, K.: Logs or Self-Reports? Misalignment between behavioral trace data and surveys when modeling learner achievement goal orientation. LAK (2023)
5. Effenberger, T., Pelánek, R.: Interpretable clustering of students' solutions in introductory programming. AIED (2021)
6. Hancer, E., Xue, B., Zhang, M.: A survey on feature selection approaches for clustering. AI Review (2020)
7. Lallé, S., Conati, C.: A data-driven student model to provide adaptive support during video watching across MOOCs. AIED (2020)
8. Lemay, D.J., Doleck, T.: Grade prediction of weekly assignments in MOOCs: mining video-viewing behavior. EIT (2020)
9. Marras, M., Vignoud, J.T.T., Käser, T.: Can feature predictive power generalize? benchmarking early predictors of student success across flipped and online courses. EDM (2021)
10. Mbouzao, B., Desmarais, M.C., Shrier, I.: Early prediction of success in MOOC from video interaction features. AIED (2020)
11. Mejia-Domenzain, P., Marras, M., Giang, C., Cattaneo, A.A.P., Käser, T.: Evolutionary clustering of apprentices' self- regulated learning behavior in learning journals. IEEE TLT **15**(5), 579–593 (2022)
12. Mejia-Domenzain, P., Marras, M., Giang, C., Käser, T.: Identifying and comparing multi-dimensional student profiles across flipped classrooms. AIED (2022)
13. Mubarak, A.A., Cao, H., Ahmed, S.A.: Predictive learning analytics using deep learning model in MOOCs' courses videos. EIT (2021)
14. Peffer, M., Quigley, D., Brusman, L., Avena, J., Knight, J.: Trace data from student solutions to genetics problems reveals variance in the processes related to different course outcomes. LAK (2020)
15. Peng, X., Li, Y., Tsang, I.W., Zhu, H., Lv, J., Zhou, J.T.: XAI beyond classification: interpretable neural clustering. JMLR (2022)
16. Swamy, V., Montariol, S., Blackwell, J., Frej, J., Jaggi, M., Käser, T.: InterpretCC: intrinsic user-centric interpretability through global mixture of experts. arXiv:2402.02933 (2024)
17. Swamy, V., Frej, J., Käser, T.: The future of human-centric explainable artificial intelligence (XAI) is not post-hoc explanations. arXiv:2307.00364 (2023)

18. Swamy, V., Marras, M., Käser, T.: Meta transfer learning for early success prediction in MOOCs. Learning@Scale (2022)
19. Swamy, V., Radmehr, B., Krco, N., Marras, M., Käser, T.: Evaluating the explainers: black-box explainable machine learning for student success prediction in MOOCs. EDM (2022)
20. Von Luxburg, U.: A tutorial on spectral clustering. Stat. Comput. **17**, 395–416 (2007). https://doi.org/10.1007/s11222-007-9033-z
21. Wan, H., Liu, K., Yu, Q., Gao, X.: Pedagogical intervention practices: improving learning engagement based on early prediction. IEEE TLT (2019)
22. Witten, D.M., Tibshirani, R., Hastie, T.: A penalized matrix decomposition, with applications to sparse principal components and canonical correlation analysis. Biostatistics (2009)

Recurrent Neural Collaborative Filtering
for Knowledge Tracing

Russell Moore[✉], Andrew Caines, and Paula Buttery

ALTA Institute, Computer Laboratory, University of Cambridge, Cambridge, UK
{rjm49,apc38,pjb48}@cam.ac.uk
http://alta.cambridgeenglish.org

Abstract. In knowledge tracing (KT), a computer system infers the skill level of a student from their interaction with coursework. This work introduces a new method to KT called recurrent neural collaborative filtering (RNCF) that can separate student learning and task difficulty traits into distinct parameter sets. Using five KT data-sets, with binary and scalar response modes, we show this method can improve upon previous predictive approaches. We illustrate the method's ability to cluster students and exercises into like groups, a result that bears promise for research into personalised learning and curriculum improvement.

Keywords: recurrent neural network · collaborative filtering · knowledge tracing

1 Introduction

Knowledge tracing (KT) is a task in which we infer a student's current state of knowledge or skill by observing their work over time. This lets us predict how well they will do on future tasks and tailor delivery accordingly. While human-like proficiency at tuition is still the goal [1], computerised systems are improving, although at a cost: content-construction is a labour intensive, iterative process, involving walkthroughs of problem-solving with subject matter experts, authoring, coding, testing and tuning of content [10,11]. Historically KT was carried out with hidden Markov models, but with the complexity of human skillsets, richer models bring better performance. Recently recurrent neural networks have been used to improve prediction over older methods.

In this paper we introduce a general-purpose knowledge tracing method called *Recurrent Neural Collaborative Filtering* (RNCF) and compare it to the related Deep Knowledge Tracing (DKT) on scalar and continuous response data. Like DKT, the RNCF method has good predictive ability and can learn about the skill composition of tasks, but it has the advantage of learning student representations too.

2 Background

The task of knowledge tracing (KT) was first addressed in the intelligent tutoring system (ITS) community, where it has been widely studied [2]. Although models

A. M. Olney et al. (Eds.): AIED 2024 Workshops, CCIS 2150, pp. 391–399, 2024.
https://doi.org/10.1007/978-3-031-64315-6_36

cannot capture every nuance of cognition from observations, nevertheless useful models can be built. In a KT setting we wish to predict a student's score on a learning exercise at time $t = n$, given their previous results from $t = 0 \cdots (n-1)$. Typically interactions form a time-series of tuples (s, q_t, y_t), meaning student s attempted exercise q, with result y (pass/fail or a score). Given s and q_n, we must predict y_n.

The historically most common method for KT is *Bayesian Knowledge Tracing* (BKT) [2] in which the student's knowledge state is treated as set of binary variables (unknown or known). Observations are treated as the emissions of a Hidden Markov Model (HMM) with knowledge as the hidden state. A key weakness is that in BKT models tasks need to be meta-tagged with their relevant concepts, as there is no way to infer this.

Deep Knowledge Tracing (DKT), using a recurrent neural network instead of an HMM, has been shown to be superior in prediction to BKT and does not need expert tagging, as it can infer the skill composition of each task. However, in general DKT does not produce a distinct representation for each user.

Here we demonstrate a method that combines DKT with *collaborative filtering* (CF). CF is traditionally found in *recommender systems* to factorise user-product preference [4] from observations (these can then be recombined to predict future preference), and more recently uses neural approaches [5] including recurrent [3]. Meanwhile, in education, CF has been applied with *matrix factorisation* [15], and with specialist neural architectures in static settings [9]. In CF, just a small number of interactions can provide more information than meta data [14], which is a key advantage for the approach.

3   Recurrent Neural Collaborative Filtering (RNCF)

In RNCF the KT task is analogous to *sequence tagging*. Each element in a sequence of interactions is a 5-tuple $[q, s, q_{(-1)}, y_{(-1)}, y]$ where q is the unique ID of the question attempted, s is the unique ID of the attempting student, and y is the observed response (this is the target for the machine learning task). The (-1) subscripts simply denote q and y from the previous timestep, which are included to enable 'teacher forcing' [8].

RNCF acts as a hybrid of collaborative filtering and knowledge tracing. To personalize the knowledge tracing aspect, it must be possible for the system to access the relevant parameters for each individual student. In other words we need to be able to 'swap out' sets of network weights when we focus on each different student. This aspect of the architecture is similar to *Neural Collaborative Filtering* [5]. The actual tracing operation is reminiscent of DKT except for the addition of these student-specific parameters: specifically a vector parameter **g** to represent the specific student's learning rate, that is, their ability to gain skill when exposed to practice.

A schematic view of the RNCF architecture is given in Fig. 1. Each timestep is processed by the *RNCF cell* which comprises two parts: the learning event unit (LEU) and the assessment event unit (AEU). More details are given in the next section. Exercises are modeled in two parts - the β vector that gives the

difficulty of the exercise, and the practice vector r that represents the amount
and type of practice the exercise imparts.

3.1 Model of Learning

We treat a student's interactions with exercises as an alternating sequence of
assessment events and learning events [7]. We make the simplifying assumption
that skill changes only during interactions with learning objects, and there is no
ambient learning or forgetting. However it is possible for learners to lose skill
during an interaction, and so the system can capture interference effects.

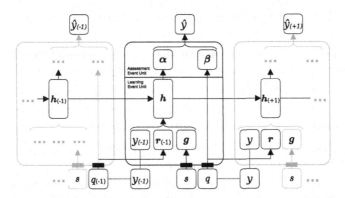

Fig. 1. Schematic of the RNCF architecture, combining learning and assessment sub-
components. The outer box (RNCF cell) encloses the operations taken in a single
timestep. The black bars show an indexing operation, where a numeric ID (s, q) is
used to look up a vector of trainable weights ($\mathbf{g}$, $\mathbf{r}$, β). Subscripts of (-1) and (+1)
denote values for the previous and next timesteps.

3.2 Internal Event Units

The *Learning Event Unit* (LEU) incorporates a recurrent neural network that
maintains for each time-step a hidden state $\mathbf{h}_t$, representing the abilities of a
given student, based on previous observations. In RNCF (unlike in DKT) this
representation is individualised to the student: the weights for the student are
loaded from a table accessed by the student's unique ID. The *Assessment Event
Unit* (AEU) simulates the act of a student attempting a question. The student's
updated vector, output by the LEU, is combined with a difficulty vector β for
the question, using an *interaction function* that can be any bivariate function
that returns a value in $[0, 1]$. We found the best function to be a pseudo-logit
subtraction passed through a weighted sum followed by a sigmoid operation:

$$\hat{y}_t = \sigma(\mathbf{W_y}(\alpha_t - \beta))$$

This forces the components of α and β into an adversarial relationship, similar to that found in compensatory logistic models. This was found to be more
performant than alternative methods during parameter tuning.

Loss Function and Regularisation. for dichotomous data, we treat the problem as binary classification, enforcing a probabilistic output by using a *binary cross-entropy* loss function:

$$\mathcal{L}_{XE}(y,\hat{y}) = -\frac{1}{N}\sum_{i=1}^{N}\left(y_i\log(\hat{y}_i) + (1-y_i)\log(1-\hat{y}_i)\right) + \Lambda$$

For continuous data we treat the problem as a regression task and use mean-squared error as the loss function:

$$\mathcal{L}_{MSE}(y,\hat{y}) = -\frac{1}{N}\sum_{i=1}^{N}(\hat{y}_i - y_i)^2 + \Lambda$$

In both cases, Λ is a regularisation term that governs activation magnitudes using ridge penalty ($P_{ridge}(x) = \lambda_1|x| + \lambda_2 x^2$ for hyperparameter penalty weights λ_1 and λ_2) on the values of α_t and β.

3.3   Hyperparameters and Architecture Search

Hyperparameter search was carried out using Bayesian optimisation with Keras tuner [12]. The parameters searched were:

Regularisation: Ridge penalty weights λ_1 and λ_2 on either exercise, student or neither, ranges 1.0×10^{-9} to 1.0; Recurrent unit choice: GRU, LSTM or RNN; Elementwise combination function: subtract, concatenate; Interaction function: sigmoid of sums, product of sigmoids, dense sigmoid, dense linear, dense ReLU; Representation vector width: 1 to 64; RNN width: 1 to 64; RNN final dropout was either 0 or 0.1.

Each neural network was trained using the Adam optimiser [6] with a batch-size of 100. To prevent overfitting, early-stopping on the validation set is used with patience=10 and best weights are kept. DKT networks were built as per the original paper with a 200-unit LSTM core and 0.1 dropout. For fairness of comparison these were updated to use Adam as the optimiser and early stopping (on a subset of held-out data) to prevent overfitting (training is terminated when validation loss climbs for 10 epochs - best weights are then restored and kept). In all cases LSTM RNN components with a subtraction interaction function and dense sigmoid output outperformed other architectures.

4   Datasets

We test the ability to predict student performance on five datasets, three for binary response prediction and two for scalar score prediction. For the binary prediction task we measure *area under the receiver operating curve* (AUC). For the scalar prediction task we measure *root mean square error* (RMSE) with all scores normalised to a unit range for comparability. We use 5-fold cross validation of *leave-one-out* prediction. Rather than splitting the dataset by student as in

Table 1. Summary of datasets in this paper. The simulated datasets are designed for pattern analysis. The ESL datasets are from production systems - these are primarily to compare performance on binary and scalar response modes. ASSISTments2009 is a classic benchmark dataset.

Name	Type	Concepts	Size	Av. Pass rate / Score
Simulated-5	Binary response, synthetic multichoice	5 concepts (1 active)	2000 students 50 tasks 20k attempts	Avg. Pass rate: 0.61 Avg. Final: 0.75
Scalar-5	Scalar response, synthetic, graded [0,1]	5 concepts (1 active) 3 student learning rate levels	2000 students 50 tasks 20k attempts	Avg. Score: 0.35 Avg. Final: 0.49
ESL-Q&A	Binary response, English language learners; multi choice Q&A game	4 basic skills 9 task types	750 students 852 tasks 26159 attempts	Avg. Pass rate: 0.69 Avg. Final: 0.68
ESL-Essays	Scalar response, English language learners; essay questions with adaptive feedback; graded [0,13]	5 brackets by item difficulty/goal; 3 by student fluency	1047 students 81 tasks 5718 attempts	Avg. Score: 0.47 Avg. Final: 0.44 (normalised to [0,1])
ASSISTments 2009	Binary response, maths questions	110 skills	4151 students 16.8k tasks 325k attempts	Avg. Pass rate: 0.62 Avg. Final: 0.63

the original DKT paper, we hold out the last timestep of each student, and train on the rest of the data. This is similar to trying to predict the current ability of a group of active students on a live system. Results are compared to DKT and to static collaborative filtering of skills (Static CF) [9] which is like RNCF but ignores temporal effects.

Simulated: We use the binary-response **Simulated-5** dataset from the original DKT paper [13] but additionally generate a scalar response dataset (**Scalar-5**) with 2,000 virtual students and 50 virtual exercises. In both cases, the simulated domain has five concepts, and each exercise has a single concept at difficulty β. Each student has a latent skill level for each concept in the domain, and acquires skill in each concept when they attempt questions aligned to that concept (Table 1).

ESL Q&A: This dataset tracks user responses on an English language learner mobile app[1]. Responses are binary valued and questions are multiple choice. This dataset has exercises testing reading, writing, listening and speaking skills.

ESL Essays: We are interested in scalar responses as well as dichotomous. This dataset measures student fluency on an essay-writing task[2]. We still use student and item identifiers for the x values but y values are now a scalar score in the range [0,13]. This corresponds to subdivisions of the *Common European Frame of Reference* (CEFR) score[3] for language fluency - each point in our scale

---

[1] https://www.cambridgeenglish.org/learning-English/games-social/exam-lift/.

[2] https://writeandimprove.com/.

[3] https://www.coe.int/en/web/common-european-framework-reference-languages/level-descriptions.

represents half a CEFR grade (traditionally written as <A1, A1, A2, B1, B2, C1, C2, >C2). These scores are normalised to [0,1].

ASSISTments 2009 : This is a classic dataset used for benchmarking KT systems, taken from the *ASSISTments* Intelligent Tutoring System. We are using the revised *skill builder* version[4] that has had duplicates removed.

5   Results

Table 2. Results for *leave-one-out* user-response prediction (scores are Area Under Curve, higher is better) for the binary response datasets. Slashes separate training score and cross-validation score. Best scores are in **bold**, next best in *italics*.

Data	Static CF.	DKT	RNCF
Simulated-5	0.685 / 0.684	*0.743 / 0.783*	**0.881 / 0.812**
ESL-Q&A	*0.822 / 0.774*	0.699 / 0.628	**0.844 / 0.782**
ASSISTments2009	*0.800 / 0.771*	0.794 / 0.764	**0.864 / 0.801**

Table 3. Results for *leave-one-out* user-response prediction (scores are RMSE normalised to a unit range, lower is better) for the scalar response datasets. Slashes separate training score and cross-validation score. Best scores are in **bold**, next best in *italics*.

Data		Static CF.	DKT	RNCF
Scalar-5		0.293 / 0.320	*0.194 / 0.211*	**0.123 / 0.163**
ESL-Essays	Prev.Best: 0.08	*0.071 / 0.052*	0.093 / 0.074	**0.036 / 0.041**

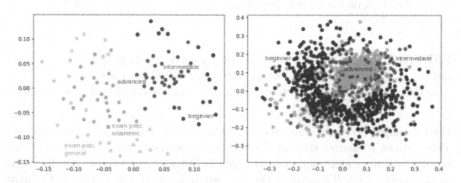

Fig. 2. MDS (distance preserving) cluster diagrams for the ESL-Essays dataset. Clustering of exercises into difficulty/goal groups (L) and of students into ability levels (R) is pronounced.

---

[4] https://sites.google.com/site/assistmentsdata/home/2009-2010-assistment-data/skill-builder-data-2009-2010.

Prediction: For the Simulated-5 data, we start by confirming that the DKT score on training is commensurate with previously reported results (0.743 AUC for us, and 0.75 in [13]). Our leave-one-out testing differs from the original experiment, but this readout measures a similar effect (of calibration on the main body of data). For the same dataset, the test score is 0.783. RNCF does notably better, with 0.881 AUC on training and 0.812 AUC on test. On Simulated-5 data, DKT and RNCF beat Static CF in both training and test (see Table 2). Ultimately RNCF outperformed DKT on all three binary response datasets, but the largest improvement was on ESL-Q&A, with a 24.5% increase in test AUC to 0.782 from 0.628. Interestingly, although DKT showed a marked improvement over the Static CF baseline on Simulated-5 (0.783 to 0.684 AUC on test data), it actually performed worse on the ESL-Q&A dataset (0.628 to 0.774 AUC on test) and on ASSISTments2009 (0.764 to 0.771 AUC on test).

A similar pattern occurs with the scalar response data (see Table 3). These regression scores are RMSE so smaller is better. For Scalar-5 DKT achieves 0.211 RMSE on test, easily beating Static CF with 0.320 RMSE. RNCF however did better than both with 0.163 RMSE. On ESL-Essays all methods display very small errors, with RNCF scoring 0.041 and Static CF second with 0.052. DKT scored 0.074. RNCF also did well on training data (0.036 to 0.071 RMSE against Static CF, with DKT scoring 0.093) which shows that the model is well calibrated across the whole body of data. ESL-Essays was used for KT previously [17] with a result of ~0.08 RMSE, which RNCF easily beats.

As shown in Fig. 2, a side-effect of factorisation is that RNCF produces representations that can cluster users and tasks quite cleanly into like groupings. Student ability groups are distinct, as are the five classes of task. This ability needs further investigation but is potentially very useful for personalised learning and organisation of curricula.

6   Discussion

In this paper we introduce *Recurrent Neural Collaborative Filtering* (RNCF) to the application of knowledge tracing in education, on simulated and real-world data for both binary and scalar responses. We show that it outperforms the related *Deep Knowledge Tracing* (DKT) and a baseline *Static CF* that cannot detect changing skill.

The collaborative filtering aspect of RNCF separates student and task representations, a feature that DKT does not support, allowing us to cluster both tasks and students into like groups. This knowledge can then be used to adapt tutoring strategies accordingly, giving us better individualisation of education.

These are early results, and we intend to widen the study of this method, across more classic and novel datasets, with new features (*e.g.* response time, answer length, lag time, day boundaries), and multiple outputs such as grammatical error categories in essays, where we believe RNCF will generalise.

We will look at the effects of extending our models with aspects such as forgetting [16] and sleep. We also intend to meta-tag a small number of tasks to see if the system can identify unknown items from these. After due diligence we hope to use the technique in production systems for assessment with active students.

This paper reports on research supported by Cambridge University Press & Assessment.

References

1. Bloom, B.: The 2 sigma problem: the search for methods of group instruction as effective as one-to-one tutoring. Educ. Res. **13**, 4–16 (1984)
2. Corbett, A.T., Anderson, J.R.: Knowledge tracing: modeling the acquisition of procedural knowledge. User Model. User-Adap. Inter. **4**, 253–278 (1994)
3. Devooght, R., Bersini, H.: Collaborative filtering with recurrent neural networks. *arXiv preprint*arXiv:1608.07400 (2016)
4. Funk, S.: Netflix update: try this at home. https://sifter.org/~simon/journal/20061211.html. Accessed 30 Sep (2020)
5. He, X., Liao, L., Zhang, H., Nie, L., Hu, X., Chua, T.S.: Neural collaborative filtering. In: WWW'17 (2017)
6. Kingma, D.P., Adam, J.B.: A method for stochastic optimization. *arXiv preprint*arXiv:1412.6980 (2014)
7. Koedinger, K.R., Corbett, A.T., Perfetti, C.: The knowledge-learning-instruction framework: bridging the science-practice chasm to enhance robust student learning. Cogn. Sci. **36**(5), 757–798 (2012)
8. Lamb, A.M., ALIAS PARTH GOYAL, A.G., Zhang, Y., Zhang, S., Courville, A.C., Bengio, Y.: Professor forcing: a new algorithm for training recurrent networks. Adv. Neural Inf. Proc. Syst. **29** (2016)
9. Moore, R., Caines, A., Elliott, M., Zaidi, A., Rice, A., Buttery, P.: Skills embeddings: a neural approach to multicomponent representations of students and tasks. Int. Edu. Data Min. Soc. (2019)
10. Murray, T.: Authoring intelligent tutoring systems: an analysis of the state of the art. In: International Journal of Artificial Intelligence in Education (IJAIED), pp. 98–129 (1999)
11. Murray, T.: An overview of intelligent tutoring system authoring tools: updated analysis of the state of the art. In: Authoring Tools for Advanced Technology Learning Environments, pp. 491–544. Springer (2003). https://doi.org/10.1007/978-94-017-0819-7_17
12. O'Malley, T., et al.: Kerastuner. https://github.com/keras-team/keras-tuner (2019)
13. Piech, c., et al.: Deep knowledge tracing. Adv. Neural Inf. Proc. Syst. **28** (2015)
14. Pilászy, I., Tikk, D.: Recommending new movies: even a few ratings are more valuable than metadata. In: Proceedings of the Third ACM Conference On Recommender Systems, pp. 93–100 (2009)
15. Thai-Nghe, N., Horvath, T., Schmidt-Thieme, L.: Context-aware factorization for personalized student's task recommendation. In: Proceedings of the International Workshop on Personalization Approaches in Learning Environments vol. 732, pp. 13–18 (2011)

16. Zaidi, A., Caines, A., Moore, R., Buttery, P., Rice, A.: Adaptive forgetting curves for spaced repetition language learning. In: Bittencourt, I.I., Cukurova, M., Muldner, K., Luckin, R., Millán, E. (eds.) AIED 2020. LNCS (LNAI), vol. 12164, pp. 358–363. Springer, Cham (2020). https://doi.org/10.1007/978-3-030-52240-7_65
17. Zaidi, A.H., Caines, A., Davis, C., Moore, R., Buttery, P., Rice, A.: Accurate modelling of language learning tasks and students using representations of grammatical proficiency (2019)

EDEN: Enhanced Database Expansion in eLearning: A Method for Automated Generation of Academic Videos

Rushil Thareja[1,3]($\boxtimes$)(ID), Deep Dwivedi[2,3](ID), Ritik Garg[2](ID), Shiva Baghel[2](ID), Mukesh Mohania[3](ID), and Jainendra Shukla[3](ID)

[1] University of South Carolina, Columbia, USA
[2] Extramarks Education, Noida, India
`deepd@iiitd.ac.in`, {`ritik.garg,shiva.baghel`}`@extramarks.com`
[3] IIIT Delhi, Delhi, India
`rushil18408@iiitd.ac.in`

Abstract. Academic video databases, integral to e-learning platforms, necessitate continual updates for freshness and diversity. In addressing this challenge, we introduce EDEN, an end-to-end method for expanding academic video databases. When presented with user-defined topics, EDEN retrieves existing videos or, if unavailable, generates new ones in the same style as the original source database, ensuring seamless integration. The system's efficacy is evidenced by its addition of 3134 videos across diverse K-12 subjects, significantly enhancing an existing database's scope. Key performance indicators, including a +6% increase in F1 BERTScore and a +9.7% rise in mean Image Visual Relevance score, demonstrate the superiority of our fine-tuned LLM and stable diffusion models over standard versions. Notably, EDEN achieves remarkable efficiency, generating one second of video content in just 0.77 s using consumer-grade GPUs.

Keywords: Online learning · Generative AI · Education Technology

1 Introduction

In the evolving domain of modern education, online learning platforms have risen to prominence, with academic videos becoming a key medium for knowledge dissemination. These videos offer significant benefits over traditional text-based materials, presenting complex concepts through visually engaging and interactive methods. However, producing such videos demands considerable labor and advanced technology. Crucially, maintaining the *style* of an existing video database poses a substantial challenge in content expansion. Recent advancements in generative AI present a solution to automate this intricate process. Our system demonstrates this capability by integrating into any existing video learning database. It can generate and add new videos aligned with the database's style for any academic topic not already covered. This approach allows the system to learn and mirror the traits, styles, and features of a pre-existing educational video database, thereby enriching it with content that harmonizes with the existing collection.

A. M. Olney et al. (Eds.): AIED 2024 Workshops, CCIS 2150, pp. 400–408, 2024.
https://doi.org/10.1007/978-3-031-64315-6_37

Envision a scenario underscoring our system's necessity: A student, pressed for time before an exam, seeks a detailed understanding of a specific topic, such as the "conservation of momentum in the collision of a neutron and a proton." Despite extensive searches on platforms like YouTube, their need remains unmet. Here, our AI system stands out as an invaluable tool. It rapidly produces custom, subject-specific educational materials, facilitating swift and comprehensive learning of various topics. Importantly, it does so while mirroring the style of an established learning platform's video database. This capability effectively simplifies a challenging task into a manageable and efficient learning process. This innovative system adeptly creates diverse content, catering to various learning styles and sustaining engagement with new material. It is particularly beneficial for remote or underprivileged communities, offering a cost-effective alternative to traditional resources through automated, on-demand video customization. Additionally, its real-time content evolution, driven by student feedback, ensures ongoing relevance and appeal. The introduction of this AI-powered method signifies a major progression in education, likely boosting engagement and expediting the widespread acceptance of e-learning. The produced videos may be organized into prerequisite graphs [11] or sorted by *effectiveness* prior to distribution to students [4].

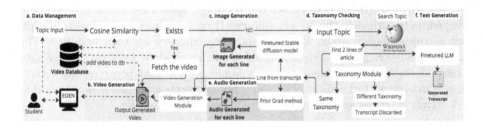

Fig. 1. End to End System Architecture

Initial efforts in video creation utilized Generative Adversarial Networks (GANs) to convert textual narratives into videos [8], followed by transformer-based models [12]. Recently, video diffusion models have shown superior performance [10]. Yet, these methods primarily produce short video clips from general descriptions and have not extensively explored generating longer animated academic content that aligns with an existing database's style. Additionally, these models demand significant computational resources and often lack consistency in temporal sequencing. Despite advancements in text, image, and audio synthesis, their integration in educational content, particularly for expanding K-12 science and math video databases, is largely unexplored. In text generation, technologies like GANs, LSTM models, and Large Language Models (LLMs) have set benchmarks, producing coherent, contextually relevant narratives. In audio generation, especially text-to-speech, conditional denoising diffusion models [7] have achieved near-human quality. Our work demonstrates a novel method to fine-tune and amalgamate these technologies, aiming to expand the video database while preserving the original style. The link for the live deployment can be found here. *Please email the primary author for data and code.*

2   Methodology

The proposed algorithm, as depicted in Fig. 1, integrates the processes of text, image, and audio generation, complemented with taxonomy prediction and video editing tasks, into a comprehensive framework for short-form academic video creation. We demonstrate this methodology on an initial database of an online learning platform, D, consisting of 2,334 academic videos. The process commences with a student's topic input, which is processed using EduBert Model [3], a BERT-base model fine-tuned on educational data, to compute cosine similarities with topics in the input database. If an existing video with a similarity score exceeding the $\lambda = 0.95$ threshold cannot be identified, our system proceeds to generate a novel video, thereby expanding the database.

In the generation phase, we initially produce the transcript using a large language model and then validate the transcript's contextual relevance to the original prompt using the Tagrec++ module [13]. This module reframes taxonomy assignment to academic text as a dense retrieval problem, introducing a cross-attention mechanism for precise categorization. Consequently, only transcripts with congruent taxonomies are retained, ensuring context preservation. Sequentially, utilizing the NLTK library [1] to split lines from the transcript, we generate spoken audio and corresponding imagery for each text line. The final stage leverages our automated video editing module, built on *MoviePy*, to consolidate the audio, image, and a dynamic caption that progressively reveals words in sync with the audio. The result is a comprehensive academic video that effectively interweaves various elements of content generation.

Table 1. Error rates for models with mean total and incorrect facts per output.

Model	Mean Total Facts	Mean Incorrect Fact	Error (%)
Falcon	13.5	4.08	30.22
Wizard	18.83	2.42	12.85
Llama2	20.5	4.75	23.17

Video Transcript Generation. To generate video transcripts for specific academic topics, our process begins with a Wikipedia search. The initial two lines from the search results are employed as prompts for our finely tuned language model, specifically developed for transcript generation. We evaluated three 7-billion parameter open-source LLMs: llama2[1], falcon[2], and wizard[3]. To determine the optimal model, we conducted experiments generating 10 video transcripts per model, prompting the model to generate 30 facts per transcript.

---

[1] https://ai.meta.com/llama/.

[2] https://falconllm.tii.ae/.

[3] https://huggingface.co/WizardLM/WizardLM-7B-V1.0.

We then manually assessed the number of facts and accuracy rate. The outcomes, detailed in Table 1, reveal that Wizard LLM exhibits the lowest error rate (12%) while generating more facts, leading us to choose this model for the transcript generation process. These 7-billion parameter models function efficiently on consumer-grade GPUs, providing inference results swiftly (under a second) on an Nvidia RTX 3060 GPU and are suitable for fine-tuning on GPUs with less than 12GB V-RAM.

Utilizing Whisper [9], an Automatic Speech Recognition (ASR) neural network, we extract 2,484 video transcripts, or over 67,695 lines, from the input database D. We enrich our training data with an additional 7,974 publicly available videos, adding 676,243 lines. This step is undertaken to enhance the model's capacity for learning, generalization, and efficient reduction of loss during the fine-tuning process. Following a fine-tuning period over 100 epochs, the model showcasing the least validation loss is selected for the task of transcript generation.

Audio Generation. We utilize Conditional Diffusion models, known for their superior performance in speech generation, as vocoders in our pipeline [6]. These models synthesize time-domain waveforms using a mel-spectrogram, a compact frequency feature representation vital for high-quality audio generation. A significant improvement, Priorgrad [7], enhances the efficiency of these models (0.104 s per on average per line on our GPU) by leveraging an adaptive prior based on data statistics and conditional information. This results in more fluent and higher quality synthesized speech. Consequently, our methodology combines Priorgrad's denoising technique with the Diffwave vocoder [6] to convert generated text into articulate, natural-sounding speech.

Image Generation. We generate illustrative images for each spoken text line by fine-tuning the CompVis Stable Diffusion model [10]. We extract over 67,000 image-text pairs from our input video database D, by considering each spoken line and finding the median image (i.e. at (line start time + line end time)/2). This enables the creation of domain-specific images resembling the input videos' animation style. The model, originally trained on a subset of the LAION-5B database [2], was fine-tuned over 10 epochs (approximately 100 GPU hours). Image quality was manually assessed using every 500 iterations, with weights saved at each 6734th iteration. We use our proposed, Image Visual Relevance score (IVR) defined subsequently to analyse the output. The model weights corresponding to approximately the 323,000th iteration were selected, where images optimally balance the training videos' animation style and pre-training photo-realism. The model can generate representative images for each line in 4.41 s (average).

2.1 Evaluation of Generated Content

The effectiveness of our fine-tuned modules for text and image generation was assessed independently. A subset of 350 academic videos from the database D were utilized for text transcript comparison, evaluated via BERTScore [15] using the EduBert Model [3]. Our fine-tuned model (P: 0.61, R: 0.69, F1: 0.65) outperformed the base model (P: 0.56, R: 0.62, F1: 0.59). Image generation, in contrast, involved creating images for 500 random transcript lines using both the base and fine-tuned stable diffusion models, and evaluated using our proposed, Image Visual Relevance (IVR) score. The IVR score, based on contextual alignment, clarity, engagement potential, and contribution to concept comprehension, was assigned by three expert teachers who were not privy to which model generated the images. The process incorporated majority voting, with the final annotations validated by a senior domain expert for quality. The fine-tuned model (mean IVR score=3.05) clearly outperformed the base model (mean IVR score=2.562), attesting to the efficacy of our fine-tuning process of both the large language and stable diffusion models for domain-specific academic applications.

To mimic student topic input for *simulation testing*, We selected academic concepts pertaining to physics and mathematics from the CMEB dataset [14]. To extract concepts for the biology and chemistry subjects, we manually extract these concepts from K12 textbooks. Following this, we utilized our pipeline to generate academic videos for each of these identified topics.

The taxonomy generation module played a crucial role in ensuring that only videos pertinent to the relevant subjects were generated, with others being discarded. This selective approach led to the addition of a considerable number of videos to the input database, enhancing its educational scope. The specifics of these additions include 896 videos in Physics totaling 15.8 h of content, 262 Mathematics videos comprising 5.5 h, 1255 videos in Biology amounting to 22.8 h, and 721 Chemistry videos contributing 13.2 h. The number of input topics for these subjects were 153 for Physics, 274 for Mathematics, 284 for Biology, and 195 for Chemistry. *The generated videos have been uploaded to the released datasets folder.* For quality assurance, each video underwent evaluation by four expert teachers, supervised by a veteran K-12 content creator. The following evaluation questions were carefully curated in consultation with the teachers and the head annotator:

- **Q1 Topic Coverage** - Considering the topic that the video covers, did it meet your expectations?
- **Q2 Factual Accuracy** - Is the information in the video factually correct?
- **Q3 Visual Relevance** - How relevant are the representational images with respect to the spoken text?
- **Q4 Educational Value** - Would this video improve a student's understanding of the topic?
- **Q5 Engagement Potential** - Would a student be interested in viewing similar videos?

The educators, possessing a deep understanding of diverse student learning styles, evaluated the content from a student's viewpoint. Their feedback effectively gauged the content's impact across a broad student demographic, eliminating individual student reviews.

Cohen's Kappa values corresponding to the five evaluation questions were 0.59 (Q1), 0.59 (Q2), 0.51 (Q3), 0.58 (Q4), and 0.47 (Q5), indicating moderate to substantial agreement among evaluators. 100 videos were annotated (95%

Fig. 2. Interface of the proposed EDEN System

confidence, 10% margin of error) [5] resulting in a total of 400 responses per question, the assessment of the videos was largely positive for topic coverage (267 affirmatives), educational value (298 affirmatives), and engagement potential (357 affirmatives), reflecting the videos' potential as beneficial educational resources. The distribution for factual accuracy was 186 partially correct, 194 totally correct, and 20 incorrect. Visual relevance saw 194 totally relevant, 174 somewhat relevant, and 32 irrelevant responses.

3 Video Generation Process

EDEN caters to K-12 students and admin users who want a comprehensive look at the video generation process. Users first encounter an authentication page (Fig. 2.1) where they can log in as either a *Student* or an *Admin*. Following successful authentication, they are directed to a search page similar to those common in e-learning platforms. On this page, users input their desired topic. The system then attempts to locate a relevant video within the database, D. If a match is found, users can select and play their chosen video from a dropdown menu (Fig. 2.2). Both user types follow this process. However, if no relevant video exists, the system activates a new video generation process (Fig. 2.3). For Student users, the system simply displays the freshly generated video upon its completion (Fig. 2.6). They also have the flexibility to request another uniquely generated video for the same topic. On the other hand, Admin users are privy to the entire video generation process. They view the initial prompts used for the video, the first two lines from a relevant Wikipedia article followed by the generated transcript and its corresponding taxonomy (Fig. 2.4). The system continuously generates transcripts until it produces one with a taxonomy that matches the prompt. Following this, the system creates spoken audio and corresponding images for each line of the validated transcript, all of which are available for admin inspection (Fig. 2.5). The final video generation stage involves the integration of the text, audio, and images into a cohesive video. Admin users can view the resulting video directly on the platform. If they wish to generate another unique video for the same topic, they can initiate the process anew. After completion, each video is incorporated into the database D, thereby expanding the system's existing catalog.

4   User Interface and Interaction

EDEN is designed with multiple user-centric interface features and interaction capabilities, aiming to optimize engagement, adaptability, and learning efficacy. A distinctive feature of this system is the real-time progress indicator. While students view an aggregated progress bar (■), admins observe separate bars for each phase of the process. The video player is enriched with numerous functionalities, including adaptable playback speed (☉) to accommodate varied learning paces, traditional play/pause controls (▶), and a video download option (⬇) that fortifies our commitment to universal and free access to education.

Furthermore, the Picture-in-Picture (PiP) mode (▣) supports simultaneous video viewing and browsing, providing students with a tool for accessing supplementary learning resources. The navigation slider (◨) affords precise control over video timelines, and the full-screen mode (⌞⌝) facilitates immersive learning experiences. Admin users are provided with additional capabilities, such as downloading audio files, copying transcript text, and viewing images corresponding to each line of the transcript directly from the platform. These functionalities, purposefully hidden from students, help maintain interface simplicity and focus. EDEN's dual-mode interface, a streamlined mode for students and a detailed mode for admins caters to diverse user needs. Incorporating a Not Safe for Work (NSFW) filter, EDEN ensures that any inappropriate images are automatically censored, preserving the platform's educational integrity.

5   Conclusion and Limitations

In this paper, we introduced EDEN, a novel methodology capable of generating short-form academic videos, thereby expanding any video content database. Amassing 57.3 h of content, with each video averaging 1.09 min and preserving the style and attributes of the input database, EDEN heralds a new era in online education and minimizing the conventional cost of content production. However, challenges with factual accuracy, content *hallucination* (Table 1), generating lesser facts and visual-to-topic alignment underscore areas for refinement and future scope for research. To mitigate these limitations, enhancements include bolstering factual verification to ensure accuracy, custom LLM fine-tuning methodologies, incorporating retrieval-augmented generation to mitigate content hallucination and add provenance, and refining visual component synchronization via custom loss functions based on detected object interactions. Additionally, the advent of advanced high-quality text-to-video engines[4] presents the possibility of utilizing generated short videos instead of static images to

---

[4] https://openai.com/sora.

accompany each spoken line. Efficiently and effectively fine-tuning such methods using videos from the source database poses a significant challenge for future research. Multi-lingual generation is also a major research direction. We remain dedicated to enhancing EDEN, enriching learning experiences and democratizing education with accessible, high-quality video content.

Acknowledgement. The primary author extends gratitude to Professors Marco Valtorta and Mark Billinghurst for their support and mentorship.

Disclosure of Interest. This research was funded and supported by Extramarks Education India.

References

1. Bird, S.: NLTK: the natural language toolkit. In: Proceedings of the COLING/ACL 2006 Interactive Presentation Sessions, pp. 69–72 (2006)
2. Cherti, M., et al.: Reproducible scaling laws for contrastive language-image learning. In: Proceedings of the IEEE/CVF Conference on Computer Vision and Pattern Recognition, pp. 2818–2829 (2023)
3. Clavié, B., Gal, K.: EduBERT: Pretrained deep language models for learning analytics. arXiv preprint arXiv:1912.00690 (2019)
4. Dwivedi, D., et al.: Effecti-NET: a multimodal framework and database for educational content effectiveness analysis. In: Proceedings of the 14th Learning Analytics and Knowledge Conference. pp. 667-677. LAK '24, Association for Computing Machinery (2024)
5. Israel, G.D.: Determining sample size (1992)
6. Kong, Z., Ping, W., Huang, J., Zhao, K., Catanzaro, B.: DiffWAVE: a versatile diffusion model for audio synthesis. arXiv preprint arXiv:2009.09761 (2020)
7. Lee, S.G., et al.: PriorGrad: improving conditional denoising diffusion models with data-driven adaptive prior. arXiv preprint arXiv:2106.06406 (2021)
8. Li, Y., Min, M., Shen, D., Carlson, D., Carin, L.: Video generation from text. In: Proceedings of the AAAI Conference on Artificial Intelligence. vol. 32 (2018)
9. Radford, A., Wu, J., Child, R., Luan, D., Amodei, D., Sutskever, I., et al.: Language models are unsupervised multitask learners. OpenAI blog 1(8), 9 (2019)
10. Rombach, R., Blattmann, A., Lorenz, D., Esser, P., Ommer, B.: High-resolution image synthesis with latent diffusion models. In: Proceedings of the IEEE/CVF Conference on Computer Vision and Pattern Recognition, pp. 10684–10695 (2022)
11. Thareja, R., Garg, R., Baghel, S., Dwivedi, D., Mohania, M., Kulshrestha, R.: Auto-req: automatic detection of pre-requisite dependencies between academic videos. In: Proceedings of the 18th Workshop on Innovative Use of NLP for Building Educational Applications (BEA 2023), pp. 539–549. Association for Computational Linguistics (2023)
12. Villegas, R., et al.: Phenaki: variable length video generation from open domain textual description. arXiv preprint arXiv:2210.02399 (2022)
13. Viswanathan, V., Mohania, M., Goyal, V.: TagRec++: hierarchical label aware attention network for question categorization. arXiv preprint arXiv:2208.05152 (2022)

14. Wang, S., et al.: Using prerequisites to extract concept maps fromtextbooks. In: Proceedings of the 25th ACM International on Conference on Information and Knowledge Management, pp. 317–326 (2016)
15. Zhang, T., Kishore, V., Wu, F., Weinberger, K.Q., Artzi, Y.: BERTScore: evaluating text generation with BERT. arXiv preprint arXiv:1904.09675 (2019)

Automatic Short Answer Grading in College Mathematics Using In-Context Meta-learning: An Evaluation of the Transferability of Findings

Smalenberger Michael[1]([✉]) [ID], Elham Sohrabi[2], Mengxue Zhang[3], Sami Baral[4], Kelly Smalenberger[5] [ID], Andrew Lan[3], and Neil Heffernan[4]

[1] Department of Mathematics and Statistics, The University of North Carolina at Charlotte, Charlotte, USA
msmalenb@charlotte.edu
[2] Department of Mathematics and Computer Science, University of South Carolina Upstate, Spartanburg, USA
[3] College of Information and Computer Science, University of Massachusetts Amherst, Amherst, USA
[4] Department of Computer Science, Worcester Polytechnic Institute, Worcester, USA
[5] Department of Mathematics and Physics, Belmont Abbey College, Belmont, USA

Abstract. Mathematics teachers use open-ended (OE) problems to inspire creativity, facilitate learning by self-explanation, and encourage transfer learning. While these types of problems are pedagogically valuable, student answers often exhibit a combination of language and mathematical expressions, and the variation in these responses can make it difficult and time-consuming to assess. There have been growing efforts to research automatic short-answer grading (ASAG) methods to support teachers with this crucial task with promising results at the K12 level. However, whether these findings transfer to other student groups or across content, specifically at the college level, remains an open question. We implement a machine learning model for ASAG developed on K12 content and student responses and evaluate the transferability of previous findings to those at the college level. Our results show that the transferability can vary significantly, buttressing the assertion that this line of investigation warrants future research.

Keywords: Automatic Short Answer Grading · Math Scoring · Machine Learning

1 Introduction

Mathematics teachers commonly use open-ended (OE) problems to inspire student creativity [15] and facilitate learning by self-explanation [8]. OE problems also lend themselves well to assessing and enhancing students' conceptual understanding [21]. However, student answers to these types of problems often exhibit a combination of language and mathematical expressions, and the variation in these responses can make it difficult and time-consuming for teachers to assess. Teachers, on average, spend about 25 percent

© The Author(s), under exclusive license to Springer Nature Switzerland AG 2024
A. M. Olney et al. (Eds.): AIED 2024 Workshops, CCIS 2150, pp. 409–417, 2024.
https://doi.org/10.1007/978-3-031-64315-6_38

of their time grading OE problems [15]. For this reason, past research showed that teachers become fatigued with grading OE problems. For example, data from ASSISTments, an online learning platform used by over 120,000 students and 3,000 K12 teachers in 2022, clearly indicate that the number assigned over a school year significantly declines [1, 11]. Furthermore, less than 14% of OE problems assigned to students were graded [1, 11], and teachers frequently resort to merely grading for completion instead of providing meaningful feedback that enhances learning [26]. Research has shown that no or delayed feedback can result in poorer student performance and disengagement [19, 24].

There have been growing efforts to research and develop methods to support teachers by automating this crucial task, often referred to in the literature as Automatic Short Answer Grading (ASAG) [1, 2, 11, 13, 22, 26]. While automating this process can be highly valuable in mitigating the obstacles mentioned earlier, it is not without risk. Grades, whether assigned by human raters or computers, can significantly affect people's lives. As Xi [25] and Madnani et al. [17] discuss in detail, decisions based on unfair student scores will likely have profound consequences.

Defining what we consider bias in ASAG is important. The key concept is the notion of a "construct," which is a set of related knowledge, skills, and other abilities that a problem measures. A fair grade on a problem assessing a given construct is one where grade differences are due only to differences in whether that construct was correctly demonstrated. Therefore, we define bias in the context of our study as any skewing or difference in grades between students that results from construct-irrelevant factors. One form of bias is model bias. It is well documented that ASAG methods may inadvertently learn to assign high weights to features irrelevant to the demonstrated mastery of a construct (e.g., longer responses receive higher grades), leading to unintended classifications, i.e., assigning an inaccurate grade to a student response [8, 10]. Hence, implementing ASAG methods that do not account for misclassification may cause significant harm by inadvertently automating bias.

Recent efforts in the ASAG literature inadequately address whether their models succumb to a potentially significant source of bias, namely the grade level of students whose responses were used to train the ASAG model. That is, many recent advances in the ASAG literature exclusively used middle and high school student responses in training their models [1, 2, 11, 13, 22, 26]. However, the literature pertaining to these advances does not address whether the performance of these models varies across students' education levels. For example, if an ASAG model was trained on middle and high school student responses to OE problems on exponents, could this model accurately grade responses to OE problems on exponents given by college students? This remains an open question that warrants investigation.

The learning science literature documents that cognitive processes change over time and that many differences exist between middle school children and young adults. The progression through neurological developmental milestones is accompanied by marked differences in mathematical processing, including retrieval, computation, reasoning, and decision-making about arithmetic relations and resolving interference between competing solutions. For example, recent research in cognitive science suggests that executive function skills, which include monitoring and manipulating information in the mind (working memory), suppressing distracting information and unwanted responses (inhibition), and flexible thinking (shifting), play a critical role in the development of mathematics proficiency [7]. Although the literature is limited in terms of disentangling the neurodevelopmental aspects of specific cognitive processes in relation to their impact on mathematics skill development, the research by Menon [18] shows that with development, as with extended practice, there is a shift from prefrontal cortex-mediated information processing to more specialized mechanisms in the posterior parietal cortex. Freeing the prefrontal cortex from computational load and thus making available valuable processing resources for more complex problem-solving and reasoning is a crucial factor in mathematical learning and skill acquisition [23].

Therefore, it is reasonable to anticipate or expect that student-generated explanations of their mathematical reasoning may change as they develop cognitively, even when discussing the same mathematics content. This provides a strong impetus for our empirical study to investigate whether the efficacy of ASAG models holds for student responses from different education levels. Hence, our research question is: When ASAG models are trained on K12 student responses to OE mathematics problems, do these models grade OE mathematics responses by K12 students similarly to responses provided by college students?

2 Related Work

One of the challenges in training ASAG models has traditionally been the need for large datasets of graded short answers to serve as training data. This stems primarily from several factors, including the complexity of natural language used in mathematics contexts with varying learning criteria, the fact that such models can be prone to overfitting, and ensuring that the training data includes a sufficiently broad range of student responses to handle partially correct answers adequately, terminologies used in an unusual manner, or unconventional but valid reasoning. For this reason, ASAG methods have begun to use meta-learning, i.e., few-shot learning methodologies to generalize to unseen mathematics problems or student responses.

A recent example of an ASAG method used in mathematics that (often significantly) outperforms existing approaches, especially for unseen mathematics problems during training, is by Zhang et al. [26]. Given this method's exceptional performance, applicability to our investigation, and availability of the data used to train the model and open source of the relevant code, we replicate this method to answer our research question. Hence, we provide the relevant details of this method next to form the basis of our replication.

3 Methodology

For a set of mathematics problems, each problem can be seen as a classification task with graded examples. One key technical stipulation to implementing this model is that we need to use a well-crafted input format to provide context to the model and help it adapt to the scoring task for each mathematics problem. Therefore, instead of only inputting the target student response we want to grade, we also include several other features as the input. These features are important to ground the model in the context of each mathematics problem, called In-context Meta-learning. Each graded example consists of multiple fields of information, including the problem text, a unique problem id, the student's response, and the teacher-assigned grade. A scoring model is trained on student responses to some problems, and its generalizability to student responses on held-out problems is evaluated. Specifically, the trained model is applied to new problems to investigate the efficacy of the model using few (or zero) scored examples for these new problems. We study two cases: i) we do not update the original model, which is called the Meta setting, and ii) we update the model using a few scored responses for new problems, which is called the Meta-finetune setting.

The scoring method is based on fine-tuning a pre-trained language model. To do this, the base model is MathBERT [22], a variant of BERT [9] pre-trained on a large mathematical corpus, and we fine-tune it on our data for downstream ASAG classification. While a significant amount of recent research tends to use the Generative Pre-trained Transformer (GPT) family of models, since our objective is ASAG, our application would not use the capability for which the GPT models have become popular, namely generation, e.g., text generation. For this reason, we investigate the BERT family of models. Nevertheless, since the performance of the GPT and BERT models tends to be similar on certain tasks, we suggest investigating the GPT family of models for ASAG in the future work section of this paper.

A key difference between Zhang et al. [26] and other ASAG methods that use BERT is fine-tuning the BERT model, i.e., updating its parameters to adapt to student-generated content effectively. That is, during training, we backpropagate the gradient on the prediction objective to both i) the classification layer, which is learned from scratch, and ii) MathBERT, which is updated from its pre-trained parameters.

4 Experiments

We use two datasets of student responses to OE problems collected from the ASSISTments online learning platform. The first dataset was used in prior work [1, 12] and consists of 141,612 total student responses from 25,069 middle and high school students to 2,042 problems, scored by 891 different graders. This dataset contains some noisy data points that increase learning difficulty. Hence we pre-processed this data as in Zhang et al. [26]. For illustrative purposes, our study calls this dataset $DK12$, but Zhang et al. [26] called this dataset $Dclean$. After pre-processing, this dataset contains 131,046 responses to 1,333 problems.

ASSISTments has almost exclusively been used at the K12 level, and the applications for which ASSISTments has been used at the college level are vastly different than the applications at the K12 level, i.e., formative assessment in mathematics. Furthermore, to our knowledge, there is no openly available dataset or datasets that can be used to compare the performance of ASAG methods between K12 and college students in mathematics. For these reasons, a second dataset was collected as part of our work, encompassing 1,014 responses from 78 distinct college students who answered 15 different OE problems from two college-level algebra courses taught by the same instructor. We call this dataset $Dcollege$.

In order to compare the two datasets, it is important to note that the mathematics topics of the problems in $Dcollege$ are within the mathematics topics covered in $DK12$, and specifically cover the topics of rational expressions, radicals, exponents, linear functions, and quadratic functions. While the problems in *Dcollege* are not in *DK*12, there are problems in *DK*12 that are isomorphic to *Dcollege*.

A significant challenge in ASAG stems from the variation in student responses and the incongruity in teacher-assigned grades. To mitigate this, we leverage subject-matter experts to ground our work in established best practices by 1) creating and implementing construct-relevant grading rubrics for student short-answer responses and 2) establishing inter-rater reliability. These measures will facilitate grades relevant to the constructs and consistent across students, leading to better ASAG model performance. To grade a mathematics construct of a particular problem assigned to the college cohort, we implement the seminal work of [4]. That is, we will evaluate each construct on a 4-part scale consisting of 1) understanding the problem, 2) planning a solution, 3) getting an answer, and 4) interpreting results. While grading rubrics will facilitate consistent grading, they do not ensure it. To ensure the student responses in the datasets for this project are consistently graded, we establish inter-rater reliability. Establishing inter-rater reliability is standard practice with a commonly accepted acceptance rate of $> 95\%$ similarity between raters. We exceeded this standard threshold by achieving an inter-rater reliability of 99%. At least two researchers in this study who hold appointments in higher education mathematics departments implemented grading rubrics to assign grades to anonymized student responses to OE problems. Each grader was assigned a set of anonymized problem responses to grade that overlapped with the set of anonymized problem responses assigned to another grader on 100 rubric items. Similarity of assigned grades on these 100 rubric items was used to establish inter-rater reliability. All but one item received identical grades. Subsequently, a consensus score was assigned.

As previously stated, the model in Zhang et al. [26] is an ASAG model capable of generalizing to previously unseen mathematics problems using a few examples. To accomplish this, a scoring model was trained on student responses to some problems, and its generalizability was evaluated using student responses to held-out problems. To test whether the efficacy of this ASAG model transfers across education levels, we again train the model using K12 student responses to some problems and investigate its performance in generalizing to college student responses to new problems isomorphic to those used in the evaluation by Zhang et al. [26]. If the model's efficacy in assigning grades to the college student responses to new problems is similar to the model's efficacy in assigning grades to the K12 student responses to new problems, then we will conclude that the finding in Zhang et al. [26] transfers across education levels. Again, we emphasize that the evaluation questions answered by the college students are isomorphic to those answered by the K12 students.

We randomly divide all problems in *DK*12 into 5 equally-sized folds in terms of the number of problems instead of the number of responses. As a result, the number of responses in each fold may vary $(26, 229 \pm 689)$ since the number of student responses to each problem is different. For each run, we use 4 folds for training and 1 fold for testing. In the test set, we make $n \in \{0, 5, 10\}$ scored responses per problem available to methods trained on the training dataset and evaluate their ability to score other responses. We emphasize that there is no overlap between training responses and test responses for these previously unseen problems.

As in [26], we use MathBERT [22] as the pre-trained model with 110M parameters as the base scoring model. We do not perform any hyper-parameter tuning and use two default settings. For the first setting, Meta, we do not further adjust the trained scoring model; instead, we only feed these responses and their scores, i.e., in-context examples, to the trained scoring model. For cases where $n < 25$, we only feed in n examples even though the method was trained with 25 examples. For cases where $n > 25$, we follow randomly sample 25 examples from the n total examples as input, following the same setting above. This experimental setting can be seen as "zero-shot" learning, where we directly test how a scoring method trained on other problems works on new problems without observing any scored responses.

For the second setting, Meta-finetune, we further fine-tune our trained method on the n new scored responses per problem. During this process, for each response as the scoring target, we use the other $n - 1$ responses as in-context examples. This experimental setting can be seen as "few-shot" learning, where we test how quickly a scoring method trained on other problems can adapt to new problems.

To assess the ASAG method, we use three evaluation measures for categorical and integer-based scores, in line with previous research [1, 11]. The first measure is the area under the receiver operating characteristic curve (AUC. We adapt the AUC calculation similar to prior approaches [14] by taking the average AUC values across all potential score categories, considering each as an independent binary classification issue. The second measure is the root mean squared error (RMSE), which treats the score categories as numerical data points. Finally, the most critical measure we use is multi-class Cohen's Kappa, a statistic commonly used for evaluating ordered categories, which aligns well with the nature of our ASAG data.

The results from Zhang et al. [26] are reported as $DK12$, and the results from the college student responses are reported as $DCollege$. As in [26], Meta ($DCollege$), the method where model parameters are not updated using in-context examples, fails to generalize to new mathematics questions without seeing training data (AUC = 0.544, RMSE = 2.123, and Kappa = 0.125 at n = 0). However, since the measures the model achieves on college student responses are within the range achieved by the model on K12 student responses when training examples are not provided, we conclude that the baseline findings on Meta transfer between education levels.

As the number of training examples increases, i.e., n = {5, 10}, the performance of Meta ($DK12$) improves, but the performance of Meta ($DCollege$) does not (AUC = 0.553, RMSE = 1.804, Kappa = 0.150 at n = 5, and AUC = 0.550, RMSE = 1.801, Kappa = 0.127 at n = 10). Furthermore, the metrics of Meta ($DCollege$) fall below the expected range of Meta ($DK12$) and hence lead us to conclude that the performance of the Meta model does not transfer between education levels.

As with Meta-finetune ($DK12$), the performance of Meta-finetune ($DCollege$) improves an n increases (AUC = 0.584, RMSE = 2.114, Kappa = 0.194 at n = 5, and AUC = 0.644, RMSE = 1.597, Kappa = 0.320 at n = 10). While the performance Meta-finetune ($DCollege$) improves more quickly than Meta-finetune ($DK12$) as n is increased 5 to 10, the performance metrics for Meta-finetune ($DCollege$) are consistently worse than those for Meta-finetune ($DK12$). While it is desirable to have a larger value of n, from this experiment, we conclude that the findings on Meta-finetune transfer between education levels. That is, our results for Meta-finetune (*DCollege*) demonstrate that this model is effective at "warm-starting" scoring models on new problems even for responses on OE problems by college students.

5 Conclusion and Future Work

In this paper, we have replicated [26] using college student responses to OE mathematics problems to assess the transferability of findings for an ASAG model. This method has two main components: a base MathBERT model pre-trained with educational content on math subjects, and a meta-learning-based, in-context fine-tuning method that promotes generalization to new problems with a carefully designed input format. Experimental results using two real-world student response datasets revealed contradicting findings: Results using the Meta method do not transfer well between education levels, but those using the Meta-finetune method do.

This conclusion buttresses the assertion that future investigation into the use of ASAG methods, particularly those used across education levels, warrants further investigation. There are ample avenues for future work. First, as is often the case, collecting a larger sample size of college student responses is desirable. From a model performance perspective, this would help to see whether Meta-finetune stabilizes and achieves comparable results to those in [26]. Similarly, with a larger college student response dataset, it may be feasible to sufficiently train Meta using *DCollege* to evaluate student responses in *DK12* subject to robustness checks. Additionally, from a study replication perspective, there is sufficient impetus to use a similar grading rubric for the K12 dataset to help investigate why the performance of Meta (*DCollege*) is consistently below that of Meta (*DK12*). Conversely, instead of investigating the efficacy of models from the BERT family, it would be interesting to investigate how other LLMs, such as GPT, would perform using these datasets.

Acknowledgments. The authors thank the National Science Foundation for their support through grants 2216036, 2215842, & 2216212. None of the opinions expressed here are those of the funders. As always, K.H.S., J.M.S., E.M.S., and W.J.S. thank you and I l. y.

Disclosure of Interests. The authors have no competing interests to declare that are relevant to the content of this article.

References

1. Baral, S., Botelho, A.F., Erickson, J.A., Benachamardi, P., Heffernan, N.T.: Improving automated scoring of student open responses in mathematics. In: International Educational Data Mining Society (2021)
2. Baral, S., Seetharaman, K., Botelho, A.F., Wang, A., Heineman, G., Heffernan, N.T.: Enhancing auto-scoring of student open responses in the presence of mathematical terms and expressions. In: International Conference on Artificial Intelligence in Education. LNCS, pp. 685–690. Springer, Cham (2022). https://doi.org/10.1007/978-3-031-11644-5_68

3. Bonham, S.W., Deardorff, D.L., Beichner, R.J.: Comparison of student performance using web and paper-based homework in college-level physics. J. Res. Sci. Teach. **40**(10), 1050–1071 (2003)
4. Charles, R.: How to Evaluate Progress in Problem Solving (1987)
5. Chen, Y., Zhong, R., Zha, S., Karypis, G., He, H.: Meta-learning via language model in-context tuning. arXiv preprint arXiv:2110.07814 (2021)
6. Chi, M.T., De Leeuw, N., Chiu, M.H., LaVancher, C.: Eliciting self-explanations improves understanding. Cogn. Sci. **18**(3), 439–477 (1994)
7. Cragg, L., Gilmore, C.: Skills underlying mathematics: the role of executive function in the development of mathematics proficiency. Trends Neurosci. Educ. **3**(2), 63–68 (2014)
8. Craig, H.K., Zhang, L., Hensel, S.L., Quinn, E.J.: African American English–speaking students: an examination of the relationship between dialect shifting and reading outcomes (2009)
9. Devlin, J., Chang, M.W., Lee, K., Toutanova, K.: BERT: pre-training of deep bidirectional transformers for language understanding. arXiv preprint arXiv:1810.04805 (2018)
10. Edwards, J., et al.: Eliciting self-explanations improves understanding. J. Speech, Lang. Hear. Res. **57**(5) (2014)
11. Erickson, J.A., Botelho, A.F., McAteer, S., Varatharaj, A., Heffernan, N.T.: The automated grading of student open responses in mathematics. In: Proceedings of the Tenth International Conference on Learning Analytics and Knowledge. Association for Computing Machinery (2020)
12. Goodfellow, I., Bengio, Y., Courville, A.: Deep Learning. MIT press (2016)
13. Gurung, A., Botelho, A., Thompson, R., Sales, A., Baral, S., Heffernan, N.: Considerate, unfair, or just fatigued? Examining factors that impact teacher. In: Proceedings of the 30th International Conference on Computers in Education. Asia-Pacific Society for Computers in Education (2022)
14. Hand, D., Till, R.: A simple generalisation of the area under the ROC curve for multiple class classification problems. Mach. Learn. **45**(2), 171–186 (2001). https://doi.org/10.1023/a:101 0920819831
15. Kwon, O.N., Park, J.H., Park, J.S.: Cultivating divergent thinking in mathematics through an open-ended approach. Asia Pac. Educ. Rev. **7**(1), 51–61 (2006). https://doi.org/10.1007/BF0 3036784
16. Leacock, C., Chodorow, M.: Comput. Humanit. **37**(4), 389–405 (2003). https://doi.org/10. 1023/A:1025779619903
17. Madnani, N., Loukina, A., Von Davier, A., Burstein, J., Cahill, A.: Building better open-source tools to support fairness in automated scoring. In: Proceedings of the First ACL Workshop on Ethics in Natural Language Processing, 41–52 (2017)
18. Menon, V.: Developmental cognitive neuroscience of arithmetic: implications for learning and education. ZDM **42**(6), 515–525 (2010). https://doi.org/10.1007/s11858-010-0242-0
19. Mestre, J., Hart, D.M., Rath, K.A., Dufresne, R.: The effect of web-based homework on test performance in large enrollment introductory physics courses. J. Comput. Math. Sci. Teach. **21**(3), 229–251 (2002)
20. Min, S., Lewis, M., Zettlemoyer, L., Hajishirzi, H.: MetaICL: learning to learn in context. arXiv preprint arXiv:2110.15943 (2021)
21. Peng, S., Yuan, K., Gao, L., Tang, Z.: PISA 2015 Assessment and Analytical Framework: Science, Reading, Mathematics, Financial Literacy and Collaborative Problem Solving (2017)
22. Shen, J.T., et al.: MathBERT: a pre-trained language model for general NLP Tasks in mathematics education. In: NeurIPS 2021 Math AI for Education Workshop (2021)
23. van Merriënboer, J.J.G., Sweller, J.: Cognitive load theory and complex learning: recent developments and future directions. Educ. Psychol. Rev. **17**(2), 147–177 (2005). https://doi. org/10.1007/s10648-005-3951-0

24. VanLehn, K., et al.: The Andes physics tutoring system: lessons learned. Int. J. Artif. Intell. Educ. **15**(3), 147–204 (2005)
25. Xi, X.: How do we go about investigating test fairness? Lang. Test. **27**(2), pp.147–170 (2010)
26. Zhang, M., Baral, S., Heffernan, N., Lan, A.: Automatic short math answer grading via in-context meta-learning. In: International Educational Data Mining Society (2022)

Enhancing Algorithmic Fairness in Student Performance Prediction Through Unbiased and Equitable Machine Learning Models

Luciano de Souza Cabral[1,2,5]([envelope]) [ID], Filipe Dwan Pereira[3,5] [ID],
and Rafael Ferreira Mello[2,4,5] [ID]

[1] Federal Institute of Pernambuco (IFPE), Jaboatão dos Guararapes, PE, Brazil
[2] Center for Excellence in Social Technologies (NEES-UFAL), Maceió, AL, Brazil
luciano.cabral@nees.ufal.br
[3] Federal University of Roraima (UFRR), Boa vista, RR, Brazil
[4] Federal Rural University of Pernambuco (UFRPE), Recife, PE, Brazil
[5] Centro de Estudos Avançados de Recife, Recife, PE, Brazil

Abstract. The rapid adoption of ML algorithms has spurred the development of educational applications aimed at enhancing teaching and learning experiences. However, contemporary research underscores ethical concerns regarding their real-world implementation. A significant challenge lies in identifying and addressing potential biases in prediction models to mitigate their impact on diverse minority groups, including ethnicities, disabilities, nationalities, and genders. This paper conducts a thorough examination of algorithmic fairness, focusing on a detailed comparative analysis of traditional machine learning methods within a Student Performance Prediction (SPP) application for CS1 programming courses. The insights derived from this investigation not only enrich the ongoing discourse surrounding algorithmic bias and fairness but also pave the way for refining the development of just and equitable ML models.

Keywords: Fairness · bias · education · learning analytics · student performance prediction

1 Introduction

In the age of prevalent machine learning, ensuring algorithmic fairness has become paramount. As artificial intelligence systems play a central role in decision-making across various domains, there is a pressing need to scrutinize their fairness and address potential biases. [2,9].

This research primarily focuses on investigating algorithmic fairness methodologies within machine learning approaches applied to student performance prediction. The objective is to uncover the nuances surrounding fairness within algorithms, through a comprehensive exploration of decision-making mechanisms.

© The Author(s), under exclusive license to Springer Nature Switzerland AG 2024
A. M. Olney et al. (Eds.): AIED 2024 Workshops, CCIS 2150, pp. 418–426, 2024.
https://doi.org/10.1007/978-3-031-64315-6_39

By utilizing traditional machine learning algorithms, the study aims to discern subtle differences between approaches and their impact on promoting fairness.

Despite ongoing evolution in methodologies, a significant portion of these studies rely on static analyses of student performance, frequently predating their university enrollment [1] [14,15]. While standardized cases are common, observable behavioral dynamics throughout the semester highlight scenarios where initial student struggles can lead to successful course completion.

The academic literature suggests that machine learning algorithms often struggle with datasets containing numerous attributes, leading to reduced performance [14,15]. To address this, dimensionality reduction techniques have been employed, aiming to enhance model performance, reduce computational costs, and improve interpretability [14,16]. These techniques fall into two main categories: aggregation, which combines original attributes through linear functions, and selection, which retains relevant attributes while discarding others [6]. Additionally, transfer learning leverages knowledge from one task to aid in a related task, particularly beneficial in scenarios with limited data [10]. However, some feature selection methods may inadvertently identify attributes highly correlated with the target class, potentially causing overfitting in the resulting model.

The work advocates for utilizing data from the Automatic Code Correction Environment (ACAC) known as CodeBench from the Federal University of Amazonas for training and evaluating machine learning models. The aim is to assess these models not only based on performance but also through the lens of algorithmic fairness. The goal is to demonstrate that these models, besides being effective, are also fair, meaning they are unbiased and promote equity.

This paper is structured: Sect. 2 presents the state of the art and related work; Sect. 3 outlines the methodology for the experiments; Sect. 4 discusses the results obtained; Sect. 5 covers final considerations, limitations, and future implications; and finally, acknowledgments and references are provided.

2   Literature Review

2.1   Predicting Student Performance in Programming Classes

In systematic review [14], the authors analyzed around 70 papers focusing on improving pedagogical processes in programming classes through machine learning models. Notably, only two works within this body of literature utilized predictions from real-class data [2,16], while others were limited to constructing models without significant real-world applications.

Some ML algorithms, including ensembles, decision trees, Bayesian networks, naive Bayes, support vector machines (SVM), and neural networks, have been applied for student performance prediction. Additionally, clustering algorithms have been used in 17 studies to categorize students based on shared profiles and assess correlations with performance metrics. Notably, there's a preference for interpretable models like decision trees and clustering, despite potential robustness compromises. Conversely, neural networks, though more robust, present challenges as black box models. Exploration of algorithmic fairness and interpretation of machine learning algorithms remain limited in this domain [2].

2.2 Algorithmic Fairness and Bias Challenge

Fairness in AI has emerged as a crucial topic in academic and industry spheres, focusing on mitigating bias and discrimination within AI systems. This is a challenging task due to the various forms of bias that may arise. Different types of fairness, such as group fairness, individual fairness, and counterfactual fairness, address various aspects of equitable AI. While fairness and bias are closely related, they differ significantly; fairness is a deliberate goal, whereas bias can be unintentional. Achieving fairness in AI requires careful consideration of context and stakeholders [8].

The primary challenge lies in creating models that account for the impact of diverse groups (e.g., varying ethnicities, disabilities, nationalities, or genders) in the final prediction model. Numerous studies in the literature utilize fairness measures to develop reliable applications, particularly in the context of automatically scoring written productions, assessing students' progress in online discussions, and categorizing the relevance of content written by students [17].

However, it is uncommon to encounter research that evaluates various ML models while also ensuring, through at least three different fairness metrics, the absence of bias or the effectiveness of fairness for one or more groups, particularly those requiring greater assurances of Diversity, Equity, and Inclusion (DEI) [3].

2.3 Algorithm Fairness Frameworks

Algorithm fairness frameworks are essential for addressing the ethical challenges posed by biased algorithms and ensuring equitable outcomes in machine learning applications. These frameworks offer methodologies, tools, and guidelines to assess, measure, and mitigate biases, thereby promoting fairness and transparency. Notable algorithm fairness frameworks, such as AIF360, and ABROCA, have been developed to tackle these issues [12] [13].

AIF360 by Bellamy et al. (2018) [5] and (2019) [4] addresses fairness concerns in machine learning models, providing an open-source toolkit to detect and mitigate biases. Offering pre-processing, in-processing, and post-processing techniques, AIF360 covers various aspects of the machine learning pipeline. Its flexibility and extensive set of algorithms contribute significantly to promoting fairness and transparency in the development and deployment of ML models.

ABROCA, short for "Adversarial Bias Reduction through Overfitting Control Algorithm," focuses on mitigating biases through adversarial training. Gardner, Brooks, and Baker (2019) [11] developed ABROCA as a framework to address both subtle and overt biases in models, aiming to balance bias reduction with maintaining model performance.

3   Method

This paper seeks to assess the performance and algorithmic fairness of various machine learning models employed in predicting student performance in programming classes, utilizing data from the CodeBench Dataset. The methodology section elaborates on the dataset, the real educational scenario in which machine learning methods were applied, and the metrics used to evaluate the models.

3.1   Research Questions

This study aims to answer the following research questions:

- RQ1: *Which ML algorithms are most effective for predicting student performance in programming classes with real data?*
- RQ2: *Can ML algorithms be demonstrated to be bias-free or fair through multiple metrics or measures?*
- RQ3: *What methods are utilized to mitigate bias and explain the models?*

3.2   Dataset

The CodeBench Dataset 1.03, accessible for free download from the UFAM project CodeBench[1]. The dataset comprises behavioral data from 1655 students across six semesters (2016.1–2018.2), capturing interactions with the automatic code correction environment (ACAC), procrastination rates, IDE usage, keyboard shortcuts, system access, and patterns of variables and operators. A second version of the dataset focuses solely on sociodemographic information to comply with the General Data Protection Law, omitting personal details. This subset includes data such as gender, age, shift, PC and internet availability at home, computer sharing practices within the family, prior programming experience, and completion of technical courses.

3.3   Educational Scenario

The prediction of student performance in the Programming Logic subject using real data, aiming to enable timely interventions for improving teaching and learning [14]. Additionally, assesses whether the ML models used for automatic prediction exhibit bias or fairness. The rationale behind this investigation lies in the high failure rate observed in programming classes and the non-normal distribution of gender and ages, highlighting the need for research to guide decision-making and promote diversity, equity, and inclusion (DEI).

---

[1] https://codebench.icomp.ufam.edu.br/dataset/.

3.4 Assessment Metrics

Confusion matrix (precision, coverage, F1-score, accuracy), along with the Quadratic Weighted Kappa (QWK) statistical metric, commonly used in Artificial Intelligence in Education (AIED) were used to measure the ML models. ABROCA, AIF360, and DALEX were used to detect fairness and bias in ML models.

4 Experiments, Results and Discussion

4.1 Performance and Bias Reduction (100% Vs 34.49% Rows)

Using the initial dataset containing two versions (behavioral and socio-demographic data) each with approximately 2000 x 20 shape, was conducted data aggregation. This process expanded the dataset to 3082 rows but introduced repetitions (data from failed students and dropouts) and inconsistencies (missing or incorrect data). Various data cleaning and size reduction strategies were then applied, resulting in 1063 x 30 shape. As demonstrated by [7], these strategies maintained the model's performance and alleviated age bias.

Table 1. Versions generated after data processing.

Version	Rows × Columns	%	Description
Merge-v1	3082 × 30	100%	With duplicates, with null data and -1 s.
Merge-v4	1063 × 30	34.49%	No duplicates, null data, -1 s, and corrected ages

Measurements to assess the impact of data reduction on the model's performance was conducted, ranging from 100% to 34.49% of the dataset. Subsequently, the decision tree classifier with basic settings (random_state = 0, max_depth = 3) was developed. The data were divided in 80% for training and remain for test using train_test_split function from sklearn.model_selection.

Table 2. Decision Tree Results Data Processing.

Version	Rows × Columns	Precision	Recall	F1-score	Accuracy	QWK
Merge_v1	3082 × 30	0.82	0.82	0.82	0.82	0.598
Merge_v4	1063 × 30	0.80	0.79	0.79	0.79	0.578
Merge_v4-FEG	1063 × 29	0.81	0.79	0.79	0.79	0.591

The feature importance before merge points to an exacerbated importance for age, after, the age was 5th feature, pointing to fisrtExamGrade as 1st.

Observing a mere 34.49% utilization, Merge-v4 version witnessed negligible shifts 0.02% dip in precision and 0.03% in recall, f1, and accuracy metrics. FirstExamGrade (FEG) remained pivotal. Meanwhile, the age disappear of the Merge-v4 feature importance, and reducing age bias between SocDem features only with data treatment, confirming [7]'s hypothesis.

Excluding FEG column slightly improved model performance, with a 0.01% increase in precision and a 0.013 boost in QWK. Key factors impacting the model include correct logical lines (lloc), systemAccess rate, procrastination (student's idle time with the application), and age, maintaining a consistent pattern of influential factors without compromising algorithmic fairness.

4.2   Algorithmic Fairness Experiments

Additional experiments using the AIF360 framework were carried out to verify whether the models used present bias or not in relation to gender and age. The Dataset used for these experiments was Merge-v4 (1063 rows × 29 columns).

Table 3. Mensuring Bias with AIF360 for DT, RF and LR Models.

Model	Metric	Value
DT	Equal Opportunity Difference	0.0
RF	Equal Opportunity Difference	-0.004291251384274664
LR	Equal Opportunity Difference	0.06076965669988921

The AIF360 framework output, 'Equal Opportunity Difference' indicator presents values 0.0 for the DT, and little bit different for the RF and LR, which indicates that the DT model is fairer than its competitors.

Experiments using ABROCA metric were conducted to assess gender and age bias in models. The dataset used was Merge-v4-FEG. The data were split into an 80% training set and the remaining for the test set using the train_test_split function from sklearn.model_selection with random_state = 0. All models were configured with parameters random_state=123, and for RF and DT, the parameter max_depth was added with values 4 and 3, respectively.

Table 4. Mensuring Bias with ABROCA for DT, RF and LR Models.

Model	Accuracy	Kappa	AUC	F1	ABROCA
DT	0.7887	0.5786	0.7905	0.7883	0: 0.0397, 1: ≥ AUC
RF	0.7511	0.5095	0.7656	0.7523	0: 0.0564, 1: ≥ AUC
LR	0.7699	0.5465	0.7849	0.7710	0: ≥ AUC, 1: 0.0204

The DT and RF models demonstrate strong discrimination capacity for gender 1 (male), which is expected given its majority representation in the dataset (Merge-v4-FEG: 1 - 673, 0 - 390; Xtrain: 1 - 536, 0 - 314; Xtest: 1 - 137, 0 - 76), while exhibiting more random results for gender 0 (female). Conversely, the LR model's results appear inconsistent with the data, raising uncertainty regarding potential bias in its performance. It returned results closer to reality for females but more random results for males, inversely proportional to data distribution.

5   Final Considerations, Limitations and Future Directions

Experiments conducted using ML algorithms in SPP with real data yield insights to address the research questions. RQ1 is answered by identifying DT and RF as the best-performing and fairest algorithms, with DT particularly notable for its transparency and explainability.

Utilizing real data from CS1 programming classes sourced from the Codebench Dataset, initial analysis involves merging behavioral and socio-demographic data to mitigate age bias in the DT model. Despite reducing the dataset to around 34% of its original size and applying appropriate data treatment, including removal of duplicates and inconsistent data, significant mitigation of age bias is achieved. Moreover, the model's performance remains comparable to previous experiments with nearly triple the data, successfully addressing RQ2.

Furthermore, three different algorithmic fairness frameworks are tested alongside DT, RF, and LR models under identical conditions, datasets, and configurations. The results reinforce earlier findings, highlighting DT as the fairest, followed by RF, while LR appears to be less fair, thus answering RQ3.

Limitations include the relatively small dataset size (1063 rows post-cleaning), potentially limiting model training adequacy. Additionally, the scope of the study could be expanded by incorporating more works and fairness frameworks, possibly in future journal publications where space constraints are less stringent.

Moreover, gender analysis and other experiments, although not included due to space limitations. The experiments scripts can be made accessible by demand, to to enhance transparency and reproducibility. Potential national or international collaborations are envisioned to extend the research in future directions.

References

1. Ahadi, A., Lister, R., Vihavainen, A.: On the number of attempts students made on some online programming exercises during semester and their subsequent performance on final exam questions. In: Proceedings of the 2016 ACM Conference on Innovation and Technology in Computer Science Education, ITiCSE '16, pp. 218-223. Association for Computing Machinery, New York, NY, USA (2016). https://doi.org/10.1145/2899415.2899452
2. Azcona, D., Hsiao, I.H., Smeaton, A.F.: Personalizing computer science education by leveraging multimodal learning analytics. In: 2018 IEEE Frontiers in Education Conference (FIE), pp. 1–9 (2018). https://doi.org/10.1109/FIE.2018.8658596
3. Barney, N.: Diversity, equity and inclusion (DEI), Technical report, TechTarget (2023), https://www.techtarget.com/searchhrsoftware/definition/diversity-equity-and-inclusion-DEI
4. Bellamy, R.K.E., et al.: AI fairness 360: an extensible toolkit for detecting and mitigating algorithmic bias. IBM J. Res. Dev. **63**(4/5), 4:1–4:15 (2019). https://doi.org/10.1147/JRD.2019.2942287
5. Bellamy, R.K.E., et al.: AI fairness 360: an extensible toolkit for detecting, understanding, and mitigating unwanted algorithmic bias. arXiv preprint arXiv:1810.01943 (2018). https://api.semanticscholar.org/CorpusID:52922804
6. Broder, R.: Performance analysis on machine learning algorithms trained on biased data, Course Completion Work (Graduation) - Bachelor's Degree in Computer Science, Technical report, Institute of Science and Technology, Federal University of São Paulo, São José dos Campos (2021). https://repositorio.unifesp.br/handle/11600/60716
7. Cabral, L., Pereira, F., Mello, R.: Avaliando influência de características de desempenho na predição resultado acadêmico em disciplinas de programação. In: Anais do II Workshop de Aplicações Práticas de Learning Analytics em Instituições de Ensino no Brasil, pp. 90–98. SBC, Porto Alegre, RS, Brasil (2023). https://doi.org/10.5753/wapla.2023.236172
8. Ferrara, E.: Fairness and bias in artificial intelligence: a brief survey of sources, impacts, and mitigation strategies. Science **6**(1) (2024). https://doi.org/10.3390/sci6010003
9. Fonseca, S., Oliveira, E., Pereira, F., Fernandes, D., Carvalho, L.: Adaptação de um método preditivo para inferir o desempenho de alunos de programação. In: Brazilian Symposium on Computers in Education (SBIE), vol. 30, pp. 1651–1660 (2019). https://doi.org/10.5753/cbie.sbie.2019.1651
10. Foresti, T.: Machine learning types: learn about the different types of machine learning, including supervised, unsupervised, semi-supervised, and reinforcement learning. Technical report, Awari (2023). https://awari.com.br/machine-learning-types-principais-tipos-de-aprendizado-em-machine-learning
11. Gardner, J., Brooks, C., Baker, R.: Evaluating the fairness of predictive student models through slicing analysis. In: Proceedings of the 9th International Conference on Learning Analytics and Knowledge, LAK19, pp. 225–234. Association for Computing Machinery, New York, NY, USA (2019). https://doi.org/10.1145/3303772.3303791
12. Hattatoglu, B.: Fairness in machine learning: ensuring fairness in datasets for classification problems. Technical report, Business Informatics Master's Thesis. Department of Information and Computing Sciences, Utrecht University, Netherlands, July 2021. https://studenttheses.uu.nl/handle/20.500.12932/40070

13. Juijn, G., Stoimenova, N., Reis, J., Nguyen, D.: Perceived algorithmic fairness using organizational justice theory: an empirical case study on algorithmic hiring. In: Proceedings of the 2023 AAAI/ACM Conference on AI, Ethics, and Society, AIES 2023, pp. 775–785. Association for Computing Machinery, New York, NY, USA (2023). https://doi.org/10.1145/3600211.3604677

14. Pereira, F., Souza, L., Oliveira, E., Oliveira, D., Carvalho, L.: Predição de desempenho em ambientes computacionais para turmas de programação: um mapeamento sistemático da literatura. In: Anais do XXXI Simpósio Brasileiro de Informática na Educação, pp. 1673–1682. SBC, Porto Alegre, RS, Brasil (2020). https://doi.org/10.5753/cbie.sbie.2020.1673

15. Cristea, F.D., et al.: Using learning analytics in the Amazonas: understanding students' behaviour in introductory programming. Br. J. Edu. Technol. 51(4), 955–972 (2020). https://doi.org/10.1111/bjet.12953

16. Quille, K., Bergin, S.: CS1: how will they do? How can we help? A decade of research and practice. Comput. Sci. Educ. 29(2–3), 254–282 (2019). https://doi.org/10.1080/08993408.2019.1612679

17. Reagan, M.: Understanding bias and fairness in AI systems. Technical report, TowardsDataScience - Medium (2023). https://towardsdatascience.com/understanding-bias-and-fairness-in-ai-systems-6f7fbfe267f3

Leveraging GPT-4 for Accuracy in Education: A Comparative Study on Retrieval-Augmented Generation in MOOCs

Fatma Miladi[✉], Valéry Psyché, and Daniel Lemire

TELUQ University, 5800 rue Saint-Denis, Montreal, QC H2S 3L5, Canada
{fatma.miladi,valery.psyche,daniel.lemire}@teluq.ca

Abstract. Large Language Models (LLMs), such as Generative Pre-trained Transformers (GPTs), have demonstrated remarkable capabilities in natural language processing (NLP). However, these models often encounter challenges such as inaccuracies and hallucinations, which can undermine their utility. Retrieval-Augmented Generation (RAG) has emerged as a promising approach to enhance model accuracy and reliability by integrating external databases. This study investigates the use of RAG to improve the accuracy of GPT models in educational settings, particularly within the realm of Massive Open Online Courses (MOOCs). Through a comparative analysis of various GPT model iterations, we observed a significant improvement in accuracy, increasing from 60% with GPT-3.5 to 80% using the RAG-augmented GPT-4. This enhancement highlights the considerable potential of RAG-augmented GPT models in improving the accuracy of content generation. Such enhanced accuracy suggests revolutionizing assessment methodologies and learning experiences, fostering an educational environment that is more interactive and tailored to individual needs.

Keywords: Generative pre-trained transformers · GPT · Evaluation · MOOC · Online learning · Exercises assessments · Retrieval augmented generation

1 Introduction

The advent of Large Language Models (LLMs), such as the generative pre-trained transformer (GPT), has revolutionized the field of artificial intelligence, particularly in natural language processing (NLP) [1,2]. These models have demonstrated remarkable performance across various domains including finance, technology, and healthcare [3–5]. However, despite their impressive capabilities, large language models are not devoid of limitations. A significant challenge they face is their tendency to 'hallucinate', producing content that may not be factually accurate [6,7]. Such hallucinations can lead to the generation of information that is sometimes opposed to established facts, posing challenges for their reliable application in critical domains.

To address this issue, researchers developed a Retrieval-Augmented Generation (RAG) approach introduced by Lewis et al. in 2020 [8]. RAG aims to enhance LLMs by integrating external knowledge sources into the generation process. This integration not only improves the model's ability to generate accurate and relevant responses, but also represents a significant advancement within the realm of LLMs, particularly for generative tasks [9,10]. Although Retrieval-Augmented Generation has shown promise in various domains, its application in educational contexts remains largely unexplored. This research gap motivated our investigation into the potential of large language models augmented by RAG to enhance content accuracy in educational environments.

In this study, we investigated the potential of retrieval augmentation techniques to enhance the accuracy of traditional GPT models. Our primary focus was on educational settings, particularly in Massive Open Online Courses (MOOCs). This investigation is motivated by our central research question: How does integrating Retrieval-Augmented Generation with GPT models impact the accuracy of content in an educational context? To address this question, we formulated the following hypothesis (H1): The GPT-4 model, when enhanced with retrieval-augmented capabilities, will surpass both GPT-3.5 and its RAG-augmented version, as well as the standard GPT-4 model, in generating accurate responses. This hypothesis paves the way for a comparative analysis to understand the additional benefits of integrating RAG techniques with advanced GPT models in education.

2   Data

Our study utilized the MOOC focused on artificial intelligence (AI), developed by University TELUQ [11]. This course is divided into four main modules, each focusing on different aspects of AI. The first module introduces general AI concepts. The second module is dedicated to symbolic AI, whereas the third module covers connectionist AI. The final module discusses the application of artificial intelligence in education. These modules are supplemented by various learning resources including videos, texts, in-depth concepts, definitions, exercises, and illustrative images. The MOOC comprises 115 formative assessment exercises, encompassing a range of formats, including 24 true/false exercises, 24 multiple-choice questions (MCQs), 13 matching exercises, and 54 fill-in-the-blank exercises.

3   Models

In this study, we compare four AI models: RAG-augmented GPT-4, RAG-augmented GPT-3.5, GPT-3.5, and GPT-4. The augmented variants incorporated retrieval-augmented generation to enhance the accuracy.

As illustrated in Fig. 1, the architecture of the augmented models utilizes a sophisticated workflow designed to enhance user interaction through a web interface. This process is initiated by the user query, which is converted into a vector representation to encapsulate its semantic meaning. We employed OpenAI's text-embedding-ada-002 [12] for this purpose, enabling an effective retrieval-augmented generation [8]. Next, the system compares query embedding and a specialized database filled with text embeddings. Following this, it identifies and selects the text segments, or 'chunks', that demonstrate the greatest cosine similarity scores. The selected segments were integrated with the original query to provide additional context, thereby enriching the prompt. This enhanced prompt is then processed using a large language model such as GPT-4 to generate a comprehensive and relevant response, drawing on domain-specific knowledge to ensure accuracy.

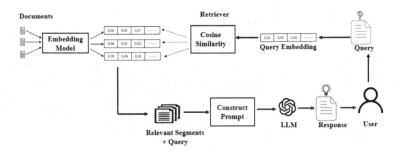

Fig. 1. Overview of the Model Architecture: From User Query Processing to Response Generation.

4 Experimental Design

To evaluate the effectiveness of the models in assessment exercises, we applied zero-shot and few-shot prompting, which are widely used in large language model studies for performance benchmarking [13,14]. Assessment questions were presented as they appear in the MOOC, utilizing prompt templates for true/false, multiple-choice, matching, and fill-in-the-blank questions, as shown in Table 1. We compared responses from the RAG-augmented GPT-4, RAG-augmented GPT-3.5, GPT-3, and GPT-4 models against correct answers, treating partially correct answers as incorrect, in line with MOOC standards. Our comparative analysis indicated that the outcomes of zero-shot and few-shot prompts were similar. Therefore, we chose zero-shot prompting as our primary evaluation method owing to its immediate practicality.

Table 1. Assessment exercises used in the performance evaluation of GPT models.

Type of Question	Sample Prompt
True/false	Indicate whether the following statement is true or false: An intelligent agent cannot adapt its actions to its environment nor act upon it. 1. True 2. False
Multiple Choice Question	Select the correct answer: According to Yann LeCun, making a machine intelligent allows it to: A. dream. B. memorize. C. learn. D. perceive.
Matching exercise	Match each definition with its corresponding term from the following: Definitions: 1. Various digital technology, mathematical, and other components that enable the design of an autonomous car. 2. The ability of a neural network to adjust itself, changing its behavior based on an environment, this ability can be used during the learning phase. 3. A robotic arm that has learned through trial-and-error manipulation to handle a Rubik's Cube. Terms: A. Artificial Intelligence B. Adaptability C. Intelligent Agent
Fill-in-the-blank	Fill in the blank: To pass the test of ..., the computer must be equipped with an artificial vision device to perceive objects and a robotic capability to manipulate objects and move.

5   Results

In this study, we evaluated the performance of the RAG-augmented GPT-4 against GPT-3.5, RAG-augmented GPT-3.5, and the standard GPT-4 model, using a dataset of 115 French exercises from a MOOC on AI. The results summarized in Tables 2, 3, 4 and 5, indicate a progressive improvement in accuracy across generations of GPT models. The GPT-3.5 model achieved a baseline success rate of 60%. This was followed by the RAG-augmented GPT-3.5 that exhibited an improved success rate of 74%. The GPT-4 model further increased the accuracy to 77% and the RAG-augmented GPT-4 achieved the highest success rate of 80%.

True/False Exercises. In the True/False exercises, as indicated in Table 2, the GPT-3.5 model demonstrated foundational capability with a 65% success rate. This was enhanced using the RAG-augmented GPT-3.5, which achieved an 85% success rate. Both the GPT-4 model and its augmented variant further improved the performance, reaching an 87% success rate, representing the highest level of accuracy among the models tested.

Multiple-Choice Questions (MCQs). As shown in Table 3, the performance on MCQs improved across the GPT model versions. The standard GPT-3.5 model began with a success rate of 60%, which was enhanced to 73% with the GPT-3.5 augmented model. Subsequently, both GPT-4 and its augmented version achieved a further increase in success rate, reaching 76%.

Matching Exercises. The GPT-3.5 model achieves a 67% success rate, which is surpassed by the RAG-augmented GPT-3.5 at 75%, as shown in Table 4. The GPT-4 model continues this trend of improvement, reaching an 81% success rate, whereas the RAG-augmented GPT-4 achieves the highest accuracy at 87%.

Fill-in-the-Blank Exercises. In the fill-in-the-blank exercises, as shown in Table 5, there was a noticeable progression in the model performance. The GPT-3.5 model begins with a success rate of 48%, which is significantly enhanced to 63% with the RAG-augmented GPT-3.5. Following this, the GPT-4 model achieved a success rate of 65%, with the RAG-augmented GPT-4 further improving performance to 72%.

Table 2. True/False exercises assessments results.

Module Topic of MOOC	True/False Exercises			
	GPT-3.5	RAG-augmented GPT-3.5	GPT-4	RAG-augmented GPT-4
General AI concepts	7/8 (87%)	8/8 (100%)	8/8 (100%)	8/8 (100%)
Symbolic AI	3/4 (75%)	3/4 (75%)	4/4 (100%)	4/4 (100%)
Connectionist AI	3/6 (50%)	5/6 (83%)	4/ 6 (67%)	5/6 (83%)
AI applications in education	3/6 (50%)	5/6 (83%)	5/6 (83%)	4/ 6 (67%)
Total	65%	85%	87%	87%

Table 3. MCQ exercises assessments results.

Module Topic of MOOC	MCQ Exercises			
	GPT-3.5	RAG-augmented GPT-3.5	GPT-4	RAG-augmented GPT-4
General AI concepts	4/7 (57%)	5/7 (71%)	5/7 (71%)	5/7 (71%)
Symbolic AI	5/7 (71%)	5/7 (71%)	5/7 (71%)	5/7 (71%)
Connectionist AI	5/8 (62%)	4/8 (50%)	5/8 (62%)	5/8 (62%)
AI applications in education	1/2 (50%)	2/2 (100%)	2/2 (100%)	2/2 (100%)
Total	60%	73%	76%	76%

6 Discussion

Our findings indicate that the RAG-augmented GPT-4 not only exhibited marked proficiency in navigating various exercises from the MOOC but also consistently outperformed the standard GPT-3.5, the RAG-augmented GPT-3.5, and the standard GPT-4 model. This superior performance aligns with our initial hypothesis (H1), substantiating the claim that the RAG-augmented GPT-4 is highly effective in producing accurate responses.

Table 4. Matching exercises assessments results.

Module Topic of MOOC	Matching Exercises			
	GPT-3.5	RAG-augmented GPT-3.5	GPT-4	RAG-augmented GPT-4
General AI concepts	2/4 (50%)	2/4 (50%)	3/4 (75%)	4/4 (100%)
Symbolic AI	2/2 (100%)	2/2 (100%)	2/2 (100%)	2/2 (100%)
Connectionist AI	2/4 (50%)	2/4 (50%)	2/4 (50%)	2/4 (50%)
AI applications in education	2/3 67%	3/3 (100%)	3/3 (100%)	3/3 (100%)
Total	67%	75%	81%	87%

Table 5. Fill in the blank exercises assessments results.

Module Topic of MOOC	Fill in the Blank Exercises			
	GPT-3.5	RAG-augmented GPT-3.5	GPT-4	RAG-augmented GPT-4
General AI concepts	9/14 (64%)	8/14 (71%)	11/14 (79%)	10/14 (71%)
Symbolic AI	8/13 (61%)	11/13 (85%)	9/13 (69%)	13/13 (100%)
Connectionist AI	5/13 (38%)	7/13 (54%)	7/13 (54%)	8/13 (62%)
AI applications in education	4/14 (29%)	6/14 (43%)	8/14 (57%)	8/14 (57%)
Total	48%	63%	65%	72%

In our analysis, including fill-in-the-blank exercises, we demonstrated the effectiveness of the GPT augmented models. By leveraging retrieval-augmented capabilities, our results are in alignment with the findings of Mao et al. [15], specifically highlighting that RAG significantly enhances the accuracy of open-domain question answering. This underscores the utility of RAG for navigating complex question formats.

Despite the promising outcomes of our study, acknowledging its limitations is crucial. Our research was conducted exclusively on a single MOOC platform and focused on assessments in French, including multiple choice, true/false, matching, and fill-in-the-blank questions. This specialization may limit the generalizability of our findings to other MOOCs, especially to those that utilize a diverse array of assessment types and languages. To mitigate these limitations, future research should include more diverse exercises that involve different MOOCs. Addressing these limitations could provide a more comprehensive and fair comparison.

Our study primarily focused on the capabilities of GPT models, particularly those enhanced by Retriever-Augmented Generation, instead of conducting a broad examination of every generative AI technology, such as Gemini or Copilot. To explore GPT-4's augmented capabilities to produce accurate and contextually appropriate responses, we sought to uncover their transformative impact on education for both educators and students.

For educators, the use of this advanced technology may become a key element in developing effective course content. By adjusting the complexity of the course materials, educators can strike a perfect balance of difficulty, ensuring that each lesson aligns with the diverse learning abilities of their students. This allows for a more engaging and interactive learning experience, in which students are neither overwhelmed by excessive challenges nor bored by tasks that fail to stimulate their intellect.

For students, the GPT-4-augmented model may transform educational experience into immersive and interactive dialogue. Serving as a sophisticated 'learning companion,' as envisaged by Chan and Baskin [16], it can provide instant clarifications and offer detailed explanations tailored to the student's current level of understanding. This personalized interaction not only encourages active learning and critical thinking, but also allows students to explore subjects at their own pace and according to their interests. Moreover, embedding a technology's ability to facilitate contextually rich interactions can further enhance retention and motivation, ultimately improving learning outcomes.

7 Conclusion and Future Work

Our study evaluated the GPT-4-augmented model by leveraging the retrieval-augmented capabilities through 115 assessment exercises. The achievement of a notable success rate of 80% highlights its potential in an educational context.

Looking forward, we aim to deepen our understanding of the GPT-4-augmented model's impact on online learning experiences. To this end, we plan to conduct case studies involving students from different countries. These studies will focus on evaluating various aspects of the learning experience, including student motivation, feelings of isolation, knowledge acquisition, and retention.

The potential effects of integrating the RAG-augmented GPT-4 into education could provide important insight regarding the future possibilities of AI-supported learning. By providing educators with advanced tools to customize the curriculum and offering students an immersive and tailored learning experience, this model has the potential to establish a new standard for educational technology. Therefore, the integration of Retrieval-Augmented Generation into GPT-4 has the potential to not only enhance the teaching and learning methods but also mark the beginning of a new era characterized by interactive and personalized education.

References

1. Vaswani, A., et al.: Attention is all you need: Advances in Neural Information Processing Systems, vol. 30 (2017)

2. Radford, A., Wu, J., Child, R., Luan, D., Amodei, D., Sutskever, I.: Language models are unsupervised multitask learners. OpenAI Blog **1**(8), 9 (2019)
3. Wu, S., Irsoy, O., Lu, S., Dabravolski, V., Dredze, M., Gehrmann, S., Mann, G.: BloombergGPT: a large language model for finance. arXiv preprint arXiv:2303.17564 (2023)
4. Chen, M., et al.: Evaluating large language models trained on code. arXiv preprint arXiv:2107.03374 (2021)
5. Yang, X., et al.: A large language model for electronic health records. NPJ Digital Med. **5**(1), 194 (2022)
6. Zhang, Y., et al.: Siren's song in the AI ocean: a survey on hallucination in large language models. arXiv preprint arXiv:2309.01219 (2023)
7. Zhou, C., et al.: Detecting hallucinated content in conditional neural sequence generation. arXiv preprint arXiv:2011.02593 (2020)
8. Lewis, P., et al.: Retrieval-augmented generation for knowledge-intensive NLP tasks. Adv. Neural. Inf. Process. Syst. **33**, 9459–9474 (2020)
9. Shi, W., et al.: REPLUG: retrieval-augmented black-box language models. arXiv preprint arXiv:2301.12652 (2023). https://doi.org/10.48550/arXiv.2301.12652
10. Liu, J., Jin, J., Wang, Z., Cheng, J., Dou, Z., Wen, J.: RETA-LLM: a retrieval-augmented large language model toolkit (2023). arXiv preprint arXiv:2306.05212. https://doi.org/10.48550/arXiv.2306.05212
11. CLOM Mots d'IA. https://clom-motsia.teluq.ca/. Accessed 17 Jan 2024
12. Neelakantan, A., et al.: Text and code embeddings by contrastive pre-training (2022). arXiv preprint arXiv:2201.10005
13. Bommarito, J., Bommarito, M., Katz, D.M., Katz, J.: GPT as knowledge worker: a zero-shot evaluation of (AI) CPA capabilities (2023). arXiv preprint arXiv:2301.04408
14. Brown, T., et al.: Language models are few-shot learners. Adv. Neural. Inf. Process. Syst. **33**, 1877–1901 (2020)
15. Mao, Y., et al.: Generation-augmented retrieval for open-domain question answering. arXiv preprint arXiv:2009.08553 (2020). https://doi.org/10.18653/v1/2021.acl-long.316
16. Chan, T.W., Baskin, A.B.: Studying with the prince: the computer as a learning companion. In: Proceedings of the International Conference on Intelligent Tutoring Systems, vol. 194200 (1988)

A Personalized Multi-region Perception Network for Learner Facial Expression Recognition in Online Learning

Yu Xiong[1], Song Zhou[1], Jing Wang[2(✉)], Teng Guo[3], and Linqin Cai[2]

[1] School of Communications and Information Engineering, Chongqing University of Posts and Telecommunications, Chongqing 400065, China
xiongyu@cqupt.edu.cn

[2] Artificial Intelligence and Smart Education Research Center, Chongqing University of Posts and Telecommunications, Chongqing 400065, China
{wangj,cailq}@cqupt.edu.cn

[3] Guangdong Institute of Smart Education, Jinan University, Guangzhou 510632, China
tengguo@jnu.edu.cn

Abstract. Learners' emotions are closely related to their learning status. Effective emotion recognition helps educators grasp learners' emotional states in a timely manner, thereby better adjusting teaching strategies and providing personalized guidance. Facial expressions, as the outward expression of inner activities, are an important basis for identifying human emotions. Therefore, facial expression recognition has become a common method for detecting learners' emotional states, especially in online learning. However, the unique facial features of individual learners lead to different appearances of the same expression, which seriously affects the accuracy of facial expression recognition. To this end, we propose a personalized multi-region perception network (PMPN) for adaptively learning individual differences in facial expressions. On the one hand, a multi-scale global perception module (MGPM) is designed to capture feature representations at different scales to adapt to individual unique facial features. On the other hand, an adaptive local attention module (ALAM) is proposed to capture more personalized local features. Experimental results show that the proposed PMPN can accurately identify learners' facial emotions in online learning by considering learners' individual differences, and achieves the most significant recognition results. This method has positive significance for timely grasping the learning status of online learners, improving the online learning experience and realizing personalized education.

Keywords: Online Learning · Learner Facial Expression Recognition · Individual Differences · Personalized Multi-region Perception

A. M. Olney et al. (Eds.): AIED 2024 Workshops, CCIS 2150, pp. 435–443, 2024.
https://doi.org/10.1007/978-3-031-64315-6_41

1 Introduction

As an important supplement to offline education, online learning has become a typical learning method in the digital era. In China, the number of registered users has exceeded 100 million in the "National Smart Education Public Service Platform" since its launch in 2022 [2]. However, it should be noted that due to the time and space separation characteristics of online learning methods, there is a lack of emotional communication and collaborative interaction between teachers and learners, which will lead to online learners feeling lonely and resulting in learning burnout [10]. The emotions during the learning process are an important factor reflecting learning status. Effective emotion recognition helps teachers provide learners with more scientific and personalized educational guidance [6]. Therefore, it is crucial for in-depth promotion of online education to accurately identify the emotional state of learners in online learning.

Facial expression recognition (FER) technology can meet the actual needs of online learning due to its non-invasive and convenience [1]. There are already some methods to use FER to detect learners' emotional state in online learning, and achieved remarkable results [1,15]. However, these methods mainly focus on eliminating the interference of irrelevant factors such as occlusion and illumination on learners' expression recognition, ignoring the individual specificity of learners. In online learning, learners' emotional expression is a spontaneous presentation of inner emotions, which is driven by the course content. As a result, their facial expressions are more personal. Due to the differences in individual characteristics such as facial features, expression habits and emotional experiences, learners' facial expressions of the same category are inconsistent in spatial appearance and expression effect [16]. This inconsistent will seriously affect the judgment of learners' true emotions.

Research in cognitive science and psychology shows that the discriminative features of different facial expressions are not evenly distributed across the face [4]. Some crucial regions, such as eyes and mouth, can well reflect the differences between individuals [18]. Additionally, the eye and mouth regions are focal points for emotional expression. Therefore, focusing on these regions can help weaken the impact of personalized characteristics on learners' emotion recognition.

In this paper, we propose a personalized multi-region perception network (PMPN) to adaptively learn global and local personalized differences of learners. PMPN is mainly composed of a multi-scale global perception module (MGPM), an adaptive local attention module (ALAM) and an emotion recognition module. In MGPM, a multi-level convolutional block is designed to capture facial features at different scales in the global perceptual domain. ALAM contains a dynamic division strategy and an attentive block. The dynamic division strategy aims to locate local key regions based on personalized facial features. Based on this, the attentive block is introduced to focus on more salient feature representations. The emotion recognition module is applied to fuse global and local expression features to obtain the final emotion category.

2   Proposed Method

Figure 1 shows the overall framework of the proposed PMPN. Different from basic expressions, academic emotions mainly include enjoyment, confusion, distraction, fatigue and neutral [1]. These emotions are more challenging due to their greater complexity and diversity in appearance.

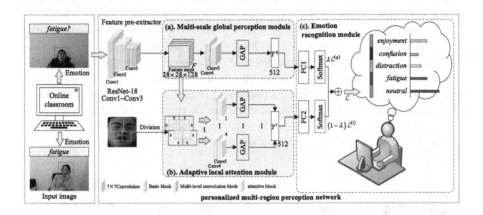

Fig. 1. The overall structure of the proposed PMPN.

2.1   Multi-scale Global Perception Module

The facial features (face outline shape, eye or mouth size, etc.) of different learners are inconsistent due to individual differences. To adapt to facial features of different scales, we design a multi-level convolution block to capture multi-scale feature representations in the global perceptual domain. Specifically, F is first evenly divided into m feature map subsets $F_i^g (1 \leq i \leq m)$, after undergoing a 3×3 convolution operation. Next, F_i^g is further input into a 3×3 convolution operation $f_i^{3 \times 3}(\cdot)$. To learn the cooperative interaction of features at different scales, the output Y_{i-1}^g of the previous convolution $f_{i-1}^{3 \times 3}(\cdot)$ is superimposed on F_i^g as the input of the convolution $f_i^{3 \times 3}(\cdot)$. Therefore, the output Y_i^g of each $f_i^{3 \times 3}(\cdot)$ can be expressed as:

$$Y_i^g = \begin{cases} f_i^{3 \times 3}\left(F_i^g\right) & i = 1 \\ f_i^{3 \times 3}\left(F_i^g + Y_{i-1}^g\right) & 2 < i \leq m \end{cases} \tag{1}$$

To capture rich facial features, we concatenate the Y_i^g along the channel axis as the output $Y^{g'}$ of a multi-level convolution block. Then a global average pooling (GAP) layer is used to obtain the global spatial feature vector Y^g.

2.2   Adaptive Local Attention Module

Dynamic Division Strategy. The dynamic division strategy consists of the coarse-grained region division stage and the dynamic refinement stage. In the first stage, we first use the highly robust Retinaface [3] to extract the five facial landmarks of the eyes, nose, and mouth, with coordinates denoted as $P_i = (x_i, y_i)$, where $0 \leq x_i, y_i \leq S, i \in$ (eye1, eye2, nose, mouth1, mouth2) and S is the largest pixel point of the image, as shown in Fig. 2(a). Next, the landmark P_{nose} of the nose, the midpoint P_{eye} of P_{eye1} and P_{eye2}, and the midpoint P_{mouth} of P_{mouth1} and P_{mouth2} are used as reference points to generate three segmentation lines $\{y_1, x_1, x_2\}$, as shown in Fig. 2(b). In the second stage, there are three adaptive division principles that need to be followed: 1) Since the center position contains more emotional information, the divided regions should be concentrated in the center of the face as much as possible; 2) Since the face anatomy meets the characteristics of "wide at the top and narrow at the bottom", the length of the eye and mouth region should be different; 3) To adapt to individual specificity, the size of local regions should be set according to the corresponding facial features.

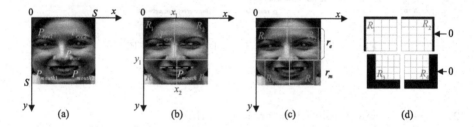

Fig. 2. Schematic diagram of the dynamic division strategy.

To fit the input feature map, each region is limited to a square region and left-right symmetrical regions have the same length. As shown in Fig. 2(c), the coordinates of the upper left corner and lower right corner corresponding to each region can be calculated, resulting the final local face region (R_1, R_2, R_3, R_4):

$$
\begin{cases}
R_1 = [((x_{eye} - r_e), (y_{nose} - r_e)) : (x_{eye}, y_{nose})] \\
R_2 = [(x_{eye}, (y_{nose} - r_e)) : ((x_{eye} + r_e), y_{nose})] \\
R_3 = [((x_{mouth} - r_m), y_{nose}) : (x_{mouth}, (y_{nose} + r_m))] \\
R_4 = [(x_{mouth}, y_{mouth}) : ((x_{mouth} + r_m), (y_{nose} + r_m))]
\end{cases}
\tag{2}
$$

where x_{eye} and x_{mouth} represent the abscissas of P_{eye} and P_{mouth} respectively. Taking region R_1 as an example, if twice the distance from P_{eye1} to P_{eye} is less than the distance x_{eye} from the left edge of the image to P_{eye}, r_e is set to twice the distance from P_{eye1} to P_{eye}, otherwise it is x_{eye}. Likewise, r_m is set in the same way. Finally, we directly map the region into F to obtain the local feature representation F_i^l, as shown in Fig. 2(d).

Attentive Block. Since different regions convey different emotional informa-tion, we introduce an attentive block [19] to adaptively perceive the key features of F_i^l. Subsequently, we first extend the attentive block to four regions to per-ceive the unique expression information under different regions. Then, the output feature maps Y_i^l of the four regions are concatenated along the spatial axes to obtain the feature $Y^{l'}$. Finally, $Y^{l'}$ is subjected to a GAP layer to obtain a local feature vector Y^l.

2.3 Emotion Recognition Module

To obtain the final learner emotion categories, we apply a decision-level fusion strategy to fuse global features Y^g and local features Y^l. Subsequently, a fully connected layer and a softmax function are added to obtain the emotion label. In the experiment, the difference between the predicted results and the true labels is measured through the cross-entropy loss function $\mathcal{L}$. The total loss $\mathcal{L}$ can be expressed as:

$$\mathcal{L} = \lambda \mathcal{L}^{(g)} + (1 - \lambda)\mathcal{L}^{(l)} \qquad (3)$$

where $\mathcal{L}^{(g)}$ represents the loss of MGPM, $\mathcal{L}^{(l)}$ represents the loss of ALAM, λ represents the balance weight of global and local features.

3 Experiments

To evaluate the performance of PMPN in recognizing learners' emotions in online learning, we first conduct experiments on a spontaneous student expression dataset OL-SFED [1] to ensure the effectiveness of academic emotion recogni-tion. Then, we compared the recognition accuracy of the proposed PMPN with the baselines on three widely used basic expression datasets (CK+ [13], JAFFE [9] and FER2013 [5]) to further verify the generalization performance of PMPN.

Table 1. Comparison of PMPN with other baseline models on the OL-SFED, CK+, JAFFE and FER2013 datasets.

Methods	Accuracy(%) OL-SFED	Methods	Accuracy(%) CK+	JAFFE	FER2013
CNN-RGB [1]	45.3	BGA-Net [17]	96.36	95.51	71.89
CNN-MBP [1]	74.1	ResNet18+SVM [11]	96.94	84.44	67.65
VGG-ADA [1]	68.7	Auto-FERNet [12]	98.89	97.14	73.78
VGG-RDA [1]	87.7	HFT-CNN [7]	98.50	94.10	-
VGG-RDA-ADA [1]	91.60	self-attention [8]	99.00	97.00	64.89
AffectXception [15]	91.87	SAN-CNN [14]	99.18	98.75	74.17
PMPN (Ours)	**96.82**	PMPN (Ours)	**99.22**	**98.75**	**74.69**

3.1   Overall Comparison

Table 1 shows the recognition results of the proposed PMPN and baseline models on four datasets. On OL-SFED dataset, PMPN achieves the best recognition result (96.82%), proving the effectiveness of the proposed method in identifying academic emotions. Compared to VGG-RDA-ADA and AffectXception, our method achieves significant improvements of 5.22% and 4.95%, respectively. This result indicates that it is beneficial to consider students' individual characteristics for academic emotions recognition. On three basic expression datasets, the proposed PMPN achieves the highest recognition accuracy of 99.22%, 98.75% and 74.69% respectively, which proves the outstanding generalization of PMPN. Since students rarely display emotions such as fear, sadness, and anger during learning, this suggests a difference between academic emotions and basic expressions. Therefore, the adaptability of PMPN shows that it can meet a wider range of academic expression recognition needs in the future.

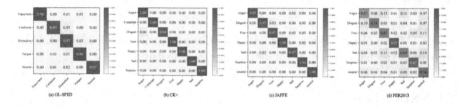

Fig. 3. The confusion matrix of PMPN on the OL-SFED, CK+, JAFFE and FER2013 datasets.

To further evaluate the recognition performance of PMPN for different expressions, we draw confusion matrices for different datasets, as shown in Fig. 3. For the OL-SFED dataset, the recognition rates of *enjoyment* and *fatigue* are higher, while the recognition rates of *confusion*, *distraction*, and *neutral* are lower. This may be due to the significant differences in facial muscle changes when students express *enjoyment* and *fatigue*. For example, students usually raise the corners of their mouths when they show *enjoyment*, while they usually open their mouths wide when they yawn due to *fatigue*. However, the appearances of *confusion*, *distraction* and *neutral* are similar, which also should be emphasized. Similar performance is also seen on three basic expression datasets. Furthermore, since the more realistic FER2013 dataset contains more interference information, the recognition accuracy of PMPN is slightly lower. Therefore, our future work will consider how to resist interference from the real world to obtain more accurate recognition results.

3.2   Ablation Study

To verify the effectiveness of MGPM and ALAM, we conduct ablation studies on four datasets, and the recognition results are shown in Table 2. On the OL-SFED, CK+, JAFFE and FER2013 datasets, MGPM is added to the backbone

Table 2. Different components in PMPN are evaluated on the OL-SFED, CK+, JAFFE and FER2013 datasets.

Dataset	MGPM	ALAM	Accuracy(%)	Dataset	MGPM	ALAM	Accuracy(%)
OL-SFED			89.93	JAFFE			95.00
	√		93.47		√		97.50
	√	√	**96.82**		√	√	**98.75**
CK+			95.31	FER2013			67.35
	√		97.66		√		71.64
	√	√	**99.22**		√	√	**74.69**

ResNet18, which improved the FER results by 3.54%, 2.35%, 2.50% and 4.29% respectively. This shows that MGPM can coordinate features of different semantics through multi-level convolution operations to learn the overall differences in expressions. On this basis, ALAM is added to increase the recognition accuracy by 3.35%, 1.56%, 1.25% and 3.05% respectively. The results demonstrated that ALAM can adaptively locate key face regions based on individual differences among students and focus on salient local features. In summary, MGPM and ALAM are able to extract personalized global and local expression features to mitigate the impact of individual specificity on students' academic emotion recognition.

4 Conclusion and Future Work

To help teachers grasp learners' learning status in a timely manner and provide personalized learning support, this paper proposed a new method, namely PMPN, to recognize learners' facial expressions for automatically detecting learners' learning emotions in online learning. PMPN adaptively learns the global and local differences in individual expressions of different students, and can effectively capture personalized emotional expressions. Experimental results show that the proposed PMPN can effectively capture personalized emotional expressions and achieve the most significant results.

In the future, we will continue to optimize the proposed method to deal with various interferences in real teaching environments. Before this technology can be used in practical applications, we must fully alert users and obtain their authorization. At the same time, users' facial data must be properly stored to ensure their privacy protection during the learning process.

Acknowledgement. This study was funded by National Natural Science Foundation of China (No.62377007), Chongqing Special Key Project for Technology Innovation and Application Development (No. CSTC2021jscx-gksbX0059) and Chongqing Key Research Project for Higher Education Teaching Reform (No.232073).

References

1. Bian, C., et al.: Spontaneous facial expression database for academic emotion inference in online learning. IET Comput. Vis. **13**(3), 329–337 (2019)
2. Dai, H.M., Teo, T., Rappa, N.A., Huang, F.: Explaining Chinese university students' continuance learning intention in the MOOC setting: a modified expectation confirmation model perspective. Comput. Educ. **150**, 103850 (2020)
3. Deng, J., et al.: RetinaFace: single-stage dense face localisation in the wild. CoRR **abs/1905.00641** (2019)
4. Friesen, E., Ekman, P.: Facial action coding system: a technique for the measurement of facial movement. Palo Alto **3**(2) (1978)
5. Goodfellow, I.J., et al.: Challenges in representation learning: a report on three machine learning contests. In: Neural Information Processing, pp. 117–124 (2013)
6. Guo, H., Gao, W.: Metaverse-powered experiential situational English-teaching design: an emotion-based analysis method. Front. Psychol. **13**, 1–9 (2022)
7. He, Y., Zhang, Y., Chen, S., et al.: Facial expression recognition using hierarchical features with three-channel convolutional neural network. IEEE Access **11**, 84785–84794 (2023)
8. Indolia, S., Nigam, S., Singh, R.: A framework for facial expression recognition using deep self-attention network. J. Ambient. Intell. Humaniz. Comput. **14**, 9543–9562 (2023)
9. Lyons, M.J., et al.: The Japanese female facial expression (JAFFE) database. In: Proceedings of Third International Conference on Automatic Face and Gesture Recognition, pp. 14–16 (1998)
10. Jagadeesh, M., Baranidharan, B.: Facial expression recognition of online learners from real-time videos using a novel deep learning model. Multimedia Syst. **28**, 2285–2305 (2022)
11. Ji, L., Wu, S., Gu, X.: A facial expression recognition algorithm incorporating SVM and explainable residual neural network. SIViP **17**(22), 4245–4254 (2023)
12. Li, S., Li, W., Wen, S., et al.: Auto-FERNet: a facial expression recognition network with architecture search. IEEE Trans. Netw. Sci. Eng. **8**(3), 2213–2222 (2021)
13. Lucey, P., et al.: The extended Cohn-Kanade dataset (CK+): a complete dataset for action unit and emotion-specified expression. In: 2010 IEEE Computer Society Conference on Computer Vision and Pattern Recognition - Workshops, pp. 94–101 (2010)
14. Putro, M.D., Nguyen, D.L., Jo, K.H.: A fast CPU real-time facial expression detector using sequential attention network for human-robot interaction. IEEE Trans. Industr. Inf. **18**(11), 7665–7674 (2022)
15. Komaravalli, P.R., Janet, B.: Detecting academic affective states of learners in online learning environments using deep transfer learning. Scalable Comput. Pract. Exper. **24**(4), 957–970 (2023)
16. Sowden, S., Schuster, B.A., Keating, C.T., Fraser, D.S., Cook, J.L.: The role of movement kinematics in facial emotion expression production and recognition. Emotion (Washington, D.C.) **21**, 1041 – 1061 (2021)

17. Tang, C., Zhang, D., Tian, Q.: Convolutional neural network-bidirectional gated recurrent unit facial expression recognition method fused with attention mechanism. Appl. Sci. **13**(22), 1–15 (2023)
18. Xia, Y., Yu, H., Wang, X., Jian, M., Wang, F.Y.: Relation-aware facial expression recognition. IEEE Trans. Cogn. Dev. Syst. **14**(3), 1143–1154 (2022)
19. Zhao, Z., Liu, Q., Wang, S.: Learning deep global multi-scale and local attention features for facial expression recognition in the wild. IEEE Trans. Image Process. **30**, 6544–6556 (2021)

Neural Automated Essay Scoring for Improved Confidence Estimation and Score Prediction Through Integrated Classification and Regression

Masaki Uto[⊠] and Yuto Takahashi

The University of Electro-Communications, Tokyo, Japan
{uto,takahashi}@ai.lab.uec.ac.jp

Abstract. Essay writing questions are a type of constructed-response question commonly used in educational assessments. However, substantial scoring costs and reduced evaluation reliability due to rater biases can be problematic, especially in large-scale assessments. To overcome these challenges, automated essay-scoring models utilizing machine learning technologies have gained significant attention. In recent years, scoring models based on deep neural networks have achieved high accuracy. However, even highly accurate neural models still suffer from scoring errors, limiting their adoption in high-stakes assessments. To address this issue, recent studies have explored scoring models that provide confidence levels along with score predictions. This study proposes improvements to the latest confidence-predicting scoring model, enhancing its performance in both confidence estimation and score prediction.

Keywords: Automated essay scoring · reliability · confidence estimation · human-in-the-loop · educational measurement

1 Introduction

Essay writing questions are a type of constructed-response question widely used in various assessment settings to measure students' practical and higher-order abilities, including expressive skills and logical reasoning [1]. Essay writing questions necessitate human raters' grading of students' written essays. However, human grading is expensive and time-consuming, particularly for large-scale assessments [6,10]. To address this, numerous studies have employed natural language processing (NLP) and machine learning to explore automated essay-scoring (AES) methods.

Recent AES methods are generally based on deep neural network models. Specifically, AES models based on pretrained transformer networks, including Bidirectional Encoder Representations from Transformers (BERT) [2], have been used popularly (e.g., [8,9,12,13]). However, even these neural AES models can produce scoring errors that would be problematic in high-stakes assessments.

© The Author(s), under exclusive license to Springer Nature Switzerland AG 2024
A. M. Olney et al. (Eds.): AIED 2024 Workshops, CCIS 2150, pp. 444–451, 2024.
https://doi.org/10.1007/978-3-031-64315-6_42

To mitigate this, some recent studies have introduced models that provide confidence levels along with score predictions. This approach allows human raters to focus only on grading those essays where the model indicates a low confidence level for its score prediction, enhancing scoring accuracy while minimizing grading costs.

A recent confidence-predicting scoring model [4] was designed as a BERT-based classifier that treats score data as nominal scale data[1]. That model calculates the probability that an input essay is associated with each score category. Then, the model uses the highest classification probability score as its prediction and the corresponding probability as its confidence level. This model is trained using a cross-entropy loss function, which treats errors between gold-standard and predicted scores equally, regardless of their magnitude.

In contrast, previous popular AES models were typically designed as regressors, treating score data as ordinal scale data. These models were trained using a squared-error loss function, which can consider the magnitude of errors between gold-standard and predicted scores. Regression-based models generally offer more accurate score predictions than classification-based models due to their consideration of the ordinal nature of scores. Moreover, enhancing score prediction accuracy would likely improve confidence estimation accuracies.

We thus propose a novel confidence-predicting neural AES model designed as a regressor. Specifically, we modify the output layer of a conventional BERT-based classifier to function as a Gaussian linear regression model. This modification allows for estimations of variance in regression-based prediction scores that reflect confidence levels. This study applied a widely used benchmark dataset to compare our model's performance with that of the conventional model. The results indicated that our model underperformed in confidence estimations as compared to conventional methods, but improved score prediction accuracy. These findings suggest that combining regression-based scoring prediction with classification-based confidence estimation might yield superior performance in both areas.

We thus further propose a dual-output neural AES model: a BERT-based model with both regressor and classifier output layers. The loss function of this hybrid model combines mean-squared error from the regressor and cross-entropy error from the classifier as a weighted linear sum. Experiments with actual data demonstrate that our hybrid model achieves equal or superior performance to models designed solely as classifiers or regressors in both score prediction and confidence estimation.

2   Conventional Method

This section describes a conventional classification-based scoring model capable of estimating confidence levels [4]. The conventional model employs BERT as the

---

[1] Note that although this model was originally proposed for automated short-answer grading, we apply it for AES because it is designed with a common architecture suitable for both scoring tasks.

base model. BERT is a multilayer bidirectional transformer model pretrained on an extensive text corpus. The pretrained BERT model can be adapted to various downstream tasks by fine-tuning with output layers using a task-specific supervised dataset. In AES applications, it is necessary to add a special [CLS] token at the beginning of each essay. BERT then condenses essays with this token into fixed-length real-valued hidden vectors, known as the distributed text representation, corresponding to the [CLS] token output. Scores are then derived by inputting the distributed text representation vector to an output layer.

2.1 Model Architecture

Conventional confidence-predicting scoring adopts an approach similar to that described above. Specifically, letting h represent the distributed text representation, the model calculates the probability that an input essay corresponds to each score $k \in \{0, 1, \ldots, K-1\}$ (where K is the number of score categories) as

$$P_k = \frac{W_k h + b_k}{\sum_{c=0}^{K-1} W_c h + b_c}, \tag{1}$$

where W_k and b_k are a weight matrix and a bias corresponding to score category k, both learned during the fine-tuning process.

Model training is conducted by minimizing the cross-entropy error using the backpropagation algorithm. Specifically, for a given training dataset comprising essay–score pairs $\mathcal{D} = \{(x_n, y_n) | n = 1, \ldots, N\}$ (where N is the number of training data entries, x_n is the n-th essay defined as a sequence of words, and y_n is the score assigned by a human rater to x_n). The cross-entropy error is defined as $-\sum_{n=1}^{N} \log P_{y_n}(x_n)$, where $P_{y_n}(x_n)$ is the classification probability corresponding to score y_n when essay x_n is input.

We predict the score of essay x using classification probabilities from the trained model as $\hat{y} = \operatorname{argmax}_k P_k(x)$.

2.2 Confidence Estimation

An earlier study [4] proposed three confidence estimation approaches based on this scoring model.

Classification Probability: This approach utilizes the classification probability $P_{\hat{y}}(x)$ as the confidence measure when the predicted score for a target essay x is $\hat{y}$.

Trust score: This approach utilizes an index called the trust score [5] as the measure of confidence.

Gaussian process regression: This approach utilizes a Gaussian process regression (GPR) model, a machine learning model capable of estimating prediction uncertainty.

2.3   Remaining Problems

As described above, the conventional model is designed as a neural classifier in which scores are treated as nominal scale data and cross-entropy error is used as the loss function for model training. Thus, the model treats errors between gold-standard and predicted scores equally, regardless of their magnitude, implying that the scores' ordinal nature is overlooked. In contrast, earlier neural scoring models were generally designed as regressors, treating score data as ordinal. These models are trained with a squared error-loss function considering error magnitudes between gold-standard and predicted scores. Regression-based models generally provide more accurate score predictions than classification-based models, because they consider the scores' ordinal nature. Moreover, enhancing score prediction accuracies improves confidence estimation accuracies.

3   Confidence-Predicting Regressor-Based Scoring Model

From the above, we propose a confidence-predicting regression-based scoring model. The proposed model is based on BERT, with an output layer designed as a Gaussian linear regression. This design enables estimations of predicted score variance, which reflects confidence levels. Specifically, the model constructs the distributed representation vector h using the BERT model for a given essay x, then passes it through output layers

$$\hat{\mu} = sigmoid(\boldsymbol{W}_m \boldsymbol{h} + b_m), \quad \hat{\sigma}^2 = exp(\boldsymbol{W}_v \boldsymbol{h} + b_v), \tag{2}$$

where $\boldsymbol{W}_m$, $\boldsymbol{W}_v$, b_m, and b_v are trainable parameters and $sigmoid(\cdot)$ and $exp(\cdot)$ denote the sigmoid and exponential functions, respectively.

3.1   Model Training

Given training data $\mathcal{D} = \{(\boldsymbol{x}_n, y_n)|n = 1, \ldots, N\}$, model training is performed by using backpropagation to minimize the following negative log-likelihood after standardizing the gold-standard scores y_n to $[0, 1]$:

$$L_{NL} = \sum_{n=1}^{N} \left\{ \frac{\|z_n - \hat{\mu}(\boldsymbol{x}_n)\|^2}{2\hat{\sigma}^2(\boldsymbol{x}_n)} + \frac{1}{2} \log\left(2\pi\hat{\sigma}^2(\boldsymbol{x}_n)\right) \right\}. \tag{3}$$

where z_n is the standardized score corresponding to y_n, and $\hat{\mu}(\boldsymbol{x}_n)$ and $\hat{\sigma}^2(\boldsymbol{x}_n)$ are the predicted mean and variance calculated through Eq. (2) given essay $\boldsymbol{x}_n$.

Here, we interpret Eq. (3). Focusing on $\hat{\sigma}^2(\boldsymbol{x}_n)$, the two terms inside the braces work in opposition to the goal of minimizing L_{NL}. Specifically, when $\|z_n - \hat{\mu}(\boldsymbol{x}_n)\|^2$ is sufficiently small, meaning the predicted score $\hat{\mu}(\boldsymbol{x}_n)$ is close to the standardized gold-standard score z_n, $\hat{\sigma}^2(\boldsymbol{x}_n)$ can be minimized to reduce the second term, thereby decreasing L_{NL}. Conversely, when $\|z_n - \hat{\mu}(\boldsymbol{x}_n)\|^2$ is large, indicating the predicted score is far from the gold-standard score, a larger $\hat{\sigma}^2(\boldsymbol{x}_n)$ increases the denominator of the first term, reducing L_{NL}. The proposed model is thus trained to estimate a smaller variance $\hat{\sigma}^2(\boldsymbol{x}_n)$ for essays where $\hat{\mu}(\boldsymbol{x}_n)$ closely matches the gold-standard score z_n, and a larger variance $\hat{\sigma}^2(\boldsymbol{x}_n)$ otherwise.

3.2   Score Prediction and Confidence Estimation

By this mechanism, the trained proposed model provides the predicted score for any essay x as $\hat{\mu}(x)$. Note that the predicted value $\hat{\mu}(x)$ must be linearly transformed back to the original K-level score scale, similar to conventional regression-based scoring models [6,10]. This transformation is necessary because the sigmoid activation function is applied to the output layer for the function $\hat{\mu}$.

Furthermore, the proposed model represents the confidence level of the predicted score $\hat{\mu}(x)$ through the variance $\hat{\sigma}^2(x)$. A larger variance signifies greater uncertainty in the predicted score. Therefore, we adopt the negative of the variance, $-\hat{\sigma}^2(x)$, as the confidence measure.

4   Evaluation Experiment

In this section, we compare the performance of both score prediction and confidence estimation between the proposed model and the conventional model, using experiments with actual data.

For our experiments, we utilized the Automated Student Assessment Prize (ASAP) dataset, a widely used benchmark dataset for AES studies. The ASAP dataset comprises essays written by English-native students in grades 7 to 10 responding to eight prompts corresponding to essay writing questions. Each essay has one gold-standard score given by human raters. Scores are allocated based on ordered categories, each with varying score ranges.

In this experiment, we evaluated the accuracies of score predictions and confidence estimations for each dataset by conducting five-fold cross-validation for each prompt. We used the quadratic weighted kappa (QWK) and correlation coefficients as metrics for score prediction evaluations. Higher values for both metrics indicate greater score prediction accuracy. For confidence estimation evaluation, we employed reversed pair proportion (RPP) [11] and area under the receiver operating characteristic curve (ROC-AUC) as metrics. Lower RPP and higher ROC-AUC values indicate higher estimating confidence.

Table 1. Evaluation results of score prediction performance

	Model	Prompt								Avg.
		1	2	3	4	5	6	7	8	
QWK	Classifier	0.737	*0.650*	**0.696**	0.802	*0.802*	0.777	0.757	0.543	0.720
	Regressor	*0.790*	**0.655**	0.662	*0.811*	0.789	*0.794*	**0.829**	*0.696*	*0.753*
	Hybrid	**0.812**	0.645	*0.683*	**0.812**	**0.804**	**0.806**	*0.822*	**0.724**	**0.764**
Correlation	Classifier	0.759	**0.667**	**0.698**	*0.807*	**0.810**	0.797	*0.771*	*0.609*	0.740
	Regressor	*0.809*	**0.667**	0.680	**0.815**	0.802	*0.808*	**0.841**	**0.746**	*0.771*
	Hybrid	**0.817**	*0.659*	*0.694*	**0.815**	*0.809*	**0.809**	**0.841**	**0.746**	**0.774**

† Bold numbers indicate best and italic numbers indicate second-best performance.

Table 2. Evaluation results of confidence estimation performance

	Model	Prompt								Avg.
		1	2	3	4	5	6	7	8	
RPP	Class. prob	*0.084*	**0.081**	*0.078*	0.074	**0.082**	**0.082**	**0.061**	0.062	*0.076*
	Trust score	0.093	0.092	0.092	*0.073*	0.093	0.086	0.069	0.079	0.085
	GPR	0.119	0.104	0.092	0.093	0.099	0.100	0.072	*0.045*	0.090
	Regressor	0.107	0.101	0.104	0.094	0.104	0.098	0.071	0.051	0.091
	Hybrid	**0.079**	*0.082*	**0.076**	**0.070**	*0.085*	*0.085*	*0.062*	**0.033**	**0.072**
ROC-AUC	Class. prob	*0.664*	*0.632*	*0.637*	0.640	**0.625**	**0.639**	**0.625**	*0.660*	**0.640**
	Trust score	0.627	0.582	0.570	*0.641*	0.575	*0.622*	*0.574*	0.566	0.595
	GPR	0.524	0.535	0.575	0.555	0.551	0.553	0.484	0.510	0.536
	Regressor	0.571	0.544	0.537	0.543	0.519	0.558	0.491	0.482	0.531
	Hybrid	**0.682**	**0.635**	**0.656**	**0.669**	*0.610*	0.621	0.550	**0.671**	*0.637*

† Bold numbers indicate best and italic numbers indicate second-best performance.

Table 1 shows the results of the score prediction accuracy evaluation, with the *Classifier* row indicating the conventional model results and the *Regressor* row indicating those of the proposed model. The *Avg.* column indicates the average performance across all prompts. The results show that the proposed model outperforms the conventional model in terms of average performance.

Table 2 shows the evaluation results for confidence estimation performance. In that table, the *Class. prob.*, *Trust score*, and *GPR* rows show the results for the conventional three-confidence-estimation methods introduced in Sect. 2.2. The *Regressor* row shows the results for the proposed method. Focusing on the average performance, the proposed method was inferior to the conventional methods in all cases. We can also see that among the conventional methods, using classification probabilities produced higher performance.

These experimental results indicate that while the regressor improves score prediction accuracy, the classifier tends to exhibit higher performance in confidence estimation, suggesting that combining regression-based scoring prediction with classification-based confidence estimation might improve performance in both areas.

5 Confidence-Predicting Hybrid Scoring Model

From the above, we also propose a dual-output neural model, namely a BERT model with regressor and classifier output layers. We refer to this model as the *hybrid model* below.

This hybrid model constructs the distributed representation vector h through a BERT, then inputs it to classification and regression output layers. Given h, the classification layer uses Eq. (1) to calculate the probability P_k that the input essay corresponds to score $k \in \{0, 1, \ldots, K-1\}$, as in the conventional classifier. The regression layer is $\hat{z} = sigmoid(W_s h + b_s)$, where W_s and b_s are trainable parameters. This regression layer is the same as the mean prediction layer of the proposed regression-based model.

Since the model has two output layers, we train it within a multitask learning framework. Multitask learning commonly designs the overall loss function as the weighted linear sum of multiple loss functions [3]. Accordingly, we define the overall loss function for the proposed hybrid model as the weighted linear sum of the mean squared error L_{reg} from the regressor and the cross-entropy error L_{cls} from the classifier. Specifically, given training data $\mathcal{D}$ along with standardized scores $\{z_n | n = 1, \ldots, N\}$, L_{reg} and L_{cls} are defined as

$$L_{reg} = \frac{1}{N} \sum_{n=1}^{N} (z_n - \hat{z}(\boldsymbol{x}_n))^2, \quad L_{cls} = -\sum_{n=1}^{N} \log P_{y_n}(\boldsymbol{x}_n), \tag{4}$$

where $\hat{z}(\boldsymbol{x}_n)$ is the predicted score given by the regression layer for essay $\boldsymbol{x}_n$. We adjust the weights for the weighted sum of these two loss functions using the loss scale balancing method [7].

The hybrid model provides the predicted score for an essay $\boldsymbol{x}$ though the regression layer as $\hat{z}(\boldsymbol{x})$. Note that as with the proposed regression-based model, $\hat{z}(\boldsymbol{x})$ should be linearly transformed to the original K-level score scale.

Furthermore, we use the classification probability obtained from the classification layer to determine the confidence level of a predicted score $\hat{z}(\boldsymbol{x})$. This is based on findings from the above experiments, which revealed that using the classification probability is the best method among the conventional approaches. Specifically, letting $\hat{y}$ be the value of $\hat{z}(\boldsymbol{x})$ linearly transformed back to the original score scale, the classification probability $P_{\hat{y}}(\boldsymbol{x})$ becomes the confidence level.

6   Evaluation Experiment for Hybrid Model

We conducted the same experiments as those described in Sect. 4 using the hybrid model. The *Hybrid* rows in Tables 1 and 2 present the results for score prediction and confidence estimation evaluation, respectively. Focusing on the averaged performance, the hybrid model produced the highest score prediction accuracies in all cases. Additionally, the hybrid model achieved the highest confidence estimation accuracy for RPP and the second-best accuracy for ROC-AUC. Note that the difference in ROC-AUC between the hybrid model and the highest-performing method was minimal, at only 0.003.

We thus conclude that the proposed hybrid model demonstrates equivalent or superior performance as compared to either a regressor or a classifier alone in terms of both score prediction and confidence estimation.

7   Conclusion

We investigated confidence-predicting AES models for improving both confidence estimations and score-prediction performance. Specifically, we first proposed a BERT-based scoring model with an output layer designed as a Gaussian linear regression. However, experiments revealed that while this model improved score prediction accuracy, its performance in confidence estimation was inferior to conventional classification-based methods. From those findings, we proposed a hybrid model that predicts scores based on a regressor and estimates confidence based on a classifier. Experiments confirmed that the hybrid model demonstrates equivalent or superior performance as compared to either a regressor or a classifier alone in terms of both score prediction and confidence estimation.

References

1. Abosalem, Y.: Assessment techniques and students' higher-order thinking skills. Int. J. Second. Educ. **4**(1), 1–11 (2016)
2. Devlin, J., Chang, M.W., Lee, K., Toutanova, K.: BERT: pre-training of deep bidirectional transformers for language understanding. In: Proc. 17th Annual Conference of the North American Chapter of the Association for Computational Linguistics, pp. 4171–4186 (2019)
3. Eigen, D., Fergus, R.: Predicting depth, surface normals and semantic labels with a common multi-scale convolutional architecture. In: Proceedings of IEEE International Conference on Computer Vision, pp. 2650–2658 (2015)
4. Funayama, H., Sato, T., Matsubayashi, Y., Mizumoto, T., Suzuki, J., Inui, K.: Balancing cost and quality: an exploration of human-in-the-loop frameworks for automated short answer scoring. In: Proceedings of International Conference on Artificial Intelligence in Education (2022)
5. Jiang, H., Kim, B., Guan, M., Gupta, M.: To trust or not to trust a classifier. In: Advances in Neural Information Processing Systems, vol. 31 (2018)
6. Ke, Z., Ng, V.: Automated essay scoring: a survey of the state of the art. In: Proceedings of International Joint Conferences on Artificial Intelligence, pp. 6300–6308. Macao, China (2019)
7. Lee, J.H., Lee, C., Kim, C.S.: Learning multiple pixelwise tasks based on loss scale balancing. In: Proceedings of IEEE/CVF International Conference on Computer Vision, pp. 5107–5116 (2021)
8. Li, X., Yang, H., Hu, S., Geng, J., Lin, K., Li, Y.: Enhanced hybrid neural network for automated essay scoring. Expert Syst. (2022)
9. Nadeem, F., Nguyen, H., Liu, Y., Ostendorf, M.: Automated essay scoring with discourse-aware neural models. In: Proceedings of Workshop on Innovative Use of NLP for Building Educational Applications, pp. 484–493 (2019)
10. Uto, M.: A review of deep-neural automated essay scoring models. Behaviormetrika **48**(2), 4459–484 (2021)
11. Xin, J., Tang, R., Yu, Y., Lin, J.: The art of abstention: selective prediction and error regularization for natural language processing. In: Proceedings of Joint Conference of the 59th Annual Meeting of the Association for Computational Linguistics and the 11th International Joint Conference on Natural Language Processing, pp. 1040–1051 (2021)
12. Xue, J., Tang, X., Zheng, L.: A hierarchical BERT-based transfer learning approach for multi-dimensional essay scoring. IEEE Access **9**, 125403–125415 (2021)
13. Yamaura, M., Fukuda, I., Uto, M.: Neural automated essay scoring considering logical structure. In: Proceedings of International Conference on Artificial Intelligence in Education, pp. 267–278 (2023)

An Intelligent System for Chinese Dance Creation Using Generative Artificial Intelligence

Luoxi Wang[1], Shuaijing Xu[1], and Yu Lu[2(✉)]

[1] The Experimental High School Attached to Beijing Normal University,
Beijing 100032, China
[2] School of Educational Technology, Faculty of Education,
Beijing Normal University, Beijing 100875, China
luyu@bnu.edu.cn

Abstract. We present an innovative and practical system for creating Chinese dance using generative artificial intelligence techniques. A full-attention cross-modal transformer is utilized to generate 3D dance motions that match the music and are rendered with realistic visual effects. Furthermore, we fine-tuned a pre-training model on a dedicated dataset to generate easily understandable narrations that enhance users' understanding of the generated dance motion sequences. The experiment was conducted to evaluate the quality of the generated content, and its outcomes suggest the prospective applicability of the proposed system.

Keywords: Generative Artificial Intelligence · Dance Creation · Multimodal Information Generation

1 Introduction

Dance is an art form that holds great cultural significance. Chinese dance is a prominent representative of Chinese civilization, with roots in ancient traditions. It incorporates diverse elements such as martial arts, acrobatics, and traditional opera. In contemporary dance creation and instruction, practitioners, including dancers and choreographers, heavily rely on personal experiences and skills. The process of creating a performance involves several stages, including music selection, choreography, stage design, etc. The high costs and time investments associated with dance limits its influence and dissemination, especially among the younger generation.

To tackle these challenges, we present a practical and innovative system for creating Chinese dance with textual explanations. The system utilizes generative artificial intelligence (GAI) techniques to produce continuous classical dance movements and complete dance passages that correspond to the given classical music. Additionally, it provides textual information to clarify the generated dance for those who are not familiar with Chinese classical dancing. The textual

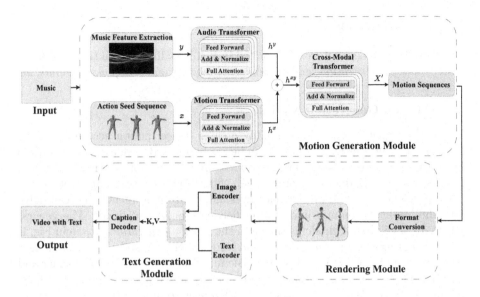

Fig. 1. Overall System Structure

information can aid learners in comprehending the significance of the generated dance movements. In short, the proposed system provides a comprehensive and diverse teaching resource for classical dance education, preserving and advancing classical dance, as well as promoting its wider dissemination among younger generations.

2   Related Work

Music-Generated-Dance is a special application of AI, involving the automatic synthesis of dance that match the rhythm and style of the music. For 2D dance generation, online dance video data is primarily utilized. Ferreira [3] introduces a Graph Convolutional Adversarial Network model, which uses a graph structure to represent joint connections, generating natural and smooth movements. Ye [13] introduces ChoreoNet, mimicking choreography by first generating a dance sketch and then refining poses. 3D dance generation are more suitable for practical applications. There is a scarcity of large-scaled 3D datasets. Li [4] establishes the AIST++ 3D music dance dataset, containing over a million frames of 3D skeletal pose data, encompassing 10 dance styles. Li and Yu [10] propose the Bailando model based on a variational autoencoder and conditional random field, capturing pose variations while ensuring temporal consistency. Li produced the FACT (full-attention cross-modal transformer) that is used for the task of 3D dance generation [5], but they do not include textual explanation. Our work is the only system that specialized in Chinese dance generation, can generate both the dance motion, and the text information for dance.

3   System Design

The overall system structure, as depicted in Fig. 1, is divided into three modules, namely *motion generation module, rendering module*, and *text generation module*.

3.1   Motion Generation Module

The motion generation module mainly creates dance motion according to the style of the input music. The music can be freely selected from external resources by the system users. The full-attention cross-modal transformer (FACT) is used for the task of 3D dance generation [5]. The FACT network can capture the correspondence between musical elements and dance movements, and can delve into the implicit semantic information between music and dance. The general structure of the FACT model is shown in the *motion generation module*in Fig. 1. The model takes as input a short motion seed sequence X and a relatively longer piece of music Y. It first encodes the music input using an audio transformer and the motion seed using a motion transformer into audio and motion embeddings h^y and h^x, respectively. The results are then sent to a cross-transformer where N feature motion sequences X' are generated. The model is self-supervised and regresses to generate predicted future motion sequences.

The FACT model use a full attention mechanism that predicts context at future timestamps based on input at the current timestamp. Compared to causal attention [9], it has a clear advantage in generating unfrozen, more realistic motion sequences. The FACT model concatenates motion and audio sequences and embeds them in a 12-layer cross-model transformer. Its sufficient depth allows to pay enough attention to the music condition and to align the generated classical dance motions with the input music sequences.

3.2   Rendering Module

The rendering module primarily serves the function of visualizing the generated dance motions. The module mainly adopts a software Blender to perform rendering. We rendered about 1200 frames of video. The rendering workflow is illustrated in Fig. 2.

3.3   Text Generation Module

The text generation module uses contrastive language-image pre-training (CLIP) [8] for the task of generating textual information for the dance video. The goal is to provide dance students or instructors with detailed explanations of actions, thereby enhancing the effectiveness of dance teaching and learning. CLIP consists of an image encoder and a text encoder. The image encoder and text encoder transforms the input image and the textual data into a high-dimensional feature vector representing the semantic information of the image through a series of convolutional, pooling, and fully connected layers. Fine-tuning was performed on a dedicated dataset CCDMM using CLIP to process classical dance

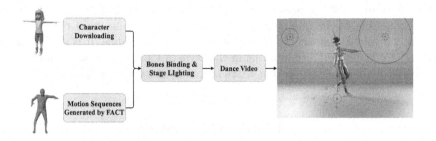

Fig. 2. The Workflow of Rendering using Blender

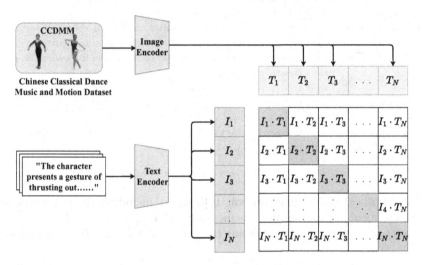

Fig. 3. CLIP Fine-tuning Structure

style action videos generated by the FACT model as input and match the textual descriptions of the actions cut from each frame of the video. The process of fine-tuning on CLIP is illustrated in Fig. 3.

4   System Evaluation

4.1   Data Preparation

We have created a dataset specifically for Chinese classical dance, known as the Chinese Classical Dance Music and Motion Dataset (CCDMM), as there are currently no publicly available datasets for this purpose. CCDMM contains 650 motion image samples and 60 audio segments of classical dance. The dataset's actions were classified into 11 dance motions, guided by professional classical dancers and referencing literature on Chinese classical dance, such as 'Tang Dynasty Music and Dance' and 'Teaching Methods of Chinese Classical Dance

Body Posture'. Table 1 presents the labels. The CCDMM dataset holds significant value due to the uniqueness of Chinese classical dance and the absence of publicly available datasets for classical dance motions online. It is utilized in our project to generate dance motions in the style of Chinese classical dance and to fine-tune models for generating textual commentary that aligns with classical dance actions.

Table 1. CCDMM Labels

Action Name	Action Description
Ti Jin	Dancer presents a posture with arms at the sides, palms pressed
Chong Zhang	Dancer presents a posture of thrusting out the palms
Shanbang Anzhang	Dancer presents a posture with arms raised parallel to the body, palms pressed
Tuozhang Shanbang	Dancer presents a posture of arms resting on the front of the head
Tuozhang Anzhang	Dancer presents a posture of arms resting on the front of the head, palms pressed
Fanshen	Dancer presents a turning posture
Da Tiao	Dancer presents a jumping posture
Daoti Zijinguan	Dancer presents a jumping posture, with the calf pressed to the back of the head
Qiao	Dancer stands on his head and legs are parted
Qinglong Tanzhang	Dancer is shown with the left leg out and the palm leaning out from the side
Shuangtuo Zhang	Dancer presents arms raised and hands above his head

We utilized data augmentation techniques including data augmentation methods such as rotation, flip, zoom, scale and random resize. To increase the quantity of available data, addressing the issue of insufficient data when applying CCDMM.

4.2   Classical Dance Motion Generation

Evaluation Metrics. We evaluate the generated dance motions on motion quality, generation diversity, and motion-music relevance.

We use Fréchet Inception Distance (FID) to evaluate motion quality by extracting meaningful features that represent motion characteristics from both real and generated videos, and then calculating the FID score based on these features by utilizing Geometric Feature Extractor, and Dynamic Feature Extractor.

For generation diversity, we measure it by computing the average Euclidean distance in the feature space for 40 generated motions on the AIST++ test set. The diversity of motions in geometric and dynamic feature spaces is denoted as $Distk$ and $Distg$, respectively.

To evaluate the relevance of motion and music, we use the Beat Alignment Score ($BeatAlign$). This score assesses the correlation between motion and music based on the similarity between their respective beats. The Beat Alignment Score is calculated as the average distance between each motion beat and its nearest music beat:

$$\text{BeatAlign} = \frac{1}{m} \sum_{i=1}^{m} \exp\left(-\frac{\min_{\forall t_j^y \in B^y} \left\| t_i^x - t_j^y \right\|^2}{2\sigma^2} \right) \tag{1}$$

where $B^x = \{t_i^x\}$ is the set of motion beats, $B^y = \{t_j^y\}$ is the set of music beats, and σ is a normalization parameter.

Table 2. Comparison of Evaluation Scores between FACT using CCD0MM for Classical Dance motion Generation and Original FACT [5]

	Motion Quality		Motion Diversity		Motion-Music Corr
	FID_k	FID_g	$Dist_k$	$Dist_g$	BeatAlign
FACT(ours)	36.9034	17.5712	6.7666	6.0198	0.281
FACT [5]	35.35	12.40	5.94	5.30	0.241

Fig. 4. Partial Display of Generated Results

Evaluation Results. Table 2 presents the test results for all evaluation metrics. The results show that our sequence has more diversity, as our score on Dist is higher. This indicates that the FACT model based on the CCDMM dataset performs well. An exemplary portion of the generated continuous classical dance motions was extracted, as shown in Fig. 4. The chosen music has a relatively slow and rhythmic pace. The generated classical dance motions exhibit a calm and gentle quality. The upper limb motions appear to be more diverse than the lower limb motions.

4.3 Text Information Generation

Evaluation Metrics. We evaluated our model through several common metrics used in video caption. *BLEU* [7] is used in machine translation. *METEOR* [2] is for evaluating text generation tasks. *ROUGE* [6] is used machine translation, automatic summarization, and question answer generation. We use the $ROUGE_L$ method. *CIDEr* [11] uses TF-IDF to assign different weights to different n-grams for video caption.

Evaluation Results. We evaluated CLIP fine-tuned on our CCDMM dataset and compared it with MS COCO [1] fine-tuned Show [12], Attend and Tell and MS COCO CLIP. The results of the comparison are shown in Table 3. We see that the fine-tuned CLIP model achieves the highest score in most of the metrics. This confirms the effectiveness of our model.

Table 3. Comparison of Evaluation Scores for CLIP Fine-tuned on CCDMM Dataset, MS COCO Fine-tuned Show, Attend and Tell, and MS COCO CLIP

Metric Name	Ours	MS COCO Show	MS COCO c5	MS COCO c40 [1]
BLEU 1	0.786	0.708	0.663	0.880
BLEU 2	0.768	0.489	0.469	0.744
BLEU 3	0.665	0.344	0.321	0.603
BLEU 4	0.521	0.243	0.217	0.471
METEOR	0.904	0.239	0.252	0.335
$ROUGE_L$	0.694	–	0.484	0.626
$CIDEr - D$	2.274	–	0.854	0.910

Fig. 5. Comparison between Before and After CLIP Fine-tuned

We also compare the results between using the original CLIP and the fine-tuned CLIP, shown in Fig. 5. The original CLIP generates descriptions such as "a close-up of a person in a purple outfit and hat". In contrast, the CLIP fine-tuned to our classical dance dataset CCDMM generates descriptions such as "The character presents a posture with arms raised parallel to the body, palms pressed in front" (Shanbang Anzhang). The original CLIP focuses primarily on the appearance of the person's skin, while the CLIP fine-tuned to our classical dance dataset CCDMM pays attention to the person's movements, thus achieving our goal of generating textual interpretations for dance movements. Figure 6 shows an exemplary generation for the system: the dance sequence is on the right, and the textual information with time steps is on the left.

Part of the Textual Information:

0.0-1.4s The character presents arms raised and hands above his head.

1.4-1.525s The character presents a turning posture.

1.525s-2.875s The figure stands on his head and legs are parted.

2.875s-2.975s The character presents a turning posture.

Fig. 6. System Generation Result

5   Discussion and Conclusion

This work proposes and implements a novel system that generates dance motion sequences based on Chinese classical music. The selected music can be transformed into 3D dance motion sequences with realistic visual effects. Textual descriptions can also be generated to offer semantic meaning and guidance for the dance. The preliminary experiment results partially validate the quality of the generated content and the effectiveness of the proposed system.

The proposed system, while exhibiting considerable potential in dance education, necessitates substantial enhancements to fulfill its promise fully. Among the imperative improvements is the introduction of a novel function designed to assist users in employing the system as a supplementary tool for both instructional and learning purposes. Furthermore, there is a pressing requirement to encapsulate the outcomes of our research and to engineer a comprehensive, end-to-end dance education integration system through software, thereby broadening its accessibility and utility for dance pedagogy.

References

1. Chen, X., et al.: Microsoft coco captions: data collection and evaluation server. arXiv preprint arXiv:1504.00325 (2015)
2. Denkowski, M., Lavie, A.: Meteor universal: language specific translation evaluation for any target language. In: Proceedings of the Ninth Workshop on Statistical Machine Translation, pp. 376–380 (2014)
3. Ferreira, J.P., et al.: Learning to dance: a graph convolutional adversarial network to generate realistic dance motions from audio. Comput. Graph. **94**, 11–21 (2021)
4. Li, R., Yang, S., Ross, D.A., Kanazawa, A.: AI choreographer: music conditioned 3D dance generation with AIST++. In: Proceedings of the IEEE/CVF International Conference on Computer Vision, pp. 13401–13412 (2021)
5. Li, R., Yang, S., Ross, D.A., Kanazawa, A.: AI choreographer: Music conditioned 3D dance generation with AIST++. In: Proceedings of the IEEE/CVF International Conference on Computer Vision (ICCV), pp. 13401–13412 (2021)
6. Lin, C.Y.: Rouge: a package for automatic evaluation of summaries. In: Text Summarization Branches Out, pp. 74–81 (2004)

7. Papineni, K., Roukos, S., Ward, T., Zhu, W.J.: Bleu: a method for automatic evaluation of machine translation. In: Proceedings of the 40th Annual Meeting of the Association for Computational Linguistics, pp. 311–318 (2002)
8. Radford, A., et al.: Learning transferable visual models from natural language supervision. In: International Conference on Machine Learning, pp. 8748–8763. PMLR (2021)
9. Radford, A., Narasimhan, K., Salimans, T., Sutskever, I., et al.: Improving language understanding by generative pre-training (2018)
10. Siyao, L., Yu, W., et al.: Bailando: 3D dance generation by actor-critic GPT with choreographic memory. In: Proceedings of the IEEE/CVF Conference on Computer Vision and Pattern Recognition, pp. 11050–11059 (2022)
11. Vedantam, R., Lawrence Zitnick, C., Parikh, D.: Cider: consensus-based image description evaluation. In: Proceedings of the IEEE Conference on Computer Vision and Pattern Recognition, pp. 4566–4575 (2015)
12. Xu, K., Ba, J., et al.: Show, attend and tell: neural image caption generation with visual attention. In: International Conference on Machine Learning, pp. 2048–2057. PMLR (2015)
13. Ye, Z., et al.: Choreonet: towards music to dance synthesis with choreographic action unit. In: Proceedings of the 28th ACM International Conference on Multimedia, pp. 744–752 (2020)

Leveraging Language Models and Audio-Driven Dynamic Facial Motion Synthesis: A New Paradigm in AI-Driven Interview Training

Aakash Garg$^{(\boxtimes)}$ ⓘ, Rohan Chaudhury$^{(\boxtimes)}$ ⓘ, Mihir Godbole$^{(\boxtimes)}$ ⓘ, and Jinsil Hwaryoung Seo$^{(\boxtimes)}$ ⓘ

Texas A&M University, College Station, TX, USA
{aakash.garg80,rohan.chaudhury,amigo2000,hwaryoung}@tamu.edu

Abstract. The paper introduces an innovative conversational AI chatbot equipped with a visual avatar, specifically tailored for enhancing skills for nursing interviews in real-time. The chatbot utilizes large language models to simulate realistic interview scenarios, enabling nurses to practice and refine their techniques in a dynamic environment. We present a unique integration of avatar animation into the chatbot system adding a significant layer of realism to these interactions, fostering more natural and engaging interactions. We utilize SadTalker, an AI model that uses 3D motion coefficients to animate still images with audio input. The incorporation of certain rendering techniques, detailed in the paper, helps us reach real-time audio-visual generation. The paper emphasizes the potential of such AI-driven tools in revolutionizing nursing education, particularly in developing critical interview skills for Courtroom Trials. Future developments will aim to explore further and expand the capabilities of this framework, investigating its potential applications across a wider spectrum of educational and professional training contexts.

Keywords: Large Language Model · Conversational AI · Real time Audio-Visual Generation

1 Introduction

The advent of conversational AI in nursing education represents a pivotal shift towards interactive and dynamic learning methodologies. With the advancements in large language models, capabilities are extending to create scenarios and personas, assisting students in conceptualizing the best course of action in diverse situations, such as providing care for AI-simulated patients or interacting with simulated family members [1]. This aspect of AI enriches the learning experience, offering practical, hands-on practice in a safe, simulated setting.

A. Garg, R. Chaudhury and M. Godbole—These authors contributed equally to this work.

R. Chaudhury—Work done while studying at Texas A&M University.

© The Author(s), under exclusive license to Springer Nature Switzerland AG 2024
A. M. Olney et al. (Eds.): AIED 2024 Workshops, CCIS 2150, pp. 461–468, 2024.
https://doi.org/10.1007/978-3-031-64315-6_44

While these advancements offer an interactive, risk-free environment for hands-on practice, they complement rather than replace the invaluable mentorship and clinical experiences provided by human educators. Studies [2] reinforce the importance of blending AI with traditional teaching methods, ensuring a comprehensive learning experience that prepares nursing students for real-world challenges.

The primary focus of this paper is to employ conversational AI in nursing education for conducting mock court testimony trials. This innovative approach aims to familiarize nursing students with legal procedures and terminology, preparing them for real-life courtroom scenarios and enhancing their proficiency in legal aspects of healthcare. For this use case, the conversational AI agent will take the persona of a prosecutor who will question the user based on a preset background setting reflecting a court trial. Here the user is a forensic nurse who will use this application as a mock practice for a real-world court trial. This initiative reflects the growing importance of integrating specialized skills, like legal nursing consultancy, into nursing education through advanced AI technologies. Our main contributions are as follows:

- A conversational AI chatbot design tailored to train forensic nurses for mock trial testimonies.
- Integration of an animated avatar generation model in the conversational AI system to enhance naturalness and simulate real-world scenarios, ensuring consistent avatar imagery across one scenario. The character can be changed as per the scenario.
- A mock prosecutor persona using large language models to conduct the mock trial testimony with the user as the forensic nurse.
- Real-time generation of speech and avatar for seamless interaction between the virtual prosecutor and the forensic nurse.

2   Related Works

2.1   Impact of Mock Trial Simulations in Nursing Education

Harding et al. [3] highlight the use of courtroom simulations to teach nursing students about the critical importance of accurate documentation in wound care, emphasizing legal and ethical implications in healthcare practices. Song et al. [4] analyze the role of mock trials in nursing education, particularly focusing on enhancing students' skills in ethical decision-making and evidence-based practice. Hanshaw et al. [5] evaluates the impact of high-fidelity simulation in nursing education on developing various levels of cognitive skills among nursing students, aligning with Bloom's Taxonomy. Tonapa et al. [6] conducts a systematic review demonstrating the effectiveness of high-fidelity simulations in improving learning outcomes for nursing students. Kim et al. [7] examine the impact of a web-based clinical simulation program, FIRST2ACT, on nurses' responses to patient deterioration, indicating the increasing relevance of technology in nursing education.

2.2   Conversational AI in Nursing Education

Applications of conversational AI agents now span the education, customer support, entertainment, and healthcare sectors. In healthcare, recent works have focused on utilizing AI to aid nursing education. Seo et al. [17] focuses on the development of a virtual reality (VR) training program using conversational AI for SBIRT skills training in nursing education. Tomasz et al. [18] mentions that conversational tutoring agents are valuable as supplementary educational tools, offering safe and practical solutions for students to engage with challenging learning scenarios. AI and ChatGPT revolutionize nursing education through customized learning experiences, simulations, easy accessibility, and enhanced efficiency, thereby boosting student engagement, and learning outcomes, and providing significant support to educators [19].

2.3   Image to Video Generation

The field of animating still images from audio has gained significant attention over time due to its wide application across various domains such as News reporting, Job Interviews, etc. Min et al. [21] synthesized realistic videos of a person talking from just a single reference image, ensuring accurate lip synchronization, head poses, and eye blinks. It utilizes a pre-trained generator and image encoder to predict latent representations that match given audio inputs. Lee et al. [22] employs a multimodal approach, integrating sound, image, and text embeddings to produce lifelike videos. To maintain the video's semantic consistency with the audio, the audio is used as the key input to provide temporal context. Similarly, Kumar et al. [23] also focus on creating talking person videos from a single image and an audio signal.

3   Conversational AI Model Development

The recent rapid developments in the field of AI have resulted in conversational AI agents interacting in almost a human-like fashion. We leverage these capabilities and create a personalized conversational AI agent using Large Language Models (LLMs), zero-shot learning, and memory components to improve contextual understanding.

3.1   Advanced AI for Enhanced Conversational Experience

The capabilities of LLMs like Open AI's GPT models have evolved in terms of better contextual understanding and accurate comprehension. Our conversational AI agent uses the GPT-4 [15] model which can produce empathetic and context-aware conversation responses, developing trust between the user and agent.

3.2 Zero-Shot Learning for Context and Flexibility

Zero-shot learning is a powerful tool that can be used to guide the model toward a desired behavior without explicit training [24]. By providing the background information about a prosecutor in a mock court setting along with custom-tailored instructions and guidelines about the interaction with the user, the agent can respond dynamically and adjust its responses, facilitating contextually appropriate interactions with users. Additionally, it strictly ensures that the AI agent does not break out of the mock testimony setting. Instructions in the prompt such as "Remember, the aim is not to intimidate but to prepare the nurse for the types of questions and scenarios they may face in a real court setting. Your questions should encourage thoughtful responses, demonstrate the nurse's competency and professionalism, and highlight the critical role of forensic nursing in the justice system." facilitate appropriate interaction with the users.

Our agent is instructed to assess professional expertise, evaluate evidence collection and handling, probe patient care and interaction, and understand interdisciplinary collaboration.

Although few-shot in-context examples can improve the quality of the generated responses, zero-shot prompting yielded excellent responses without any need for additional examples.

3.3 Natural and Professional Interactions

The AI model representing the prosecutor is designed to simulate a realistic cross-examination process for forensic nurses. The design ensures a professional interaction with the nurse as the AI agent questions the user (nurse) in a structured manner. Since the model is created to train the nurses, it is also designed to be encouraging and thoughtful, enabling natural conversations between the user and the agent.

3.4 Memory Components for Contextual Understanding

We extend and utilize the work by Garcia-Pi, B. et al. [16] on the use of memory components for enhanced memory retention. Three memory components are used: Background Memory, Initial Memory, and Latest Memory. The prompt and the background of the prosecutor are stored together in the Background memory component which is passed into the "content" for the role of "system" in the Chat Completions API call. Based on that as the conversation moves forward we store the initial few exchanges (threshold based on user input) between the role of "user" and "assistant" in the Initial Memory. These are stored in a database and get added to the entire message list when it is passed to the API call. When the initial memory threshold is reached, then a queue data structure (queue size is based on user input) stores the incoming conversation. The queue gets updated as the conversation moves forward and the queue threshold is reached, hence this queue stores the latest conversation between the user and the virtual patient. We call this queue – Latest Memory. Now whenever a new input is received we combine the contents in Background Memory, Initial Memory, and Latest Memory, with the new user input and pass it to the API call to get the response from the GPT-4 model. This enables us to keep the text input to the model within the context length of the model (as the two threshold values can be quite small, around 3 or 4) while ensuring that the model retains most of the important information it requires to continue the conversation. This bears a resemblance to human interaction patterns, particularly when encountering new individuals. We typically recall the initial exchange where personal details such as names are shared. Subsequent conversations tend to focus on recent discourse, often disregarding the historical context between the initial encounter and the present discussion.

4 Avatar Development

Recently, many studies have shown the positive impacts of the use of avatars in the context of education and developing nursing skills [9]. Talking head avatars are crucial in bridging the gap between theoretical and practical clinical skills in pediatric nursing courses [10]. In this section, we delve into the specifics of implementing these avatars and their integration with the conversational AI agents.

4.1 Process of Developing Talking AI Avatar

In the process of creating the static AI avatar image for a prosecutor, Adobe Firefly is used, guided by the prompt: "40-year-old prosecutor in court with a white background". This prompt was meticulously chosen to yield an avatar that visually resonates with the intended professional demeanor and setting. The AI avatar is shown in Fig. 1. The next critical step involved animating this static AI-generated avatar image to synchronize with the generated audio. For this purpose, we took inspiration from the work done by [12], and rendered the talking avatar video which is embedded into our Conversational AI platform. The work focuses on animating still images based on audio input and specializes in syncing facial movements and expressions in an image to correspond with the spoken words in an audio clip.

4.2 Avatar Rendering, Integration, and Testing Details

Zhang et al. [12] generates 3D motion coefficients from audio (including head pose and facial expressions) and then uses a 3D-aware face render to create the talking head videos, which are then enhanced using GFPGAN [11]. We created our model using two face rendering techniques facevid2vid [14], and pi-render [13], and have provided a brief comparison between the two (Please check out the demo videos and the associated code here[1]). We used pi-render for our

---

[1] https://github.com/aakash-garg/Leveraging-Language-Models-and-AudioDriven-Dynamic-Facial-Motion-Synthesis-For-AI-Interview-Training/.

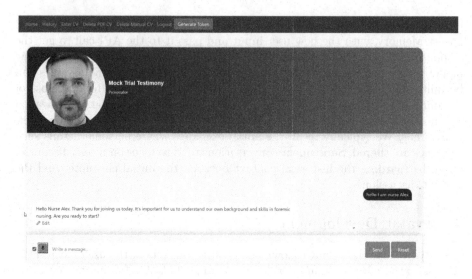

Fig. 1. AI avatar in the Mock trial demo.

application as it has a faster inference and is well-suited for our use case, as we need real-time outputs for seamless conversation. Due to this reason, we didn't use GFPGAN enhancement as it will make the algorithm slower. Our application is written in Flask, the avatar module is written in PyTorch, and our Flask app communicates with the PyTorch module through a Restful API to get the avatar video by taking the audio sample and avatar image.

5   Pipeline Overview

Figure 2 shows the entire pipeline of our application where the input data in either typed text or transcribed speech format is passed on to the Language Model combined with the memory component elements. The Language model then generates response text representing a response from the prosecutor. A text-to-speech model generates speech from the response text. The avatar generation module is called to take the generated speech and the avatar image as input and generate the talking avatar as output, which gets rendered in real-time on the chatbot application.

6   Conclusion and Future Work

This paper presents a conversational AI chatbot, equipped with a talking AI avatar of a prosecutor, specially designed for nursing education for conducting mock court testimony trials. We utilize state-of-the-art large language models and video rendering techniques to create a seamless interface for conversation, preparing nursing students for potential legal scenarios in their careers. Future

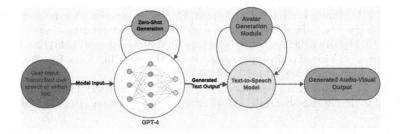

Fig. 2. System Flowchart

work includes integrating emotional intelligence into the AI avatar to allow for more nuanced conversations, as well as analyzing larger applications in domains such as law enforcement and healthcare communication training.

However, such advances raise ethical concerns, such as data privacy, consent, and the possibility of biases. Transparent data usage regulations and informed approval are critical to upholding ethical norms. Particularly in the legal sector, hallucinations in the model can lead to a wrong understanding of the law and associated legal terms. Hence, the associated AI must be grounded and guided to ensure correctness. Furthermore, the potential for AI to replicate or magnify biases necessitates constant monitoring and modifications to AI behavior to ensure fairness and impartiality in simulated interactions. Moving ahead, we will focus on appropriately leveraging AI's potential to improve learning while navigating the ethical environment in education and beyond.

References

1. Daily Nurse: ChatGPT and Its Potential in Nursing Education (2023). https://dailynurse.com/chatgpt-and-its-potential-in-nursing-education/
2. Lee, J., Kim, H.: Analysis of the effect of an artificial intelligence chatbot educational program on non-face-to-face classes: a quasi-experimental study. BMC Med. Educ. **21**(1), 1–11 (2021). https://doi.org/10.1186/s12909-021-02567-5
3. Harding, M., Troyer, S., Bailey, M.: Using courtroom simulation to introduce documenting quality wound care to beginning nursing students. Nurse Educ. **39**(6), 263–264 (2014). https://doi.org/10.1097/NNE.0000000000000078.PMID: 25330261
4. Song, C., Jang, A.: Mock trial as a simulation strategy allowing undergraduate nursing students to experience evidence-based practice: a scoping-review. PLoS ONE **18**(8), e0289789 (2023). https://doi.org/10.1371/journal.pone.0289789
5. Hanshaw, S.L., Dickerson, S.S.: High fidelity simulation evaluation studies in nursing education: a review of the literature. Nurse Educ. Pract. **46**, 102818 (2020). https://doi.org/10.1016/j.nepr.2020.102818. Epub 9 June 2020. PMID: 32623148
6. Tonapa, S.I., Mulyadi, M., Ho, K.H.M., Efendi, F.: Effectiveness of using high-fidelity simulation on learning outcomes in undergraduate nursing education: systematic review and meta-analysis. Eur. Rev. Med. Pharmacol. Sci. **27**(2), 444–458 (2023). https://doi.org/10.26355/eurrev_202301_31040. PMID: 36734697

7. Kim, J.A., Jones, L., Terry, D., Connell, C.: An exploration of nurses' experience following a face-to-face or web-based intervention on patient deterioration. Healthcare **12**(11), 3112 (2023). https://doi.org/10.3390/healthcare11243112

8. Siobhan, O.: Virtual reality and avatars in health care. Clin. Nurs. Res. **28**, 523–528 (2019). https://doi.org/10.1177/1054773819845824

9. Irwin, P., Coutts, R.A., Graham, I.W.: Looking good sister! The use of a virtual world to develop nursing skills. 33–45 (2019). https://doi.org/10.1007/978-981-32-9582-7_3

10. Flood, J.L., Commendador, K.: Avatar case studies: a learning activity to bridge the gap between classroom and clinical practice in nursing education. Nurse Educ. **41**(1), 3–4 (2015). https://doi.org/10.1097/NNE.0000000000000195

11. Wang, X., Li, Y., Zhang, H., Shan, Y.: Towards real-world blind face restoration with generative facial prior. In: The IEEE Conference on Computer Vision and Pattern Recognition (CVPR)

12. Zhang, W., et a.: SadTalker: learning realistic 3D motion coefficients for stylized audio-driven single image talking face animation (2022)

13. Ren, Y., Li, G., Chen, Y., Li, T.H., Liu, S.: PIRenderer, Controllable Portrait Image Generation via Semantic Neural Rendering (2021)

14. Wang, T.-C., Mallya, A., Liu, M.-Y.: One-Shot Free-View Neural Talking-Head Synthesis for Video Conferencing (2021)

15. OpenAI Report: GPT-4 technical report. arXiv:2303-08774 (2023)

16. Garcia-Pi, B., et al.: AllyChat: developing a VR conversational AI agent using few-shot learning to support individuals with intellectual disabilities. In: Abdelnour Nocera, J., Kristín Lárusdóttir, M., Petrie, H., Piccinno, A., Winckler, M. (eds.) INTERACT 2023. LNCS, vol. 14145, pp. 402–407. Springer, Cham (2023). https://doi.org/10.1007/978-3-031-42293-5_43

17. Seo, J., et al.: Development of virtual reality SBIRT skill training with conversational AI in nursing education. In: Wang, N., Rebolledo-Mendez, G., Matsuda, N., Santos, O.C., Dimitrova, V. (eds.) AIED 2023. LNCS, vol. 13916, pp. 701–707. Springer, Cham (2022). https://doi.org/10.1007/978-3-031-36272-9_59

18. Tomasz, S., Moh'd, A., Thomas, K., Kristina, Y.: Development of a conversational agent for tutoring nursing students to interact with patients, pp. 171–182 (2022). https://doi.org/10.1007/978-3-031-32883-1_15

19. Navigating the Pedagogical Landscape: Exploring the Implications of AI and Chatbots in Nursing Education (2023, preprint). https://doi.org/10.2196/preprints.48530

20. Prajwal, K.R., Mukhopadhyay, R., Namboodiri, V.P., Jawahar, C.V.: Learning individual speaking styles for accurate lip to speech synthesis. In: The IEEE/CVF Conference on Computer Vision and Pattern Recognition (CVPR) (2020)

21. Min, D., Song, M., Hwang, S.J.: StyleTalker: one-shot style-based audio-driven talking head video generation. arXiv [Cs.CV] (2022). http://arxiv.org/abs/2208.10922

22. Lee, S.H., et al.: Sound-guided semantic video generation. arXiv [Cs.CV] (2022). http://arxiv.org/abs/2204.09273

23. Kumar, N., Goel, S., Narang, A., Hasan, M.: Robust one shot audio to video generation. arXiv [Cs.CV] (2020). http://arxiv.org/abs/2012.07842

24. Brown, T.B., et al.: Language models are few-shot learners. arXiv preprint arXiv:2005.14165 (2020)

Potential Pitfalls of False Positives

Indrani Dey[1]([✉]), Dana Gnesdilow[1]([✉]), Rebecca Passonneau[2],
and Sadhana Puntambekar[1]

[1] University of Wisconsin-Madison, Madison, WI 53706, USA
{idey2,gnesdilow}@wisc.edu, puntambekar@education.wisc.edu
[2] The Pennsylvania State University, State College 16801, USA
rjp49@psu.edu

Abstract. Automated writing evaluation (AWE) systems automatically assess and provide students with feedback on their writing. Despite learning benefits, students may not effectively interpret and utilize AI-generated feedback, thereby not maximizing their learning outcomes. A closely related issue is the accuracy of the systems, that students may not understand, are not perfect. Our study investigates whether students differentially addressed false positive and false negative AI-generated feedback errors on their science essays. We found that students addressed nearly all the false negative feedback; however, they addressed less than one-fourth of the false positive feedback. The odds of addressing a false positive feedback was 99% lower than addressing a false negative feedback, representing significant missed opportunities for revision and learning. We discuss the implications of these findings in the context of students' learning.

Keywords: AI Accuracy · Automated Feedback · Science Writing

1 Introduction

Studies on artificial intelligence (AI) systems are increasingly exploring how to support classroom activities, by automating routine parts of teachers' work, and allowing them to provide more meaningful support to students [1, 2]. However, AI systems are often imperfect, resulting in frustration and trust issues among users [3], possibly from an incomplete understanding of how AI works and its capabilities. To thrive in a world increasingly permeated by AI, which includes educational spaces, students need to develop a critical understanding of what AI is and how it works so they can competently interact with it and utilize AI-generated output [4]. While it is important that the output is accurate for students to use it effectively, AI systems are not infallible. Therefore, we need to better understand how students respond to errors in AI-generated output, to inform best practices for scaffolding the use of AI in educational contexts. This study investigates the extent to which students addressed false positive and false negative errors in AI-generated feedback in the context of revising their science writing.

AI technologies, particularly Natural Language Processing (NLP) techniques are increasingly being used to assess students' writing, particularly essays that are hard to

A. M. Olney et al. (Eds.): AIED 2024 Workshops, CCIS 2150, pp. 469–476, 2024.
https://doi.org/10.1007/978-3-031-64315-6_45

assess in a timely manner [5], such as written explanations of scientific phenomena that students often struggle with. This automated feedback, tailored to individual student needs, allows for ongoing formative assessment [6]. Further, teachers can use the AI-generated feedback to identify gaps in students' understanding and drive instruction, while students can use the feedback to critically evaluate and refine their work, thereby deepening their understanding and improving their learning outcomes.

The accuracy of automated feedback on writing is an ongoing challenge. Building realistic expectations and understanding of AI-generated outputs in end-users plays a role in their user experience and acceptance of AI technologies [7], which, in turn, can impact their engagement and learning. False positive errors, where an incorrect answer is marked as correct, can be particularly challenging to address, as students may not be able to identify the errors. Studies have found that students often ignore feedback from a false positive output as they may not have identified it as missing or incorrect [8], and may be reluctant to correct something they thought they already had right [9], thus, missing opportunities to revise and learn. On the other hand, while false negative errors (where correct answers are marked as wrong) can lead to more engagement as students spend more time reviewing their answers [9], these errors more readily contribute to students' dissatisfaction and perceptions of unfairness [10], leading to trust issues with the system and a lower propensity for future tool use [3, 11].

While studies have explored students' perceptions and/or revision behaviors using automated feedback on short answers [9, 10] or writing quality on English essays [3, 8, 11], few have investigated student revision behaviors using automated feedback on the science content in essays; fewer have focused on the extent to which students address AI assessment and feedback errors when revising scientific essays. Our previous study found that although students included significantly more science ideas in their revised essays, the NLP system we used made errors about 25% of the time [12]. This study further investigates the nature of the errors – i.e., false positive versus false negative– and whether students addressed the feedback by revising ideas in their essays (or not) when they received erroneous feedback. The research questions guiding our study are:

1. What were the rates of false positive and false negative feedback errors generated by our NLP system?
2. When students received erroneous feedback, did they address the false positive or false negative feedback in similar ways?
3. To what extent did addressing erroneous feedback impact students' learning?

2 Methods

2.1 Study Design and Context

A total of 238 students from three 8th-grade public middle school science classrooms in the midwestern US participated in this study ($n_1 = 96$, $n_2 = 80$, and $n_3 = 62$). Students conducted experiments using a virtual roller coaster simulation to explore relationships between height, mass, and energy. They wrote an essay using data from their trials to explain the scientific phenomena behind their roller coaster design. They submitted these essays to our NLP system, PyrEval (described below), which assessed the essays and provided automated feedback. Students were supposed to use this feedback to identify missing ideas, revise, and submit their revised essays to PyrEval for re-assessment. Students were aware they were receiving AI-generated feedback that may not be 100% accurate. Students took the same multiple-choice test before and after the unit.

2.2 PyrEval Assessment and Feedback

PyrEval, the NLP system that provided automated feedback on students' writing in this study, uses a wise-crowd model to identify weighted vectors of key content ideas [13], known as content units or CUs; more important ideas are more highly weighted. For this unit, PyrEval identified 6 high-weighted CUs aligned with the important ideas or relationships students should have included in their essays (see Fig. 1).

Once students submitted their essays, PyrEval parsed each essay into separate sentences and examined each sentence for the presence or absence of each of the 6 highly weighted CUs. If PyrEval detected a certain CU, it would produce a vector score of 1, whereas a vector score of 0 indicated that PyrEval did not detect that CU in the essay. These vector scores were presented to the students in the form of a feedback chart (see Fig. 1), indicating which ideas may have been present or missing from their essays. The chart also provided a "My Confidence" column, indicating PyrEval's estimated accuracy in detecting or not detecting a particular CU in the essay. Students were asked to also attend to this column when considering what feedback to address when revising.

Feedback		My Confidence
Height and Potential Energy	✓	Medium
Relation between Potential Energy and Kinetic Energy	?	High
Total energy	?	Low
Energy transformation and Law of Conservation of Energy	?	High
Relation between initial drop and hill height	✓	Medium
Mass and energy	✓	High

Fig. 1. Sample feedback chart, showing each CU, if it was detected (green check mark) or not (orange question mark) in the essay, and PyrEval's approximate accuracy about the detection. (Color figure online)

The vector scores for each essay were recorded in the backend in the following (example) format: [1, 0, 0, 1, 0, 1], with a 1 or 0 indicating the presence or absence of an idea in that essay, respectively. Thus, in the above example, PyrEval detected CUs 0, 3, and 5 in a particular essay, but did not detect CUs 1, 2, and 4. These vector scores were then used to assess student's performance for each essay as well as PyrEval's accuracy.

2.3 Data Sources and Analyses

Out of the students who submitted both an initial and revised essay, we randomly chose 20 students from each teacher's classes, for a total of 60 students and 120 corresponding essays (60 × (1 initial essay + 1 revised essay)). For an in-depth look into how students responded to erroneous automated feedback, we first determined the percent of false positive and false negative feedback errors in their initial essays and then examined whether students addressed the feedback by making revisions. As false positive errors can result in missed learning opportunities, we also used students' scores on the pre to post content knowledge test and their responses to erroneous feedback to understand how students' responses (or lack thereof) may have impacted their science learning.

PyrEval Accuracy. To assess PyrEval's accuracy, two researchers independently coded 20% of the 60 students' initial and revised essays for each content unit, and compared their codes to PyrEval's vector scores. If PyrEval determined the presence of a CU in an essay when it was absent, we considered it a *false positive error*. If PyrEval failed to detect a CU present in the essay, we considered it a *false negative error*. The two coders achieved substantial inter-rater reliability with Kappa = 0.768 [14]. Discrepancies were resolved through discussion and one researcher coded the remaining data. We calculated the percent error by adding the total false positive and false negative errors, dividing it by 360 (60 essays x 6 CUs per essay), and multiplying by 100.

Addressing Feedback. To assess whether students addressed the automated feedback or not, we parsed each student's initial and revised essay into separate sentences and then compared each sentence to examine the revisions. We then coded which CU was addressed for each revision, i.e., each update to an existing idea or an addition of a new idea. Two researchers independently coded 15% of the 60 students' initial and revised essays to determine which CUs students addressed in their revisions, achieving almost perfect inter-rater reliability (Kappa = 0.911). One researcher then coded the remaining data. We used this data in a Chi-square test of independence to explore whether there were differences in the proportion of students who addressed false positive versus false negative feedback. We also used this data to perform a logistic regression to determine whether the type of feedback errors and students' initial essay scores would predict whether they would address erroneous feedback or not.

Physics Content Knowledge Test. Students took a multiple-choice content knowledge test assessing their understanding of the relationships between the height of the initial drop and mass of the roller coaster car and the amount of energy and speed on the ride. It also assessed their understanding of energy transformation and conservation. Students could score a maximum score of 11 points on the test. We used this data to explore the extent to which addressing erroneous feedback may have impacted students' science content learning, conducting a multiple linear regression analysis.

3 Results

We will first provide the descriptive statistics of PyrEval's performance in detecting CUs in the initial essays. We will then present the extent to which students addressed the PyrEval feedback, followed by the results of the Chi-square and regression analyses.

PyrEval Accuracy. PyrEval searched for a total of 360 CUs in students' initial essays. We found that PyrEval was 74.7% accurate in correctly identifying whether the students included or did not include the CUs. On the other hand, we found that PyrEval made errors 25.2% of the time, with a total of 91 errors. Of these errors, 73.6% were false positive errors and 26.4% were false negative errors. This means that PyrEval was nearly three times as likely to make a false positive rather than a false negative error.

Nature of Students' Revisions in Response to Automated Feedback. Of our 60-student sample, 96.7% revised by updating one or more existing ideas or adding new ideas. Two students made no revisions and resubmitted their initial essays. PyrEval identified a total of 105 CUs that were missing from all initial essays (including 24 false negatives), of which 76.1% were addressed in revisions. This indicated that the majority of students revised based on PyrEval's feedback that a CU was missing from their essays. However, there were 67 instances where PyrEval gave students false positive feedback. In this case, we found that only 15 of these false positive errors were addressed.

To examine whether students were significantly more likely to address false negative or false positive feedback errors using a chi-squared analysis of independence, we created a contingency table (Table 1) for the 91 total errors by PyrEval, indicating whether errors were false positive or false negative and if students addressed them or not.

Table 1. Contingency table showing the frequency of whether students addressed or did not address false positive or negative errors made by PyrEval.

	Addressed feedback	Did not address feedback	Total
False positive	15 (0.22)	52 (0.78)	67
False negative	22 (0.92)	2 (0.08)	24
Total	37	54	91

Table 1 shows that while 92% of the students addressed false negative feedback, only 22% addressed false positive feedback. We found that a significantly higher proportion of students addressed false negative feedback than false positive feedback ($X^2_{(1, N = 91)}$ $= 35.150, p < 0.0001$) when revising their essays.

To ensure these findings were not based on students' performance on their initial essays or the teacher they had, we performed a logistic regression. In the model, the likelihood of a feedback being addressed or not was considered the outcome variable, the error type as the independent variable, and students' initial essay CU score and their teacher as the covariate. The logistic regression output revealed a significant coefficient for addressing false positive feedback (-3.62233), when addressing false negative feedback was the baseline category ($p < 0.05$). The odds of a student addressing a false positive feedback was 99% lower than addressing a false negative feedback, if all other variables are constant, given an odds ratio of $\exp(-3.62233) = 0.026$.

Thus, students were overwhelmingly more likely to address a false negative error than a false positive error, leading us to investigate whether there was a relationship between failing to address a false positive error and students' science learning. We focused only on false positive feedback errors for two reasons: first, students addressed most of the false negative feedback (see Table 1); second, studies found that failing to address false positive feedback on students' short answers negatively impacted learning outcomes [9], and we wanted to explore this for students' longer writing pieces.

Relationship Between Not Addressing False Positive Feedback and Science Learning. Of the 60 students in our sample, 43 received at least one false positive feedback. Of these forty-three students, 34 had taken both the pre and post-tests. Thus, we had the complete data for these 34 students, which was then used for the multiple linear regression. We calculated the percentage of false positive feedback students did not address by dividing the total number of false positives each student addressed by the total number of false positives received, then multiplied by 100.

We conducted a multiple linear regression, using students' post-test scores as the dependent variable, the percent of false positive errors addressed as the independent variable, and their pre-test score and initial essay score as covariates. The model explained 17.9% of the variance ($R^2 = 0.1786$, $F_{(3,30)} = 2.175$, $p = 0.1$). While there was a negative relationship between the percent of false positive errors not addressed and post-test scores, it was not statistically significant ($x = -0.5844$, $p = 0.3415$). The model also indicated slight positive relationships with pre-test scores ($x = 0.1783$, $p = 0.1581$) and essay scores ($x = 0.3659$, $p = 0.0692$), but neither were statistically significant.

4 Discussion

While automated writing assessments can support students' learning, students may not effectively interpret and utilize the AI-generated feedback, especially when the system provides inaccurate feedback. Our study investigated whether students responded differently in addressing AI-generated feedback with false positive or false negative errors, to understand if students were critically examining the AI feedback to make targeted revisions based on what was truly missing in their essays.

We found that PyrEval made errors about the presence or absence of science ideas in essays about one-fourth of the time and that it was three times more for a false positive error than a false negative error. Not only was the potential for receiving a false positive error much higher than a false negative error, students were significantly more likely to address a false negative error than a false positive error, thus magnifying the potential for missed learning opportunities.

The majority of students had addressed at least some of PyrEval's feedback, suggesting that they were not averse to making revisions, as also seen in other studies [3, 11]. Therefore, false positive feedback errors represent lost opportunities for students to address missing ideas. Although we found that students who did not address false positive feedback had lower learning outcomes, unlike other studies, it was not significant [9]. This may be due to a few reasons. First, there may be differences in students' revisions on shorter writing assignments [9] versus long essays and in different subjects. Second, our sample size was limited. Third, since the pre and post-tests were identical, recall bias may have further affected the internal validity. Another potential confound could be that students participated in other science activities apart from writing their essays in this unit, which may have influenced their conceptual learning. Further, students may differ in their understanding of latent relationships between CUs, the examination of which was beyond the scope of this study (e.g., students do not mention CU2 if they mention CU3). Future research may address and expand on these issues.

Although students addressed all false negative errors in our study, it could create trust issues with the feedback. This may lead to disengagement with the AI systems integrated with schoolwork, which in turn, could affect learning outcomes. Other studies investigating user experiences with AI systems also found differences in users' trust, participation with, or acceptance of the AI system, such as a conversational agent [15] or a scheduling assistant [7]. These studies highlight imperfections in AI outputs and discuss the trade-off between favoring a false positive versus negative error. Despite reducing the error in potentially more harmful contexts, there is still a reliance on technology that may cause other unanticipated issues in students' participation or learning.

Thus, helping learners understand how AI output may contain potential errors and how to identify and address them, may help mitigate some of the effects of both false positive and false negative errors, as well as provide students with more agency over their learning [4]. Instead of passively following automated feedback, students can be encouraged to work in partnership with the technology to mindfully engage with the feedback and critically apply it to improve and develop understanding [16]. Thus we recommend that developing these competencies should be considered as a part of AI-literacy as well as AI-related curricula. Our future work will be to investigate students' perceptions and actions on using AI-generated output and help inform how to help learners more effectively use automated feedback to maximize their learning.

Acknowledgments. We thank the students and teachers who participated in this study.

References

1. Holstein, K., McLaren, B.M., Aleven, V.: Co-designing a real-time classroom orchestration tool to support teacher-AI complementarity. J. Learn. Anal. **6**, 27–52 (2019)
2. Holstein, K., McLaren, B.M., Aleven, V.: Designing for complementarity: teacher and student needs for orchestration support in AI-enhanced classrooms. In: Isotani, S., Millán, E., Ogan, A., Hastings, P., McLaren, B., Luckin, R. (eds.) AIED 2019. LNCS, vol. 11625, pp. 157–171. Springer, Cham (2019). https://doi.org/10.1007/978-3-030-23204-7_14
3. Roscoe, R.D., Snow, E.L., McNamara, D.S.: Feedback and revising in an intelligent tutoring system for writing strategies. In: Lane, H.C., Yacef, K., Mostow, J., Pavlik, P. (eds.) AIED 2013. LNCS, vol. 7926, pp. 259–268. Springer, Heidelberg (2013). https://doi.org/10.1007/978-3-642-39112-5_27
4. Markauskaite, L.: Rethinking the entwinement between artificial intelligence and human learning: what capabilities do learners need for a world with AI? Comput. Educ. Artif. Intell. **3**, 100056 (2022)
5. Ramesh, D., Sanampudi, S.K.: An automated essay scoring systems: a systematic literature review. Artif. Intell. Rev. **55**(3), 2495–2527 (2022)
6. Ai, H.: Providing graduated corrective feedback in an intelligent computer-assisted language learning environment. ReCALL **29**(3), 313–334 (2017)

7. Kocielnik, R., Amershi, S., Bennett, P. N.: Will you accept an imperfect AI? Exploring designs for adjusting end-user expectations of AI systems. In: Proceedings of the 2019 CHI Conference on Human Factors in Computing Systems, pp. 1–14 (2019)

8. Dodigovic, M., Tovmasyan, A.: Automated writing evaluation: the accuracy of grammarly's feedback on form. Int. J. TESOL Stud. 3(2), 71–87 (2021)

9. Li, T. W., Hsu, S., Fowler, M., Zhang, Z., Zilles, C., Karahalios, K.: Am I wrong, or is the autograder wrong? Effects of AI grading mistakes on learning. In: Proceedings of the 2023 ACM Conference on International Computing Education Research, vol. 1, pp. 159–176 (2023)

10. Hsu, S., Li, T. W., Zhang, Z., Fowler, M., Zilles, C., Karahalios, K.: Attitudes surrounding an imperfect AI autograder. In: Proceedings of the 2021 CHI Conference on Human Factors in Computing Systems, pp. 1–15 (2021)

11. Roscoe, R.D., Wilson, J., Johnson, A.C., Mayra, C.R.: Presentation, expectations, and experience: sources of student perceptions of automated writing evaluation. Comput. Hum. Behav. 70, 207221 (2017)

12. Gnesdilow, D., et al.: The impact of middle school students' writing quality on the accuracy of the automated assessment of science content. In: Proceedings of ISLS (2024)

13. Gao, Y., Chen, S., Passonneau, R. J.: Automated pyramid summarization evaluation. In: Proceedings of the 23rd Conference on Computational Natural Language Learning (CoNLL), pp. 404–418. Association for Computational Linguistics (2019)

14. Stemler, S.: An overview of content analysis. Pract. Assess. Res. Eval. 7(17), 137–146 (2001)

15. Do, H.J., Kong, H.K., Tetali, P., Lee, J., Bailey, B.P.: To err is AI: imperfect interventions and repair in a conversational agent facilitating group chat discussions. Proc. ACM Hum.-Comput. Interact. 7, 1–23 (2023)

16. Salomon, G., Perkins, D.N., Globerson, T.: Partners in cognition: extending human intelligence with intelligent technologies. Educ. Res. 20(3), 2–9 (1991)

Predictive Modelling with the Open University Learning Analytics Dataset (OULAD): A Systematic Literature Review

Lingxi Jin[1] , Yao Wang[2], Huiying Song[3], and Hyo-Jeong So[1(✉)]

[1] Department of Educational Technology, Ewha Womans University, Seoul, South Korea
{jinlingxi,hyojeongso}@ewha.ac.kr
[2] National Institute of Education, Nanyang Technological University, Singapore, Singapore
nie23.wy3836@e.ntu.edu.sg
[3] Department of Computer Science, Ewha Womans University, Seoul, South Korea
shy114@ewha.ac.kr

Abstract. Higher education has experienced an unparalleled digital transformation, driven by the widespread adoption of online learning with massive users, which has risen to an explosive growth in the generation and analysis of student-related data. Within this transformation, predictive modeling has emerged as a useful tool for predicting critical indicators in the learning process, encompassing students' academic performance, class retention, and dropout rates. With this backdrop, this study aims to conduct a systematic review of recent publications focused on predictive modeling, with a specific emphasis on the Open University Learning Analytics Datasets (OULAD). Following the PRISMA process, we identified 17 research articles published from 2017 to 2024, concentrating on OULAD in higher education. For our analysis, we categorized the purpose of predictive modeling into three types: (a) predicting students' performance, (b) identifying at-risk students, and (c) predicting student engagement. The central focus lies on the identification of algorithms predominantly employed in these studies, including machine learning, deep learning, and statistical models. By investigating the methodologies and algorithms employed, this review informs researchers in learning analytics and educational data mining of the potential opportunities and challenges associated with predictive modeling using OULAD in higher education.

Keywords: Predictive modelling · Open University Learning Analytics Dataset (OULAD) · Educational data mining (EDM)

1 Introduction

With the widespread adoption of Massive Open Online Courses (MOOCs), the scale of student data collection has reached unprecedented levels. Educational institutions now have access to vast amounts of data on students' learning patterns, preferences, and progress. This wealth of data enables educators to personalize instruction, ultimately enhancing student engagement and success. This digital revolution has not only transformed the delivery methods of instruction but has also given rise to the field dedicated to

A. M. Olney et al. (Eds.): AIED 2024 Workshops, CCIS 2150, pp. 477–484, 2024.
https://doi.org/10.1007/978-3-031-64315-6_46

examining learner data. Learning analytics [4], academic analytics and educational data mining [7] are three prominent fields in the domain of leveraging learner-generated data and analytical models to uncover valuable insights into learning processes. In particular, learning analytics focuses on utilizing these data to unpack learning processes and social connections while making predictions and recommendations about learning outcomes [18].

Within this context, the Open University Learning Analytics Dataset (OULAD) which is publicly accessible, becomes a valuable resource for predictive modeling. OULAD includes a wealth of learning-related data, such as study duration, access frequency, and course grades, offering researchers valuable insights into student learning behaviors [13]. The data of OULAD is utilized to model training that is authentic and reliable, supporting the development of future virtual learning environments and enabling other educational institutions to use the data to innovate and develop their instructional methods. However, several challenges persist in the practical implementation of predictive modeling, including the quality and integrity of data, the efficacy of feature selection, as well as the complexity and interpretability of models [13].

With this backdrop, this paper aims to provide a systematic literature review of predictive modeling research using OULAD. By investigating the methods and algorithms employed, this review seeks to inform researchers in learning analytics and educational data mining of the potential opportunities and challenges associated with predictive modeling using OULAD in higher education. The research questions that guided this study are as follows: **RQ1**: What are the research trends of the predictive modeling using OULAD? **RQ2**: What are the research themes in predictive modeling using OULAD? **RQ3**: What prediction models are used in OULAD in the literature?

2 Methods

2.1 Data Sources and Search Strategy

This systematic review followed the PRISMA (Preferred Reporting Items for Systematic Reviews and Meta-analyses) 2020 guidelines to ensure the rigor in reporting [16]. A wide range of databases was used as our primary source, including IEEE Xplore, Scopus, ScienceDirect, and Web of Science. ERIC and ProQuest were included as additional sources to expand the literature in the educational context relevant to our study. We selected these databases since most publication channels led by the Society of Learning Analytics Research (SoLAR), the International Educational Data Mining Society (IEDMS), and the International Artificial Intelligence in Education Society (IAIED) are available in these databases. The literature search was conducted in Dec 2023 using two sets of search terms as shown in Table 1. The first set of search terms consists of different variations of the term OULAD. The second set includes terms related to predictive modelling. In total, 3,062 papers that contain the two sets of search terms within the title and abstract were identified in the initial search process.

Table 1. Search terms used in each category.

Categories	Search terms in the category
Open University Learning Analytics Dataset	OULAD, OUL*, OULA* dataset, Open University, Learning analytics
Predictive modelling	Prediction model, Predictive analytics, Educational Data Mining

2.2 Inclusion and Exclusion Criteria

The overview of the search protocol is summarized in Fig. 1. This protocol adhered to the PRISMA process [16] and was reconstructed utilizing the template provided by Page et al. [16]. To identify suitable studies for this review, we employed the following inclusion and exclusion criteria. First, we limited the publication period from the beginning of public access to the OULAD in 2017 to January 2024 to ensure the relevance of the literature. Next, we limited the search to peer-reviewed journal articles, case studies, and conference proceedings for quality assurance. Duplicate copies of the studies were removed. Based on this criterion, books and review articles were also excluded from the search. The search was further limited to papers published in English to match the authors' language proficiency, reducing the search to 1,021 papers. Next, the titles and abstracts were scanned to ensure that the chosen articles were relevant to the main goal of this research. By this criterion, we excluded studies such as the use of non-OULAD for predictions (e.g., the use of the International University of La Rioja educational dataset to predict learner performance by [10] and non-education contexts [22]. This reduced the number of shortlisted articles to 29. The full text of each study was reviewed and 12 articles that did not use OULAD for predictive modelling were further removed. This PRISMA process led to the identification 17 articles eligible for the final analysis.

2.3 Open University Learning Analytics Dataset

Open University Learning Analytics Dataset includes 22 modules with 32,593 Open University registered students. It mainly contains three types of information: demographic data, student performance data, and learning behavior data [3]. First, the demographic data covers basic information about students, including age, gender region, and previous education background. These data are crucial for understanding students' learning backgrounds and characteristics. Second, student performance data mainly reflects students' learning achievements and results during their studies at OU, including their evaluation results. These data help researchers understand students' learning progress and mastery levels. Third, learning behavior data documents students' online learning activities, such as daily clicks, which can reveal students' learning habits and patterns.

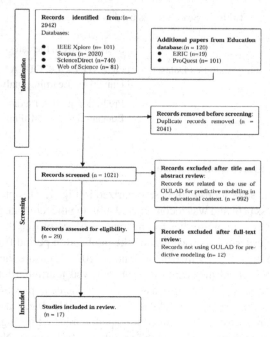

Fig. 1. PRISMA Flowchart for Presented Systematic Review (Framework adapted and modified from Page et al.).

3 Findings

3.1 Bibliometric Analysis

The results reveal a trend in the publication timeline, with all 17 papers collected being published after 2021. This indicates a surge in interest in using the OULAD for predictive modelling in recent years. The year 2022 stood out as the highest number of papers to the field (n = 7). Regarding the type of publication, 17 articles were evenly distributed, with 9 published in academic journals and 8 published in conference proceedings.

3.2 Predictive Models

Three overarching themes emerged as the main focus of predictive models: (a) student performance, (b) students at risk and (c) student engagement, as illustrated in Table 2. In this section, we describe the diverse aspects of predictive modelling based on the OULAD employed in the research and the results obtained.

Student Performance Prediction. The first theme, student performance, focuses on the classification of final grades and the prediction of students' pass or fail outcomes in particular courses. The studies employed various models and techniques including the BART model [8], RNN + LSTM [12], Logistic Regression [14, 15], K-Nearest neighbour [15, 17], random forest [8, 26], Gradient-Boosted Tree [15], 1D-CNN [15],

Table 2. Aspect of the predictive modelling based on OULAD.

Theme	Code	Authors	Years	Dataset Features	Algorithms	Best Accuracy
Student Performance	C1. Student Academic Performance	Almahdi et al. [2]	2023	Engagement, Demographic, Performance	MLP, ANN, NB, SVM	MLP (98.94%)
		Drousiotis et al. [8]	2021	Engagement, Performance	BART, RF, DT	BART (92%)
		Al-Tameemi et al. [3]	2021	Interaction	DL	-
		Kukkar et al. [12]	2024	Behavior, Engagement	RNN, LSTM, ML	RNN + LSTM + RF (96.78%)
		Liu et al. [15]	2022	Behavior, Performance	LSTM, 1D-CNN, RF, GBT, K-NN, LR	LSTM (90.25%)
		Lakshmi et al. [14]	2023	Relationship, Performance	GCN, SVM, LR, FFN	PRISM+GCN (almost 82%)
		Qiu et al. [17]	2022	Engagement, Demographic, Performance and Emotion	SVC(R), SVC(L), Naïve Bayes, KNN(U), KNN(D), and Softmax	BCEP Model (97.4%)
		Wang et al. [26]	2022	Demographic, Performance, Behavior	Light GBM, XGBoost, GBDT (Gradient Boosting DT, RF, AdaBoost)	-
	C2. Early Prediction	Tonghui. [24]	2023	Demographic, Performance	ML	-
	C3. Risk Students Performance	Souai. [20]	2022	Sequential trajectories, Behavior	BLSTM	96.90%
	C4. MOOC Performance	Hao et al. [11]	2022	Performance Behavior	LSTM, 1D-CNN	LSTM (90.25%)
At-risk student	C5. Student Dropouts	Adnan et al. [1]	2021	Engagement, Performance, Behavior	RF, SVM, ET, AdaBoost, Gradient Boost	RF (91%)
		Shafiq et al. [21]	2022	Demographic, Behavior	LSTM, RTV-SVM, RF, ANN, J48, DT, XGBoost, Naïve Bayes	LSTM (97.25%)
		Ali et al. [5]	2021	Engagement, Performance	Ada Boost, RF, SVM, LR, KNN	-
	C6. Learning Trajectories	Gupta et al. [9]	2022	Engagement, Demographic, Behavior	HMM, LR	HMM (94.83%)
	C7. Predict Student Success	Waheed et al. [27]	2023	Engagement, Demographic, Performance	LSTM, GBT, SVM, ANN, LR, DT, KNN	LSTM (89.77%)
Student Engagement	C8. Student Participation	Raj et al. [18]	2022	Behavior, Performance, Interaction	RM, KNN, DT, LR	RF (94.12%)

GCN [14], and Light GBM [26] to predict student academic performance based on diverse datasets such as behavioral data, demographic data, and interaction logs. For instance, Almahdi and Sharef [2] proposed an MLP network model, while Drousiotis et al. [8] employed the BART model, both of which effectively processed complex datasets including behavioural and demographic data to predict student performance. These models successfully predicted whether students passed or failed courses. Additionally, Kukkar et al. [12] demonstrated the superiority of LSTM algorithms in predicting student academic performance by integrating RNN + LSTM feature extraction techniques and various machine learning models.

At-risk Student Prediction. The second theme focuses on identifying at-risk students, understanding the factors that affect student performance, and early prediction of dropouts and withdrawals during or after a course. Regarding student dropouts, Adnan et al. [1] examined the challenges faced by online learners, emphasizing high dropout rates and low engagement levels. Their proposed predictive model, leveraging machine learning and deep learning algorithms, identified at-risk students, thereby enabling effective intervention strategies. Similarly, Shafiq et al. [21] introduced a machine learning-based predictive model focused on identifying students prone to attrition in virtual learning environments. Additionally, Ali et al. [5] examined the issue of dropout rates in MOOCs by employing sequential classification approaches and comparing various classification models to predict the likelihood of student dropout rates.

Student Engagement. The final theme, student engagement, indicates the degree of interaction and active participation of students during the learning process. For instance, Raj et al. [18] predicted students' engagement in a course by analyzing data on their behaviours in a virtual learning environment (VLE). The researcher utilized various features, including student activities, interactions, and scores in the VLE, to construct a robust model that accurately predicts student engagement. Through experimentation, it was determined that the Random Forest classifier yielded optimal results within the current setup. Leveraging the Random Forest classification algorithm resulted in an impressive accuracy rate of 95%, precision of 95%, and correlation coefficient of 98%.

4 Discussion and Conclusion

This paper systematically reviews recent research in predictive modelling using the OULAD to identify current trends and advances in the field. The findings indicate that most of the existing publications use predictive modelling to assess student performance and predict those students who are at risk of failing or dropping out of school. At the same time, less attention has been paid to student engagement. In terms of the techniques used for prediction, machine learning models including decision trees, random forest, support vector machines (SVM), logistic regression, gradient boosting algorithms, and k-nearest neighbours (KNN) have been used with significant frequency. Statistical models such as linear regression and generalized linear models have also been used in some studies to explore the relationship between predictor variables and outcomes of interest. Deep learning models including neural networks convolutional neural networks (CNNs), recurrent neural networks (RNNs), and auto encoders have gained popularity in recent years for their ability to handle large-scale data and extract meaningful patterns from complex data. The results of these studies vary depending on the chosen model, the characteristics of the dataset, and the specific research question. Generally, machine learning and deep learning models demonstrate good predictive performance in most cases, especially when dealing with complex and non-linear relationships in the data. However, it is essential to consider the limitations of each model and to evaluate their performance using appropriate evaluation methods.

Based on the key findings, we present some areas for future research directions. First, researcher need to consider the potential difference of predictive power and accuracy due to the limited availability of training data and the lack of interpretability of the model. Future research should not only focus on the accuracy and performance of the model, but also examine the interpretability and explain ability of the model. This will help increase trust in the model's predictions and make it easier for educators to understand and accept the model's results. At the same time, efforts should be made to improve the quality and quantity of training data to reduce the performance differences of models in different educational scenarios. Second, it is necessary to strengthen the integration of educational background and theoretical knowledge with machine learning models to improve the model's ability to understand and solve educational problems. As Baker and Yacef [6] aptly pointed out, understanding students' educational backgrounds, applying relevant theories, and considering various phenomena in educational settings are crucial for effective performance prediction. Therefore, future research should comprehensively consider these factors to enhance the accuracy and practicality of predictive models. Third, it is important to note that the literature surveyed lacks empirical testing of predictive models in real instructional contexts and other virtual learning environments. Hence, future research should address this gap by conducting empirical studies to evaluate predictive models' effectiveness in diverse educational settings.

References

1. Adnan, M., et al.: Predicting at-risk students at different percentages of course length for early intervention using machine learning models. IEEE Access. **9**, 7519–7539 (2021)
2. Almahdi, A.A., Sharef, B.T.: Deep learning based an optimized predictive academic performance approach. In: 2023 International Conference on IT Innovation and Knowledge Discovery, pp. 1–6. IEEE (2023)
3. Al-Tameemi, G., et al.: A deep neural network-based prediction model for students' academic performance. In: 2021 14th International Conference on Developments in eSystems Engineering, pp. 364–369. IEEE (2021)
4. Avella, J.T., Kebritchi, M., Nunn, S.G., Kanai, T.: Learning analytics methods, benefits, and challenges in higher education: a systematic literature review. Online Learn. **20**(2), 13–29 (2016)
5. Ali, H.A., Mohamed, C., Abdelhamid, B., El Alami, T.: Prediction MOOC's for student by using machine learning methods. In: 2021 XI International Conference on Virtual Campus, pp. 1–3. IEEE (2021)
6. Barber, R., Sharkey, M.: Course correction: using analytics to predict course success. In: Proceedings of the 2nd International Conference on Learning Analytics and Knowledge, pp. 259–262. ACM (2012)
7. Campbell, J.P., DeBlois, P.B., Oblinger, D.G.: Academic analytics: a new tool for a new era. EDUCAUSE Rev. **42**(4), 40 (2007)
8. Drousiotis, E., Shi, L., Maskell, S.: Early predictor for student success based on behavioural and demographic indicators. In: Cristea, A.I., Troussas, C. (eds.) ITS 2021, LNCS, vol. 12677, pp. 161–172. Springer, Cham (2021). https://doi.org/10.1007/978-3-030-80421-3_19
9. Gupta, A., Garg, D., Kumar, P.: Mining sequential learning trajectories with hidden Markov models for early prediction of at-risk students in e-learning environments. IEEE Trans. Learn. Technol. **15**(6), 783–797 (2022)
10. Hidalgo, A.C., Ger, P.M., Valentin, L.D.L.F.: Using Meta-Learning to predict student performance in virtual learning environments. Appl. Intell. **52**(3), 1–14 (2022)
11. Hao, J., Gan, J., Zhu, L.: MOOC performance prediction and personal performance improvement via Bayesian network. Educ. Inf. Technol. **27**(5), 7303–7326 (2022)
12. Kukkar, A., Mohana, R., Sharma, A., Nayyar, A.: A novel methodology using RNN + LSTM + ML for predicting student's academic performance. Educ. Inf. Technol. (2024). https://doi.org/10.1007/s10639-023-12394-0
13. Kuzilek, J., Hlosta, M., Zdrahal, Z.: Open university learning analytics dataset. Sci. Data **4**(1), 1–8 (2017)
14. Lakshmi, P. R., Geetha, A. V., Priyanka, D., Mala, T.: PRISM: predicting student performance using integrated similarity modeling with graph convolutional networks. In: 2023 12th International Conference on Advanced Computing, pp. 1–7. IEEE (2023)

15. Liu, Y., Fan, S., Xu, S., Sajjanhar, A., Yeom, S., Wei, Y.: Predicting student performance using clickstream data and machine learning. Educ. Sci. **13**(1), 17 (2022)
16. Page, M.J., et al.: The PRISMA 2020 statement: an updated guideline for reporting systematic reviews. Int. J. Surg., 88 (2021)
17. Qiu, F., et al.: Predicting students' performance in e-learning using learning process and behaviour data. Sci. Rep. **12**(1), 453 (2022)
18. Raj, N.S., Renumol, V.G.: Early prediction of student engagement in virtual learning environments using machine learning techniques. E-Learn. Digital Media **19**(6), 537–554 (2022). https://doi.org/10.1177/20427530221108027
19. Ranjeeth, S., Latchoumi, T.P., Victer Paul, P.: A survey on predictive models of learning analytics. Procedia Comput. Sci. **167**, 37–46 (2020). https://doi.org/10.1016/j.procs.2020.03.180
20. Souai, W., et al.: Predicting at-risk students using the deep learning BLSTM approach. In: 2022 2nd International Conference of Smart Systems and Emerging Technologies, pp. 32–37. IEEE (2022)
21. Shafiq, D.A., Marjani, M., Habeeb, R.A.A., Asirvatham, D.: A conceptual predictive analytics model for the identification of at-risk students in VLE using machine learning techniques. In: 2022 14th International Conference on Mathematics, Actuarial Science, Computer Science and Statistics, pp. 1–8. IEEE (2022)
22. Spadon, G., et al.: Pay attention to evolution: time series forecasting with deep graph-evolution learning. IEEE Trans. Pattern Anal. Mach. Intell. **44**(9), 5368–5384 (2021)
23. Tonghui, X.: Using data mining models to predict students' academic performance before the online course start. J. Educ. Online **20**(1), 222–235 (2023). https://doi.org/10.9743/jeo.2023.20.1.15
24. Torres Martín, C., Acal, C., El Homrani, M., Mingorance Estrada, Á.C.: Impact on the virtual learning environment due to COVID-19. Sustainability **13**(2), 582 (2021)
25. Ujkani, B., Minkovska, D., Nakov, O.: Understanding student success prediction using SHapley Additive exPlanations. In: 2023 International Scientific Conference on Computer Science, pp. 1–4. IEEE (2023)
26. Wang, C., Chang, L., Liu, T.: Predicting student performance in online learning using a highly efficient gradient boosting decision tree. In: Shi, Z., Zucker, J. (eds.) IIP 2022, LNCS, vol. 643, pp. 508–521. Springer, Cham (2022)
27. Waheed, H., Hassan, S.-U., Nawaz, R., Aljohani, N.R., Chen, G., Gasevic, D.: Early prediction of learners at risk in self-paced education: a neural network approach. Expert Syst. Appl. **213**, 118868 (2023). https://doi.org/10.1016/j.eswa.2022.118868

GAMAI, an AI-Powered Programming Exercise Gamifier Tool

Raffaele Montella[1(✉)] [iD], Ciro Giuseppe De Vita[1] [iD], Gennaro Mellone[1] [iD],
Tullio Ciricillo[1] [iD], Dario Caramiello[1] [iD], Diana Di Luccio[1] [iD], Sokol Kosta[2] [iD],
Robertas Damasevicius[3] [iD], Rytis Maskeliunas[3] [iD], Ricardo Queiros[4] [iD],
and Jakub Swacha[5] [iD]

[1] University of Naples "Parthenope", Naples, Italy
raffaele.montella@uniparthenope.it
[2] Aalborg University, Copenhagen, Denmark
[3] Kaunas University of Technology, Kaunas, Lithuania
[4] INESC TEC, Porto, Portugal
[5] University of Szczecin, Szczecin, Poland

Abstract. This paper presents GAMAI, an AI-powered exercise gamifier, enriching the Framework for Gamified Programming Education (FGPE) ecosystem. Leveraging OpenAI APIs, GAMAI enables the teachers to leverage the storytelling approach to describe the gamified scenario. GAMAI decorates the natural language text with sentences needed by OpenAI APIs to contextualize the prompt. Once the gamified scenario has been generated, GAMAI automatically produces the exercise files for the FGPE AuthorKit editor. We present preliminary results in AI-assessed gamified exercise generation, showing that most generated exercises are ready to be used with none or minimum human effort needed.

Keywords: Gamification · Programming Education · Software for Teachers · Artificial Intelligence

1 Introduction

Incorporating theoretical frameworks such as Self-Determination Theory [2], modern learning systems aim to empower learners by providing them with meaningful choices and autonomy within gamified exercises [9]. In programming teaching, learners often select different paths to solve a problem, allowing them to exercise autonomy and engage more deeply with the material [4,6]. Narrative transportation theory serves as another cornerstone of such approaches, as it seeks to immerse learners in captivating storylines within gamified scenarios [11], as learners may find themselves embarking on a virtual adventure where they must apply programming concepts to progress through the storyline, thereby becoming emotionally attracted to learning process [16]. Other approaches leverage the Extended Elaboration Likelihood Model (E-ELM) principles to strategically incorporate persuasive narratives within exercises [3].

A. M. Olney et al. (Eds.): AIED 2024 Workshops, CCIS 2150, pp. 485–493, 2024.
https://doi.org/10.1007/978-3-031-64315-6_47

Although several reports have confirmed the benefits of applying gamification to learning programming (see [13] and works cited therein), its practical use is constrained by the limited availability of gamified programming exercises, and creating new gamified exercises is a severe challenge to teachers [8]. By integrating OpenAI's GPT-based models, such as GPT-3, GAMAI exploits such models and enhances the coherence and relevance of generated narratives, ensuring that prompts and feedback are tailored to each learner's context [5]. This lets teachers employ a storytelling approach to articulate the gamified scenario, taking the burden of automatically generating gamification specification files compatible with the FGPE ecosystem [8]. In the subsequent sections, we delve into the design, implementation, and evaluation of GAMAI.

2   Related Work

While GAMAI is unique in its capability of generating *gamified* programming exercises, exploiting AI to generate exercises is not new. Kurdi [7] conducted a systematic review of automatic exercise generation in many domains, including programming. His study of 93 papers concludes that there need to be more tools for generating exercises of controlled difficulty and with facilities like improving presentation and generating feedback. Zavala [15] presented a tool that uses Automatic Item Generation (AIG) to address the problem of creating many similar programming exercises to avoid plagiarism yet ensure consistency in testing many students with questions of the same difficulty level. The solution, Goliath, automatically generates Python programming exercises using a template system. Agni [1] serves as a dynamic code playground tailored for learning JavaScript, featuring a component powered by the ChatGPT API that automates the exercise creation process by generating statements, solution code, and test cases. ExGen [14] leverages LLMs to generate many exercises from which those are chosen that suit a specific difficulty level and concept the student is working on. TESTed- [12] is an educational testing framework that supports the creation of programming exercises with automated assessment capabilities in a programming-language-independent manner. Sarsa [10] explored OpenAI Codex (deprecated since March 2023) to create new programming exercises and code explanations.

3   Design and Implementation

Gamified programming exercises can only be used within a technical milieu, allowing for the incorporation of gamification in programming education. The target environment for GAMAI is FGPE (Framework for Gamified Programming Education), an open, programming-language-agnostic set of exercise formats, exemplary exercises, and the supporting software [8]. The key components of the FGPE ecosystem include the editor for both programming exercises and gamification rules which provides access to GAMAI, the AI-powered Programming Exercise Gamifier tool (FGPE AuthorKit), gamification service which processes game rules and manages the overall game state (FGPE GS), and the interactive learning environment (FGPE PLE), which lets students access gamified exercises, solve them, and receive graded feedback. In contrast, teachers can organize exercise sets, grant students access, and monitor their learning progress. Figure 1 provides a comprehensive visualization of the overarching FGPE architecture, emphasizing the innovative contribution of the proposed AI-powered exercise gamification solution.

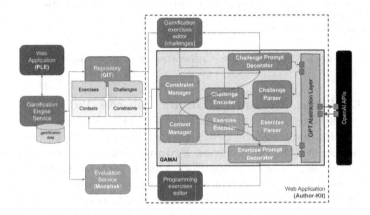

Fig. 1. AI-powered exercises gamifier.

GAMAI uses generative pre-trained transformer-based large language model to help the creators implement exercises and challenges. Designing, implementing, training, and reinforcing a dedicated LLM from scratch is out of the scope of this specific research, so we designed GAMAI with leveraging OpenAI API usage in mind. However, the **GPT Abstraction Layer** component abstracts the interface to OpenAI API, enabling the FGPE developers to test GAMAI with diverse and different large language model services. **AI-assisted gamified scenarios creation** involves creators with diverse skills, including those in digital humanities, communications, and storytelling. Prompts are constructed before each interaction with the GPT service, focusing on the critical aspects of the "acting as" statement. For instance, a creator proficient in fantasy saga writing could define characters and plotlines mapped onto gamified exercises with branching storylines and twists. The **Context Manager** and **Constraint Manager** components automatically set the context using data from the project repository, ensuring consistency in the generated content. The **Exercise Prompt Decorator** component enhances the creator-generated prompt by providing additional information about the type of exercise needed. Figure 2 illustrates the prompt generated by this component. Line 1 retrieves the exercise creator's skill from the project setup or exercise collection level. Line 2, written by the exercise creator, specifies the quantity and type of exercises to generate. Line 3 is assembled automatically using the GAMAI exercise prompt user interface or details provided by the exercise creator. Line 4 is automatically appended to retrieve results in JSON format for simplified parsing.

```
1   Acting as a research assistant in computer science.
2   Generate 10 gamified programming exercises of increasing
        difficulty for a first-year introduction to computing
        programming using the C language for bachelor students in
        a computer science degree class.
3   The answer to the exercise's correct solution has to be chosen
        from a set of 5.
4   Return the result in JSON.
```

Fig. 2. Gamified exercise generation prompt example (1).

The **Challenge Prompt Decorator** enhances the creator-generated prompt by detailing the gamification scenario, main plot, and types of quests with related plot branching and twists. Figure 3 illustrates the prompt produced by this component. Like the **Exercise Prompt Decorator**, line 1 is retrieved from the project-level contexts repository. Line 2, authored by the content creator, serves as the primary statement for AI-assisted world generation. Line 3 specifies whether students compete individually or in teams. Line 4 determines how quests conclude and if participants can all win awards or only one. Line 5 outlines the quest generation statement using the GAMAI challenges prompt user interface. Line 6 describes how the plot branches from one quest to the next, often sequential in this example. Line 7 formats results in JSON for the GPT provider. Finally, line 8 defines connections to the exercise repository, exercise types (e.g., multiple choice), and difficulty levels.

```
1   Acting as an expert fantasy storyteller.
2   Create a world where young boys and girls in high school
        attend wizarding schools.
3   Each student has to compete alone to solve a quest.
4   Each student solving the final quest wins the wizard
        certificate.
5   Generate 10 different magic quests of increasing difficulty.
6   The wizard student who solves the first quest can continue
        with the second one, and so on.
7   Return the result as json.
8   Add to each quest the field:
9   "exercise": {
10      "repository":"__world__",
11      "kind":"multiple_choises",
12      "level": "__iteration___"
13  }
```

Fig. 3. Gamified scenario generation prompt example.

Once the **Exercise Prompt Decorator** and **Challenge Prompt Decorator** complete their invocations of the GPT service and receive the related results, the **Exercise Parser** and **Challenge Parser** interpret the AI-generated content via the **GPT Abstraction Layer**. Both parsers map GPT-generated data onto FGPE entities, ensuring semantic consistency. Despite variations in GPT-generated exercises and scenarios, semantic consistency within the FGPE ecosystem is maintained. The **Exercise Encoder** serves as a data sink for the **Exercise Parser**, generating exercise representations compliant with the FGPE gamified exercise schema. The Authors-Kit Exercise Editor enables comprehensive exercise management, allowing creators to modify exercise descriptions and incorporate diverse input data set providers or custom correctness checkers. Similarly, the **Challenge Encoder** processes data from the **Challenge Parser**, transforming it into gamified scenario representations adhering to the FGPE gamified scenario schema. The Authors-Kit Challenge Editor empowers creators to customize scenarios, adding or removing components as needed to refine the gamified content. The **Challenge Encoder** and Authors-Kit Challenge Editor form a cohesive unit within the FGPE architecture. While the former ensures adherence to the established scenario schema, the latter provides a user-friendly interface for creators to tailor scenarios to their pedagogical goals and creative preferences. This integrated approach emphasizes standardization and flexibility in gamified exercise design, enhancing the overall effectiveness and adaptability of the framework.

4 Evaluation and Discussion

Currently, our research focuses on evaluating exercise quality and discussing results to estimate time savings for creating gamified exercises and prompts. Below, we show one such exercise generated from the prompt at Fig. 4:

GAMAI's maturity level prohibits direct performance evaluations between GPT services, prompting our qualitative approach. While generating exercises is quicker, they often require additional human effort to meet expectations.

Considering as exercise *text* all the human language content, assigning a score, in parentheses, for each needed fix, we classify the effort as follows: (i) the exercise can be used as is (0); (ii) the text is unclear, it needs rearrangement (+1); (iii) the text must be completely rewritten, but the exercise is formerly correctly generated (+2); (iv) the text is correct, but there are issues from the technical point of view (+3); (v) the exercise, although correct in the text part, is too much easy or too much complex than expected (+4); (vi) the exercise cannot be used as is; it must be dropped or completely rewritten; for example, the code snippet is entirely missing (+5).

Then, we choose to generate 10 exercises for each one of the following prompts as part of line three of the Listing 4:

- first-year introduction to computing programming using the C language for bachelor students in a computer science degree class (1CCS).

```
{
  "question": "Which of the following is the correct way to declare an
  integer variable named 'count' in C?",
  "options": [
    {"value": "int count;", "correct": true},
    {"value": "integer count;", "correct": false},
    {"value": "count int;", "correct": false},
    {"value": "declare count as int;", "correct": false},
    {"value": "variable count is int;", "correct": false}
  ]
}
```

```
1  Acting as a seasoned full professor in computer science.
2  Generate 100 gamified programming exercises of increasing
       difficulty for a
3  ...
4  The answer to the exercise's correct solution has to be chosen
       from a set of 5.
5  Return the result in json.
```

Fig. 4. Gamified exercise generation prompt example (2).

- first-year introduction to computing programming using the Python language for bachelor students in a computer science degree class (1PCS).
- first-year introduction to computing programming using the Python language for bachelor students in an environmental science degree class (1PES).
- first-year introduction to computing programming using the Python language for bachelor students in a law degree class (1PLS).
- third-year object-oriented computing programming using the Java language for bachelor students in a computer engineering degree class (3JCE).
- third-year object-oriented computing programming using the Java language for bachelor students in a computer science degree class (3JCS).

Table 1. Creating gamified exercises experiment results

Prompt	0	1	2	3	4	5	Score
1CCS	5	4	0	1	0	0	7
1PCS	4	0	0	4	2	0	20
1PES	6	0	0	4	0	0	12
1PLW	3	1	2	3	0	1	19
3JCE	4	3	0	3	0	0	12
3JCS	6	1	0	0	3	0	13
%	40.0	21.67	3.33	25.0	8.33	1.67	

The experiment results (Table 1) indicate that most generated exercises require minimal human intervention (61.7%), though some high-difficulty exercises need manual adjustments (28.3%). Useless generations are rare (10.0%). Quality varies with programming language; C exercises are more ready-to-use than Python ones. While improvement in exercise generation and word creation is straightforward, deploying FGPE and GAMAI in real-world settings warrants discussion. Generated text may be inaccurate, biased, or raise intellectual property (IP) concerns. While verifying exercise quality is manageable, ensuring gamification word quality is more complex due to potential copyright issues stemming from GPT's training data and popular culture influence. For example, despite in the used prompt (Fig. 3), no mention has been made of "wizarding_schools", the names of three different schools of wizardry in the fantasy world of Eldoria are present in the generated exercise (listed below with the ellipsis denoting omissions).

```
{
"world": {
   "name": "Eldoria",
   "description": "A magical realm where wizarding schools are scattered
   across mystical landscapes, and young sorcerers engage
   in solo quests to prove their magical prowess."
},
"wizarding_schools": [
{
   "name": "Arcane Haven Academy",
   "location": "Nestled between ancient forests and majestic ..."
}, ... ],
"quests": [
{
   "name": "The Enchanted Amulet",
   "description": "Retrieve the Enchanted Amulet hidden within the
   Forbidden Forest, guarded by mystical creatures and
   protected by a powerful enchantment.",
   "exercise": { "repository": "__world__", "kind": "multiple_choice",
   "level": "__iteration__" }
},
...
```

5   Conclusion

This paper introduces the use of a large language model for an automatic generation of gamified exercises for programming education. It thus allows to introduce gamification to programming courses of any kind without much effort and skills. Its key novelty lies in enhancing teachers' storytelling to create contextualized prompts leading to better gamified programming exercises. An obvious limitation of the proposed approach is that the generated exercises may require some adjustments to achieve the expected quality level, however, the preliminary evaluation results indicate that it is not needed at all for more than 60% of generated exercises (Sect. 4).

Future work includes developing a standardized test suite to evaluate exercise gamification performance, serving as a benchmark for assessing challenges' effectiveness. We also plan to introduce a metric for measuring student interactions with these challenges in computer and non-computer science classes.

Acknowledgements. This research was co-funded by the European Union, grant number 2023-1-PL01-KA220-HED-000164696.

References

1. Bauer, Y., Leal, J.P., Queirós, R.: Can a content management system provide a good user experience to teachers? In: 4th International Computer Programming Education Conference (ICPEC 2023). Open Access Series in Informatics (OASIcs), vol. 112, pp. 4:1–4:8. Schloss Dagstuhl – Leibniz-Zentrum für Informatik, Dagstuhl, Germany (2023)
2. Botte, B., Bakkes, S., Veltkamp, R.: Motivation in gamification: constructing a correlation between gamification achievements and self-determination theory. In: Marfisi-Schottman, I., Bellotti, F., Hamon, L., Klemke, R. (eds.) GALA 2020. LNCS, vol. 12517, pp. 157–166. Springer, Cham (2020). https://doi.org/10.1007/978-3-030-63464-3_15
3. Brooks, J.J.: Touching Tales for Touchy Topics? Engaging Contentious Issues Through Narrative Persuasion. Ph.D. thesis, Northwestern University (2023)
4. Hosseini, R., et al.: Improving engagement in program construction examples for learning python programming. Int. J. Artif. Intell. Educ. **30**(2), 299–336 (2020)
5. Jacobsen, L.J., Weber, K.E.: The promises and pitfalls of ChatGPT as a feedback provider in higher education: an exploratory study of prompt engineering and the quality of AI-driven feedback (2023). https://osf.io/cr257/download. Accessed 25 Mar 2024
6. Kao, G.Y.M., Ruan, C.A.: Designing and evaluating a high interactive augmented reality system for programming learning. Comput. Hum. Behav. **132**, 107245 (2022)
7. Kurdi, G., Leo, J., Parsia, B., Sattler, U., Al-Emari, S.: A systematic review of automatic question generation for educational purposes. Int. J. Artif. Intell. Educ. **30** (2019)
8. Paiva, J.C., Queirós, R., Leal, J.P., Swacha, J., Miernik, F.: Managing gamified programming courses with the FGPE platform. Information **13**(2), 45 (2022)
9. Saleem, A.N., Noori, N.M., Ozdamli, F.: Gamification applications in e-learning: a literature review. Technol. Knowl. Learn. **27**(1), 139–159 (2022)
10. Sarsa, S., Denny, P., Hellas, A., Leinonen, J.: Automatic generation of programming exercises and code explanations using large language models. In: Proceedings of the 2022 ACM Conference on International Computing Education Research, vol. 1, pp. 27–43. ICER 2022, Lugano, Switzerland (2022)
11. Schmidt-Kraepelin, M., Thiebes, S., Warsinsky, S.L., Petter, S., Sunyaev, A.: Narrative transportation in gamified information systems: the role of narrative-task congruence. In: Extended Abstracts of the 2023 CHI Conference on Human Factors in Computing Systems, pp. 1–9 (2023)

12. Strijbol, N., et al.: Tested - an educational testing framework with language-agnostic test suites for programming exercises. SoftwareX **22**, 101404 (2023)
13. Swacha, J., Szydlowska, J.A.: Does gamification make a difference in programming education?: Evaluating FGPE-supported learning outcomes. Educ. Sci., 1–11 (2023). https://doi.org/10.3390/educsci13100984
14. Ta, N.B.D., Nguyen, H.G.P., Swapna, G.: ExGen: ready-to-use exercise generation in introductory programming courses. In: Proceedings of the 31st International Conference on Computers in Education Conference, pp. 1–10. Matsue, Shimane, Japan (2023)
15. Zavala, L., Mendoza, B.: On the use of semantic-based AIG to automatically generate programming exercises. In: Proceedings of the 49th ACM Technical Symposium on Computer Science Education, pp. 14–19. SIGCSE '18, Association for Computing Machinery, New York, NY, USA (2018)
16. Zeng, J., Parks, S., Shang, J.: To learn scientifically, effectively, and enjoyably: a review of educational games. Hum. Behav. Emerg. Technol. **2**(2), 186–195 (2020)

Generating Contextualized Mathematics Multiple-Choice Questions Utilizing Large Language Models

Ruijia Li[1], Yiting Wang[3], Chanjin Zheng[1,2,4], Yuan-Hao Jiang[2,4], and Bo Jiang[2,4(✉)]

[1] Faculty of Education, East China Normal University, Shanghai, China
[2] Lab of Artificial Intelligence for Education, East China Normal University, Shanghai, China
[3] Software Engineering Institute, East China Normal University, Shanghai, China
[4] Shanghai Institute of Artificial Intelligence for Education, East China Normal University, Shanghai, China
bjiang@deit.ecnu.edu.cn

Abstract. Applying mathematics to solve authentic question play important roles in math-ematics education. How to generate high-quality multiple-choice questions that have authentic context is a great challenge. By combining multiple iterations of large language model dialogues with auxiliary external tools and the LangChain framework, this work presents a novel method for automatically generating contextualized multiple-choice mathematics questions. To check the quality of generated questions, 30 questions were randomly selected and 13 human experts were invited to rate these questions. The survey result indicates that the questions produced by the proposed method exhibit a significantly higher quality compared to those generated directly by GPT4, and are already quite comparable in performance to questions that are meticulously crafted by humans across multiple dimensions. The code is available on the project home page: https://github.com/youzizzz1028/MCQ-generation-Chain.

Keywords: Automatic Question Generation · LangChain · ChatGPT · Core Literacy · Prompt Engineering

1 Introduction

1.1 Research Background

Creating multiple-choice questions (MCQs) manually requires considerable time and resources. Despite the provision of question banks, manual labor is still required for various tasks including question formatting, tagging, and integration. Moreover, an issue of inconsistent question quality persists. A critical aspect of learning system development is the investigation of methods to generate multiple-choice questions automatically while ensuring quality.

A. M. Olney et al. (Eds.): AIED 2024 Workshops, CCIS 2150, pp. 494–501, 2024.
https://doi.org/10.1007/978-3-031-64315-6_48

Another complexity arises from the additional demand for mathematics education. In mathematics, the Compulsory Education Mathematics Curriculum Standards released by Ministry of Education of China (2022 Edition) place significant emphasis on competency. Context is a crucial element in linking theoretical knowledge and practical skills to competency. It plays a vital role in connecting theory and practice, enabling the application of knowledge and abilities in a meaningful way. To enhance students' ability to apply their acquired knowledge to authentic situations, it is crucial to design math questions that incorporate authentic contexts. This presents a significant challenge to the ability of LLMs to utilize real-world knowledge.

The objective of this research is to investigate the potential application of large language models (LLMs) in deriving mathematical questions. Our method is to enhance the LLM's capability in item generation by incorporating code interpreters for logical operations and external knowledge bases for knowledge, with the aim of improving the quality and diversity of contextualized math questions [1].

1.2   Research Questions

This study aims to identify an efficient approach for the automated creation of context-specific multiple-choice questions (MCQs) in primary and middle schools. The objective of this study is to assess the quality of questions generated automatically, as well as those generated by our platform, questions crafted by human experts, and questions generated by ChatGPT. The research questions for this study are as follows:

1. What is the feasible approach, based on LLM, for generating automated multiple-choice questions in a contextualized manner?

2. What is the quality of questions generated using this solution?

2   Related Work

Kurdi [4] presents a classification and description of Automatic Question Generation (AQG) systems from seven dimensions: (a) the purpose and utilization of question generation, (b) the knowledge domain being examined, (c) the source of knowledge, (d) the generation method, (e) the type of questions, (f) the formatting of question structure, and (g) the evaluation of question quality. Cur-rent question generation technologies primarily aim to create quiz questions, support online learning platforms, or assist students in independent learning, with a focus on disciplines such as language, mathematics, humanities, and medicine. Classic methods of question generation primarily fall into three categories: (a) grammar-based, which identifies key concept words using techniques such as grammar trees and POS (Part-of-speech tagging), and generates questions by replacing these key concepts; (b) semantic-based, which primarily identifies the part of speech and dependencies between words using natural language processing techniques, and generates questions based on the extracted dependencies;

and (c) template-based, which involves the creation of preset question templates and the utilization of random number generation or keyword extraction to fill in the templates. Subsequently, with the emergence of deep learning technology, neural networks [7], reinforcement learning [2], and pre-trained language model shave been integrated into AQG systems. Additionally, some research has begun to focus on the utilization of LLMs for question generation [5].

Owing to the stringent requirements of mathematical precision and logical rigor, the utilization of connectionist artificial intelligence algorithms is not advisable for addressing mathematical challenges. Instead, it often necessitates the employment of symbolic methodologies for resolution [9]. Numerous technical solutions within AQG systems often fail to apply to the generation of mathematical problems. Presently, the most commonly used method remains the generation of random numbers based on predefined templates, which can result in challenges such as inflexible question formats, limited practical relevance, and limited scalability. Grévisse [3] delved into the quality of MCQs generated and discovered that despite the fluent language expression of GPT-generated questions, numerous issues persist, including excessively long question stems, absence of correct answers among the options, ambiguous expressions, and low relevance between the questions and knowledge points. Our experimental exploration concurs with these findings - the direct utilization of GPT models and prompts for question generation, particularly for MCQs, is challenging to achieve satisfactory out-comes. Therefore, it is imperative to introduce external knowledge bases as knowledge inputs and targeted teaching strategies. It is also necessary to introduce technologies such as code interpreters and PoT (Program-of-Thoughts) [1], adjust the parameters of LLM dialogue and control the question generation process.

3   Contextualized Mathematical Multiple-Choice Question Generation Methodology

The workflow of creating and generating multiple-choice questions is typically divided into the following six stages [3]: (a) Background Input (b) Preprocessing (c) Question Generation (d) Answer Generation (e) Distractor Generation (f) Formatting. Following this workflow, this study proposes a solution for automatically generating mathematical multiple-choice questions based on the LangChain framework, which integrates multi-round large language model dialogues and pluggable interfaces (such as SerpAPI, llm-math, etc.). This solution includes four core modules: Domain Agent, QG Question Generation Agent, Problem Solving Agent, and Option Generation Agent (Fig. 1). Each module is based on the GPT-4 model, and they are closed linked to generate mathematical multiple-choice questions with a certain quality assurance and real-world scenarios.

3.1   Domain Agent

This agent uses search engines to obtain quiz examples and problem-solving ideas highly related to specific knowledge point from extensive web resources, to

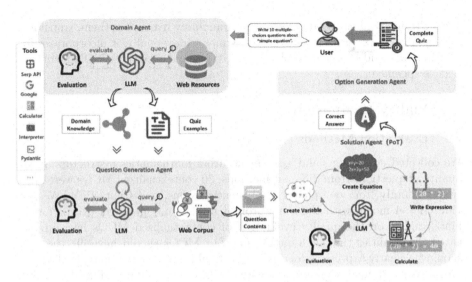

Fig. 1. Automated Solution for Producing Contextualized Multiple-Choice Questions

facilitate the establishment of domain foundation and minimize hallucinations and knowledge errors.

3.2   Question Generation Agent

This agent receives a range of parameters from preceding steps (such as domain knowledge, quiz examples, question quantity, difficulty level, etc.). By engaging in multiple iterations of experiments, a prompt is gradually crafted. Additionally, this agent utilizes SerpAPI as a supportive instrument to examine and validate the authenticity of the question context by leveraging extensive corpus data from the Internet. This approach ensures that the questions are grounded in reality, enriching contextualized characteristics.

3.3   Solution Agent

This agent designed for problem-solving utilizes llm-math interface and Python interpreter to conduct numerical computations. During analysis, the LLM provides Python expressions that can accurately resolve them. After gradual execution of program, the correct answer with natural language format is generated.

3.4   Option Generation Agent

This agent offers a standardized template of multiple-choice questions. Questions and corresponding correct answers are embedded in carefully designed prompt.

Then the agent generates distractors and randomly rearranges them, transforming the question's structure into MCQ format. Additionally, it incorporates formatting cues and coercion checks of data type, and returns a JSON formatted text as specified.

4   Quality Evaluation

4.1   Evaluation Methods

We collected evaluations and assessments from mathematics instructors on the quality of questions from different methods. 30 contextualized MCQs were gathered randomly from our platform, authentic exams, and ChatGPT4. We provide GPT-4 model with the same prompt used in our platform except format hint. Studies on the quality evaluation of contextualized MCQs have investigated the qualities that high-quality and fair MCQs should have [9]: the stem should be clearly expressed, focused on particular knowledge points, leading to a uniquely determined response, and employing positive phrases. The alternatives should be concise and straightforward, without the use of ambiguous adverbs or expressions like "all of the above are correct, all of the above are wrong", and with consistent length and language structure. Previous research has often used expert rating techniques to evaluate the quality of automatically produced questions [10]. Some studies merely gather survey responses for a single component of "question quality", but others collect expert views on dimensions such as answer-ability, knowledge relevance, and question complexity [4].

Based on the preceding research, this study argues that assessing the quality of questions may be divided into three dimensions: rationality of the stem, rationality of options, and contextual appropriateness. In addition, this research looks at experts' perspectives on the producing potential to see whether there are subjective distinctions between produced questions and authentic questions. The following Table 1 provides detailed explanation and definition of each dimension.

Table 1. Definitions of Research Dimensions and Scoring Standards

Dimension	Definition and Scoring Standards
Rationality of the Stem	The expression should be clear and concise; closely related to knowledge points; lead to a uniquely determined answer; use positive expressions, such as "the following is correct..." rather than "the following is incorrect...".
Rationality of options	Contain the only correct answer; be concise and clear, avoid the use of vague adverbs (such as "often", "occasionally"); avoid the expression of "all of the above are correct" or "all of the above are wrong"; have consistent length and grammatical structure.
Contextual appropriateness	It conforms to real life; the expression is clear, and the structure is simple; it meets the cognitive level of students; the contextual background is integrated with the knowledge learned.

4.2 Evaluation Results

The Cronbach's alpha value of the consistency among scorers is 0.898, indicating that the scoring criteria of the scorers are consistent and their understanding and scoring methods for the questions are similar. Independent sample T-tests were conducted on the questions generated by this platform (Ours) and the authentic questions (AQ), as well as on the questions generated by this platform and directly generated by GPT4. The results are shown in the following Tables 2 and 3.

Table 2. Comparison of Evaluation Results between Ours and AQ

Dimension	Source	AVG	S.D	p-value
Contextual appropriateness	AQ	4.417	0.4387	0.439
	Ours	4.267	0.4924	
Rationality of options	AQ	4.55	0.2611	1.000
	Ours	4.55	0.3680	
Rationality of Stem	AQ	4.592	0.2353	0.457
	Ours	4.508	0.2999	
Producing Potential	AQ	2.039	0.4226	0.084
	Ours	2.433	0.6243	

Table 3. Comparison of Evaluation Results between Ours and GPT4

Dimension	Source	AVG	S.D	p-value
Contextual appropriateness	GPT4	3.983	0.3817	0.236
	Ours	4.267	0.4924	
Rationality of options	GPT4	4.050	0.6058	0.043
	Ours	4.55	0.3680	
Rationality of Stem	GPT4	4.017	0.3920	0.009
	Ours	4.508	0.2999	
Producing Potential	GPT4	3.417	0.5231	0.004
	Ours	2.433	0.6243	

Upon analysis of the data, it is evident that the questions produced by this platform exhibit minimal variance compared to authentic questions across various metrics. Notably, when compared to questions directly generated by GPT4, the platform's questions exhibit statistical significance in three dimensions. While the contextual appropriateness does not reach statistical significance, its numerical value surpasses GPT4-generated questions. This indicates that the technological approach employed by our platform enhances the scientific and accuracy of the questions based on GPT4, without significant deviation from real-world questions. Therefore, it is anticipated that this platform will be suitable for practical applications in the future.

5   Discussion and Conclusion

This research presents a technique for automatically creating mathematical MCQs with context using the LangChain framework. It combines multi-round LLM dialogue with plug-gable external tools. Questions created by this solution may be guaranteed to a certain degree, with a much greater quality than those directly generated by GPT4, and already comparable to questions prepared by human experts in some respects. By using the semantic expression capabilities of LLM and external searchable knowledge base, the produced questions are more closely connected with real-life scenarios, in accordance with the goals of developing and evaluating the fundamental mathematical literacy skills.

Various research has indicated that ChatGPT should advise and collaborate on educational exams to reduce teachers' workload rather than replace them [11]. This study intends to improve question quality and recognize the need of sensitivity in education. It suggests using machine-generated exercises as teaching materials rather of giving them to students. Our platform design gives educators easy tools for creating and managing questions and banks (Fig. 2). It may also output files in docx, excel, JSON, and other formats, simplifying conventional exams and enabling compatibility with other educational platforms.

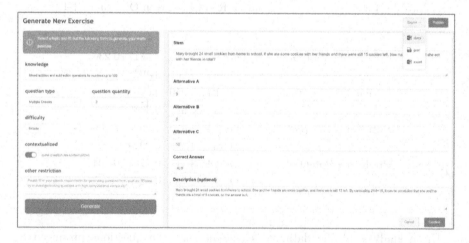

Fig. 2. Automated Solution for Producing Contextualized Multiple-Choice Questions

Nevertheless, there are still some limitations in this method. Initially, the sup-port for various question kinds is still incomplete yet, this solution specifically focuses on MCQs in the field of mathematics. However, it does not yet provide any ways for creating fill-in-the-blank questions or true/false questions. Further-more, the approach used to manage the complexity of question generated is quite rudimentary. The system merely modifies the difficulty of questions by modifying prompt and depending on the LLMs' comprehension, without using exact quantitative techniques. Currently, the produced questions are restricted to numerical and algebraic concepts and do not provide support for geometry questions or statistics chart questions in many formats. For future work, researchers could attempt to migrate the solution proposed by this study to other subjects and question types. The researchers could also optimize on the control of difficulty and the source of the input knowledge points, or embed more precise retrieval-enhanced algorithms to improve the quality of the generated questions.

Acknowledgements. This work was partially supported by the National Natural Science Foundation of China under Grant 61977058, and Natural Science Foundation of Shanghai under Grant 23ZR1418500.

References

1. Chen, W.: Program of thoughts prompting: disentangling computation from reasoning for numerical reasoning tasks. arXiv preprint arXiv:2211.12588 (2022)
2. Chen, Y.: Reinforcement learning based graph-to-sequence model for natural question generation. arXiv preprint arXiv:1908.04942 (2019)
3. Grévisse, C.: Comparative quality analysis of GPT-based multiple choice question generation. Int. Conf. Appl. Inform., 435–447. Springer Nature Switzerland (2023). https://doi.org/10.1007/978-3-031-46813-1_29
4. Kurdi, G.: A systematic review of automatic question generation for educational purposes. Int. J. Artif. Intell. Educ. **30**, 121–204 (2020)
5. Lee, U.: Few-shot is enough: exploring ChatGPT prompt engineering method for automatic question generation in English education. Educ. Inf. Technol., 1–33 (2023). https://doi.org/10.1007/s10639-023-12249-8
6. Siddiq, F.: Taking a future perspective by learning from the past-a systematic review of assessment instruments that aim to measure primary and secondary school students' ICT literacy. Educ. Res. Rev. **19**, 58–84 (2016)
7. Zhou, Q., Yang, N., Wei, F., Tan, C., Bao, H., Zhou, M.: Neural question generation from text: a preliminary study. In: Huang, X., Jiang, J., Zhao, D., Feng, Y., Hong, Yu. (eds.) NLPCC 2017. LNCS (LNAI), vol. 10619, pp. 662–671. Springer, Cham (2018). https://doi.org/10.1007/978-3-319-73618-1_56
8. Zong, M.: Solving math word problems concerning systems of equations with GPT-3. In: Proceedings of the AAAI Conference on Artificial Intelligence, pp. 15972–15979 (2023)
9. Boland, R.J., Lester, N.A., Williams, E.: Writing multiple-choice questions. Acad. Psychiatry **34**(4), 310–316 (2010). https://doi.org/10.1176/appi.ap.34.4.310
10. Das, B., Majumder, M., Phadikar, S., Sekh, A.A.: Automatic question generation and answer assessment: a survey. Res. Pract. Technol. Enhanced Learn. **16**(1), 1–15 (2021). https://doi.org/10.1186/s41039-021-00151-1
11. Lo, C.K.: What is the impact of ChatGPT on education? A rapid review of the literature. Educ. Sci. **13**(4), 410 (2023). https://doi.org/10.3390/educsci13040410

Generative AI in K-12: Opportunities for Learning and Utility for Teachers

Kristjan-Julius Laak[(✉)] and Jaan Aru

Institute of Computer Science, University of Tartu, Tartu, Estonia
{kristjan-julius.laak,jaan.aru}@ut.ee

Abstract. The goal of education is to develop natural intelligence. How are teachers using generative artificial intelligence (GenAI) to support this goal? The aim of this study is to understand how teachers have changed their teaching practices due to GenAI, their real use of GenAI in learning activities, and how they have utilised these tools to make their work more efficient. A nationwide survey was conducted among all Estonian K-12 schools. A total of 908 teachers completed the survey - a representative sample of more than 5% of Estonian K-12 teachers. The results show that 49% of teachers have already modified their teaching processes by including tasks that encourage critical thinking, eliminating written homework, and allowing the use of GenAI to generate new ideas, to name a few. In addition, 74% of teachers are using GenAI to make their work more efficient (e.g. answering emails from parents). Finally, there are various concerns and challenges, and the majority of educators are looking for professional development on the practical integration of GenAI into teaching practice.

Keywords: Generative AI · K-12 · ChatGPT · Teachers

1 Introduction

A recent review article on the use of ChatGPT in higher education raised a problem of the paucity of empirical studies, indicating that 73% of studies in the field have been based on opinions [2] and reviews [1,10,15,17], and noted that only 2 out of 69 papers examined the use of ChatGPT by academics. While most of the discussions occur in higher education [2,12,16], there have also been some studies exploring the use of ChatGPT among K-12 teachers [5,7,14,18]. However, these studies were conducted in the months following the release of ChatGPT and focused on teachers' opinions on the implications of generative AI (GenAI). In addition, they focused on ChatGPT instead of GenAI more broadly, had a small sample size, or were limited to a certain group of teachers (e.g. physicists). Thus, a larger and more representative sample to reliably understand the current adoption of GenAI in K-12 education is needed.

Here, we aimed to collect data from a large representative sample to get an accurate overview of the current usage of GenAI 1.5 years after the release

A. M. Olney et al. (Eds.): AIED 2024 Workshops, CCIS 2150, pp. 502–509, 2024.
https://doi.org/10.1007/978-3-031-64315-6_49

of ChatGPT. A study on the current use of GenAI among K-12 teachers has not yet been conducted in Estonia. Specifically, we investigate teachers' two-fold hands-on usage of GenAI: integration with learning activities and usage for individual tasks.

This study focuses on several key research questions centred around using GenAI by Estonian K-12 teachers.

1. RQ1: How have teachers changed their teaching practices because of GenAI?
2. RQ2: For which learning-related tasks and assignments is GenAI used?
3. RQ3: For which job-related tasks are teachers utilising GenAI?
4. RQ4: Which challenges, fears and questions do they have regarding GenAI?

2 Methods

The survey was run on the university's self-hosted LimeSurvey environment, a cloud-based platform for conducting online surveys. The respondents are teachers in Estonian general education schools with a population of 16,942 by 2022. The survey was an online questionnaire, and participants from all Estonian K-12 (515) and vocational (31) schools were invited via e-mail. The survey was open for two weeks in March 2024. The study was approved by the Ethics Committee of the University of Tartu, participation was voluntary, and no compensation was available.

The study was an online survey of both closed- and open-ended questions. The first section collected demographic data, including the type of school, grade level taught, how many years the respondent had been teaching, subjects being taught, age, and gender. The participants who answered that they had heard about GenAI were shown the next section, and others were directed to the last section. The second section investigated how the teachers have accommodated their teaching activities because of these tools. Participants who said they are actively using GenAI applications were shown the third section about specific usage. The last section that was shown to all participants asked about the main fears, questions and challenges regarding AI.

The survey was designed to take 10–15 min to complete to get as many answers as possible and consisted of four sections. The researchers formulated the questionnaire based on the existing scientific literature and previous studies by the Estonian Education Forum. We conducted a descriptive analysis of the results.

3 Results

A total of 1139 teachers participated in the study, of whom 908 fully completed the questionnaire (5.4% of all Estonia's K-12 teachers). The median age of respondents was 49 years (SD 12, min 20, max 76), and the median years being a teacher were 18 years (SD 14, min 0, max 52). The average survey completion time was 7min 30sec, and 95% of respondents finished within 22 min.

The respondents came from a wide range of school types (both small and large, city and countryside, private and public) and subject areas (from all 16 main subject categories as well as more than 100 different other subjects).

98% had heard about GenAI chatbots such as ChatGPT. Nevertheless, ten of the remaining 19 teachers commented that they had heard about them but were not using them, and only 9 participants (1%) said they did not have much information about them. The average age and years in school were identical to the sample. Some of the key results are shown in Fig. 1 below.

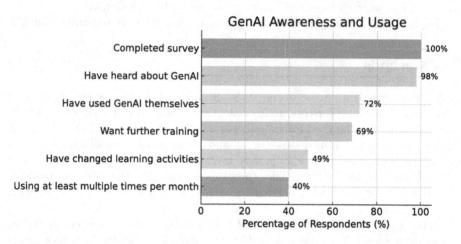

Fig. 1. Key results of the nationwide survey completed by 908 K-12 teachers, a representative sample of 5% of the population.

3.1 RQ1: How Have Teachers Changed Their Teaching Practices Because of GenAI?

To examine this question, we allowed respondents to mark down multiple answers regarding their practices. To begin with, half of the respondents (50%) had not modified or changed their teaching practices. Some teachers try to create AI-proof assignments (16%). For example, eight respondents said they had omitted written homework and only did it in the classroom. One teacher mentioned having completely abandoned homework, and one had banned the unsupervised use of smartphones in the classroom. In addition, a small subset of respondents have changed assessment criteria (10%), e.g. stopped assessing written homework. Interestingly, 25% of teachers have integrated GenAI into their lessons. For example, for letting students analyse the grammar and sentence structure of the texts created by GenAI. Moreover, a large proportion (36%) have introduced and tried out different tools together with learners and raised classroom discussions about using them, and 16% have given instructions on how to use them for homework. Lastly, 36% of teachers are now more focused on developing critical thinking in children, for example, by asking thought-provoking questions and stressing the need for critically assessing and finding reliable sources.

3.2   RQ2: For Which Learning-Related Tasks and Assignments Is GenAI Used?

Next, we investigated how the teacher regulates the usage of GenAI. The majority, 57% of teachers, have not regulated the usage of GenAI. While 17% say they allow the usage, 8% do not allow learners to use GenAI. The rest of the free-text responses (17%) mainly express a more nuanced approach depending on the task types and learning goals. For example, GenAI tools are allowed for tasks such as collecting ideas, gathering new information, and analysing texts. Nevertheless, in these situations, learners must check the grammar, critically analyse the results, and refer to the tool used.

Illustrating the need for critical thinking, one respondent said: "I often allow to use it [GenAI] and sometimes even oblige, but students already know that a chatbot is useless without critical thinking and background knowledge. Especially when the work is about Estonian language, literature and culture". Another teacher added: "I have taught students that ChatGPT is a very good option for generating ideas, but it's not a substitute for thinking for themselves. They need to be critical of the information they get from the chatbot and ask the chatbot further questions to verify the accuracy of the information it is giving. For me, the student must understand why they are learning in school and for whom they are learning".

Some teachers create specific assignments that need to be worked on with GenAI but do not allow for other tasks, ensuring each task clearly indicates whether GenAI is allowed to be used. For example, some instructions ask the learner to solve the task independently and then compare it with the GenAI solution. Teachers also allow the usage on mutual agreement or following agreed usage criteria. Some teachers use GenAI with learners in the lesson, e.g., to create images and compare learner- and AI-generated texts.

In summary, learning-related tasks and situations where learners are allowed to use GenAI are distributed as follows: to learn about AI (43%), to develop critical thinking (39%), and to get answers to questions (33%), translation (25%), text generation (18%) and summarising (11%).

A few other distinct themes were revealed from these answers. First, GenAI was claimed to be useless for specific lesson types such as handcraft, music, physical education, and others that fall outside of GenAI's capabilities. Second, GenAI is said to have no usage in primary school (1–6th grade), where the learning goals and approaches are simpler, differing from secondary school, e.g., requiring more sensory input. Third, some respondents mentioned that these tools are not usable by children with special needs and intelligence deficits. However, one teacher specifically allows a student with Asperger syndrome who struggles in writing to use these tools to ask questions and write texts from the answers.

3.3   RQ3: For Which Job-Related Tasks Are Teachers Utilising GenAI?

Investigating teachers' use of GenAI, the results showed that 72% of teachers have used GenAI for work-related tasks, of whom 93% have used ChatGPT 3.5, 10% GPT Plus (GPT-4), 22% Microsoft Copilot (previous Bing Chat), and 14% Gemini. 15% of teachers said they had also used other GenAI tools (a total of 55 different tools were mentioned), of which the most popular were Perplexity, MagicSchool, and Canva. More than 55% of teachers use GenAI more than a few times per month, and 28% use these tools at least every week.

The most popular tasks teachers use GenAI for are creating assignments (63%) and generating tests and questions (52%), sometimes based on a given website or document. Other popular functions include text summarising (38%), lesson planning (31%), getting recommendations for teaching methods (34%), and task differentiation (21%). Generating (formative) feedback (15%) and assessing student work (6%) were less common tasks given to GenAI.

A prominent task mentioned was managing teacher-parent email correspondence, e.g. creating an outline for parent's letters, especially when the parent does not speak the same language. Specific tasks mentioned were discovering multiple solutions to problems, compiling lists of necessary prior knowledge, formulating learning outcomes, generating topic introductions, worksheets, and reading recommendations, scheduling their own working time, creating workshop outlines and presentations, discovering online educational games on specific topics, and creating scenarios for a Christmas concert. One teacher uses GenAI to find ways to support students with special needs regarding specific assignments they struggle with, e.g. writing a poem.

3.4   RQ4: Which Challenges, Fears and Questions Do They Have Regarding GenAI?

Some respondents said that they do not fear anything and that these tools have made their lives easier. At the same time, some teachers mention not having enough time to get to know these tools, exemplified by this answer: "Chatbots are evolving very fast. The fear is the lack of time to familiarise yourself with these innovations. Students are ahead of teachers in this area. I often let the children teach and show me the new solutions". Besides various issues, our qualitative analysis revealed three main themes.

First, teachers raise concerns about the impact of GenAI on student learning. Teachers fear that "learners stop thinking with their own head". As one of the respondents put it: "Pupils give up looking for things themselves and even simple Googling disappears. It is easy to write your question to ChatGPT and then copy the answers to your work. Students do not check the answers they get but believe that if the answer comes from the computer, it is the truth". Another teacher added: "People are no longer using the potential of their brains, and there is a decline in our skills and knowledge". In short, teachers fear the decline of analytical thinking skills.

Second, many teachers stressed the importance of original thinking, worrying that AI could stifle learners' creativity. As one teacher said, "I am a little concerned about the development of children's creativity". Gen complicates assessment of creative work: "I have always tried to teach students to think for themselves, to express their opinions and to respect other people's opinions and views. But now, when I come across texts written by AI in my 9th-grade homework (I know my students and their expressive skills), how do I evaluate them? There is no point in banning chatbots, but there is a risk that pupils' free opinions and creativity will be increasingly diminished".

Third, some teachers expressed concerns about the reliability of the information and the accuracy of the facts provided by GenAI. Worryingly, students trust AI output: "Superficiality and blind trust - for example, students' criticality of an AI-created text tend to be rather weak, they take the answer as the truth without further analysis". This could lead to learning incorrect facts: "A text robot often gives completely wrong answers. Displays facts with references that do not exist in reality. If you do not check critically and do not have enough knowledge to distinguish between wrong and right answers, you can learn completely wrong things".

Other answers revealed that some teachers lack the knowledge of how large language models work. 69% of the respondents said they would like further training: "Since I haven't had proper training, I do not know the benefits". Another teacher said that they need training to properly use GenAI tools to make their job easier, adding that "when teachers start using these tools regularly, they can also use them with their pupils and teach them how to use them".

4   Discussion

The aim of this study was to discover teachers' current use of GenAI, compared to previous studies that have explored early opinions on ChatGPT (e.g., [9]). Our results show that nearly all Estonian K-12 teachers are aware of GenAI-based tools like ChatGPT (98%), and 49% have already changed their teaching processes to facilitate the new reality. Moreover, 25% of teachers have integrated GenAI into their teaching, in contrast to the early day study of Missouri K-12 teachers, only 5% of whom had done the same in 2023 [7]. This real-life adaptation differs from the opinions of Brazilian K-12 teachers, who assumed that ChatGPT would not have a significant impact on assessment methods [14]. Modifying assignments is critical, as even experienced teachers cannot distinguish between ChatGPT and student-generated texts and are overconfident in detecting GenAI-generated texts [6].

Our results show that 43% of teachers have already regulated the use of GenAI in student learning. While some subjects are more immune to GenAI, teachers raised concerns about the unregulated use of chatbots for mindless text generation without deeper analysis, potentially leading to a decline in (critical) thinking skills. Encouragingly, 36% of teachers have already started to support the development of critical thinking skills, and this trend is in line with the early prediction of Hong Kong teachers that critical thinking and other general competencies are essential for learning with ChatGPT [5]. The majority of teachers allow the use of these tools for specific tasks with proper referencing and critical analysis of the output, but they prohibit them for other tasks, demonstrating their key role in supporting learning in the GenAI era [5].

While 18% of Missouri teachers had used ChatGPT for their job-related tasks in 2023 [7], a massive 72% of Estonian teachers are using GenAI for their tasks in 2024, demonstrating the potential of these tools to reduce their workload. Respondents actively use 55 different GenAI tools to create assignments and tests, plan lessons, and differentiate tasks.

Some teachers are concerned that students do not check the output of these tools and blindly copy the answers into their homework. This concern is compounded by results showing that GenAI-generated misinformation is harder to detect than semantically equivalent human-written errors [4]. The failure to detect factual errors in ChatGPT's output was attributed to students' trust in the tool's answers, even when warned of the possible confirmation bias and poor quality of the output [8].

The inappropriate trust in GenAI could be explained by the innate human tendency to learn from confident agents and the fact that GenAI models do not signal uncertainty in the way that humans do (e.g. using "hmm" when talking or "I think") [11]. This bias leads children to assume certainty where there is none and further emphasizes the need for critical thinking skills to assess and correct the output.

In addition, teachers' fears highlight the need to modify their assignments and homework to be AI-proof. Indeed, without changing the tasks and assessments, students could use GenAI to do their thinking. As one respondent put it: "The rise of chatbots will lead to a further decline in functional literacy, which is already poor; vocabulary will become more limited, thinking more superficial". Overall, teachers are aware that the tools make factual and conceptual errors, a known characteristic of AI chatbots [3,13].

While only 13% of Missouri teachers were looking for support and professional development in 2023 [7], 69% of Estonian teachers said they would like to receive training. Professional development programmes should include practical examples of the use of these tools for job-related tasks, AI literacy, and the integration of GenAI into learning tasks. The real-life examples presented in this study of how teachers have integrated these tools to both improve learning and make their work more efficient could be encouraging for all educators.

Acknowledgement. This study was funded by the Estonian Research Council (grant number PSG728). We thank Kärt-Katrin Johanson for her insightful comments on the questionnaire.

References

1. Annuš, N.: Chatbots in education: the impact of artificial intelligence based Chat-GPT on teachers and students. Int. J. Adv. Natl. Sci. Eng. Res. **7**(4), 366–370 (2023)
2. Ansari, A.N., Ahmad, S., Bhutta, S.M.: Mapping the global evidence around the use of ChatGPT in higher education: a systematic scoping review. Educ. Inf. Technol., 1–41 (2023)
3. Chauncey, S.A., McKenna, H.P.: A framework and exemplars for ethical and responsible use of AI chatbot technology to support teaching and learning. Comput. Educ. Artif. Intell. **5**, 100182 (2023)
4. Chen, B., Zhu, X., et al.: Integrating generative AI in knowledge building. Comput. Educ. Artif. Intell. **5**, 100184 (2023)
5. Chiu, T.K.: The impact of generative AI (GenAI) on practices, policies and research direction in education: a case of ChatGPT and Midjourney. Interact. Learn. Environ., 1–17 (2023)
6. Fleckenstein, J., Meyer, J., Jansen, T., Keller, S.D., Köller, O., Möller, J.: Do teachers spot AI? evaluating the detectability of AI-generated texts among student essays. Comput. Educ. Artif. Intell. **6**, 100209 (2024)
7. Hays, L., Jurkowski, O., Sims, S.K.: ChatGPT in k-12 education. TechTrends **68**(2), 281–294 (2024)
8. Hill, B.: Taking the help or going alone: ChatGPT and class assignments. HEC Paris Research Paper Forthcoming (2023)
9. Kaplan-Rakowski, R., Grotewold, K., Hartwick, P., Papin, K.: Generative AI and teachers' perspectives on its implementation in education. J. Interact. Learn. Res. **34**(2), 313–338 (2023)
10. Kasneci, E., et al.: Chatgpt for good? On opportunities and challenges of large language models for education. Learn. Individ. Differ. **103**, 102274 (2023)
11. Kidd, C., Birhane, A.: How AI can distort human beliefs. Science **380**(6651), 1222–1223 (2023)
12. Labadze, L., Grigolia, M., Machaidze, L.: Role of AI chatbots in education: systematic literature review. Int. J. Educ. Technol. High. Educ. **20**(1), 56 (2023)
13. Lo, C.K.: What is the impact of ChatGPT on education? A rapid review of the literature. Educ. Sci. **13**(4), 410 (2023)
14. Monteiro, F.F., Souza, P.V.S., da Silva, M.C., Maia, J.R., da Silva, W.F., Girardi, D.: ChatGPT in Brazilian k-12 science education. Front. Educ. **9**, 1321547. Frontiers Media SA (2024)
15. Rahman, M.M., Watanobe, Y.: ChatGPT for education and research: opportunities, threats, and strategies. Appl. Sci. **13**(9), 5783 (2023)
16. Rudolph, J., Tan, S., Tan, S.: ChatGPT: Bullshit spewer or the end of traditional assessments in higher education? J. Appl. Learn. Teach. **6**(1), 342–363 (2023)
17. Verma, M.: The digital circular economy: Chatgpt and the future of stem education and research. Int. J. Trend Sci. Res. Dev. **7**(3) (2023)
18. Zhang, P., Tur, G.: A systematic review of ChatGPT use in k-12 education. Eur. J. Educ. (2023)

Author Index

A. M. Olney et al. (Eds.): AIED 2024 Workshops, CCIS 2150, pp. 511–515, 2024.
https://doi.org/10.1007/978-3-031-64315-6

Printed in the United States
by Baker & Taylor Publisher Services